Ernest G. Cravalho
Joseph L. Smith, Jr.

Massachusetts Institute of Technology

Engineering
THERMODYNAMICS

Pitman

Boston · London · Melbourne · Toronto

Pitman Publishing Inc.
1020 Plain Street
Marshfield, Massachusetts 02050

Pitman Books Limited
39 Parker Street
London WC2B 5PB

Associated Companies
Pitman Publishing Pty Ltd., Melbourne
Pitman Publishing New Zealand Ltd., Wellington
Copp Clark Pitman, Toronto

Library of Congress Cataloging in Publication Data

Cravalho, Ernest G.
Engineering thermodynamics.

1. Thermodynamics. I. Smith, Joseph L. II. Title.
TJ265.C88 621.402′1 81-240
ISBN 0-273-01604-0 AACR2

Manufactured in the United States of America

10 9 8 7 6 5 4 3 2 1

Contents

Preface

The fundamental and pervasive nature of the science of thermodynamics leads to a variety of expositions on the subject. The large number of textbooks on thermodynamics demonstrates the many different means that can be employed to introduce students to this important and challenging field. In this text, we have adopted an approach that has proven to be successful in working with students of diverse backgrounds and at various stages in their academic careers. In general, however, the text is intended for a one-semester first course in thermodynamics for students studying mechanical, electrical, oceanic, aerospace, nuclear, and civil engineering. The absence of material on chemical thermodynamics makes the text unsuitable for students in chemical engineering or materials science.

Our treatment has been developed and refined in classroom instruction involving several thousand students and many instructors, principally at the Massachusetts Institute of Technology. The material has been developed to be an integral part of the modern undergraduate engineering curriculum with much use made of the modeling techniques of system dynamics in the earlier portions of the treatment. The distinguishing objectives of this new method of presentation are as follows:

1. To present a development of classical thermodynamics that is closely related to the student's experience with real physical situations and previous subjects.
2. To illustrate the physical principles first with the simplest case, preferably one within the student's experience, and then to expand and to generalize to more complex situations.
3. To develop thermodynamics as a logical extension of the energy concepts of mechanics to more complex cases in which the thermal variables are coupled with the mechanical variables. To present

reversible and irreversible processes as an extension of conservative and dissipative processes in mechanics.

4. To develop the second law of thermodynamics as a conservation of entropy principle within the same framework that is normally used for conservation of mass and conservation of energy. The concept of entropy transfer is introduced and used with the second law in the same way that heat transfer and work transfer are used with the first law.
5. To emphasize the idealization and modeling of real physical situations for analysis by thermodynamic methods.
6. To limit the scope of the subject so that the student can develop a working understanding of the principles necessary for the more common situations requiring thermodynamic analysis.
7. To provide experience in extending simple concepts to include new and more complex areas.

Material is presented in the following order, although the instructor is certainly free to use the text in any manner appropriate to the situation at hand. Chapter 1 addresses only the basic definitions that are essential to start the development. Other definitions are given in context when needed. Especially important is the "thermodynamic method" that we have developed as an orderly means for dealing with thermodynamic situations. Chapter 2 takes up the first law, the property energy and the energy transfer interactions, heat transfer, and work transfer. Chapter 3 amplifies these concepts in a major way for uncoupled thermodynamic systems and introduces modeling concepts for the first time. Chapter 4 extends these notions to coupled thermodynamic systems with heavy emphasis on the ideal gas system model. Chapter 5 introduces reversibility and irreversibility as extensions of conservation and dissipation followed by the essence of the second law of thermodynamics. Chapter 6 considers first and second law implications of coupled thermodynamic systems in communication with more than one heat reservoir and develops the idea of a heat engine. Chapter 7 extends these concepts to develop entropy and entropy transfer, including the second law limits on work transfer and heat transfer. Chapter 8 develops the pure substance model as a higher order model of coupled thermodynamic systems and introduces the student to tabulated thermodynamic properties. Chapter 9 derives the relations that must exist among thermodynamic properties in order to satisfy the first and second laws of thermodynamics. (Much of this chapter is usually omitted in the one-semester course given at M.I.T.) Chapter 10 presents the first and second laws for open systems using a traditional control volume approach, while Chapter 11 applies these principles to components of thermodynamic plants in a manner similar to the approach used for uncoupled systems in Chapter 3. Chapter 12 integrates the material of previous chapters in the context of thermodynamic plants, both energy conversion plants and refrigeration plants. In this chapter we also address (at a simple level) the major issues confronting the power industry. All chapters contain numerous examples worked out in detail to illustrate the approach as well as many problems to be worked out by the student. These latter problems reproduce closely the situations faced by the practicing engineer.

This method of presentation, ostensibly the work of the two authors, has been influenced by the work of others, most notably Dr. G. N. Hatsopoulos and the late Professor J. H. Keenan. The reader will also note the influence of Professors L. Tisza and H. B. Callen, particularly in Chapter 9. The work on uncoupled thermodynamic systems has benefited greatly from the work of Professor H. H. Richardson and co-workers.

The authors also acknowledge the assistance of their many colleagues who have taught from the text and have shared with us their penetrating comments and suggestions. Especially helpful in this regard have been Professors J. A. Keck, J. B. Heywood, A. A. Sonin, A. Bejan, H. Paynter, W. Unkle, D. Pratt, M. Crawford, P. Thullen, K. Diller, and A. H. Shapiro. Doctors A. Rios and W. Toscano were most helpful in the early development of the material, particularly with the problems at the ends of the chapters. We also express our gratitude to S. Oliver, B. Larsen, K. Lubinger, and L. Shapiro, who assisted in the preparation of the manuscripts that led to the final form of the text. And finally, we offer a special note of thanks to the many students who labored through this material in its various forms. Their input and feedback were invaluable in the development of this approach.

CHAPTER 1

Introductory Concepts and Basic Definitions

1.1 Definition and Scope of Thermodynamics

Thermodynamics is the study of the thermal, mechanical, electromagnetic, and chemical behavior of matter and the methods by which such behavior may be exploited for the benefit of mankind. Thermodynamics is concerned with energy, energy transfer, and energy conversion with emphasis on combining thermal behavior with mechanical, electromagnetic, and chemical phenomena.

When two quantities of matter are in communication and interact, the effects are coupled so that thermal effects cannot be associated strictly with thermal interactions nor can mechanical effects be attributed solely to mechanical interactions, and it is this coupling that makes the science of thermodynamics distinctly different from other sciences. Consequently, the study of this coupling is a major concentration in the science of thermodynamics.

Since nearly every physical situation involves thermal and/or mechanical interactions, thermodynamics is very general in its application. Every physical situation, from a nuclear power plant to a clock spring, must satisfy the laws of thermodynamics.

For this reason, the science of thermodynamics has great utility in engineering practice. By applying the laws of thermodynamics in a careful, systematic manner, practicing engineers can establish the maximum amount of useful work that a given energy source can provide under a specific set of circumstances. They can also determine the amount of thermal pollution that a particular power plant delivers to the environment, or the minimum amount of a particular fuel necessary to power a spaceship from one planet to another. The list of applications of thermodynamics is endless and includes situations of every size and configuration imaginable. In short, no physical situation can escape the laws of thermodynamics.

Since the laws of thermodynamics are so encompassing, how are they formulated? Rather simply, these laws are a summary of the experience of mankind in dealing with physical situations. They are phenomenological in that they are generalized statements based on observations of physical situations within our experience. The laws of thermodynamics cannot be derived from any set of basic principles–they are a set of basic principles.

Stating the laws of thermodynamics in their most general terms will not satisfy the engineer. These laws must be stated in terms that will permit their application to physical situations in a quantitative manner. With these laws expressed in mathematical terms, the engineer must formulate a similar mathematical representation of the situation of interest. This representation is known as the model of the physical situation. Thermodynamic behavior is determined by application of these laws to the model. In essence, the model is an idealization of the physical situation consistent with the laws of thermodynamics. This only includes those physical characteristics relevant to the behavior of interest in the specific situation. All other characteristics are neglected. The precise manner of formulating the model to be used in a given situation requires some experience; hence, this text will devote considerable attention to modeling. Therefore, the model is an indispensable adjunct to the laws of thermodynamics, and the physical situation must be modeled before we can successfully apply the laws of thermodynamics. The significance of the model in the science of thermodynamics must not be underestimated.

1.2 Basic Definitions

Thermodynamics has its foundations in a set of basic definitions that constitute a unique language. Traditionally, these definitions assign precise, technical, meanings to common words and considerable care must be exercised in stating them in order to minimize any ambiguities. The importance of proper understanding and use of these terms in the study of thermodynamics cannot be overemphasized.

The definitions will be introduced as the text develops. Many important definitions will be first stated in simple, restricted, terms and later generalized as more complex situations are considered.

1.2.1 System

A *system* is defined as any quantity of matter or region of space to which attention is directed for purpose of analysis.

Clearly this is a general definition and can apply to any physical system, thermodynamic or otherwise. Generally, when attempting to determine the behavior of a system, we would expect to consider all the possible physical phenomena that can involve the system in any way. The task of enumerating all the possibilities is a formidable one and becomes virtually impossible if one is to account for all the coupling which might exist among the phenomena. Fortunately, it is often apparent from the physical situation, that many physical phenomena are of little or no consequence in determining the relevant behavioral aspects of the system. Thus, by judiciously eliminating all nonessential phenomena, we can substantially reduce the effort associated with the analysis of the system's behavior. In fact, this is precisely the point of view employed in the study of mechanical, thermal, and electrical systems. In each instance, all effects but one are neglected thereby simplifying the analysis considerably.

1.2.2 Boundary

The quantity of matter or region of space which forms the system is delineated by a *boundary*. The boundary is a surface, real or imaginary, which is prescribed in a rather arbitrary fashion and may possess special features. The most significant among these features is the boundary's ability to permit the transfer of matter into or out of the system. For example, if a system is defined as a particular quantity of matter, the system will always contain the same matter, and there can be no transfer of mass through the boundary. Such a system is called a *closed system* and is surrounded by a boundary *impermeable* to mass transfer. However, if a system is defined as a region of space without restriction to mass, it is enclosed by a boundary *permeable* to mass transfer. Systems of this type are called *open systems*.

Boundaries possess other characteristics in addition to permeability. For example, they can be rigid or deformable. This can influence the nature of mechanical interactions that the system can experience or can restrict the nature of allowable thermal interactions. At this time it is not possible to describe the details of boundary characteristics; however, in each case, the boundary and its characteristics must be carefully specified to avoid ambiguities and misunderstandings.

1.2.3 Environment

Everything outside the system boundary is referred to as the *environment*.

1.2.4 State and Property

The term *state* is used to signify the condition of a system at a specific instant. The state of the system is characterized by a collection of certain observable, macroscopic quantities, called properties. A *property* is one of those observable macroscopic quantities which are definable at a particular instant without reference to the system's history. A complete description of a system's state includes the values of all its properties.

By definition, the collection of properties that specify the state does not provide any information regarding the manner in which that system's state was achieved. In fact, any quantity which depends upon the system's history is not acceptable as a property. It cannot be used as part of the description for the state of that system.

The concepts of state and property are fundamental to thermodynamics and the distinction between properties and non-properties is of utmost importance. As an example of this difference, let us consider a system consisting of two vehicles. If the two vehicles, A and B, are located at given points in two different cities, then the straight line distance from A to B is a property of the system. This is because the value is fixed only by the position of the vehicles at the specified time and not by the manner in which they arrived in their positions. On the other hand, the distance traveled by vehicle A in reaching vehicle B depends on how the vehicle A travels. The distance traveled is known only if we know the history of the trip, that is, the route followed. Thus, the distance traveled cannot be a property of the system. However, we can say that the distance traveled is greater than or equal to the straight line distance from vehicle A to vehicle B. We can also compute the straight line distance (a property) from the complete details of the history of the trip (route followed). In thermodynamics we have an analogous situation in which a property (energy) is related to non-properties (heat transfer and work transfer).

As a second example, consider an automobile tire as a system. The depth of the tire tread is a property, but the number of miles traveled is not. There is no direct correlation between the number of miles traveled and the observable condition of the tire at a given time. A tire driven a given number of miles on a rough road at high speed under a heavy load will be in a different condition than one driven the same distance on a smooth road at low speed under a light load.

Some of the common thermodynamic quantities which are *properties* are: pressure, mass, volume, density, temperature, internal energy, enthalpy, entropy, position, velocity, charge, voltage, and force.[1]

The common quantities in thermodynamics which are *non-properties* are: heat transfer and work transfer.[2]

[1,2] The precise thermodynamic definitions of these terms will be given in subsequent chapters.

In general, properties may be grouped into a number of different categories. For example, properties may be classified as *independent* or *dependent*. A property is independent of another if, for a given system, it varies in an arbitrary fashion without producing any change in the other property. Conversely, one property is dependent upon another if a change in the latter produces a change in the former.

The selection of properties to be taken as independent is somewhat arbitrary; however, the prevailing physical situation usually dictates which properties are the natural choices. Often the choice is related to the ease with which certain properties can be controlled or measured. For example, when gas is being compressed in the cylinder of an automobile engine, the motion of the piston is determined by the crank mechanism so the volume of the gas is naturally taken as one of the independent properties. Should the physical situation be altered in some manner, the set of properties taken to be independent can be transformed into another set of independent properties in much the same manner as one transforms from one set of coordinates to another in mathematics. In any event, a complete set of independent properties is sufficient to establish the state of the system. However, until the system is properly identified, the dependent properties remain unspecified and the description of the state of the system is incomplete.

A system is identified and described by means of mathematical expressions known as the *constitutive relations*. These relations express the dependent properties in terms of the independent properties. The constitutive relations of a system depend upon the characteristics of the physical materials that comprise the system and their geometric arrangement within the system. Hence, for a system with a given set of constitutive relations, the complete state of the system can be described once the values for the independent properties are established.

In the constitutive relation there are often quantities which have the appearance of properties; however, these quantities are not true properties. Rather, they represent certain characteristics of the system and can therefore be used to identify the system. A typical example of this is the spring constant of a linear spring. A linear spring system has two properties: the force on the spring and the relative displacement of its two ends.

These properties are sufficient to describe the state of the spring. Since the two properties are related through the constitutive relation, only one of them is independent. However, because the relationship between the two properties is a linear one, a new quantity appears–the spring constant. The spring constant is not a true property. It cannot describe the state of the system in terms of force-displacement behavior and thus serves to identify the system. In fact, for purposes of analysis, the spring constant and the free length are sufficient to define a linear (and massless) spring.

In a geometrical sense, the complete set of independent properties of a given system can be used to form a multidimensional space. In that space, the constitutive relations will form a hypersurface such that every possible state of the system must lie somewhere on the surface. Thus, the independent thermodynamic properties can be regarded as thermodynamic coordinates.

1.2.5 Interaction

Two systems interact when a change of state in one system causes a change of state in the other. The *interaction* is the phenomenon at the contiguous boundary between the two systems that causes the transfer of mutual influence. In thermodynamics the energy effects of the interactions are of primary importance. Two systems are in *communication* when interactions are possible, and are *isolated* when interactions are not possible.

1.2.6 Process

A *process* occurs when a system undergoes a change of state with or without interactions with its environment. During the change of state, the system passes through a succession of states which forms the *path* of the process. The complete description of a process requires four specifications: the initial state, the final state, the path, and the interactions between the system and its environment during the change of state.

There exists a special class of processes, known as *cycles,* in which the initial and final states are identical. Quite often the cycle is a series of simple processes that represent changes of states in engineering systems. Thus, a cycle can be regarded as a series of processes that produces no net change of state in the system.

1.2.7 Equilibrium

Consider now two thermodynamic systems which are free to interact in any possible manner. If the state of one system is changed by some external agent, the two systems will interact with one another. Eventually, they will reach a particular state, known as the *equilibrium* state, determined by the physical situation. In this state, the interaction between systems ceases. Once this state has been established, the two systems cannot be made to interact any further unless the imposed circumstances are altered. In this sense, the two systems are in a state of *mutual equilibrium.* On the other hand, the two interacting systems may actually be part of a larger system. When all subsystems of this larger system reach a state of mutual equilibrium, the composite system is said to be in a state of *internal equilibrium.* That is, any one part of the composite system is in equilibrium with any other part.

It is worth noting that the underlying principle of thermodynamics is this tendency for all systems to seek the equilibrium state as defined by the particular physical circumstances. On the basis of this principle, it is possible to derive the relations of thermodynamics.[3]

Although it is not specifically mentioned, the variable time does play an important role in the definition of equilibrium. Perhaps this point can be clarified by rephrasing the definition: "When a system is in an equilibrium state, it is impossible for the system to experience a change of state without also experiencing an interaction of some sort". In a manner of speaking, an equilibrium state is a state of "rest"; nothing is happening. We note, however, that this static nature of the equilibrium state exists only on a macroscopic scale. If we observe

[3] Hatsopoulos, G. N. and Keenan, J. H., *Principles of General Thermodynamics,* John Wiley & Sons, 1965.

any system on a microscopic scale, we see that the system is in a state of constant fluctuation and is far from being at "rest". The magnitudes of these microscopic fluctuations are so minute that in most cases they are undetectable on a macroscopic scale. However, the possibility does exist for a large number of these minute fluctuations to "happen" simultaneously so that a spontaneous macroscopic fluctuation is observed. If this possibility exists for every system, then how can we talk about equilibrium states in such absolute terms? This question can be answered by noting that the probability of this occurring in most systems is so small that the event would probably happen only once during a time period comparable with the age of the universe. Therefore, a system not experiencing any interactions when allowed to do so can be regarded as being in equilibrium.

The point is that this equilibrium is relative rather than absolute. We have just stated (although not explicitly) that if the probability of a system executing a spontaneous change of state is small during the time scale of a typical observation, then the system is assumed to be in equilibrium. Simply, if a system is undergoing a change of state which is "slow" relative to the time scale of the physical phenomenon which dominates the overall physical picture, such a system can be treated as though it were in equilibrium.

In the present treatment there are four types of equilibrium which will concern us. These types are characterized by the nature of the interaction which is necessary for the system to reach equilibrium. Accordingly they are called *mechanical equilibrium, thermal equilibrium, chemical equilibrium,* and *diffusion equilibrium*. Detailed definitions of these various types of equilibrium will be presented at the appropriate points in our development.

By now it should be clear that classical thermodynamics is incapable of determining the time required for the attainment of the equilibrium state and that macroscopic thermodynamics is simply a study of the equilibrium states of matter. For this reason, some authors choose to call macroscopic thermodynamics by the name thermostatics.[4] Although this term is probably more appropriate, we choose to retain the name thermodynamics for purely historical reasons.

[4] Tribus, M., *Thermostatics and Thermodynamics,* D. Van Nostrand Co., N.Y., 1961.

1.3 The Thermodynamic Model and Method of Analysis

Two of the most important duties of an engineer are the design of engineering systems and the analysis of the behavior or performance of these systems. In general, systems vary greatly in their complexity and thermodynamic systems are no exception. Fortunately, the general techniques for analyzing thermodynamic systems are much the same regardless of the complexity of the system. These techniques can be cast in the form of steps which, if performed in sequence, will lead to an adequate solution of the problem. Whether the solution is a good one depends upon the skill of the individual formulating the problem and performing the analysis. A good solution will provide the necessary engineering information about the situation within the time available for analysis and with an economy of effort. An analysis which is more complex than necessary is time consuming and wasteful. The main steps involved in the analysis are:

1. Select the system, system boundary and environment.
2. Model possible behavior
   a. System properties and constitutive relations
   b. Possible boundary phenomena and interactions
   c. Environment properties and constitutive relations
3. Model the specific processes of system and environment
   a. Initial and final states
   b. Path of process
   c. Interactions
4. Apply laws of thermodynamics
   a. Space continuity
   b. Conservation of mass
   c. First law of thermodynamics
   d. Second law of thermodynamics

The first step is an extremely important one. It separates the system of interest from the environment and thereby limits the scope of the analysis. In most cases, this step will establish the manner in which the laws of thermodynamics are applied to the system. It should be noted that exercising a little judicious care in the choice of the system boundary can considerably reduce the effort expended in the analysis. A thorough understanding of the information to be gained from such an analysis can often provide insight in the selection of the system boundary. As a general rule the system boundary should be as global as possible and consistent with the required details. In many cases, more than one system must be defined and analyzed to obtain the required information.

The second step of the thermodynamic analysis models the relevant behavior of the system, the system boundary, and the environment. A *model* is a mathematical representation of a complicated physical situation. In formulating this model, the engineer selects the properties of the system which are important in determining the required information. Possible system behavior is thus modeled by the constitutive relations between the relevant properties.

The possible behavior of the system boundary is modeled by specification of the interactions which result in energy transfer across the boundary. These interactions produce changes in the state of the system and its environment. The energy transfer interactions are: heat transfer, and various work transfers (these will be discussed in detail in subsequent chapters). Energy can also be transferred by mass permeating the boundary of the system.

The possible behavior of the environment is modeled by the constitutive relations between the relevant properties of the environment.

The third step of the analysis models the specific processes of the system and the processes of the environment as they interact and transfer energy across the boundary of the system. The model for the process includes specification of: the initial state, the path of the processes, the final state, and the interactions across the boundary.

The fourth and final step of the analysis is the application of the laws of thermodynamics and other relevant physical principles to the model. Continuity of space requires that geometric displacements of the system and the environment are the same at the system boundary. Thus, a volume increase of the system requires an equal volume decrease of the environment. Conservation of mass requires that a closed system have constant mass. The first law of thermodynamics relates changes in the energy of the system to the energy transfer interactions, while the second law imposes restrictions on the possible interactions as will be developed in subsequent chapters.

In formulating the model and executing the analysis, the objective is to idealize the physical interaction to a level of approximation consistent with the required answers. In constructing the model, complex real phenomena are reduced to a simplified description amenable to analysis. It requires skill and ingenuity to accomplish the engineering task of modeling with a minimum expenditure of time and effort. Modeling skill is developed only through experience, but this ability marks excellence in an engineer.

Subsequent chapters are organized to give practice in formulating appropriate thermodynamic models for increasingly complex physical situations. The basics of thermodynamics are developed by adding phenomena to the thermodynamic model in steps. The problems associated with each chapter have been selected to display the modeling process and to emphasize the thermodynamic method of analysis.

1.4 Scope of the Present Treatment

The essentials of the formalism of thermodynamics will be further developed in the following chapters. We have begun with the concepts and definitions of this chapter. In the course of our development, we shall first state, without rigorous proof or discussion, the first law of thermodynamics. We will then follow with a formulation of the detailed concepts of energy and energy interactions using a simple class of thermodynamic systems in which no coupling occurs among the various energy storage modes. Within this simple class of thermodynamic systems, we shall first consider those systems which are uncoupled and have states that can be specified by a single independent property. Then we will consider those systems which are uncoupled but have states that require more than one independent property for their specification. We then extend the treatment to coupled systems with two or more independent properties, thereby completing the details of the first law of thermodynamics.

Once the concepts of stored energy and energy interactions are understood, we will then re-examine the nature of the thermodynamic processes which take place during interactions between a system and its environment. This re-examination is accomplished by first considering systems in communication with a single heat reservoir and then systems in communication with any number of heat reservoirs. Based on these results, we shall conclude that there are definite limitations imposed upon the thermodynamic behavior of a system and that these limitations are related to the number of heat reservoirs in communication with the system. In order to provide some quantitative measure of these limitations, which constitute the second law of thermodynamics, we shall introduce a new concept and a new property, the entropy.

Following the introduction of the second law, we will turn our attention to the nature of a pure substance. At this point we will formulate the various models which are useful in representing the thermodynamic behavior of real substances. Finally, we shall conclude by considering typical engineering systems and evaluating their performance by applying the first and second laws to suitable models of the actual physical situation.

Problems

1.1 Which of the following represent a thermodynamic system? For each system, describe its boundaries.

(a) An explosion
(b) A bicycle pump
(c) Three pounds of air
(d) A wave on the surface of a lake
(e) A force
(f) An automobile
(g) The volume inside an evacuated tank
(h) Ten miles of copper wire
(i) A flow through a pipe

1.2 Which of the following are properties of the specified system? Which are not?

(a) *System: The filament of an incandescent lamp*
the mass
the diameter
the number of hours of operation
the electrical resistance
the total watt-hours consumed

(b) *System: A dry cell flashlight battery*
the volume
the mass
the voltage
the total watt-hours used
the mass of each chemical compound or element in the battery

(c) *System: A clock spring*
the torque on the output shaft
the volume
the energy transferred to the spring by the input shaft
the energy transfer required to wind the spring to its present torque from the unwound condition

CHAPTER 2

The First Law of Thermodynamics

The first law of thermodynamics plays a significant role in the analysis of thermodynamic systems, and for this reason we shall devote considerable attention to its development. However, it is useful for us to outline briefly the essential features of the first law of thermodynamics as it applies to closed systems. With this information, we will be better prepared to organize and interpret the details of the succeeding development.

In essence, the first law of thermodynamics is a generalization of the observed facts about the energy interactions between a system and its environment. Specifically, it relates the various energy interactions between a system and its environment to changes of state experienced by that system during these interactions. All physical experience has confirmed that the generalizations embodied in the first law of thermodynamics must be satisfied to describe realistically a physical situation.

There are many ways in which the first law of thermodynamics can be stated, and some rather complex arguments are required to prove the equivalence of these statements. However, the simplest of these statements can be formulated by considering a system which executes a cycle. By definition, the system experiences no net change of state. Therefore, there can be no net energy interaction between the system and its environment. For our purposes it will be sufficient to consider only two forms of energy interaction between a system and its environment: work transfers[1] W, and heat transfers[2] Q. Thus, we may state the first law of thermodynamics in a formal way as:

> The net energy interaction between a system and its environment is zero for a cycle executed by the system.

The mathematical equivalent of this statement is

$$\oint \delta Q - \oint \delta W = 0 \tag{2.1}$$

The integral sign indicates the algebraic summation of each infinitesimal heat transfer δQ or work transfer δW over the complete cycle. The negative sign is introduced because historically in thermodynamics, work transfer *from* a system is taken as *positive* which is the inverse of the sign convention used in mechanics.

The major difficulty in applying the first law to physical situations lies in the determination and formulation of the heat transfers and work transfers. This will soon become apparent as we take the basic concepts of work transfer directly from mechanics and electromechanics, and reformulate them from a thermodynamics point of view. However, as we deal with progressively more complex systems, it may be necessary to extend these definitions of work transfer to avoid any ambiguity. Since heat transfer is unique to thermodynamics, it is not possible to rely on other disciplines for its formulation. In fact, one of the most difficult aspects of the formalism of thermodynamics is a rigorous definition of heat transfer. Therefore, we shall delay its discussion until later in our development. Presently, it is sufficient to regard heat transfer simply as one method of energy interaction, distinct from work transfer, between a system and its environment.

As we shall see later, it will become increasingly important for us to distinguish between the energy transfer interactions Q and W and the property which changes by virtue of these in-

[1,2] At this point we have defined work transfer and heat transfer as the two forms of energy interaction which change the energy of a system. As our development progresses, more complete definitions will be given on pages 26, 38, 39 92, and 193.

teractions. This latter quantity is defined as the energy, E, of the system at a given state. It is simply the stored energy of the system. The *change in the stored energy* of a system due to a change of state of the system from state 1 to state 2 is defined by

$$\int_1^2 \delta Q - \int_1^2 \delta W = E_2 - E_1 \qquad \textbf{(2.2)}$$

We may now use the first law of thermodynamics in the form of equation (2.1) to show that the energy change $E_2 - E_1$ is the change of a property of the system. To show this, let us consider the situation depicted in Figure 2.1. The system is one for which two independent properties are adequate to specify the state. Thus, each point in the plane of Figure 2.1 represents a state of the system. (Note: our proof is not restricted to systems of this type. We have used this example simply because it is easier to present in graphical form.) The state of the system is

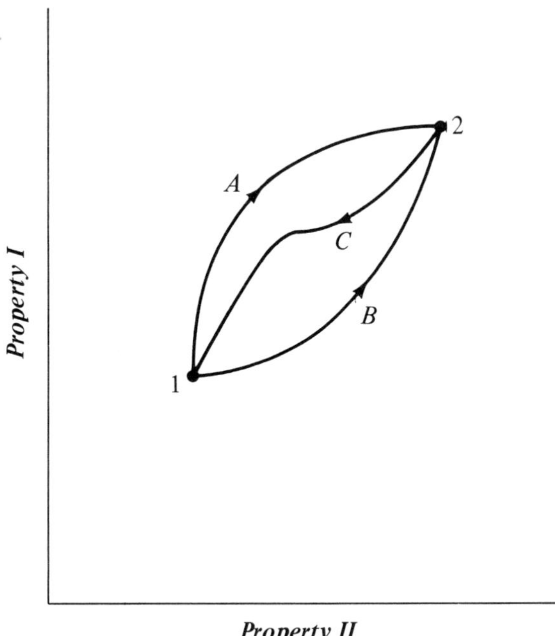

Figure 2.1 Change of state in a system with two independent properties

changed from state 1 to state 2 by two different processes, A and B. From equation (2.2) we have

$$\left[\int_1^2 \delta Q - \int_1^2 \delta W\right]_{\text{process } A} = (E_2 - E_1)_{\text{process } A} \tag{2.3}$$

and

$$\left[\int_1^2 \delta Q - \int_1^2 \delta W\right]_{\text{process } B} = (E_2 - E_1)_{\text{process } B} \tag{2.4}$$

Now we can find some other process to return the system from state 2 to state 1, say process C, such that

$$\left[\int_2^1 \delta Q - \int_2^1 \delta W\right]_{\text{process } C} = (E_1 - E_2)_{\text{process } C} \tag{2.5}$$

If we apply the first law of thermodynamics to one cycle consisting of process A and process C and to a second cycle consisting of process B and process C, we have from equation (2.1)

$$\left[\int_1^2 \delta Q - \int_1^2 \delta W\right]_{\text{process } A} + \left[\int_2^1 \delta Q - \int_2^1 \delta W\right]_{\text{process } C} = 0 \tag{2.6}$$

and

$$\left[\int_1^2 \delta Q - \int_1^2 \delta W\right]_{\text{process } B} + \left[\int_2^1 \delta Q - \int_2^1 \delta W\right]_{\text{process } C} = 0 \tag{2.7}$$

Thus, from equations (2.6) and (2.7) it follows that the net energy interaction for *any two processes between state 1 and state 2* must have the same value

$$\left[\int_1^2 \delta Q - \int_1^2 \delta W\right]_{\text{process } A} = \left[\int_1^2 \delta Q - \int_1^2 \delta W\right]_{\text{process } B} \tag{2.8}$$

Thus, when we substitute the definition of the energy change, equation (2.2), we show that

$$(E_2 - E_1)_{\text{process } A} = (E_2 - E_1)_{\text{process } B} \tag{2.9}$$

Since processes A and B are arbitrary, we have shown that the change of energy between two states of a system is independent of the process (the method by which the change of state is carried out). Therefore, this change must depend only on the states of the system and must be a change in a property. If we choose one state as an arbitrary datum for energy, then the energy relative to this state depends only on the state of the system. Thus, the energy of a system relative to a reference 0 is

$$E_1 = E_0 + \int_0^1 \delta Q - \int_0^1 \delta W \tag{2.10}$$

where the integrals are the heat transfer and work transfer interactions respectively for *any* process between state 0 and state 1. For convenience, it is customary to assign the energy of the reference state the value zero; thus $E_0 = 0$.

Note that the individual energy transfers, integral of δQ and the integral of δW, are individually related to specific changes in the stored energy of the system only in especially simple cases. In more complex cases, the integrals will each depend upon the process used to change the system from state 1 to state 2 so that only their combination is related to changes in stored energy through equation (2.2). For example, a complex system could have the same change of state as the result of a heat transfer process without work transfer or as the result of a work transfer without a heat transfer.

As a convenient notation, changes in properties like $E_2 - E_1$ are given the symbol ΔE_{1-2} for finite changes of state and dE for infinitesimal changes of state. Thus,

$$E_2 - E_1 = \Delta E_{1-2} = \int_1^2 dE \tag{2.11}$$

On the other hand, the energy transfer quantities, heat transfer and work transfer, that are not generally properties of the system, are given the symbols Q_{1-2} and W_{1-2} for processes of finite extent and the symbol δQ and δW for processes of infinitesimal extent. Thus,

$$Q_{1-2} = \int_1^2 \delta Q \tag{2.12}$$

and

$$W_{1-2} = \int_1^2 \delta W \tag{2.13}$$

Then combining equations (2.2), (2.12), and (2.13), we can write the first law of thermodynamics in the form

$$Q_{1-2} - W_{1-2} = E_2 - E_1 \tag{2.14}$$

For an infinitesimal change of state, equation (2.14) becomes

$$\delta Q - \delta W = dE \tag{2.15}$$

To apply the first law of thermodynamics to physical situations, it is now necessary for us to elucidate the details associated with the evaluation of the heat transfer, work transfer, and change in stored energy for engineering systems. The following chapter accomplishes this task by first considering a very simple class of systems known as uncoupled systems. For these systems there is a separate aspect of behavior and energy storage mode associated with each type of interaction (heat transfer or work transfer). We use the term "uncoupled" because each stored energy form can be affected only by one specific interaction. These uncoupled systems are modeled as one or more pure system elements. Each pure system element models a single interaction and a single stored energy form. By virtue of these special characteristics, we are able to formulate precise definitions of work transfer, heat transfer, and temperature. Obviously, the first law analysis of this class of systems is relatively simple.

Systems in which an individual stored energy can be affected by both heat transfer and work transfer are the logical extensions of the analysis in Chapter 3. These systems are called coupled systems and are considered in Chapter 4. Naturally, the first law analysis of coupled systems is more complicated than that of uncoupled systems; however, these coupled systems are the most interesting from a thermodynamic point of view and the most useful for engineering applications.

2E.1 Sample Problem: Consider a thermodynamic system whose state can be specified by two independent properties. The states of such a system can be represented conveniently in graphical form as shown in Figure 2E.1.

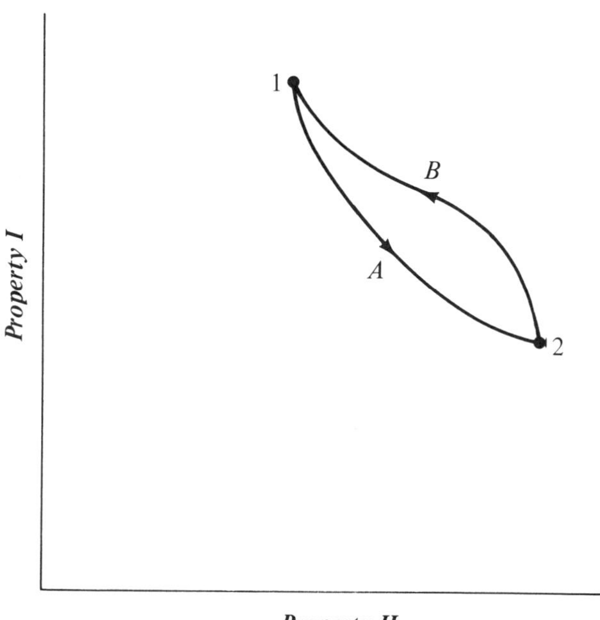

Figure 2E.1 States of a system with two independent properties

The states 1 and 2 are states specified by the values of the properties (I_1,II_1) and (I_2,II_2), respectively. Consider a process *A* by which the state of the system changes from state 1 to 2 and experiences work transfer $(W_{1-2})_A = -23$ kJ and a heat transfer $(Q_{1-2})_A = 11$ kJ. The energy in state 1 referenced to some datum state is $E_1 = 32$ kJ. If the state of the system is subsequently restored to state 1 by means of a process *B* in which there was a work transfer $(W_{2-1})_B = -46$ kJ, calculate the heat transfer $(Q_{2-1})_B$.

Solution: From the first law it follows that for process *B*

$$(Q_{2-1})_B = E_1 - E_2 + (W_{2-1})_B$$

where E_2 must be determined from the application of the first law to process A. Thus,

$$E_2 = E_1 + (Q_{1-2})_A - (W_{1-2})_A$$

or

$$E_2 = 32 + 11 + 23 = 66 \text{ kJ}$$

Then

$$(Q_{2-1})_B = 32 - 66 - 46 = -80 \text{ kJ}$$

Problems

2.1 The water in a steam power plant experiences a cyclic change of state as it circulates through the plant. Each kilogram of water during each cycle of its circulation experiences a positive heat transfer of 1200 kJ in the boiler. The net mechanical work transferred from each kilogram during every cycle is 400 kJ. What is the heat transfer between the water in the plant and the water in the river that is used as coolant?

2.2 A given closed system is changed from state 1 to state 2 by a process, A. For process A the heat transfer for the system is $+15$ kJ, and the work transfer is $+4$ kJ.

(a) A second process, B, also changes the system from state 1 to state 2. The work transfer for process B is -19 kJ. What is the heat transfer for process B?

(b) What work transfer would be required to change the system from state 1 to state 2 without any heat transfer?

(c) What must be the relation between the heat transfer and the work transfer for any process which changes the system from state 1 to state 2?

2.3 A system consists of an elastic coil spring. When the spring is compressed 20 mm from the relaxed (zero force) state, the work transfer for the system is -20 Joules. Is it possible for the spring system to return to its original relaxed state without heat transfer or work transfer?

2.4 As shown in the state diagram of Figure 2P.4, a closed system with two independent properties undergoes a process which changes its state from a to b along the path a-c-b. During this process the heat transfer is 30 kJ and the work transfer is 25 kJ.

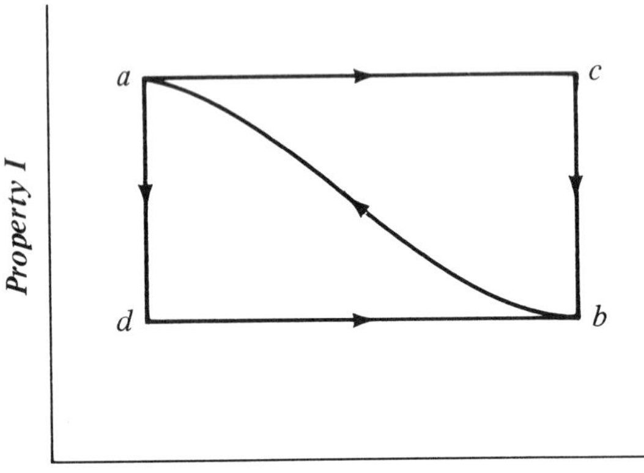

Figure 2P.4

(a) If the system follows the path *a-d-b* while experiencing a work transfer of 7.6 kJ, determine the heat transfer over the same path.

(b) If the system now returns from state *b* directly to state *a* over the path shown while experiencing a work transfer of −18.5 kJ, what is the heat transfer for this process along *b-a*?

(c) Determine the heat transfer for the paths *a-c* and *c-b* if the energy of the system at state *a* is 7.7 kJ, the energy of the system at state *c* is 46.0 kJ, and processes *a-d* and *c-b* are zero work transfer processes.

CHAPTER 3

Thermodynamic Behavior of Uncoupled Systems

3.1 Introduction

The thermodynamic behavior of a system is a consequence of the thermal heat transfer interaction and the mechanical work transfer interactions experienced by the system. Generally, both heat transfer and work transfer interactions influence all of the properties of the system. There are no separate thermal aspects attributed solely to heat transfer and there are no separate mechanical aspects attributed solely to work transfer. Because of this coupling, the analysis of the thermodynamic behavior of a system is considerably more complicated than the analysis of purely mechanical, electrical, magnetic, or thermal behavior of a system. Fortunately, there exist many physical situations in which the coupling between thermal and mechanical aspects is vanishingly small. For these uncoupled systems it is sufficient to model them as a collection of pure system elements, each capable of representing a particular aspect of system behavior. Since the behavior of each system element is independent of the behavior of all the other elements, the thermodynamic behavior is simply the collective behavior of the individual system elements. Of course, this particular modeling procedure is limited to uncoupled behavior and breaks down when the thermal and mechanical aspects become coupled.

As discussed in Chapter 1, even though the thermodynamic behavior of a coupled system is markedly different from that of an uncoupled system, the techniques for analyzing the two are much the same. The only difference arises in the detailed manner in which the techniques are applied. Thus, much insight can be gained by detailed study of the thermodynamic nature of the simpler uncoupled system. Once these details are mastered, the complexity associated with the coupled system is measurably reduced. The detailed study of the thermodynamic behavior of the coupled system then follows as the logical extension of the uncoupled case. We propose to execute this study by first enumerating the typical pure system elements which comprise the uncoupled thermodynamic system model and then describing their individual behavior. In concluding, we will assemble these elements to form a typical uncoupled thermodynamic system.

The pure system elements necessary to describe uncoupled thermodynamic behavior fall into three categories: (1) pure conservative system elements, (2) pure thermal system elements, and (3) pure dissipative system elements.

These system elements are pure in that each exhibits a single phenomenon, and at most a single independent property fixes the state of a pure system element. This makes the changes of state simple and directly related to the interactions across the system boundary.

3.2 Pure Conservative Mechanical System Elements and the Concept of Work Transfer

This class of system elements has a single independent mechanical property and can experience only mechanical work transfer with the environment. The work transfer arises from the forces of the environment on the system and the motion of the system boundary. We define this work transfer in the following manner[1]:

[1] More general, complete, definitions of work transfer are given on pages 92 and 193.

The infinitesimal work transfer δW *from* a system due to a concentrated boundary force **F** *exerted on the system* by its environment is given by the negative scalar product of the force vector **F** and the infinitesimal displacement vector $d\mathbf{r}$ of *the boundary* at the instantaneous point of application of the boundary force. Thus

$$-\delta W = \mathbf{F} \cdot d\mathbf{r} \tag{3.1}$$

The negative sign appearing in equation (3.1) is simply a matter of thermodynamic convention introduced in the early development of the subject. Also in equation (3.1) we have used the symbol δ rather than d to distinguish between an infinitesimally small interaction and the infinitesimally small change of state arising from the interaction.

The integration of equation (3.1) for a finite displacement of the boundary of the system is relatively simple when the force always acts through the same boundary point. In this case the resulting work transfer W_{1-2} from the system is simply the integral of equation (3.1),

$$- W_{1-2} = \int_{\mathbf{r}_1}^{\mathbf{r}_2} \mathbf{F} \cdot d\mathbf{r} \tag{3.2}$$

where **r** is in this case simply the displacement of the one boundary point through which the force **F** is continuously applied. The work transfer depends upon the details of how the force **F** varies as the boundary is displaced. Thus, the work transfer cannot be determined from the initial and final states of the system. This is a trivial point in the case of a pure conservative mechanical system element, but as we shall see shortly, it is an important point in the case of a pure dissipative mechanical system element.

In the more complex case in which the concentrated force **F** moves from point to point along the boundary while at the same time the boundary is moving, the integration of equation (3.1) requires special care. In this case, the displacement of the point of application of force **F** is *not* the same as the instantaneous displacement of the system boundary where **F** is applied. The displacement $d\mathbf{r}$ is the displacement of the mass point of the boundary during the time that the force **F** is applied to that point. Thus, in a finite process the displacement vector **r** is not related to the displacement of a single point on the boundary of the system. The classic example of a process of this type is the force applied to a system boundary through a rolling contact point.

The pure conservative mechanical system element experiences only work transfer interactions; thus, the first law of thermodynamics, equation (2.2), gives the change in the stored energy which results from the work transfer

$$-W_{1-2} = E_2 - E_1 = \int_{\mathbf{r}_1}^{\mathbf{r}_2} \mathbf{F} \cdot d\mathbf{r} \qquad \textbf{(3.3)}$$

Again, it is important to note that the thermodynamic sign convention for work transfer requires that the *work transfer* from the system *decreases the stored energy E.* From equation (3.3) it is clear that when a pure conservative mechanical system element is restored to the initial state after having experienced a change of state, the work transfer associated with that return must be in the opposite direction of the work transfer associated with the original change of state. Further, the magnitudes of the two work transfers are identical. It is because of this "recoverability" of work transfer that the pure conservative mechanical system element is described as "conservative". That is,

$$\oint \delta W = 0 \qquad \textbf{(3.4)}$$

for any cycle of a pure conservative mechanical system element.

The constitutive relation of a pure conservative mechanical system element provides the relation between the boundary force and the boundary displacement. Thus, the change in the stored energy is determined by carrying out the integration indicated in equation (3.3). Representative pure conservative mechanical system elements are the pure translational spring, the pure gravitational spring, and the pure translational mass. The thermodynamic models for these systems are given in Figures 3.1, 3.2, and 3.3.

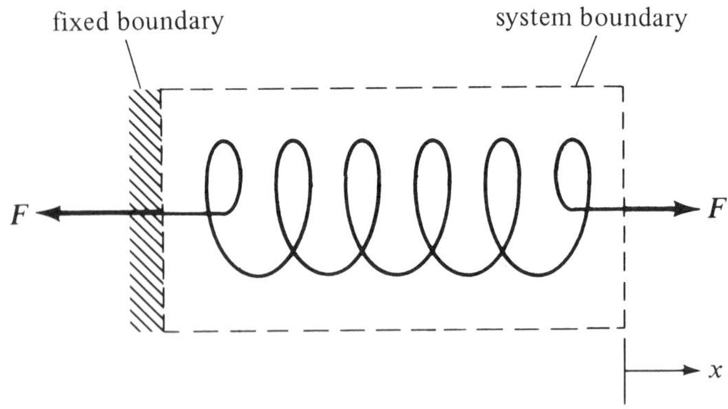

Properties:	F = the through force *on* the system boundary, positive for tension x = extension relative to $F = 0$ state, positive for elongation
Constitutive relation:	$F = kx : k$ = spring constant
Interaction:	$-W_{1-2} = \int_{x_1}^{x_2} F\,dx$
Stored elastic energy change:	$E_2 - E_1 = \frac{1}{2}kx_2^2 - \frac{1}{2}kx_1^2$

Figure 3.1 Pure translational spring (linear)

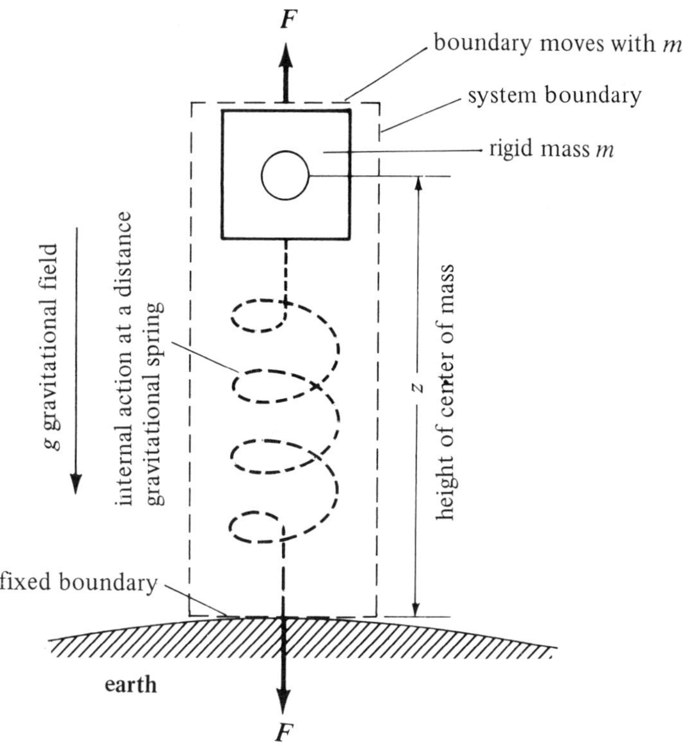

Properties:	F = through force on the system boundary, positive for tension z = height of the center of mass m, positive upward
Constitutive relation:	$F = mg$ g = acceleration of gravity $g_{std} = 9.80665 \text{ m/s}^2$
Interaction:	$-W_{1-2} = \int_{z_1}^{z_2} F\,dz$
Stored gravitational energy change:	$E_2 - E_1 = mgz_2 - mgz_1$

Figure 3.2 Pure gravitational spring

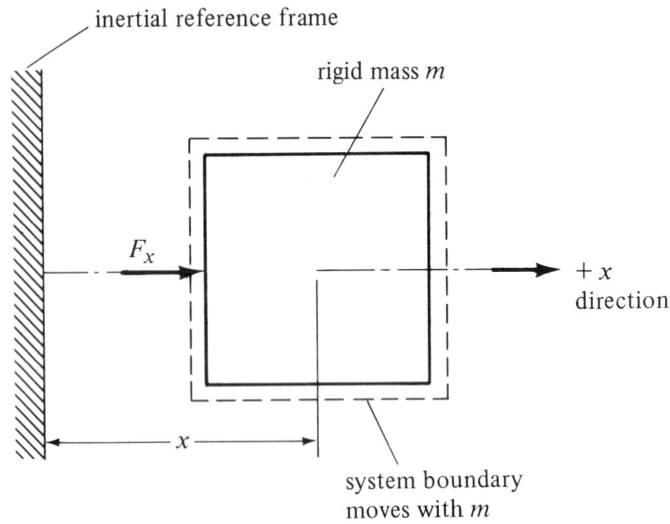

Properties: F_x = net external force *on* the system boundary, positive in the positive x direction

$$\vartheta_x = \frac{dx}{dt} = \text{velocity of the center of mass } m, \text{ relative to inertial frame}$$

Constitutive relation:

$$F_x = m\,\frac{d\vartheta_x}{dt}$$

Interaction:

$$-W_{1-2} = \int_{x_1}^{x_2} F_x\,dx = \int_{x_1}^{x_2} m\vartheta_x\,d\vartheta_x$$

Stored kinetic energy change:

$$E_2 - E_1 = \frac{1}{2}\,m\vartheta_{x_2}^2 - \frac{1}{2}\,m\vartheta_{x_1}^2$$

Figure 3.3 Pure translational mass

3E.1 Sample Problem: Compute the work transfer and change in energy when the extension of a pure translational spring is changed from -0.16 m to $+0.16$ m. The spring constant k is 2000 N/m.

Solution: The system and the process are defined and modeled as in Figure 3.1. The energy change for extension from $x_1 = -0.16$ m to $x_2 = +0.16$ m is

$$\begin{aligned} E_2 - E_1 &= \frac{1}{2} k x_2^2 - \frac{1}{2} k x_1^2 \\ &= \frac{1}{2}(2000 \text{ N/m})(+0.16 \text{ m})^2 - \frac{1}{2}(2000 \text{ N/m})(-0.16 \text{ m})^2 \\ &= 0 \end{aligned}$$

The first law gives the work transfer

$$-W_{1-2} = E_2 - E_1 = 0$$

Note that the total work transfer is zero; however, the work transfer from $x_1 = -0.16$ m to $x = 0$ is $+25.6$ J while the work transfer from $x = 0$ to $x_2 = +0.16$ m is -25.6 J.

3E.2 Sample Problem: Compute the work transfer required to lift 100 kg a vertical distance of 1.2 m in a gravitational field of 9.80 m/s^2.

Solution: The system and the process are defined and modeled as a pure gravitational spring as in Figure 3.2. Application of the first law gives

$$-W_{1-2} = E_2 - E_1 = mgz_2 - mgz_1$$

If we take $z_1 = 0$ then $z_2 = 1.2$ m, and the first law becomes

$$\begin{aligned} -W_{1-2} &= E_2 - E_1 \\ &= (100 \text{ kg})(9.80 \text{ m/s}^2)(1.2 \text{ m}) - 0 \\ &= 1176 \text{ J} \end{aligned}$$

The negative sign of the work transfer shows that the work transfer is from environment to the system.

3E.3 Sample Problem: Compute the work transfer required to increase the translational velocity of a 10 kg mass by 1 m/s when at an initial velocity of 1 m/s. Repeat for an initial velocity of 10 m/s.

Solution: The system and processes are modeled as a pure translational mass as in Figure 3.3. Application of the first law gives

$$-W_{1-2} = E_2 - E_1$$

$$= \frac{1}{2} m\vartheta_2^2 - \frac{1}{2} m\vartheta_1^2$$

$$-W_{1-2} = (1/2)(10 \text{ kg})(2 \text{ m/s})^2 - (1/2)(10 \text{ kg})(1 \text{ m/s})^2$$

$$= 15 \text{ J}$$

For an initial velocity of 10 m/s

$$-W_{1-2} = (1/2)(10 \text{ kg})(11 \text{ m/s})^2 - (1/2)(10 \text{ kg})(10 \text{ m/s})^2 = 105 \text{ J}$$

The work transfer is from the environment to the system to produce the acceleration.

3.3 Pure Conservative Electrical System Elements and Electrical Work Transfer

This class of system elements has a single independent electrical property and can experience only electrical work transfer with the environment. The electrical work arises from the external transfer of electric charge between two points on the system boundary that are different in electrostatic potential. The electric work transfer for a two terminal electrical system element is

$$-W_{1-2} = \int_{q_1}^{q_2} \nu \, dq \qquad \textbf{(3.5)}$$

where ν is the electric potential difference between the two terminals when infinitesimal charge dq is transferred. The integration of equation (3.5) for a finite charge transfer, $q_2 - q_1$, requires the relation between the voltage v and the charge q as specified by the constitutive relation for the system. In SI metric units v is in volts and q is in ampere-seconds (Coulombs) for work transfer in Joules. The minus sign that appears in equation (3.5) is due to the thermodynamic sign convention.

Equation (3.5) is derived from equation (3.1) by evaluating the mechanical work transfer required to move mechanically the infinitesimal charge dq between the two points with potential difference v. The mechanical force is determined from the electrostatic field equations.

As with the mechanical systems, the first law of thermodynamics, equation (2.2), gives the change in stored energy which results from the work transfer

$$-W_{1-2} = E_2 - E_1 = \int_{q_1}^{q_2} v\, dq \tag{3.6}$$

The constitutive relation gives the voltage, v, as a function of the charge, q, for the integration indicated in equation (3.6). Representative pure conservative electrical system elements are the pure electrical capacitance and the pure electrical inductance. The models for these systems are given in Figures 3.4 and 3.5.

3E.4 Sample Problem: Compute the electrical work transfer required to charge a 0.001 farad capacitor to a voltage of 1000 volts from the discharged state.

Solution: The system and process are modeled as a pure electrical capacitance as in Figure 3.4. Application of the first law gives

$$-W_{1-2} = E_2 - E_1 = (1/2C)q_2^2 - (1/2C)q_1^2$$

If the constitutive relation is used to eliminate q, we have

$$\begin{aligned} -W_{1-2} &= (1/2)Cv_2^2 - (1/2)Cv_1^2 \\ &= (1/2)(0.001 \text{ As/V})(1000 \text{ V})^2 - 0 \\ &= 500 \text{ AsV} = 500 \text{ J} \end{aligned}$$

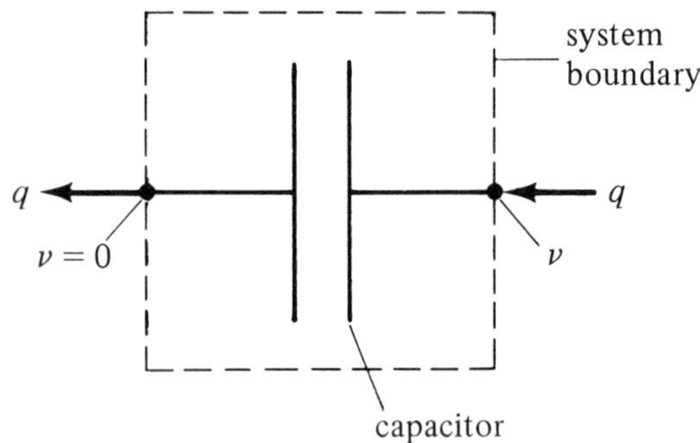

Properties:

v = electric potential difference across the boundary

q = electric charge transferred through the boundary, positive for positive charge into system at positive potential

Constitutive relation: $v = \frac{1}{C} q$

C = capacitance in farads, (ampere second)/volt

Interaction: $-W_{1-2} = \int_{q_1}^{q_2} v \, dq$

Change in stored energy: $E_2 - E_1 = \frac{1}{2C} q_2^2 - \frac{1}{2C} q_1^2$

Figure 3.4 Pure electrical capacitance

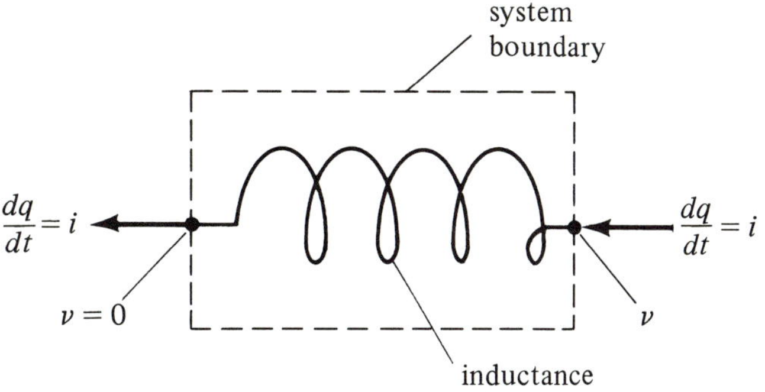

Properties: v = electric potential difference across the boundary

$i = \frac{dq}{dt}$ = charge flowing through the boundary

Constitutive relation: $v = L\frac{di}{dt}$

L = inductance in henries, (volt second)/ampere

Interaction: $-W_{1-2} = \int_{q_1}^{q_2} v\,dq = \int_{i_1}^{i_2} Li\,di$

Change in stored energy: $E_2 - E_1 = \frac{1}{2}Li_2^2 - \frac{1}{2}Li_1^2$

Figure 3.5 Pure electrical inductance

3.4 Pure Thermal System Elements and the Concepts of Heat Transfer

The pure system element which models the thermal behavior of uncoupled systems is the pure thermal system element. The state of this system element is completely described by one independent thermal property and the only energy transfer interaction is heat transfer. The pure thermal system element cannot experience work transfer. However, it is a good model for the thermal behavior of liquids and solids that do not change volume significantly as the result of heat transfer interaction or changes in thermal state. The model for this system is shown in Figure 3.6.

At this early stage of our development of thermodynamics, we do not have the background to express the heat transfer interaction in terms of an interaction integral as we did with the work transfer. Thus, we cannot use the first law of thermodynamics to derive an expression for changes in stored energy. At this point we consider the stored thermal energy, U, as a function of the independent property temperature to be the given

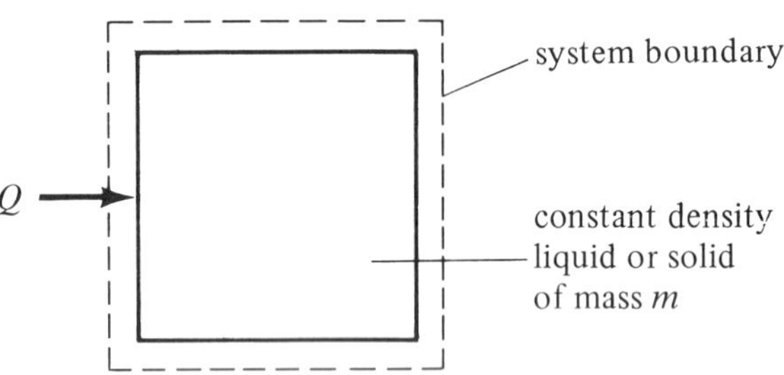

Property: T = temperature

Constitutive relation: U = function of T

Interaction: Q_{1-2} = heat transfer

Change in stored thermal energy: $U_2 - U_1 = mc(T_2 - T_1) = Q_{1-2}$

c = specific heat of system

Figure 3.6 Pure thermal system with constant specific heat

constitutive relation which describes the possible behavior of the pure thermal system.

In the remainder of the present treatment, the symbol U is commonly used to denote the stored energy that can be influenced by heat transfer. As we shall see later, in more complex systems U is not simply stored thermal energy since in the coupled case U can be increased or decreased by work transfer as well as by heat transfer.

For a pure thermal system element, it is often convenient to express the constitutive relation in terms of the change in energy for unit change in temperature. Thus, we define the *heat capacity* of a pure thermal system element as the derivative

$$C = \frac{dU}{dT} \tag{3.7}$$

The heat capacity per unit mass of the system element is known as the *specific heat c* where

$$c = \frac{C}{m} = \frac{1}{m}\frac{dU}{dT} = \frac{du}{dT} \tag{3.8}$$

and u is the stored thermal energy per unit mass of the system. In general the specific heat, and heat capacity, are functions of temperature. When the specific heat of the pure thermal system is modeled as constant (as in Figure 3.6) the thermal energy becomes

$$U_2 - U_1 = m(u_2 - u_1) = mc(T_2 - T_1) = C(T_2 - T_1) \tag{3.9}$$

The *heat reservoir* is a pure thermal system that is very useful in thermodynamics. The temperature of a heat reservoir has a fixed value independent of the thermal energy U. Thus, a heat reservoir experiences all heat transfer and changes in energy U at a fixed specified temperature. Also note from equation (3.9) that a heat reservoir is the model for a system with a heat capacity so large that the temperature change is negligible for the energy change involved.

In formulating the pure thermal system element model we have in effect the first limited definition of heat transfer[2]:

> *Heat transfer* is the energy transfer interaction that occurs between pure thermal system elements.

Although the concept of temperature is familiar from the sense of touch, we can now give a simple definition of temperature[3]:

> *Temperature* is the property of a system which indicates the potential for heat transfer with other systems.

We may also define the inequality of temperature:

> Two systems (or a system and its environment) in communication are *unequal in temperature* when they experience a heat transfer interaction at a finite rate.

We may define equality of temperature and thermal equilibrium:

> Two systems (or a system and its environment) in communication are *equal in temperature* when they reach a state of mutual *thermal equilibrium* for which all heat transfer interactions cease.

The utility of the temperature as a unique measure of the potential of a system for heat transfer is guaranteed by the generalization called the *zeroth law of thermodynamics* which can be stated as follows:

> If two systems are each equal in temperature to a third system, then the temperatures of the two systems themselves are equal.

The sign convention for heat transfer and the sign convention for temperature change are such that:

> For a heat transfer in the absence of work transfer, the heat transfer is positive for the low temperature system which increases in energy and negative for the high temperature system which decreases in energy. For the system with the positive heat transfer, the temperature change is positive or in the limiting case zero.

---

[2] A more complete definition of heat transfer will be given on page 193.

[3] Equation (9.4), page 276, may be taken as a more formal definition of temperature.

These characteristics of heat transfer and temperature are closely related to the second law of thermodynamics, which generalizes the tendency of all systems to come to a mutual equilibrium (both thermal and mechanical equilibrium) when the systems are brought into communication.

In the considerably more complex situations which we shall consider in subsequent chapters, it is usually more convenient to identify heat transfer interactions by means of the property temperature.

> An interaction between two closed systems is a heat transfer if the interaction is solely the result of the temperature difference between the two systems.

The terms *adiabatic* and *diathermal* are used to model the heat transfer characteristics of the boundary of a system. The property known as the thermal conductivity is used as an indication of the heat transfer characteristics of a material. A material with a high value for the thermal conductivity will allow a more rapid heat transfer as the result of a given temperature difference than a material with a low value for the thermal conductivity. When the material in the vicinity of a system boundary has a low value for the thermal conductivity, the boundary itself is often modeled as a complete barrier to heat transfer despite the temperature difference across the boundary. The boundary is then said to be *adiabatic*. Processes within systems surrounded by adiabatic boundaries are termed *adiabatic processes*. A system must be surrounded by an adiabatic boundary to be *thermally isolated* from the environment. In short, the term adiabatic implies: modeled as zero heat transfer. Very few real boundaries or processes are absolutely free from heat transfer, but the adiabatic model is used when the heat transfer is insignificant compared to the other aspects of the situation.

When the boundary of a system is within an environment which has a high value for the thermal conductivity, the boundary may be modeled as a perfect conductor such that heat transfer can occur with a vanishingly small temperature difference. The boundary of the system is then said to be a *diathermal boundary*. The temperature difference across a diathermal boundary is always zero or of infinitesimal magnitude.

Having defined equality of temperature and a method for ordering temperatures from low to high, we can now discuss temperature scales, temperature measurement, and thermometers. One of the serious difficulties in developing thermodynamics is that the definition of the fundamental thermodynamic temperature scale depends upon the second law of thermodynamics; however, the second law of thermodynamics demands the concept of temperature for its development.

We will use the thermodynamic Kelvin scale of temperature throughout the development. After the development of the second law of thermodynamics we will be able to show the special significance of a thermodynamic temperature scale. The traditional practical centigrade and Fahrenheit temperature scales are linearly related to the Kelvin scale within a negligible margin of error. The modern Celsius scale is defined in terms of the thermodynamic Kelvin scale. The relations between the temperature scales are

$$\mathrm{C} = \mathrm{K} - 273.15 \tag{3.10}$$

$$F = (9/5)\mathrm{K} - 459.67 \tag{3.11}$$

A *thermometer* is a system which samples and measures the potential of systems for heat transfer. In order to effect the sampling and measurement, the thermometer should be capable of convenient thermal communication with other systems *without experiencing any work transfer.* When a thermometer is placed in thermal communication with a system for temperature measurement, a heat transfer occurs until the thermometer and the system are in thermal equilibrium and temperature equality has been established. If the heat transfer does take place in the absence of work transfer, a change of state of the thermometer occurs which is manifested in a change of an observable (mechanical or electrical) property of the thermometer. When thermal equilibrium prevails, the observable property will remain constant. The magnitude of the heat transfer necessary to achieve thermal equilibrium between the thermometer and the system must be such that it produces an undetectable change of state but also a significant change of state in the thermometer. These requirements are usually satisfied by careful thermometer design. A typical example of such a design is the mercury-in-glass thermometer in which the change of state of the thermometer is revealed by a change in the volume occupied by the mercury. In this case, the observable property is the height of the mercury in the glass capillary tube.

The historic Fahrenheit and centigrade temperature scales were established and defined by bringing the mercury-in-glass thermometer into thermal equilibrium with water at its atmospheric pressure boiling point. The height of the mercury column in the constant bore capillary tube was arbitrarily assigned the value 100 C or 212 F. (Any other numbers would have been just as suitable.) The thermometer was then brought into thermal equilibrium with water at its atmospheric pressure freezing point, and the height of the mercury arbitrarily assigned the value 0 C or 32 F. The change of state of the thermometer was assumed to be linear. Hence, the distance on the capillary tube between these two points was divided into 100 equal parts or "degrees" to define the centigrade scale and 180 equal parts to define the Fahrenheit scale. The scales were then linearly extrapolated above and below the fixed points by using the same distance for each degree change in temperature.

3.5 Pure Dissipative System Elements

In the previous development we considered pure system elements for which the change in stored energy is directly related to the single energy transfer across the boundary during the change of state. We now consider pure dissipative system elements which have two energy transfer interactions across the boundary but no stored energy. The unique characteristic is work absorption in that these system elements act as "sinks" for work transfer because they are incapable of any positive work transfer. Since pure dissipative system elements do not have stored energy, the first law of thermodynamics requires a heat transfer which is equal to the work transfer rather than being related to the change of state of the system. Thus, the behavior of the pure dissipative system elements is quite different from the behavior of the pure conservative system elements. All work transfer to a pure dissipative system element is "dissipated" and appears as a heat transfer from the system. These system elements "generate" a heat transfer as the result of the dissipation.

Consider the mechanical system consisting of the spring and the mass shown in Figure 3.7. The composite system is initially at rest with the mass shown in Figure 3.7 at some equilibrium position $x_1 = 0$. If the mass is now displaced to some new location x_2 by means of an external work transfer, the stored mechanical energy of the system is changed. The mass is now released from this position (external force removed) and the system is allowed to oscillate.

One of the simplest models that we can use to represent this physical situation would consist of a pure translational mass and a pure translational spring in a vacuum. For this model the initial work transfer to the composite system would be stored within the spring subsystem as elastic energy. When the mass is released from position x_2, there is a work transfer from the spring subsystem to the mass subsystem. By means of this work transfer, the stored kinetic energy of the mass will increase at the expense of the stored elastic energy of the spring. Thus, the work transfer for the spring is positive while that for the mass is negative. The work transfer for the composite system is, of course, zero. When the mass reaches the position x_1 again, the stored elastic energy of the spring has been reduced to its original zero value; however, the kinetic energy of the mass has increased by an amount equal to the change in the stored elastic energy of the spring. At this position, the mass begins to experience a positive work transfer while the spring experiences a negative work transfer. This process continues indefinitely with the work transfer between the two constituent subsystems reversing sign at the limits of travel of the mass and at the position x_1. The nature of the oscillation is such that the total stored mechanical energy of the composite system is a constant; it does not change with time. Further, this mechanical energy stored within the composite system is equal to the energy transferred by means of the original work transfer to the composite system.

If we were to compare the performance of this model with the performance of a typical physical system, we would find that there is a disparity. The total stored mechanical energy of the physical system is not constant; it decreases with the passage of time. For a given spring and mass, at a given time, the magnitude of the disparity will depend rather strongly upon the circumstances, ranging from quite small to quite large. In all cases, the

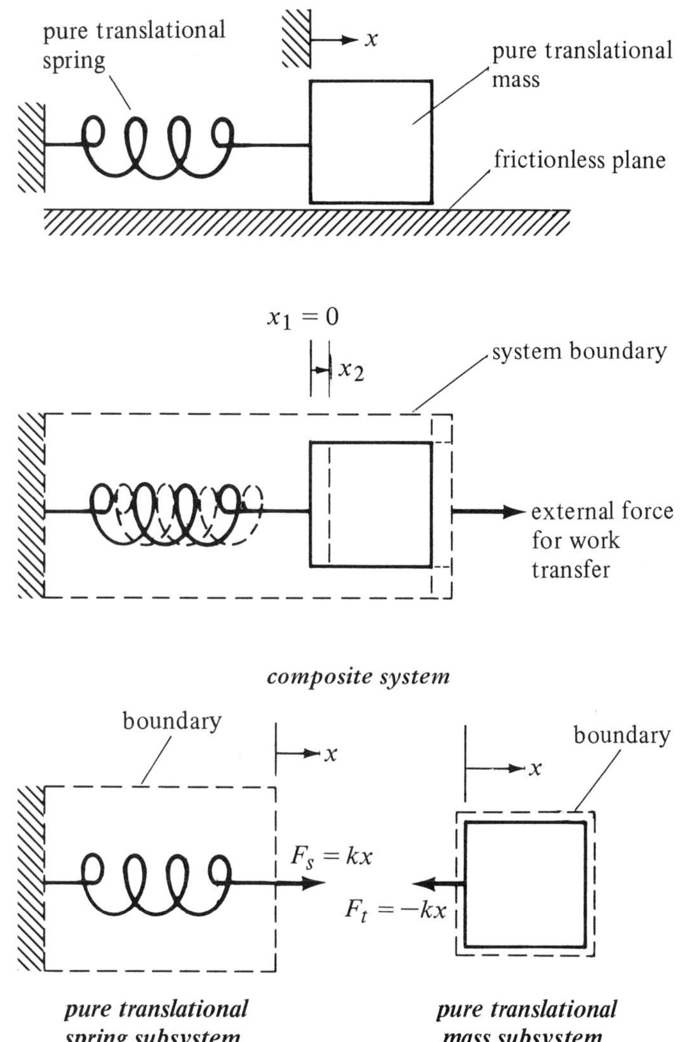

Figure 3.7 Mechanical system

disparity increases in magnitude with time. Finally, the actual physical system exhibits an additional characteristic very different from the model described above. After some period of time, the physical system ceases to oscillate and attains a state of internal equilibrium. In the physical situation just described, the final state of mechanical equilibrium is very nearly identical to the initial state of mechanical equilibrium. As a consequence, the system cannot be returned to its original state by a process that will recover the original work transfer. A system which behaves in this manner is said to be non-conservative.

It is then apparent, that the system model described previously is an inadequate representation of the actual physical system, especially after a long time has passed. In fact, closer inspection of the situation reveals that any model composed of pure conservative system elements is insufficient to describe the behavior of the actual physical system. (This is not to imply, however, that there are not those special situations for which the model composed of pure conservative system elements is sufficient.) Accordingly, we would like to incorporate into our system model a system element which will permit us to represent the non-conservative nature of physical systems. The system element with this capability is called a pure dissipative system element.

In order to be able to describe the nature of pure dissipative system elements it is worth considering some examples of dissipative phenomena in actual physical systems that can be modeled as uncoupled. In a simple Newtonian fluid the rate of dissipation of mechanical energy is proportional to the square of the rate of deformation within the fluid. The fluid property viscosity is the constant of proportionality in the consecutive relation characterizing the dissipation, and in this manner the viscosity serves as a measure of the dissipative nature of the fluid.

The dissipation associated with the sliding of the surface of one rigid solid over the surface of a second rigid solid is known as friction. One example is called Coulomb friction. The friction coefficient (analogous to the fluid property viscosity) is the proportionality factor between the dissipative response (the friction force) and the system input (the load force normal to the interface between the surfaces). However, unlike viscosity, the coefficient of friction is a property of the two solid surfaces involved and not just the solids themselves. Thus, the friction coefficient describes the dissipative nature of the interface. Dissipation of this type is important in the selection of construction material for bearings or brakes.

Perhaps more closely related to the viscous dissipation of a fluid is the bulk dissipation of a solid. Normally, this type of dissipation in solid systems is quite small and is usually neglected. However, there are those instances in which such a dissipation is significant. For example, the selection of the material of construction of a bell or the frame of a machine tool must consider dissipation of this type. Unfortunately, this form of dissipation in solid systems is more complex than in the case of a fluid. Other complexities arise in the deformation of a solid material beyond its elastic limit and into its plastic range where it acquires a permanent deformation after the loading forces are relaxed. Also in this category is the deformation of a visco-elastic material. This can be seen in some commercial plastics which initially assume a deformed state but gradually return to the original state when the load is removed.

From the foregoing examples, it is apparent that there exist in nearly all mechanical systems dissipative effects which, for convenience, we group under the general heading "friction". However, dissipative phenomena are not peculiar to purely mechanical systems alone. Other types of systems also exhibit dissipative phenomena, but whether these dissipative effects are of any consequence in a given physical situation depends upon the prevailing circumstances. In those cases for which dissipation is significant, we would like, if possible, to lump all the dissipative effects into a single pure system element, thereby greatly simplifying the analysis of the system behavior. The pure system element with the desired behavior is a *pure dissipative system element.*

A pure system element of this type possesses two particular characteristics. First, the net work transfer for these system elements is always negative. The pure dissipative system element is incapable of experiencing a net positive work transfer with another system. That is, work transfer is always *into* a pure dissipative system element. Second, the detailed manner in which the pure dissipative system element interacts with its environment is extremely important. That is, the kinetics of the work transfer process have a significant influence on the behavior of the system so that the interaction cannot be described in terms of the values of system properties before and after the interaction occurred. The detailed nature of pure dissipative system elements can be best illustrated by means of two well known examples.

A pure translational damper shown in Figure 3.8 is a one-dimensional pure system element of zero mass which can experience only translational displacements along its axis. The two significant properties of the damper are ϑ, the finite velocity of one end of the damper relative to the other end, and F, the axial boundary force exerted on each end of the damper. Since the damper has no mass, the magnitude of the boundary force must be the same at both ends of the boundary. Thus, the constitutive relation for a pure translational damper gives the force as a function of the velocity.

It is a characteristic of the pure translational damper that the boundary force acts in the same direction as the velocity of the boundary relative to the other end. Therefore, the boundary force F and the relative velocity ϑ always have the same sign[4]. When the relative velocity changes sign, the boundary force reverses direction. Because of this relationship between the boundary force and the relative velocity, the work transfer for the pure translational damper is always negative. This most important point can be clarified by the mathematical definition of work transfer, equation (3.1).

$$\delta W = -F\,dx \tag{3.12}$$

But $dx = \vartheta\,dt$. Thus,

$$\delta W = -F\vartheta\,dt \tag{3.13}$$

Since F and ϑ always have the same sign and dt is always positive, equation (3.13) shows that the work transfer for a pure translational damper is always negative. Thus, the work transfer associated with the displacement of the system element boundary is always *into* the system element.

Equations (3.12) and (3.13) reveal another significant feature of the pure translational damper. The explicit calculation of the work transfer requires the *time* history of the state of the pure translational damper; consequently, the work transfer

[4] Note that when the relative velocities of the two ends of the damper are apart the damper is in tension and both F and ϑ are positive.

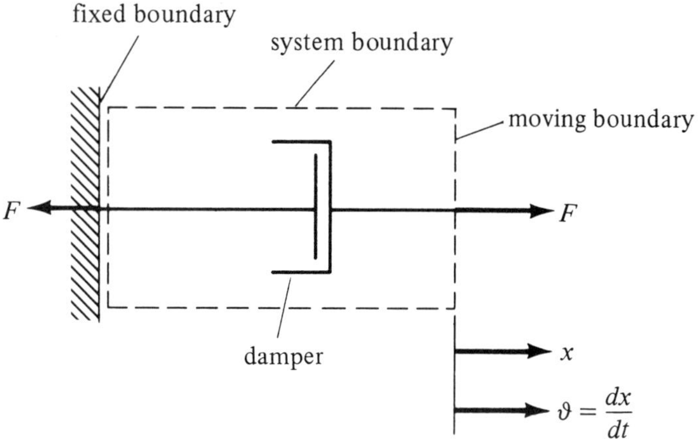

Properties: F = through force on the system boundary

ϑ = rate of extension or $\dfrac{dx}{dt}$

Constitutive relation: $F = b\vartheta$

b = damping coefficient

Interactions: $$-W_{1-2} = \int_{x_1}^{x_2} F\,dx = \int_{t_1}^{t_2} b\vartheta^2\,dt = -Q_{1-2}$$

Stored energy change: $E_2 - E_1 = 0$

Figure 3.8 Pure translational damper (linear)

cannot be represented solely in terms of the system properties before and after the work transfer interaction. We conclude, then, that the work transfer for a pure translational damper is *not a property* as it is in the case of a pure conservative mechanical system element.

For the pure translational damper of Figure 3.8, the work transfer is obtained by substituting the constitutive relation into equation (3.13) and integrating

$$-W_{1-2} = \int_{t_1}^{t_2} b\vartheta^2\,dt \qquad \textbf{(3.14)}$$

Since the pure translational damper has no stored energy change, the first law of thermodynamics requires a heat transfer interaction equal to the work transfer

$$-Q_{1-2} = -W_{1-2} \tag{3.15}$$

The evaluation of the integral in equation (3.14) requires more detailed information than simply the end states of the work transfer process. For the pure translational damper, we must know the state of the system element at every *instant of time* during the interaction before we can determine the work transfer and the heat transfer. In distinct contrast to the pure conservative mechanical system element, it makes a considerable difference whether the work transfer process for a pure translational damper was carried out slowly or rapidly.

The electrical system analog of the pure translational damper is the pure electrical damper, more commonly known as the resistor. The boundary force (the voltage) is determined by the rate at which charge crosses the boundary (current). The general constitutive relation gives the potential difference v (voltage) as a function of the current i. For the linear resistor

$$v = iR \tag{3.16}$$

where the proportionality constant R is the electrical resistance. Using equation (3.6), the work transfer is

$$-W_{1-2} = \int_{q_1}^{q_2} v\,dq = \int_{q_1}^{q_2} iR\,dq \tag{3.17}$$

But $dq = idt$. Then,

$$-W_{1-2} = \int_{t_1}^{t_2} i^2R\,dt \tag{3.18}$$

Since the pure electrical damper has no stored energy change, the first law of thermodynamics requires

$$-Q_{1-2} = -W_{1-2} \tag{3.19}$$

Again, the work transfer can be determined only when the current is known as a function of time for the process. Equation (3.18) clearly illustrates the fact that the work transfer and the heat transfer are both negative since i^2, R, and dt are all positive quantities.

3E.5 Sample Problem: Compute the work transfer and heat transfer for a dashpot damper which is extended at a constant rate of 1.2 m/s m/s for 0.1 second. The damping coefficient b is 17 Ns/m.

Solution: The system and process are modeled as a pure translational damper as in Figure 3.8. The work transfer is

$$-W_{1-2} = \int_{t_1}^{t_2} b\dot{\vartheta}^2\,dt = b\dot{\vartheta}^2(t_2 - t_1)$$

$$-W_{1-2} = (17\ \text{Ns/m})(1.2\ \text{m/s})^2(0.1\ \text{s} - 0) = 2.45\ \text{J}$$

Since the stored energy is zero, the first law requires

$$-Q_{1-2} = -W_{1-2} = 2.45\ \text{J}$$

3.6 History of First Law of Thermodynamics

In passing, it is worth noting that some of the terminology of thermodynamics currently in use appears contradictory and illogical in light of the accepted definitions. This terminology is a product of the evolutionary development of thermodynamics and has remained in use for historical reasons. Although the origin of thermodynamics can be traced to the discovery of fire, it was not until the eighteenth century that the basic concepts of thermodynamics were formalized. At that time, the science of mechanics developed rapidly from the early works of Newton, and even today non-relativistic mechanics still appear in much the same form.

During that same period of time, the effects associated with temperature and heat transfer were receiving considerable attention from Gabriel D. Fahrenheit, Anders Celsius and others who together formulated what was then known as the caloric theory of heat. In keeping with the conservation principles which were being concurrently developed in mechanics, heat or caloric was regarded as a quantity which was conserved and flowed from one system to another. In this form, the simple concepts of the caloric theory are adequate for the pure thermal system element and for interactions between these system elements. However, despite considerable effort and ingenuity, these simple concepts were just not adequate to model more complex physical situations.

In spite of the deficiencies of the caloric theory, much of its terminology remains in use in thermodynamics today. For example, the terms "specific heat" and "heat capacity" originated from the concepts of stored heat and are now actually misnomers for the energy derivatives with respect to temperature, equations (3.7) and (3.8). The term "latent heat" (to be defined in Chapter 8) is another example of a term still in use today. This term was originally devised for the more complex process associated with a heat transfer to a boiling liquid. The adjective "latent" was used for this particular type of heat transfer since it did not produce the increase in temperature normally associated with heat transfer to a liquid.

Since the caloric theory preceded the formulation of the first law of thermodynamics, it was not recognized that in a coupled thermodynamic system the heat transfer and work transfer are related. Thus, a separate system of units was used for each of these interactions. The heat transfer units were originally defined in terms of the heat transfer required to produce a given change of state of a given mass of water. The calorie was defined as the heat transfer necessary to change the temperature of one gram of water at atmospheric pressure from 15 C to 16 C. The British thermal unit (Btu) was defined as the heat transfer required to change the temperature of one pound of water at atmospheric pressure from 59.5 F to 60.5 F.

During the first half of the nineteenth century, the first law of thermodynamics was formulated to relate properly the mechanical and thermal behavior of systems. One of the early workers in this regard was Count Rumford (Benjamin Thomson) who, after making some observations of the heat transfer process resulting from the boring of cannon barrels, tried to relate work transfer and heat transfer in the context of the caloric theory. For a number of years this interrelation remained the subject of considerable controversy and confusion. The matter remained unsettled until the careful measurements by James Prescott Joule during the period 1840–1848 established a firm experimental foundation for the first law of thermodynamics.

Joule's experiment consisted of a brake or paddle wheel surrounded by a bath of water which was thermally isolated from the environment. The experiment can be modeled as a pure dissipative system element in thermal communication with a pure thermal system element. Joule carefully measured the work transfer to the dissipative element and the resulting change of state (temperature change) of the pure thermal system element. As illustrated by equation (3.15), Joule was able to show that the work transfer into the dissipative element necessary to produce a change of state in the pure thermal system element was always proportional to the heat transfer required to produce the same change of state. The two were proportional rather than equal because Joule measured work transfer and heat transfer in independent units. The constant of proportionality measured by Joule is referred to as the "mechanical equivalent of heat" and was given the symbol J. In reality, the constant is merely the factor for converting from one set of energy units to a second. Thus, the numerical value of J is different for each combination of work transfer and heat transfer units. On this basis it is possible to formulate the modern definition of the calorie as

$$1 \text{ calorie} = 4.1868 \text{ Joules} \qquad \textbf{(3.20)}$$

The modern definition of the British thermal unit (Btu) is

$$1 \text{ Btu} = 778.169 \text{ ft-lbf} \qquad \textbf{(3.21)}$$

These modern definitions were formulated so that the old definitions from the caloric theory are still true.

The real significance of Joule's measurements was that the measurements led to the eventual conclusion that both work transfer and heat transfer are energy transfer interactions and that either can change the energy of the system. Thus, work transfer and heat transfer interactions do not involve two separate quantities which are individually conserved.

Although the first law of thermodynamics requires that heat transfer and work transfer be equivalent in their measurement, these two interactions are not equivalent in every way. In our subsequent development of the second law of thermodynamics we will derive specific numerical limits on the interchangeability of both heat transfer and work transfer.

3.7 Uncoupled Thermodynamic Systems

Heretofore, we have focused on pure systems elements, each with a simple one-dimensional behavior. We will now use these system elements to model the multidimensional behavior of more complex, but uncoupled thermodynamic systems.

In summary, the three general classes of pure system elements are: (1) pure conservative system elements, (2) pure thermal system elements, and (3) pure dissipative system elements. We now review these characteristics.

The characteristics of pure conservative system elements are:

1. The states are fixed by a single independent property. The constitutive relations relate the dependent properties to the single independent property. All changes of state follow the same path defined by the succession of values of the independent property.
2. The only interaction is work transfer which is stored by changing the energy of the system.
3. The first law of thermodynamics becomes

$$-W_{1-2} = E_2 - E_1 \tag{3.22}$$

The characteristics of pure thermal system elements are:

1. The state is fixed by a single independent property, either temperature or the stored thermal energy.
2. The only interaction is heat transfer which is stored by changing the stored thermal energy of the system.
3. The first law of thermodynamics becomes

$$Q_{1-2} = U_2 - U_1 \tag{3.23}$$

The characteristics of pure dissipative system elements are:

1. The instantaneous rate of interaction is fixed by the state of the system. The system does not store energy. The interaction during a change of state depends on the path of the change of state and is not specified by the end states.
2. The work transfer and the heat transfer are always negative.
3. The first law of thermodynamics becomes

$$-W_{1-2} = -Q_{1-2} \tag{3.24}$$

As an introduction to the modeling of the multidimensional behavior of more complex, but uncoupled thermodynamic systems, we will consider systems without dissipation. These uncoupled systems can be modeled as an assembly of pure conservative system elements and pure thermal system elements as shown in Figure 3.9. Since the conservative system elements interact only by work transfer and the thermal system elements only by heat transfer, the thermal elements cannot interact with or influence the mechanical elements and vice versa. Thus, the pure thermal system elements model the *thermal aspects* of the physical situation. They are completely divorced from the *mechanical aspects* of the physical situation which are modeled by the pure conservative system elements. Systems which can be modeled in this manner are called *uncoupled systems* because of this separation of thermal and mechanical behavior.

The first law of thermodynamics for an uncoupled system is the sum of the first law expressions for each of the system elements included in the model. When these expressions are added, all internal interactions between system elements cancel. Then the net external interactions equal the sum of the energy changes for all system elements included in the model. The internal interactions cancel because each appears twice; once with a positive sign and once with a negative sign. The relation between the thermodynamic system and the component system elements is illustrated in Figure 3.9. Figure 3.9 (a) shows the total system with external interactions Q and W. Figure 3.9 (b) shows the pure system elements as interacting subsystems. The *interaction nodes* are used to interrelate the interactions at the contiguous boundaries of the subsystems. The net interaction at every node must be zero.

We can now develop the first law of thermodynamics for the uncoupled system by summing the first law for each pure system element which is a subsystem. For the three pure conservative system elements A, B, and C, respectively

$$-W_A = (E_2 - E_1)_A \tag{3.25}$$
$$-(W_{Bi} + W_{Bii}) = (E_2 - E_1)_B \tag{3.26}$$
$$-W_C = (E_2 - E_1)_C \tag{3.27}$$

For the two pure thermal system elements

$$Q_I = (U_2 - U_1)_I \tag{3.28}$$
$$Q_J = (U_2 - U_1)_J \tag{3.29}$$

The three interaction node equations are

$$W_A + W_{Bi} = 0 \tag{3.30}$$
$$W_{Bii} + W_C - W = 0 \tag{3.31}$$
$$-Q_I - Q_J + Q = 0 \tag{3.32}$$

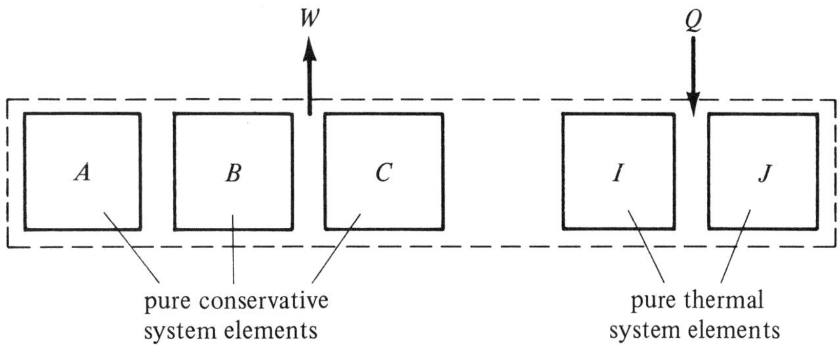

(a) Total system model

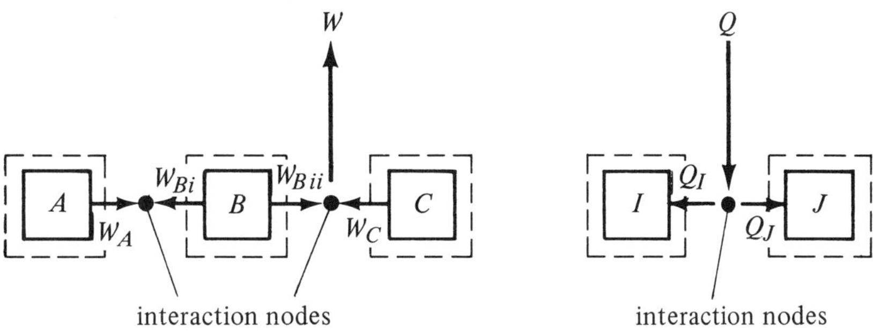

(b) Subsystem models and interaction nodes

Figure 3.9 Model for uncoupled thermodynamic system without dissipation

When we sum equations (3.25, 3.26, 3.27, 3.30, and 3.31), we obtain

$$-W = (E_2 - E_1)_A + (E_2 - E_1)_B + (E_2 - E_1)_C \qquad \textbf{(3.33)}$$

and from summing equations (3.28, 3.29, and 3.32), we obtain

$$Q = (U_2 - U_1)_I + (U_2 - U_1)_J \qquad \textbf{(3.34)}$$

Note that the internal interactions W_A, W_{Bi}, W_{Bii}, W_C, Q_I and Q_J each appear once in a subsystem first law and once in an interaction node equation with the opposite sign. Thus, all internal interactions add to zero in the sum for the first law for the total system.

For an uncoupled thermodynamic system without dissipation, the first law of thermodynamics is in two independent and uncoupled parts. The conservative, work transfer related part, is

$$-W_{1-2} = \sum_i (E_2 - E_1)_i \tag{3.35}$$

where the energy change summation is for all conservative system elements. The heat transfer related part is

$$Q_{1-2} = \sum_i (U_2 - U_1)_i \tag{3.36}$$

where the summation is for all thermal system elements. It is this separation of the first law with separate forms of stored energy that characterizes the system with uncoupled properties. The use of the uncoupled system model without dissipation is illustrated in the following sample problem.

3E.6 Sample Problem: A lead shot with a mass of 0.1 kg free falls in a vacuum tank while transferring heat to the cooler tank walls, Figure 3E.6. The initial height is 30 m, the initial velocity is zero, and the initial temperature is 400 C. The final height is 0 m and the final temperature is 200 C. The specific heat of lead is 133 J/kgC. The acceleration of gravity is 9.80 m/s^2.

Determine the final velocity of the lead shot and the values of the heat transfer and work transfers involved in the process.

Solution: Step 1; Select the system: The system is the lead shot in the gravitational field. The system boundary is defined in Figure 3E.6 (a). The boundary is moving, but the boundary force is zero; thus, the work transfer W is zero. The heat transfer across the boundary by thermal radiation is Q.

Step 2; Model behavior: The lead shot is modeled as an assembly of a pure translational mass, a pure gravitational spring, and a pure thermal system shown in Figure 3E.6 (b).

Step 3; Model the processes: The interaction between the pure translational mass and the pure gravitational spring is an internal interaction within the system. The system has no external work transfer. The external heat transfer for the system is Q from the environment to the system. The mechanical processes

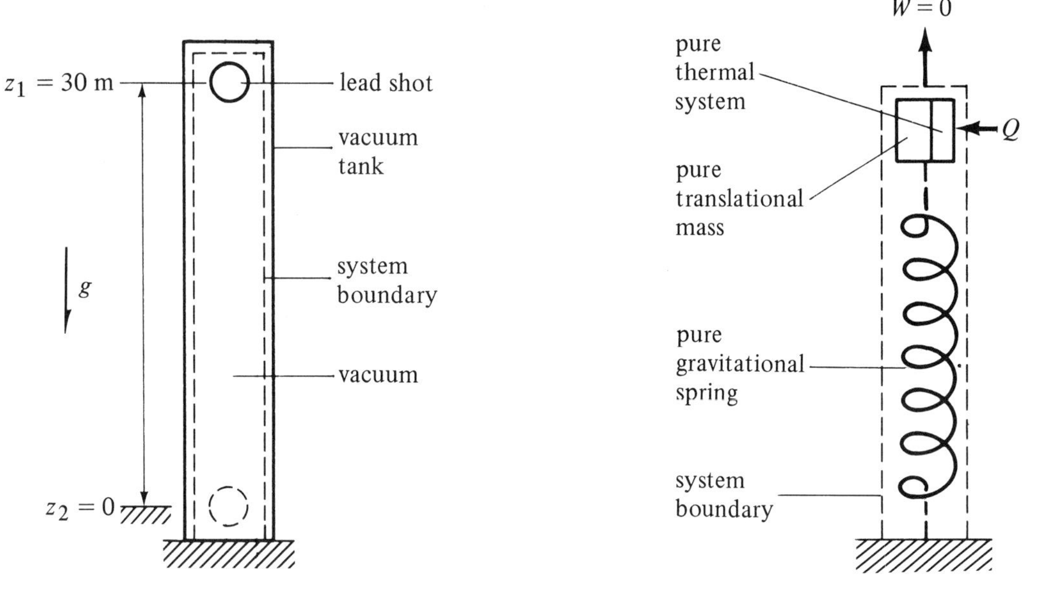

(a) Apparatus

(b) Thermodynamic model

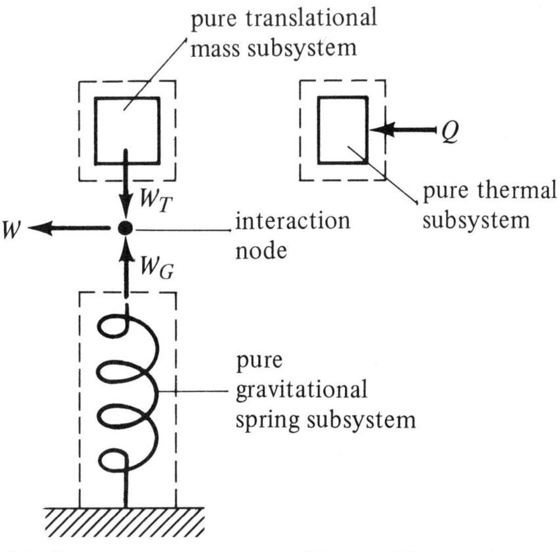

(c) Component systems and internal interactions

Figure 3E.6 Uncoupled thermodynamic system without dissipation

are without any dissipation. From the constitutive relations for the pure system elements we have

$$(E_2 - E_1)_{\text{kinetic}} = (1/2)m\vartheta_2^2 - (1/2)m\vartheta_1^2$$
$$(E_2 - E_1)_{\text{grav}} = mgz_2 - mgz_1$$
$$(U_2 - U_1)_{\text{therm}} = mc(T_2 - T_1)$$

Step 4; Apply the laws of thermodynamics: For an uncoupled system without dissipation, the first law is

$$-W_{1-2} = (E_2 - E_1)_{\text{kinetic}} + (E_2 - E_1)_{\text{grav}}$$
$$Q_{1-2} = (U_2 - U_1)_{\text{therm}}$$

since W_{1-2} is zero

$$(E_2 - E_1)_{\text{kinetic}} = -(E_2 - E_1)_{\text{grav}}$$
$$(1/2)m\vartheta_2^2 - 0 = -(0 - mgz_1)$$
$$\vartheta_2 = (2gz_1)^{1/2} = [2(9.80 \text{ m/s}^2)(30 \text{ m})]^{1/2} = 24.2 \text{ m/s}$$

Note that the scalar first law does not establish the direction of the velocity.

$$Q_{1-2} = mc(T_2 - T_1)$$
$$Q_{1-2} = (0.1 \text{ kg})(133 \text{ J/kgC})(200 \text{ C} - 400 \text{ C}) = -2660 \text{ J}$$

The negative sign indicates the heat transfer is from the system to the environment.

For the system defined in Figure 3E.6 (a) and modeled in Figure 3E.6 (b), the work transfer is zero so the total mechanical energy of the system remains constant. The gravitational energy is converted to kinetic energy by internal interactions. The negative heat transfer decreases the stored thermal energy of the system.

Additional understanding of this problem is obtained by determining the internal interactions between the component subsystems. The subsystems and interaction nodes are shown in Figure 3E.6 (c). When the first law is applied to the subsystems, we have for the pure thermal system

$$Q_{1-2} = (U_2 - U_1)_{\text{therm}} = -2660 \text{ J}$$

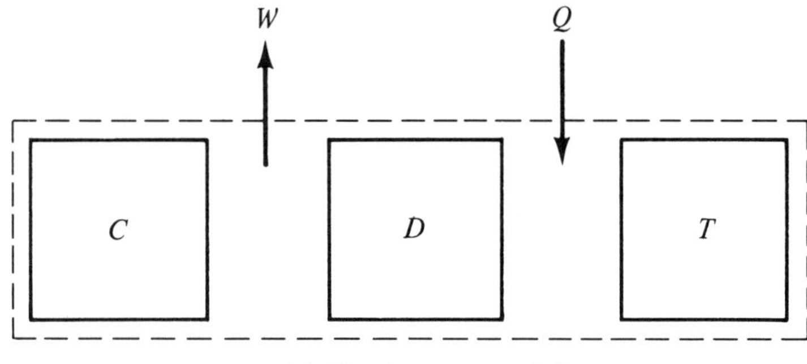

(a) Total system model

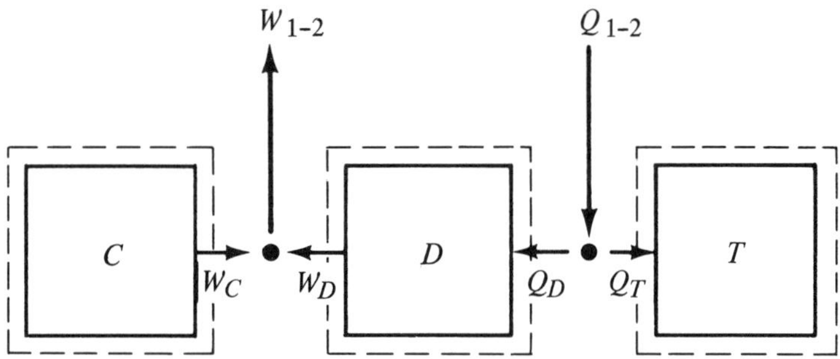

(b) Subsystem models and interaction nodes

Figure 3.10 Model for uncoupled thermodynamic system with dissipation

For the pure translational mass

$$-W_T = (E_2 - E_1)_{\text{kinetic}} = (1/2)m\vartheta_2^2 - 0$$
$$-W_T = (1/2)(0.1 \text{ kg})(24.25 \text{ m/s})^2 = 29.4 \text{ J}$$

For the pure gravitational spring

$$-W_G = (E_2 - E_1)_{\text{grav}} = 0 - mgz_1$$
$$-W = -(0.1 \text{ kg})(9.80 \text{ m/s})^2(30 \text{ m}) = -29.4 \text{ J}$$

Thus, the gravitational spring subsystem transfers work to the translational mass subsystem. However, we must remember that we can identify this work transfer across the boundary between subsystems only in our model Figure 3.10 (b) and (c).

In the real physical system Figure 3.10 (a) there is no geometric boundary between the gravitational and inertial effects.

Uncoupled thermodynamic systems with dissipation are more complex than those without dissipation because the dissipation provides a means for work transfer to influence the thermal aspects of the system. However, the dissipation is unidirectional. Thermal processes cannot influence the mechanical aspects of the system. Dissipation only allows mechanical processes to influence the thermal aspects.

The uncoupled thermodynamic system with dissipation is modeled as an assembly of pure system elements with the mechanical aspects modeled by appropriate pure conservative system elements. The thermal aspects are modeled by pure thermal system elements and the dissipative aspects modeled by pure dissipative system elements, as shown in Figure 3.10.

The first law of thermodynamics for the uncoupled system with dissipation is the sum of the first law expressions for each system element. As before, all internal interactions between subsystems are eliminated in the addition. Figure 3.10 shows the relation between the total system and the subsystems comprising the pure system elements. The first law expressions for the subsystems are

$$-W_C = (E_2 - E_1)_C \tag{3.37}$$

$$Q_D - W_D = 0 \tag{3.38}$$

$$Q_T = (U_2 - U_1)_T \tag{3.39}$$

The interaction node equations are

$$W_C + W_D - W_{1-2} = 0 \tag{3.40}$$

$$-Q_T - Q_D + Q_{1-2} = 0 \tag{3.41}$$

By direct summation of equations (3.37–3.41), the first law for the system is

$$Q_{1-2} - W_{1-2} = (E_2 - E_1)_C + (U_2 - U_1)_T \tag{3.42}$$

When an uncoupled system undergoes a process involving dissipation, the first law does not separate into two independent equations because the dissipation provides an internal interaction between the mechanical and thermal aspects of the system. If we combine equations (3.37) and (3.40), we have

$$-(W_{1-2} - W_D) = (E_2 - E_1)_C \tag{3.43}$$

and if we combine equations (3.39) and (3.41), we have

$$Q_{1-2} - Q_D = (U_2 - U_1)_T \tag{3.44}$$

But we know that Q_D and W_D are negative so we can conclude

$$(E_2 - E_1)_C < -W_{1-2} \tag{3.45}$$
$$(U_2 - U_1)_T > Q_{1-2} \tag{3.46}$$

When dissipation occurs during a process in an uncoupled system, the increase in stored mechanical energy is less than the mechanical work transfer into the system. Further, the increase in stored thermal energy is greater than the heat transfer into the system, as expressed by equations (3.45) and (3.46). In effect the dissipation "creates" the thermal energy at the expense of the "loss" of an equal amount of mechanical energy as required by the first law, equation (3.42).

The use of the uncoupled system model with dissipation is illustrated in the following sample problem.

3E.7 Sample Problem: A spring is connected within an oil filled piston-cylinder apparatus as shown in Figure 3E.7. The spring is slowly extended by pulling on the string. The string is cut and the spring returns to the unextended length. Determine the final temperature.

The initial conditions, state 1, are: The spring is at zero force. The oil, spring, and piston are at 20 C. During the stretching process, the spring extends 0.1 m to state 2. After the string is cut the spring returns to the unextended length, state 3. The spring constant is 1000 N/m. The combined heat capacity of the spring, oil and piston is 4000 J/C.

Solution: Step 1; Select system: The system for analysis is the components within the boundary drawn at the cylinder wall.

Step 2; Model the behavior: The system is modeled as a pure translational spring, a pure translational damper, and a pure

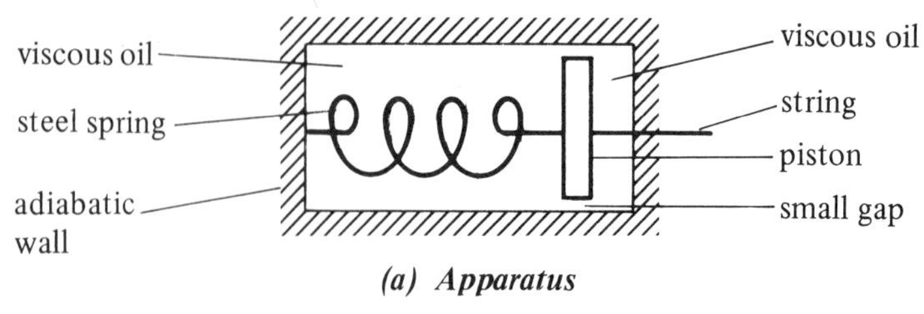

(a) Apparatus

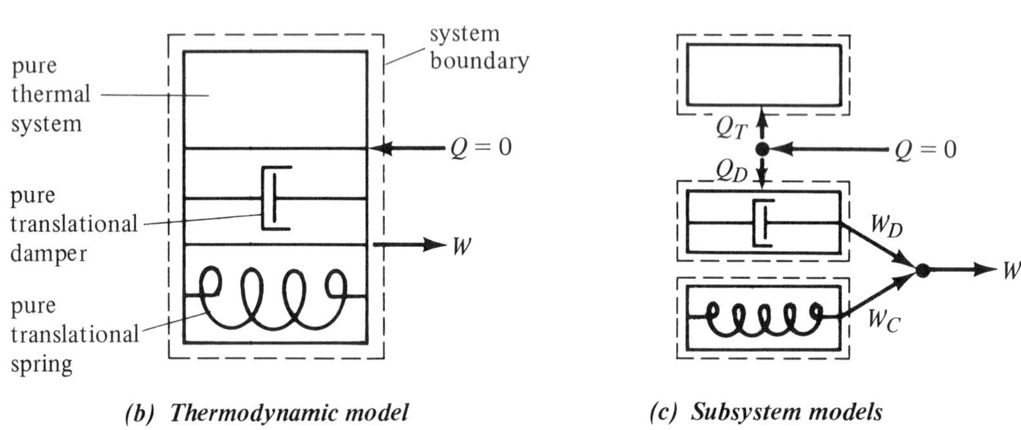

(b) Thermodynamic model *(c) Subsystem models*

Figure 3E.7 Uncoupled thermodynamic system with mechanical dissipation

thermal system representing the thermal aspects of the spring, the oil, and the piston. The cylinder wall is modeled as adiabatic.

Step 3; Model the process: The process 1–2 is modeled as a conservative process since the extension is assumed slow enough so that the damping force may be taken as zero. This process has a zero heat transfer since the system is surrounded by an adiabatic wall. The process 2–3 is modeled as a zero work transfer process with dissipation. The work transfer is zero since the motion at the system boundary is zero. This process is also adiabatic.

Step 4; Apply the laws of thermodynamics: The first law of thermodynamics for the system defined in Figure 3E.7 (b) is

$$Q - W = (E_f - E_i)_E + (U_f - U_i)_T$$

where the energy change is the sum of the energy changes for each uncoupled system element included in the model. For the conservative process 1–2 the first law separates.

$$Q_{1-2} = (U_2 - U_1)_T = 0 = C(T_2 - T_1)$$
$$-W_{1-2} = (E_2 - E_1)_E = (1/2)kx_2^2 - (1/2)kx_1^2$$
$$-W_{1-2} = (1/2)(1000 \text{ N/m})(0.1 \text{ m})^2 - 0 = 5 \text{ J}$$

For the dissipative process 2–3 the first law becomes

$$Q_{2-3} - W_{2-3} = (E_3 - E_2)_E + (U_3 - U_2)_T = 0$$

Since $Q_{2-3} = W_{2-3} = 0$,

$$C(T_3 - T_2) = -(1/2)kx_3^2 + (1/2)kx_2^2$$

but $x_3 = x_1 = 0$, and $T_2 = T_1$. Thus

$$T_3 - T_1 = \frac{k}{2C} x_2^2$$
$$T_3 = 20 + \frac{1000 \text{ N/m}}{2(4000 \text{ J/C})}(0.1 \text{ m})^2 = 20.00125 \text{ C}$$

It is important to note that, for the system defined by the boundary shown in Figure 3E.7 (b), the temperature rise is the result of internal dissipation and not the result of a heat transfer.

This problem can also be solved by analysis of the individual subsystems shown in Figure 3E.7 (c). The analysis here is more involved; however, it provides additional information about the internal interactions involved in each process. For the conservative process 1–2 we have assumed the damping force is zero so

$$(W_D)_{1-2} = (Q_D)_{1-2} = 0$$

The heat transfer interaction node equation requires

$$-Q_T - Q_D + Q = 0$$
$$(Q_T)_{1-2} = Q_{1-2} - (Q_D)_{1-2} = 0 - 0 = 0$$

The first law for the thermal subsystem is

$$0 = (U_2 - U_1)_T = C(T_2 - T_1)$$
$$T_2 = T_1$$

The first law for the spring subsystem is

$$-(W_C)_{1-2} = (1/2)kx_2^2 - (1/2)kx_1^2 = 5\text{ J} - 0 = 5\text{ J}$$

The interaction node equation requires

$$W_D + W_C - W = 0$$
$$-W_{1-2} = -(W_C)_{1-2} - (W_D)_{1-2} = 5\text{ J} - 0 = 5\text{ J}$$

For the dissipative process 2–3, the first law for the spring requires

$$-(W_C)_{1-2} = \frac{1}{2}kx_3^2 - \frac{1}{2}kx_2^2 = 0 - 5\text{ J} = 5\text{ J}$$

Since $W_{2-3} = 0$, the interaction node equation requires

$$(W_D)_{2-3} = -(W_C)_{2-3} = -5\text{ J}$$

The first law for the damper requires

$$-(W_D)_{2-3} = -(Q_D)_{2-3} = 5\text{ J}$$

Since $Q_{2-3} = 0$, the interaction node equation requires

$$(Q_T)_{2-3} = -(Q_D)_{2-3} = 5\text{ J}$$

The first law for the thermal subsystem is

$$(Q_T)_{2-3} = (U_3 - U_2)_T = C(T_3 - T_2)$$

But $T_2 = T_1$, thus

$$T_3 = T_1 + \frac{(Q_T)_{2-3}}{C} = 20\text{ C} + \frac{5\text{ J}}{4000\text{ J/C}} = 20.00125$$

3E.8 Sample Problem: A resistive heating element is used to heat an oil bath contained within an imperfectly insulated container, Figure 3E.8 (a). The element is supplied with 16 amperes

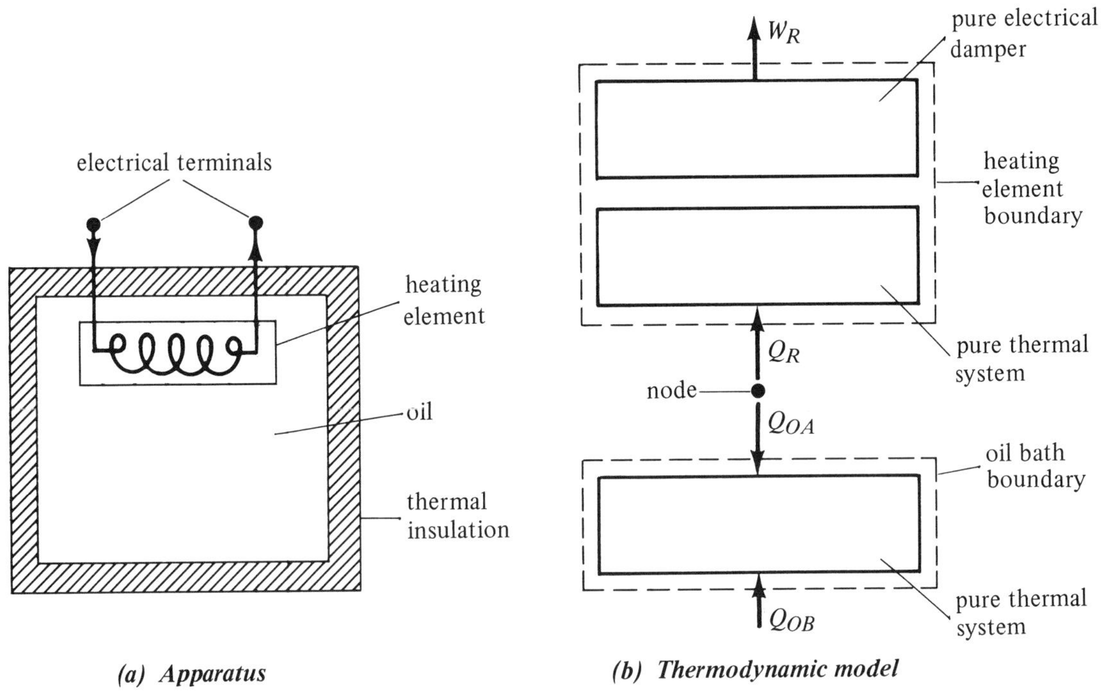

Figure 3E.8 Uncoupled thermodynamic system with electrical dissipation

of current at a potential of 220 volts for 60 seconds. The mass of the heating element is 0.5 kg and the mass of the oil is 2 kg. The specific heat of the heating element is 700 J/kgC and of the oil is 2100 J/kgC. The initial temperature of the oil and heating element is 17 C and the final temperature is 21.12 C. What is the heat transfer to the oil from the heating element? What is the heat transfer to the insulation around the oil container?

Solution: Step 1; Select systems: Two systems are defined as shown in Figure 3E.8 (b). System R is the heating element and system O is the oil bath, with the boundaries as shown. System R has electrical work transfer W_R at the terminals and heat transfer Q_R from the oil bath. System O has heat transfer Q_{OA} from the heater and Q_{OB} from the insulation.

Step 2; Model behavior: System R is modeled as a pure electrical damper and a pure thermal system. System O is modeled as a pure thermal system.

Step 3; Model processes: The process in system R is dissipative with a negative work transfer, an increase in thermal energy U, and a negative heat transfer. The process in system O is thermal with two heat transfers and a change in thermal energy U. For the electrical damper of system R

$$-W_R = \int_{t_1}^{t_2} i^2R\,dt = \int_{t_1}^{t_2} vi\,dt$$

$$-W_R = (220\text{ V})(16\text{ A})(60\text{ s}) = 21.12\text{ kWs} = 21.12\text{ kJ}$$

where the constitutive relation for the damper has been used to eliminate R. For the thermal system of system R

$$(U_2 - U_1)_R = (mc)_R(T_2 - T_1)_R$$

For system O

$$(U_2 - U_1)_O = (mc)_O(T_2 - T_1)_O$$

Step 4; Apply the laws of thermodynamics: For system R the first law gives

$$\begin{aligned} Q_R - W_R &= (U_2 - U_1)_R = (mc)_R(T_2 - T_1) \\ Q_R &= W_R + (mc)_R(T_2 - T_1) \\ &= 21.12\text{ kJ} + (0.5\text{ kg})(0.7\text{ kJ/kgC})(21.12\text{ C} - 17\text{ C}) \\ &= -19.64\text{ kJ} \end{aligned}$$

For system O the first law gives

$$Q_{OA} + Q_{OB} = (U_2 - U_1)_O = (mc)_O(T_2 - T_1)$$

The interaction node equation is

$$Q_R + Q_{OA} = 0$$

Thus

$$\begin{aligned} Q_{OB} &= (mc)_O(T_2 - T_1) + Q_R \\ Q_{OB} &= (2\text{ kg})(2.1\text{ kJ/kgC})(21.12\text{ C} - 17\text{ C}) + (-19.64\text{ J}) \\ Q_{OB} &= -1.96\text{ kJ} \end{aligned}$$

The heat transfer to the oil from the heater is $Q_{OA} = -Q_R =$ 19.64 kJ. The heat transfer to the insulation is $-Q_{OB} = 1.96$ kJ.

Problems

3.1 A string-mass system is assembled as shown in Figure 3P.1. The pulley and string may be assumed massless. The support string restrains the assembly in the position shown. When the support string is cut, the rider of mass m_r and the mass m_2 descends while the mass m_1 rises. The mass m_2 will pass through the hole in the bracket but the rider will not.

(a) Determine the maximum distance s which the mass m_2 descends from the starting position.

(b) Describe the behavior of the rider in terms of its various stored energy modes.

(c) Is there a work transfer involved? If so, where?

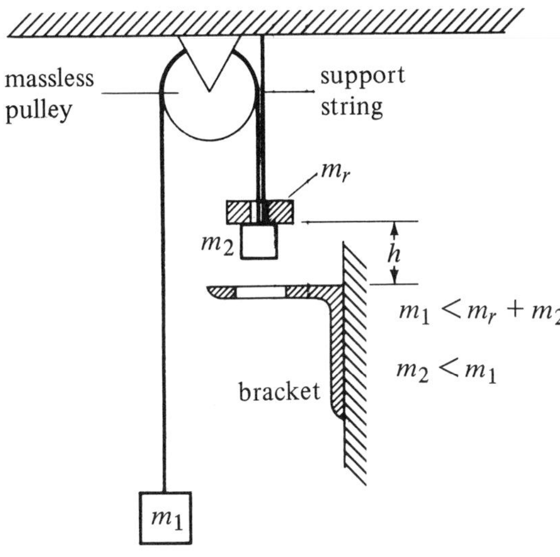

Figure 3P.1

3.2 A mass M hangs from a steel cable, stretching the cable a length x_1 beyond its free length $x_0 = 0$. The mass is lifted just enough to relieve the tension in the cable and is then dropped.

(a) Describe the model that you would use to determine the maximum force on the cable after the mass is dropped.

(b) What is the maximum force in the cable?

(c) What are the limitations of your model?

3.3 A device used to measure the power output (the rate of positive work transfer) of a machine is called a dynamometer. One type of dynamometer is an absorption dynamometer in which all of the power produced by the machine is dissipated. There are techniques available for measuring the rate at which the energy is dissipated so that it is possible to measure power in this manner. One technique for making such a measurement is shown in Figure 3P.3. The dynamometer is supported

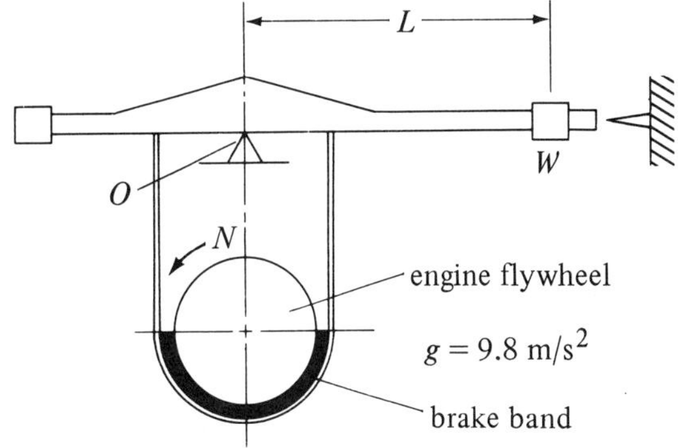

Figure 3P.3

on the knife edge O and the brake band is tightened, while the engine is running, until the arm balances the weight of the attached mass W. With the weight removed and the brake band loose, the arm is in equilibrium. On a particular test $W = 25$ kg, $L = 600$ mm and $N = 1640$ revolutions per minute.

(a) Determine the power output at this speed for the engine being tested.

(b) If the test is conducted in the ambient atmosphere of a typical laboratory, determine the work transfers, heat transfers, and changes in stored energy for the brake and the environment as separate systems. Also determine these interactions for the composite system consisting of the brake band and the environment.

(c) Repeat part (b) above for a test conducted in a perfectly evacuated chamber.

(d) An alternate dynamometer design would be to couple a D.C. generator to the engine being tested. The electrical output from the generator is then passed through a bank of resistors. Describe the nature (positive, negative, or zero) of the work transfers, heat transfers, and changes in stored energy for the system consisting of:

(i) the generator, resistors, and environment.
(ii) the generator and resistors.
(iii) the resistors.
(iv) the resistors and the environment.
(v) the environment.

3.4 As part of a heat treating operation, a steel part with a mass of 1 kg is to be quenched from a temperature of 1000 C to a temperature of 100 C. Available as a quenching medium is a water bath at 21 C enclosed in an adiabatic chamber. The properties of the two media are:

specific heat of steel = 0.473 kJ/kgK
specific heat of water = 4.187 kJ/kgK

(a) Describe a model suitable to determine the amount of water necessary to perform the quenching process. State all the assumptions associated with this model.

(b) What is the necessary amount of water?

(c) Is there any heat transfer for the system consisting of the steel part and the water?

3.5 An electron beam welder, Figure 3P.5, is a device which employs a beam of high energy electrons (20 kV) in a high vacuum chamber to melt a very localized spot on a metal target. In welding, the parts to be joined are used as the target while the beam traverses the seam to be welded. In a certain application, two parts are to be welded in a manner such that the temperature of a particular area, remote from the weld itself, must not exceed 205 C. The problem is to determine, in a very simple way, a safe limit for the welding time for each current setting of the welder.

The data for the welding machine are:
acceleration voltage 20 kV
beam current 0 to 300 mA.

The data for the parts are:
total mass 0.2268 kg
specific heat 2082 J/kgK
maximum initial temperature 33 C.

Carefully define and model the system for the solution of this problem in thermodynamic terms. Give values for the work transfer, heat transfer, and changes in energy for your system.

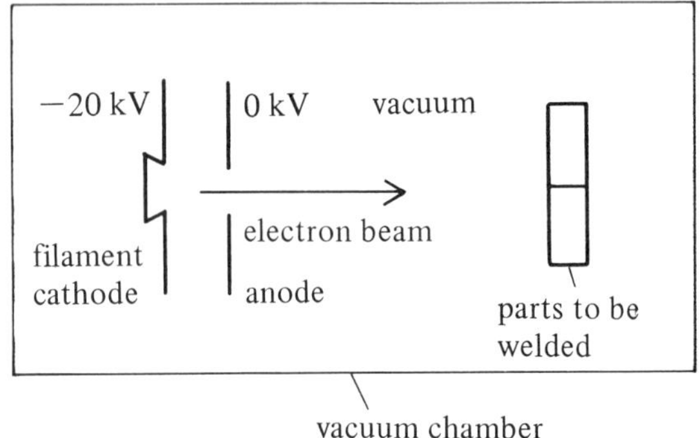

Figure 3P.5

3.6 A system consists of an externally insulated vessel filled with water. Immersed in the water is a paddle wheel which is driven by an electric motor which delivers a constant 200 watts. When the vessel contains 5 kg of water, it is found that if the motor operates for a time period of 1130 seconds, the temperature of the complete system will eventually increase 6 C. If the mass of the water in the vessel is increased to 10 kg of water, it is found that 1758 seconds are required to change the temperature of the system 6 C with the same power input rate of 200 watts.

(a) Describe a model to determine the heat capacity of the vessel and paddle wheel combined and the specific heat of the water.

(b) What is the heat capacity (in kJ/K) of the vessel and paddle wheel combined?

(c) What is the specific heat of the water in kJ/kgK?

(d) Make a table similar to the one here showing whether the heat transfer, work transfer, and change in stored energy of the paddle wheel, the vessel and the water is positive, negative, or zero for a given change of temperature. Do the same for the composite system consisting of paddle wheel, vessel, and water.

TABLE 3P.6

	Q_{1-2}	W_{1-2}	ΔE
paddle wheel			
vessel			
water			
composite system			

3.7 Consider the mechanical system shown in Figure 3P.7 in which a mass M is suspended from a string in a gravity field g. Below the mass is a rigid platform of mass ($m \ll M$) which is mounted on two identical springs with spring constants k. The string supporting the mass M is cut, and the mass falls from a height H striking the platform midway between the two springs.

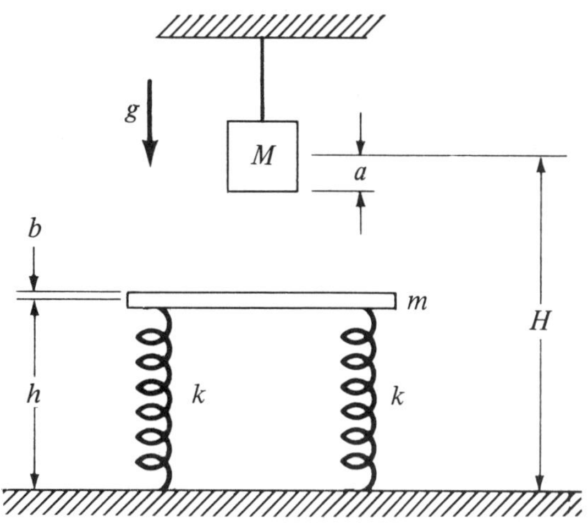

Figure 3P.7

(a) Define the system, in terms of the simplest model composed of pure conservative mechanical system elements, that will enable you to determine the minimum height that the platform attains after being struck by the falling mass. What is this minimum height?

(b) What is the equilibrium height of the platform?

(c) In view of the system definition of (a) above, how is this equilibrium height attained?

(d) Describe the work transfer and energy transfer of each system element of the system as defined in (a) above.

(e) Suppose that mass M is replaced by a smaller mass at a proportionally larger free fall height. If the mass is reduced by a factor of 8 and the free fall height $(H - a - b - h)$ is increased by a factor of 8, what effect does this have on the minimum height and the equilibrium height? Note that the dimension a of mass M remains constant.

3.8 A mass, a spring, and a damper are connected as shown in Figure 3P.8. The com-

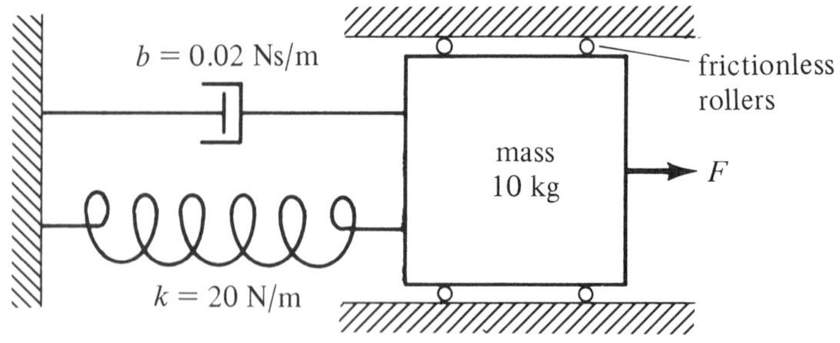

Figure 3P.8

plete assembly is mounted on a table in a laboratory. Initially the system is in equilibrium both internally and with its environment. An external force now moves the mass very slowly, extending the spring 20 mm. The force is then suddenly removed thereby allowing the system to oscillate and subsequently return to equilibrium. Define an appropriate model for the composite system and use this model to determine:

(a) The net work transfer during the initial stretching of the spring.

(b) The change in stored energy of the composite system during the initial stretching of the spring.

(c) The net heat transfer during the initial stretching of the spring.

(d) Repeat parts (a), (b), and (c) above for the process after the external force is released (the process by which the spring returns to equilibrium).

3.9 A freight handling system consists of the equipment shown in Figure 3P.9. Boxes of freight move down the roller conveyor from the second floor to the first floor of a building where they are readied for shipment. The boxes are brought to a stop by sliding them across the horizontal loading platform. The sliding friction coefficient between the platform and the boxes is 0.5.

(a) Define the system and an appropriate model to determine the minimum length of the platform necessary to stop the boxes. What is this length?

(b) Does the necessary length depend upon the mass of the box?

(c) For a system consisting of the box, the conveyor, and the platform, determine the work transfer, heat transfer, and the change in energy for the process by which the box slides down the conveyor and comes to equilibrium.

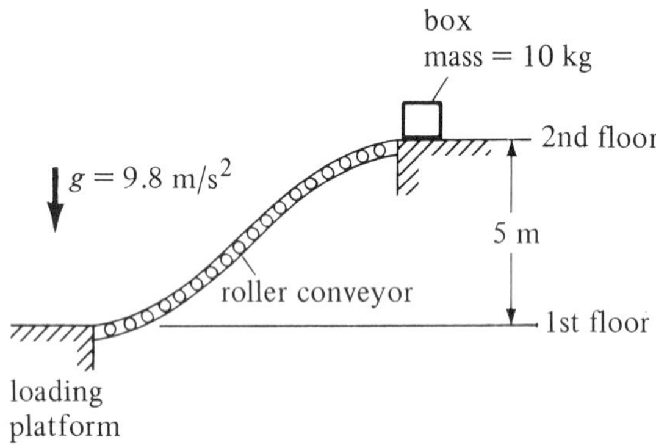

Figure 3P.9

3.10 The track shown in Figure 3P.10 consists of a frictionless circular portion smoothly joined to a rough inclined plane with an angle of 37 degrees to the horizontal. A block is released from rest at the position shown and slides along the track with its center of gravity following the trajectory shown in Figure 3P.10. The coefficient of friction between the block and the inclined plane is 0.3.

(a) What fraction of the translational kinetic energy of the block at the bottom of the track is dissipated by friction as the block slides up the plane? Be sure to describe the model used for the solution.

(b) How high will the center of gravity of the block move up the plane? Measure the height from the low point of the trajectory.

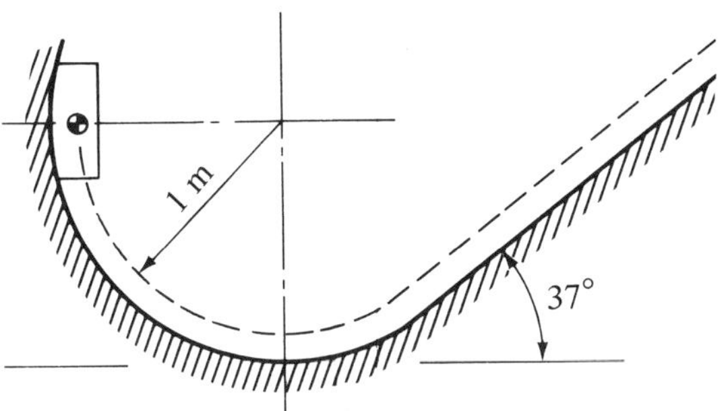

Figure 3P.10

3.11 A casting of 45 kg is taken from an annealing furnace at a temperature of 500 C and is plunged into an insulated tank containing 350 kg of oil at a temperature of 25 C. When the casting and oil reach thermal equilibrium, the uniform temperature is 40 C. The specific heat of the oil is 2.1 kJ/kgK.

(a) Describe a model suitable for finding the specific heat of the casting. (Neglect the heat capacity of the tank.)

(b) What is the specific heat of the casting?

(c) What is the heat transfer experienced by the casting?

(d) If the casting and the bath are now allowed to come to thermal equilibrium with the environment at 27 C, what is the heat transfer for this composite system?

3.12 An insulated copper container with a mass of 5 kg contains 40 kg of water. The water is stirred by a very light paddle wheel connected to an electric motor as shown in Figure 3P.12. By virtue of this stirring action, the uniform temperature of the system consisting of the vessel and the water rises at a rate of 0.05 K/s. Make a table showing the rates of heat transfer and work transfer and the rate of change of stored energy of the system consisting of:

(a) The water.

(b) The container.

(c) The water and the container.

(d) The water, the container, and the motor.

Note: The specific heat of copper is 0.39 kJ/kgK while that of water is 4.186 kJ/kgK.

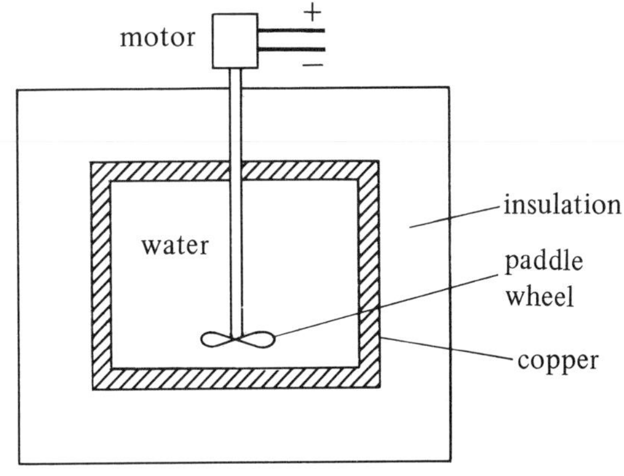

Figure 3P.12

3.13 A self contained electric hoist consists of an electric motor, a winch with a cable, and a brake. The cable is attached to a fixed support

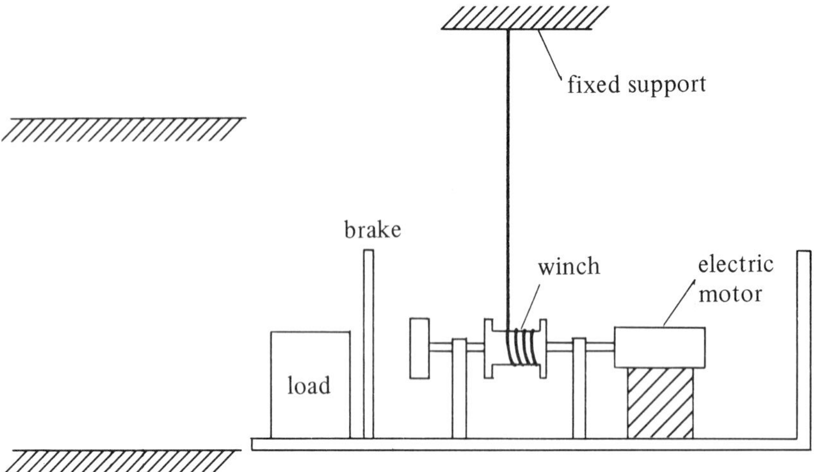

Figure 3P.13

above as in Figure 3P.13. The electric motor lifts the hoist and load. The brake is used to control the rate of descent of the hoist with the motor off.

(a) Construct a model suitable for a simple description of the energy changes and energy transfer associated with lifting a load with this hoist and returning the empty hoist back to the lower level.

(b) Use the model to describe the energy transfers and energy changes during this operation.

(c) How would the model and the energy transfer and changes be modified if the electric motor were powered by batteries carried on the hoist?

CHAPTER 4

Thermodynamic Behavior of Coupled Systems

4.1 Tests for Coupled Behavior

In Chapter 3 we considered systems with particularly simple thermodynamic behavior. From their very nature, it was obvious that the systems were uncoupled and therefore possessed separate and distinct thermal and mechanical aspects. Suppose now, however, we are confronted with a system with an unknown internal structure. In order to analyze the behavior of this system, we must construct a suitable thermodynamic model. To do this, we must first determine whether the system is coupled or uncoupled. Thus, we require a test for coupling in terms of the external behavior of the system without knowledge of the internal structure.

For example, if we are presented with an electro-mechanical "black box", we could determine whether the system as represented by the box is electro-mechanically coupled or uncoupled if we knew the internal structure of the box. For instance let us assume the system was composed of two pure conservative system elements—one a pure translational spring, the other a pure electrical capacitance. This system, as shown schematically in Figure 4.1, is capable of mechanical

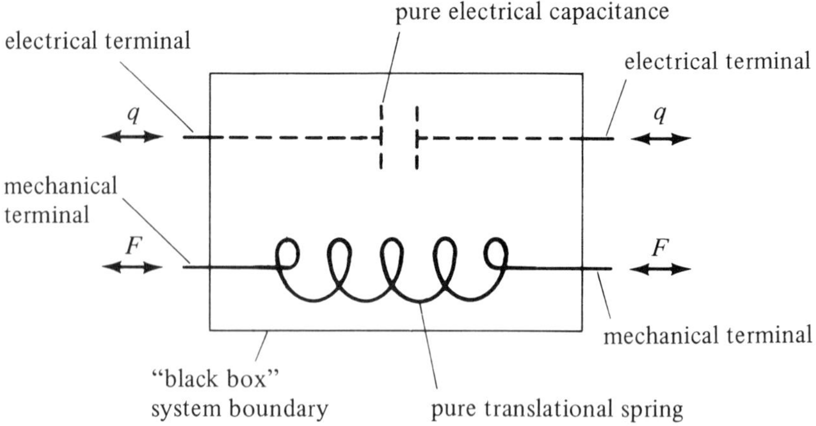

Figure 4.1 An uncoupled electro-mechanical system

work transfer associated with the motion of its boundary and electrical work transfer associated with charge transfer from one of its terminals to the other. For this internal construction we know the constitutive relations to be

$$F = kx \tag{4.1}$$

and

$$\nu = \frac{q}{C} \tag{4.2}$$

Had we not known the internal construction, equations (4.1) and (4.2) could have been determined by experiment. If we apply the first law of thermodynamics, equation (2.2), and the definitions of work transfer, equation (3.2), we have

$$E_2 - E_1 = \int_{q_1}^{q_2} \nu \, dq + \int_{x_1}^{x_2} F \, dx \tag{4.3}$$

The constitutive relations for this uncoupled system, equations (4.1) and (4.2), show that mechanical work transfer affects only the mechanical properties F and x and not the electrical properties ν and q. Conversely, electrical work transfer affects only the electrical properties ν and q and not the mechanical properties F and x. As a result, the paths of the integrals ap-

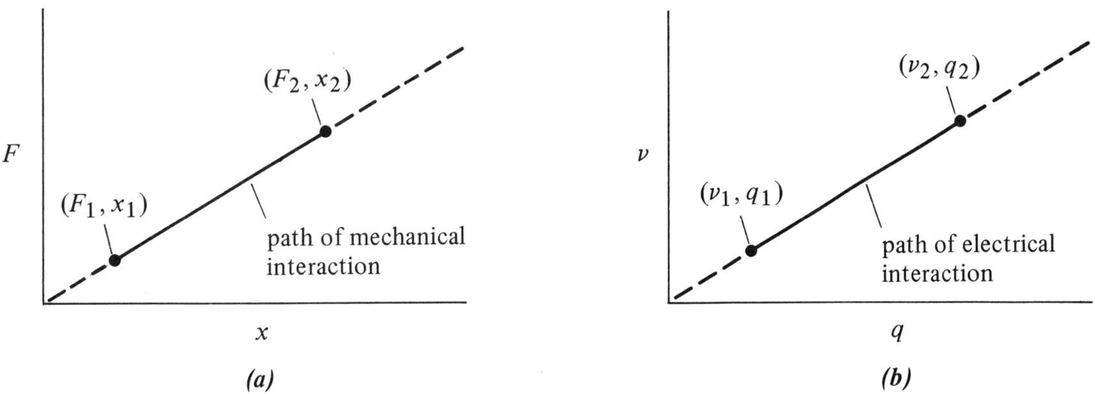

Figure 4.2 Paths of mechanical and electrical work transfer processes in the electro-mechanically uncoupled system

pearing in equation (4.3) take on the simple form shown in Figure 4.2. It follows that all paths of all processes executed by this system must be somewhere along the lines shown on the F–x plane and v–q plane of Figure 4.2. Then for any cycle executed by the system, not only do we have

$$\oint \delta W_{\text{net}} = \oint dE = 0 \tag{4.4}$$

but also

$$-\oint \delta W_{\text{mechanical}} = 0 \tag{4.5}$$

and

$$-\oint \delta W_{\text{electrical}} = 0 \tag{4.6}$$

as well. From the definition of a property, it follows that the total stored energy of equation (4.3) is actually the sum of two separate energy forms. These are: a stored mechanical energy associated with the mechanical interaction resulting from a change of the mechanical properties of the spring and a stored electrical energy associated with the electrical interaction resulting from a change of the electrical properties of the capacitance. Thus, as we previously discussed in Sections 3.2, 3.3, and 3.7.

$$E_2 - E_1 = (E_2 - E_1)_{\text{mechanical}} + (E_2 - E_1)_{\text{electrical}} \tag{4.7}$$

where

$$(E_2 - E_1)_{\text{mechanical}} = (k/2)x_2^2 - (k/2)x_1^2 \tag{4.8}$$

and

$$(E_2 - E_1)_{\text{electrical}} = (1/2C)q_2^2 - (1/2C)q_1^2 \tag{4.9}$$

It is apparent from equations (4.5) and (4.6) that during the cycle the system experiences no net mechanical interaction and no net electrical interaction. In this sense, the system can be regarded as electro-mechanically uncoupled. In fact we can use equations (4.5) and (4.6) to formulate a generalized test for an uncoupled system.

> For a given system, the properties associated with two different energy transfer interactions are uncoupled if for *all* cycles executed by the system, in the absence of *any dissipative effects,* the net interaction for each mode of energy transfer is individually zero. Thus,

$$\oint \delta(\text{interaction}) = 0 \tag{4.10}$$

Each group of properties can then be used to characterize a separate, independent, aspect of the behavior of the system. Note that as a consequence of equation (4.10), it is only possible to define a stored energy form associated with a single given interaction, and this can only be altered by the one particular type of nondissipative interaction. All other types of nondissipative interactions will not affect it.

Thus, for a system of unknown internal structure which is capable of two possible energy interactions, we can determine whether the properties associated with each of these two energy interactions are uncoupled by simply applying the above test. If the properties are uncoupled, it is then possible as far as these interactions are concerned, to construct a model for the system

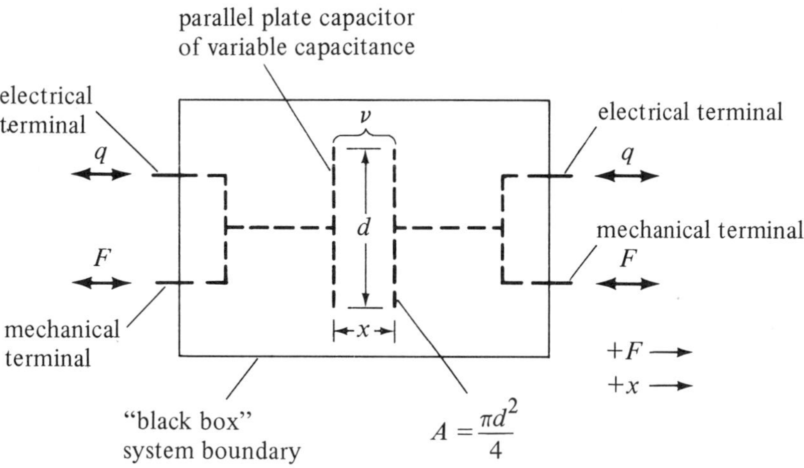

Figure 4.3 A coupled electro-mechanical system

composed solely of pure conservative system elements. Because the system model can be formulated in this manner, it follows that the system will have separate constitutive relations for each aspect of system behavior.

Suppose now that the internal structure of our "black box" does not consist of the pure system elements of Figure 4.1, but of a parallel plate capacitor in which two plates of area A are separated by a mechanically variable vacuum gap of thickness x shown in Figure 4.3. Such a system is capable of experiencing a work transfer by a displacement of its mechanical terminals or by an electrical charge transfer from one electrical terminal to the other. However, this situation is quite different from the previous one. Specifically, if the potential difference between the plates is v and the charge on the capacitor is q, the constitutive relations for the system can be determined by experiment to be

$$v = \frac{qx}{\epsilon_0 A} \tag{4.11}$$

and

$$F = \frac{vq}{2x} = \frac{q^2}{2\epsilon_0 A} \tag{4.12}$$

where ϵ_0 is the permittivity of free space and F is the external force required to hold one plate in a static position. Thus, if we apply the first law of thermodynamics to this system, we obtain for an infinitesimal change of state

$$-\delta W_{\text{net}} = -\delta W_{\text{mechanical}} - \delta W_{\text{electrical}} = dE \tag{4.13}$$

From the definitions of work transfer

$$-\delta W_{\text{mechanical}} = F\,dx \tag{4.14}$$

and

$$-\delta W_{\text{electrical}} = \nu\,dq \tag{4.15}$$

and for a finite change of state from state 1 to state 2, the first law of thermodynamics becomes

$$E_2 - E_1 = \int_{q_1}^{q_2} \nu\,dq + \int_{x_1}^{x_2} F\,dx \tag{4.16}$$

Equation (4.16) is identical to equation (4.3). However, in the present case, it is clear from the constitutive relations (4.11) and (4.12) that the various changes of state associated with the interactions are more complex than those of the previous case.

Let the system experience the cycle *ABCD* shown in Figure 4.4. Because of the coupled constitutive relations, the paths of a cycle of a coupled system may enclose an area such as in Figure 4.4, whereas the paths of a cycle of an uncoupled system must be lines as in Figure 4.2.

At fixed charge q_1 and fixed force F_1, the system experiences a mechanical interaction characterized by an increase in gap thickness from x_1 to x_2 ($A \rightarrow B$). At the fixed gap x_2, the system experiences an electrical interaction characterized by an increase in charge from q_1 to q_2 ($B \rightarrow C$). At the fixed charge q_2 and fixed force F_2, the system experiences a mechanical interaction characterized by a decrease in gap thickness from x_2 to x_1 ($C \rightarrow D$). At the fixed gap x_1, the system experiences an electrical interaction characterized by a decrease in charge from q_2 to q_1 ($D \rightarrow A$).

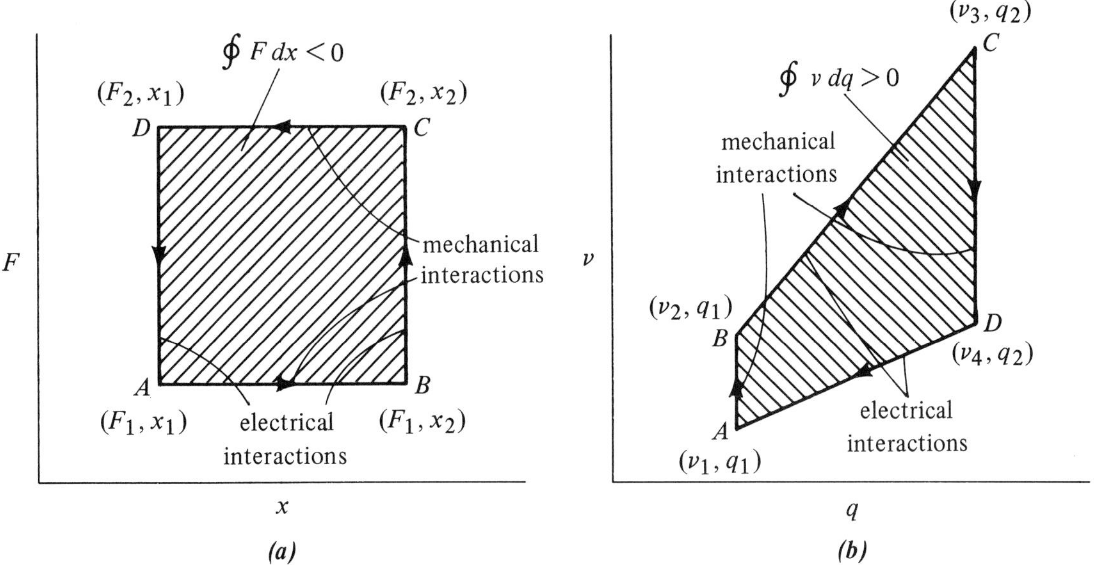

Figure 4.4 Cycle of a parallel plate capacitor

If we apply the first law to each process in turn, we obtain the following equations:

$$E_B - E_A = \int_{q_A}^{q_B} v\,dq + \int_{x_A}^{x_B} F\,dx = F_1(x_2 - x_1) \tag{4.17}$$

$$E_C - E_B = \int_{q_B}^{q_C} v\,dq + \int_{x_B}^{x_C} F\,dx = \frac{x_2}{\epsilon_0 A}\left[\frac{q_2^2}{2} - \frac{q_1^2}{2}\right] \tag{4.18}$$

$$E_D - E_C = \int_{q_C}^{q_D} v\,dq + \int_{x_C}^{x_D} F\,dx = F_2(x_1 - x_2) \tag{4.19}$$

$$E_A - E_D = \int_{q_D}^{q_A} v\,dq + \int_{x_D}^{x_A} F\,dx = \frac{x_1}{\epsilon_0 A}\left[\frac{q_1^2}{2} - \frac{q_2^2}{2}\right] \tag{4.20}$$

$$-\oint \delta W_{\text{mechanical}} = F_1(x_2 - x_1) + F_2(x_1 - x_2) < 0 \tag{4.21}$$

$$-\oint \delta W_{\text{electrical}} = \frac{x_2}{2\epsilon_0 A}(q_2^2 - q_1^2) + \frac{x_1}{2\epsilon_0 A}(q_1^2 - q_2^2) > 0 \tag{4.22}$$

$$-\oint \delta W_{\text{net}} = -\oint \delta W_{\text{mechanical}} - \oint \delta W_{\text{electrical}} = 0 \tag{4.23}$$

Thus during this cycle, energy enters the system by means of an electrical work transfer and leaves by means of a mechanical work transfer. The net work transfer and the energy change for the cycle are zero, and the first law is satisfied.

Equation (4.21), the net mechanical interaction, is represented in Figure 4.4 (a) by the negative area enclosed by the curve *ABCD*. Equation (4.22), the net electrical interaction, is represented in Figure 4.4 (b) by the positive area enclosed by the curve *ABCD*. The magnitudes of these areas are such that the net interaction for the cycle as represented by their algebraic sum is zero.

Thus, for this system there exists at least one cycle for which the net mechanical interaction is nonzero and the net electrical interaction is nonzero. That is,

$$\oint \delta W_{\text{mechanical}} \neq 0 \tag{4.24}$$

and

$$\oint \delta W_{\text{electrical}} \neq 0 \tag{4.25}$$

Of course, the first law of thermodynamics requires that

$$\oint \delta W_{\text{net}} = 0 = \oint \delta W_{\text{mechanical}} + \oint \delta W_{\text{electrical}} \tag{4.26}$$

or

$$\oint \delta W_{\text{mechanical}} = -\oint \delta W_{\text{electrical}} \tag{4.27}$$

Equation (4.27) shows that the properties associated with the two types of interactions are interrelated, that is, coupled to one another. Actually this fact was known from the very outset since the constitutive relations themselves were coupled. There was no separate mechanical constitutive relation or separate electrical constitutive relation.

On the basis of the information gained from this example, we can formulate a general test for coupled properties:

> For a given system, the properties associated with two different energy transfer interactions are coupled if for at least one cycle executed by the system, in the *absence of any dissipative* effects, the net interaction for each mode of energy transfer is individually nonzero. Thus,

$$\oint \delta(\text{interaction}) \neq 0 \tag{4.28}$$

To satisfy the first law of thermodynamics, the net interaction summed for all modes of energy transfer must be zero for the cycle.

As a consequence of equation (4.28), the coupled system does not have a separate aspect of behavior associated with a given interaction. The system simply possesses a coupled state which cannot be traced to some particular interaction.

Any one interaction is capable of altering this state and changing all of the properties (mechanical or electrical) of the system. Specifically, the energy of the system is a property and is fixed by the values for a set of independent properties of the system. For example, we can obtain an expression for the energy change of the system in Figure 4.3 by integrating equation (4.16) for a change of state from x_1 and q_1 to a second state x_2 and q_2 where x and q are selected as the independent properties.

$$E_2(x_2 q_2) - E_1(x_1 q_1) = \frac{1}{2\epsilon_0 A}(x_2 q_2{}^2 - x_1 q_1{}^2) \tag{4.29}$$

Clearly the energy of the system is electro-mechanical and not the sum of a separate mechanical energy and a separate electrical energy.

An additional consequence of the coupling is that the system model cannot be constructed of pure system elements. The system model must now have constitutive relations which include all the coupled properties. In this sense, the system is said to be coupled through the constitutive relations. Of course some of the properties of a system can be coupled while simultaneously other properties remain uncoupled. In such cases, the system model will consist of several types of elements with pure system elements modeling the uncoupled aspects.

On the basis of a given example, we have been able to formulate positive tests for the coupling that may exist among the properties that define the state of a thermodynamic system. In the presented example, the coupling was known *a priori,* since the constitutive relations were known. However, when the constitutive relations are not known, the nature of the coupling can be established by applying the tests formulated here. The coupling that occurs between the thermal and mechanical aspects of a thermodynamic system is of greatest interest to us in our present treatment, and we shall consider it in detail.

4.2 Coupled Thermodynamic Systems in the Absence of Dissipation

Consider a thermodynamic system that can experience both work transfer and heat transfer. For the sake of simplicity, let us restrict the work transfer to a simple displacement of the boundary. If such a system were uncoupled, we would have according to the test for uncoupled system behavior the following

$$\oint \delta Q = 0 \text{ and } \oint \delta W = 0 \tag{4.30}$$

For this system, in the absence of dissipation, we can model the thermal aspect as a pure thermal system element. The mechanical aspect can also be modeled as a pure conservative mechanical system element. Although uncoupled thermodynamic systems do exist, they are not usually considered in detail in thermodynamics because their thermal aspects are very simple. However, their mechanical aspects can be quite complicated and for this reason the sciences of mechanics and fluid mechanics consider uncoupled systems in considerable detail.

In contrast to equation (4.30) we have for the coupled thermodynamic system

$$\oint \delta Q \neq 0 \text{ and } \oint \delta W \neq 0 \tag{4.31}$$

but in order to satisfy the first law of thermodynamics

$$\oint \delta Q - \oint \delta W = 0 \tag{4.32}$$

Equations (4.31) and (4.32) are two of the most powerful in thermodynamics. The very basis of the science of energy conversion is that a thermodynamically coupled system can have a net positive heat transfer and a net positive work transfer for a cycle. Energy conversion is possible only when thermodynamic coupling is present.

The coupled thermodynamic system does not possess separate thermal and mechanical aspects but only a thermodynamic state. The constitutive relations are functions of both the thermal and mechanical properties of the system. Any interaction, regardless of whether it is a heat transfer or a work transfer, will result in a change of *both* the mechanical and the thermal properties of the system. An adiabatic work transfer from the system involves a change in the temperature (usually a decrease) of the system as well as a change in the geometry and forces. A heat transfer into the system at constant geometry (no work transfer) will change the forces (usually increase) on the system as well as the temperature of the system. Under these circumstances, it is not possible to relate the changes in mechanical properties (force and displacement) to a property of the system which one would like to define as the stored mechanical energy of the system. Also, it is not possible to relate the temperature of the system to a property which one would like to define as the stored thermal energy of the system. The *internal energy, U*, is the only property which can be defined in terms of the coupled properties of the system as the result of the first law of thermodynamics. This property is, in general, a function of both the thermal and mechanical properties of the system. The first law of thermodynamics for a coupled thermodynamic system then becomes

$$Q_{1-2} - W_{1-2} = U_2 - U_1 \qquad \textbf{(4.33)}$$

For example, if a coil spring is made of a material with a thermal expansion, the response of the spring to a work transfer will also depend upon the heat transfer and temperature. The work transfer for a given displacement will be different under adiabatic conditions than for constant temperature conditions.

Some physical situations involve both coupled and uncoupled behavior. Certain interactions satisfy the test for uncoupled behavior while other interactions meet the coupled requirements. An example is the motion of a rigid system in the conservative gravitational field. Since the position can be changed while all other properties remain constant, the system has a separate gravitational aspect characterized by its position and its gravitational potential energy. In this instance the first law becomes

$$Q_{1-2} - W_{1-2} = U_2 - U_1 + (E_2 - E_1)_{\text{gravitational}} \qquad \textbf{(4.34)}$$

where the work transfer W_{1-2} includes the work transfer which changes the gravitational state of the system.

Generally, a system with uncoupled aspects has a separate uncoupled constitutive relation between the uncoupled properties. For such systems, the uncoupled energy separates and the first law of thermodynamics becomes

$$Q_{1-2} - W_{1-2} = U_2 - U_1 + \sum \Delta E_{\text{uncoupled}} \qquad \textbf{(4.35)}$$

where the work transfer is for both the uncoupled and coupled aspects of the system.

In coupled thermodynamic systems, it is the loss of identity of the energy associated with work transfer or heat transfer which is the source of complexity of thermodynamics over and above mechanics. However, this loss of identity can be exploited to produce useful work transfer by means of heat transfer. That is, the stored internal energy of a thermodynamic system may be increased by means of a heat transfer into the system. This stored internal energy may then be converted into a useful work transfer. The device which permits us to accomplish this transformation of energy is called a heat engine. In order to make the most effective use possible of the heat engine, the working substance should be as strongly coupled as possible. Since gases are among the most strongly coupled substances, they are widely used in heat engines.

4.3 A Simple Coupled Thermodynamic System—The Ideal Gas Model

Matter exists in essentially three forms: solid, liquid, and gaseous. Although solids and liquids may have important coupling, gases are the most strongly coupled and hence are of special interest to the thermodynamicist. Therefore, we now develop a simple model that includes the essential features of the thermodynamic coupling in gases. This model involves the thermal property, temperature, and the mechanical properties, pressure and volume.

From experimental observations of the behavior of gases at low pressures, thermodynamicists have developed the ideal gas model. This model describes the behavior of gases well at moderate to low pressures and exactly in the limit of zero pressure. Its constitutive relation can be written

$$PV = n\overline{R}T \tag{4.36}$$

Here, P is the absolute pressure that the gas exerts on the walls of its container (see appendix to this chapter), V is the volume of space occupied by the gas, n is the number of kg-moles of the gas, $\overline{R}$ is a constant of proportionality known as the universal gas constant ($\overline{R}$ = 8.3134 kJ/kg-mole K), and T is the ideal gas absolute temperature. Since the number of moles is equal to the mass of the gas, m, divided by its molecular weight, M, the constitutive relation, equation (4.36), can be written

$$PV = \frac{m}{M}\overline{R}T = mRT \tag{4.37}$$

Here, R is known as the gas constant and has a value which depends upon the chemical composition of the gas and the system of units. Equations (4.36) and (4.37) are also known as the ideal gas equation of state.

Although not immediately obvious, the ideal gas constitutive relation can be used to define a temperature scale known as the ideal gas temperature scale. The ratio of any two temperatures T_1 and T_2 on this scale is defined as

$$\frac{P_2 V_2}{P_1 V_1} = \frac{T_2}{T_1} \tag{4.38}$$

Here, $P_2 V_2$ is the pressure-volume product when the ideal gas system is at equilibrium at the temperature T_2 and $P_1 V_1$ is the pressure-volume product when the same ideal gas system is at equilibrium at the temperature T_1.

In order to establish numerical values on the ideal gas temperature scale, an arbitrary value must be assigned to one temperature. By convention, this temperature is selected as the freezing point of pure water at 611.3 N/m^2. This is the triple point for water which we will define in a later chapter.[1] The Kelvin temperature is an ideal gas temperature scale for which the assigned value is 273.16 K for the triple point of water. The number 273.16 K was selected for the Kelvin scale so that the boiling point of water at one atmosphere would be 100 K above the freezing point of water at this pressure. Thus, this makes a temperature change of 1 K on the Kelvin scale the same as a change of 1 C on the old centigrade or the Celsius scale. Thus, the Kelvin ideal gas temperature is related to the Celsius temperature by the familiar relation:

$$T(\mathrm{K}) = t(\mathrm{C}) + 273.15 \qquad \textbf{(4.39)}$$

A part of the generality of the ideal gas temperature scale results from the fact that equation (4.38) is true for any ideal gas system regardless of the size of the system or the chemical composition of the gas. This generality may be illustrated by considering two ideal gas systems which are initially in mutual thermal equilibrium. If these two systems are isolated and each cooled to the state at which the pressure-volume product for each system is one-half the initial value of each system, then equation (4.38) shows that each will be at a temperature exactly one-half of the initial temperature. Thus, the systems are in mutual equilibrium and if placed in thermal communication, no heat transfer would occur.

Another significant feature of the ideal gas model is the especially simple relation for the change in internal energy. In fact the change in internal energy of an ideal gas can be shown to depend only upon the temperature. Thus,

$$U_2 - U_1 = m[f(T_2) - f(T_1)] \qquad \textbf{(4.40)}$$

Under certain circumstances which we shall describe subsequently, equation (4.40) can assume a special form in which the

[1] Section 8.6, p. 238.

TABLE 4.1

Ideal gas model	*Composition*	*Molec. weight*	*R, kJ/kgK*	c_v, *kJ/kgK*
Carbon Dioxide	CO_2	44.01	0.189	0.662
Oxygen	O_2	32.00	0.260	0.657
Air		28.97	0.287	0.716
Nitrogen	N_2	28.01	0.297	0.741
Helium	He	4.00	2.078	3.153
Hydrogen	H_2	2.02	4.116	10.216

functional dependence of the internal energy upon the temperature is linear. Then equation (4.40) assumes the form

$$U_2 - U_1 = mc_v(T_2 - T_1) \tag{4.41}$$

where c_v is the constant of proportionality known as the specific heat at constant volume[2]. The reason for this name will become apparent as we proceed. Typical engineering values for c_v of various ideal gas models are listed in Table 4.1. We are not yet in a position to show that equation (4.40) is a direct result of the constitutive relation, equation (4.36), and the first law of thermodynamics, because the second law of thermodynamics is needed for the proof.[3]

The expression for the internal energy change of the ideal gas, equation (4.41), is the same as that for the pure thermal system, but this similarity is deceptive because the internal energy of the ideal gas may be changed by work transfer or by heat transfer. This may easily be shown by combining equations (4.37) and (4.41) to eliminate the temperature and obtain

$$U_2 - U_1 = \frac{c_v}{R}(P_2 V_2 - P_1 V_1) \tag{4.42}$$

Thus, the internal energy is related to the mechanical properties P and V as well as to the thermal property T.

[2] The general definition of specific heat c_b is given by equation (9.53), page 287.
[3] See Chapter 9, Section 9.6.2, page 296.

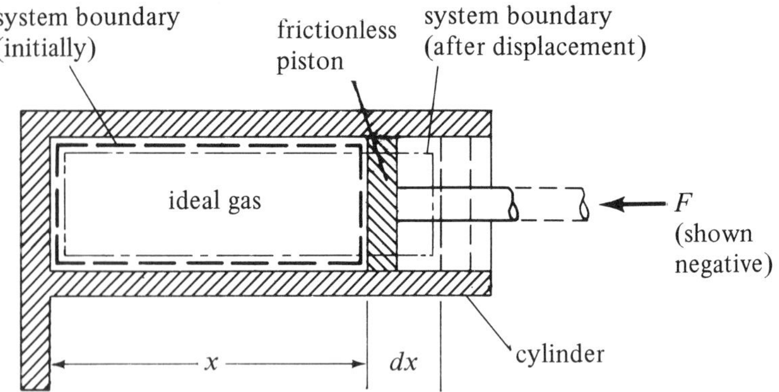

Figure 4.5 Work transfer in an ideal gas system

An ideal gas can experience a work transfer by means of a motion of the system boundary against the pressure force normal to the boundary. Figure 4.5 shows an ideal gas system with a uniform pressure P experiencing an infinitesimal work transfer as the system changes state an infinitesimal amount. The x-direction force exerted by the piston on the gas system is $F = -PA$. Here, A is the area of the piston and the minus sign indicates that the force is compressive for a positive pressure. The x-direction displacement of the boundary of the gas is dx, where dx is positive for extension or expansion of the system. Equation (3.1) gives the net infinitesimal work transfer for the gas as a system. Since the forces and displacements are colinear, equation (3.1) becomes

$$\delta W = -F\,dx \tag{4.43}$$

or

$$\delta W = PA\,dx\,. \tag{4.44}$$

Since the net change in volume is

$$dV = A\,dx \tag{4.45}$$

the work transfer can be written

$$\delta W = P\,dV \tag{4.46}$$

For a finite change of state (from state 1 with volume V_1 to state 2 with volume V_2), equation (4.46) can be integrated to give

$$W_{1-2} = \int_{V_1}^{V_2} P\, dV \tag{4.47}$$

In carrying out the integration indicated in equation (4.47), it should be remembered that P is the uniform pressure of the gas during the volume change dV. For a finite change in volume P may change as V changes, so that the integral $\int P\, dV$ can be evaluated only after the path (the series of states) connecting the initial and final states is known.

Heat transfer at a finite rate to or from an ideal gas occurs if the temperature of the gas is different from the temperature of the environment. Unfortunately, the evaluation of the heat transfer directly in terms of the temperature difference across the system boundary is a very complicated process and itself forms a special branch of engineering science. Because work transfer and change in internal energy are usually easier to evaluate directly, we shall content ourselves in the present treatment with evaluating the heat transfer indirectly from a knowledge of these two quantities and the application of the first law of thermodynamics. As stated previously, by convention the system has a positive heat transfer across its boundary if the temperature of the gas is lower than the temperature of the environment.

An ideal gas system cannot truly be isolated from its container. In some cases, it is necessary to consider both the container and the gas when evaluating the behavior of the gas. Generally, the gas will be in thermal communication with the wall, and if the containing wall is quite strong, it is satisfactory to consider the wall completely rigid so that changes in the elastic energy of the wall caused by changes in the pressure of the gas can be neglected. Then the container may be modeled as a pure thermal system in communication with the gas inside, and the gas plus the container may be isolated from the environment.

We have introduced a coupled thermodynamic system, the ideal gas model, but have not applied the test for coupling to verify this. We have proceeded in this manner because certain aspects of processes in thermodynamic systems must be discussed in detail before we can construct a suitable test cycle. For the present, we must content ourselves with the fact that the constitutive relation for the ideal gas, equation (4.37), involves both mechanical properties (P and V) and a thermal property (T) and that this is evidence enough to regard the ideal gas as a coupled system.

In subsequent chapters we shall take up the question of cycles in coupled thermodynamic systems, and at that time it will become obvious that these systems do indeed satisfy the test for coupling.

4E.1 Sample Problem: Consider a room 3 m × 4 m × 3 m filled with air. If the pressure of the air in the room is 1.013×10^5 Pa and the temperature is 300 K, calculate the mass of the air in the room if the air can be modeled as an ideal gas.

Solution: From the ideal gas equation of state,

$$m = \frac{PV}{RT}$$

where from Table 4.1, $R = 0.287$ kJ/kgK. Then

$$m = \frac{1.013 \times 10^5 (\mathrm{N/m^2})(3\ \mathrm{m})(3\ \mathrm{m})(4\ \mathrm{m})}{0.287(10^3\ \mathrm{Nm/kgK})(300\ \mathrm{K})}$$

$$m = 42.36\ \mathrm{kg}$$

4E.2 Sample Problem: Helium is contained in a piston-cylinder apparatus fitted with a spring as shown Figure 4E.2. The spring has a spring constant of $k = 10^5$ N/m and maintains the gas in mechanical equilibrium. The gas experiences a heat transfer at a

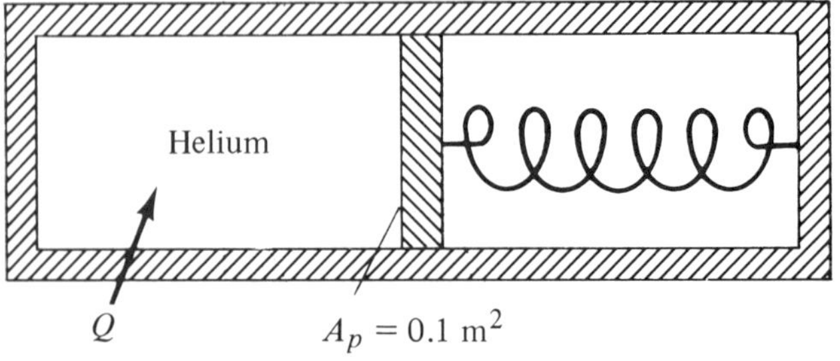

Figure 4E.2 Piston-cylinder apparatus

slow rate so that the pressure and temperature are always uniform throughout the gas. Calculate the work transfer experienced by the gas as its pressure changes from $P_1 = 1.013 \times 10^5$ Pa to $P_2 = 3.039 \times 10^5$ Pa.

Solution: The work transfer is given by the integral $\int P\,dV$ where the value of P must be known as V changes, that is, the integral can be evaluated only when we know the functional relationship between P and V or $P = f(V)$. In this situation the relationship between P and V is determined by the condition of equilibrium for the piston. Thus,

$$PA_p = kx$$

Here x is measured relative to the unstressed condition of the spring. Note that since

$$V = V_0 + A_p x$$

where V_0 is the volume of the gas when the spring is in its unstressed condition,

$$V = V_0 + \frac{PA_p^2}{k}$$

and

$$dV = \frac{A_p^2}{k}\,dP$$

Then the work transfer is

$$W_{1-2} = \int_{P_1}^{P_2} P\,dV = \frac{A_p^2}{k}\int_{P_1}^{P_2} P\,dP = \frac{A_p^2}{k}\left[\frac{P_2^2}{2} - \frac{P_1^2}{2}\right]$$

and

$$W_{1-2} = \frac{(0.1\text{ m}^2)^2}{10^5\text{ N/m}}\left[\frac{(3.039 \times 10^5\text{ N/m}^2)^2}{2} - \frac{(1.013 \times 10^5\text{ N/m}^2)^2}{2}\right]$$

$$W_{1-2} = 4.105\text{ kJ}$$

Thus, we have determined the work transfer for the gas by evaluating the line integral $\int P\,dV$ along the path determined by the mechanical equilibrium of the piston. We also could have determined this result from an application of the first law of thermodynamics to the spring.

4.4 Work Transfer in Coupled Thermodynamic Systems

In dealing with uncoupled thermodynamic systems without dissipative effects, there was never any doubt that the system had experienced a work transfer or a heat transfer during some particular process. If the question as to the mode of the energy transfer did arise, the answer could be derived simply by determining which form of stored energy was changed by the process. If the stored mechanical energy was altered in any way, the system experienced a work transfer. If the stored thermal energy was changed, and there was no dissipation, the system experienced a heat transfer.

However, from the definition of internal energy, it is clear that the question of the mode of energy transfer cannot be so simply answered in the case of the coupled system or the system with dissipative effects. The loss of identity of stored energy due to coupling and dissipation precludes the use of the stored energy in identifying the energy transfer mode. What is needed, then, is a positive test to determine whether the system has experienced a work transfer or heat transfer. Clearly a test or definition for either mode is sufficient since there are only two modes possible for a system near equilibrium. Hence, in keeping with the primary objective of thermodynamics, we propose the following definition of work transfer[4].

> A thermodynamic system experiences a positive work transfer during a given process if the only effect external to the system *could be* the increase in the stored energy of a pure translational spring. A thermodynamic system experiences a negative work transfer during a given process if the only effect external to the environment *could be* the increase in the stored energy of a pure translational spring.

Since there are many systems (electrical and magnetic) in which work transfer is not directly manifested as a mechanical force displacing the system boundary through a distance, it is sometimes necessary to employ some sort of transducer which

[4] A more complete definition of work transfer is given in Section 7.6, page 193.

converts energy from one form to another without dissipation and with no net change of state of the transducer. This device then permits us to measure the equivalent amount of mechanical work produced by these "non-mechanical" systems.

Note that we chose not to define negative work transfer in terms of the loss of stored energy of a pure translational spring. For example, suppose the system environment is replaced by a pure translational spring and a pure translational damper in thermal communication with the system. Let the spring now decrease its stored energy by experiencing a work transfer with the damper. By definition, the damper cannot store energy so the energy associated with the work transfer into the damper is transferred to the system of interest by means of a heat transfer. Thus, the sole effect external to the system was the loss of stored energy of a spring; however, the system experienced a heat transfer, not a work transfer. It follows that, the loss of stored energy of a spring is inadequate to define a negative work transfer. For this reason, the negative work transfer of a system is defined in terms of the positive work transfer of the system environment which appears as a negative work transfer when viewed from the system.

Also note that positive work transfer cannot be confused with negative heat transfer. In order to increase the stored energy of a pure translational spring by means of heat transfer, some other system, such as a transducer, must be involved. This is because a pure conservative mechanical system is itself incapable of experiencing heat transfer. Clearly the transducer involved must be a coupled thermodynamic system in order to convert heat transfer to work transfer. Although such a transducer makes it possible to increase the stored energy of a spring by means of positive heat transfer, *this is not the sole effect* external to the system of interest. The total stored energy of the transducer was unchanged during the conversion process, but there were compensating changes in the thermal and mechanical properties of the transducer. Since these changes could be detected by direct observation, they would appear as effects external to the system of interest. Admittedly the preceding argument is somewhat heuristic. It can be made more definite but only with reference to the second law of thermodynamics.

To use this definition of work transfer as a test of an interaction, the interacting systems are replaced, one at a time with a pure translational spring. If the same interaction is possible after each replacement, and only the spring and the interacting system have a net change of state, the interaction is a work transfer. For electrical work transfer or magnetic work transfer, a transducer must be employed with the spring. However, the transducer must not have a net change of state after the interaction under test.

CHAPTER 4 APPENDIX

Pressure

In dealing with pure mechanical system elements, we determined the work transfer for such systems from a knowledge of the forces acting on the system boundary. For fluid systems (gaseous or liquid systems) the forces exerted on the system boundary are transmitted throughout the system and are distributed over the system boundary.

The boundary force is then expressed in terms of the pressure, where the pressure is the boundary force per unit area of the boundary. In the general case, the pressure is different over different regions of the system boundary. The pressure at a point is then defined as the force per unit area for a small area enclosing the point. Specifically, if ΔA is a small area enclosing the point, and ΔA_c is the smallest area over which we can consider the fluid as a continuum (as matter distributed in space in a continuous manner rather than as discrete molecules) and ΔF_n is the component of the boundary force on ΔA normal to ΔA, we define pressure P at a point as

$$P = \lim_{\Delta A \to \Delta A_c} \frac{\Delta F_n}{\Delta A} \qquad \textbf{(4A.1)}$$

For a fluid in an equilibrium state, the force at the boundary is normal to the boundary and the pressure completely describes the boundary forces. If the fluid is not in an equilibrium state (subject to a rate of shear), the boundary force has components in the plane of the boundary and a stress tensor is required to describe the boundary force. For a fluid in equilibrium the pressure at a point is the same regardless of the orientation of the boundary which includes the point. That is, the pressure force at a point is the same in all directions.

In thermodynamics the pressure is measured relative to the zero force condition or a complete vacuum. This pressure is often termed absolute pressure to distinguish it from a pressure difference which is measured by common pressure gages. The absolute pressure of the atmosphere is termed atmospheric pressure. For example, a gas contained in a flexible, unstressed, partially collapsed bag would be at atmospheric pressure with the pressure forces of the gas inside the bag exactly balanced by the pressure of the atmosphere. A fluid system at equilibrium at a pressure different from the pressure of its environment must be confined within a mechanical structure (a tank) which can provide the forces arising from the pressure difference. For example, consider the system consisting of the gas sealed in a cylinder by a leak-free frictionless piston as in Figure 4A.1. At equilibrium, the force exerted on the piston face by the gas pressure balances the gravitational force on the piston and the weight plus the atmospheric pressure force on the top of the piston. If the pressure on the top of the piston is changed, the piston will move until the gas pressure inside the cylinder adjusts to a new equilibrium value. The pressure above the piston could be changed by placing the apparatus inside a tank and evacuating or pressurizing the tank.

Since most common pressure gages measure differential pressure relative to atmospheric pressure, the term gage pressure is widely used. Thus, the absolute pressure is the sum of the gage pressure measured with the standard pressure gage and the atmospheric pressure. Common vacuum gages measure pressures less than atmospheric pressure relative to atmospheric

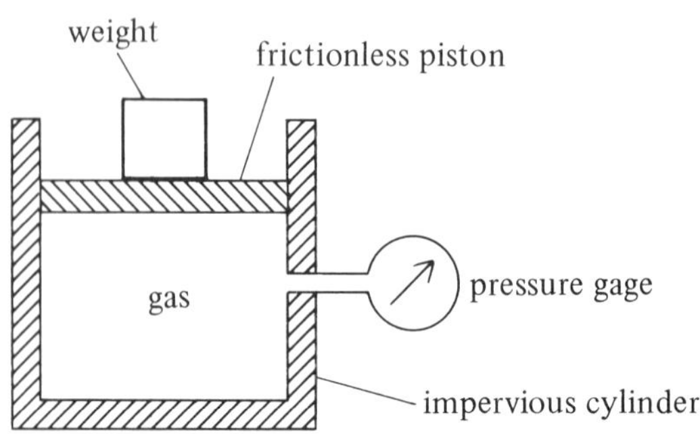

Figure 4A.1 Piston-cylinder apparatus

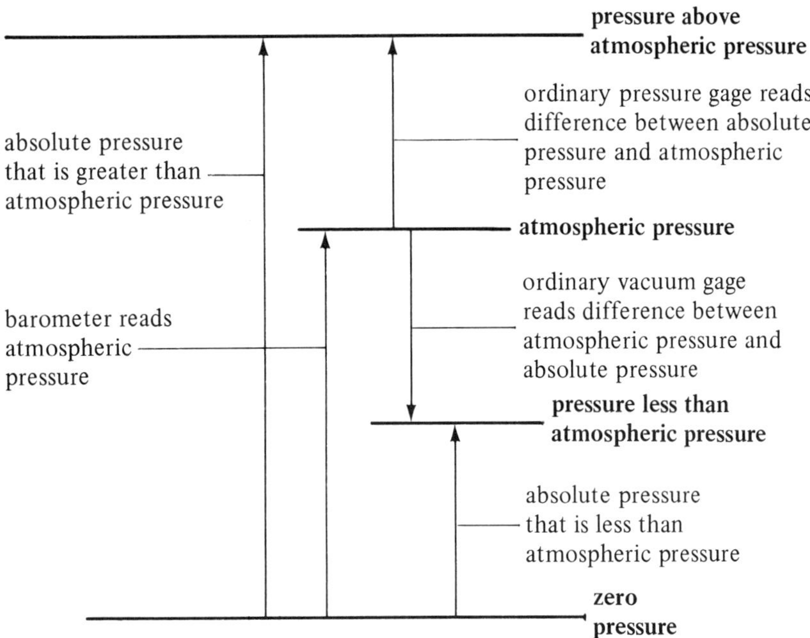

Figure 4A.2 Terms used in pressure measurement

pressure. Hence, an increasing vacuum reading is actually a decreasing absolute pressure. In Figure 4A.2 the relations between pressure measurements are shown.

Most instruments used in measuring pressure or pressure difference operate by measuring the pressure force on a sensing element. In the most common instruments, the pressure force is balanced by gravity or by the stress in an elastic element. In a manometer, the weight of a column of liquid balances the difference in pressure forces. In a dead weight pressure balance, the calibrated weights on a piston of known area balance the difference in pressure. In a bourdon tube pressure gage (the common dial type gage) the pressure difference is balanced by the elastic stresses in a curved tube of an oval cross section. A rack and pinion movement transforms the small deflection of the bourdon tube into indicating needle rotation. In the diaphragm pressure gage, the pressure difference is balanced by the stresses in an elastic diaphragm. The deflection of the diaphragm

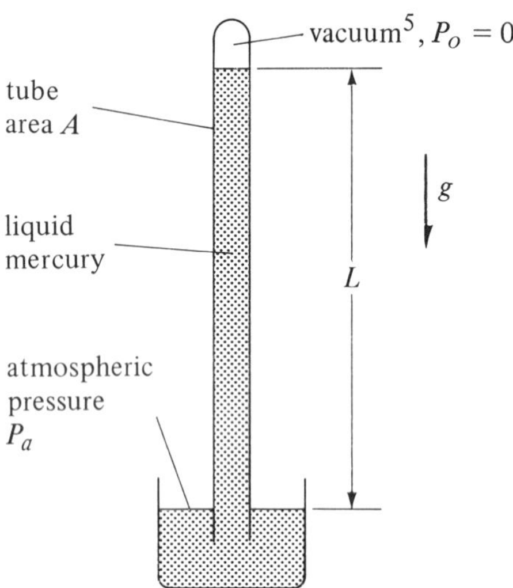

Figure 4A.3 Mercury barometer

is indicated mechanically or electrically. Pressure gages which produce an electric signal are called pressure transducers.

Any differential pressure gage becomes an absolute pressure gage when one of the pressures applied to the instrument is a complete vacuum, that is, zero pressure. An absolute pressure gage for measuring the absolute pressure of the atmosphere is called a barometer. A mercury filled manometer with a zero pressure applied to one leg is commonly used as a barometer. Such a mercury barometer is shown in Figure 4A.3. A force balance on the mercury column requires

$$P_aA = \rho LAg + P_0A \quad \textbf{(4A.2)}$$

$$P_a = \rho gL \quad \textbf{(4A.3)}$$

Thus, the column height is proportional to the atmospheric pressure. Mercury is used in barometers since its high density gives a reasonable column height and the evaporation of the liquid into the vacuum space is negligible but still measurable.

[5] The pressure is not exactly zero because of the pressure of mercury vapor which at 20 C is only 0.16 N/m^2.

The unit of pressure in international metric units is the pascal, Pa, where:

$$1 \text{ Pa} = 1 \text{ N/m}^2$$
$$1 \text{ kPa} = 10^3 \text{ N/m}^2$$
$$1 \text{ MPa} = 10^6 \text{ N/m}^2$$
$$1 \text{ mPa} = 10^{-3} \text{ N/m}^2$$
$$1\ \mu\text{Pa} = 10^{-6} \text{ N/m}^2$$

A variety of pressure units have been widely used with many of the units related directly to a specific pressure instrument. In the English units the common pressure units are pounds force per square inch, lbf/in^2; pounds force per square foot, lbf/ft^2; and inches of mercury column, in Hg (for standard density mercury in standard gravity). Before SI units were adopted, the metric units for pressure were kilogram force per square centimeter, kgf/cm^2; Bar and millimeters of mercury column, mm Hg, or Torr. The standard atmosphere was used in both the English and the metric system. Table 4A.1 gives equivalents of various pressure units.

TABLE 4A.1
Equivalents of Pressure Units

1 Pa = 1 N/m^2
1 lbf/in^2 = 144 lbf/ft^2 = 6.895×10^3 N/m^2
1 psia = 1 lbf/in^2 absolute pressure
1 psig = lbf/in^2 gage pressure
1 in Hg (32 F) = 0.491 lbf/in^2 = 3.386×10^3 N/m^2
1 mm Hg (0 C) = 1 Torr = 10^3 micron Hg = 1.333×10^2 N/m^2
1 std atm = 14.696 lbf/in^2 = 29.921 in Hg (32 F)
1 std atm = 760.00 mm Hg (0 C) = 1.0133×10^5 N/m^2
1 Bar = 10^5 N/m^2
1 kgf/cm^2 = 9.807×10^4 N/m^2

Problems

4.1 Two vacuum bell jars are interconnected as shown in Figure 4P.1. Each bell jar is connected to a separate vacuum pump. The mercury manometers read as indicated in Figure 4P.1. The atmospheric pressure is 750 mm Hg.

(a) What are the readings on the pressure gages A, B, and C?

(b) What is the absolute pressure in each of the bell jars?

(c) The bell jars are sealed to the base plate by rubber gaskets which can be modeled as pure ideal springs. How much does the gasket for bell jar II deflect after being pumped down for a gasket spring constant of 3000 N/mm?

4.2 A room 3 meters wide and 4 meters long has a ceiling 2.5 meters high. The room is to be filled with a gas to a pressure of 1 atmosphere (1.013×10^5 N/m^2). The temperature of the gas is 21 C. Assume that the gas can be modeled as an ideal gas.

(a) What is the mass of gas required if the gas is (1) hydrogen, (2) helium, or (3) air?

(b) The room is now sealed off, and the temperature of the gas is increased to 40 C by means of a heat transfer. What is the heat transfer necessary for each of the gases listed in part (a)?

4.3 A submarine filled with air at an atmospheric pressure of 98.3 kPa is sealed shut, and dives to a depth of 150 m below the surface of the ocean. The submarine is instrumented with various pressure gages as shown in Figure 4P.3. Also a barometer is installed in the submarine. If the density of sea water is 1041 kg/m^3, determine the readings of the various gages in Pa and the height, h, of the barometer in mm Hg.

4.4 Jacques Cousteau, the famed aquanaut, has designed a test cell to study the adaptability of human beings to living underwater for prolonged periods of time. The cell consists of a cylindrical steel vessel (dimensions shown in Figure 4P.4) with a circular opening at one end to permit the movement of divers in and out of the cell. There is no door covering this

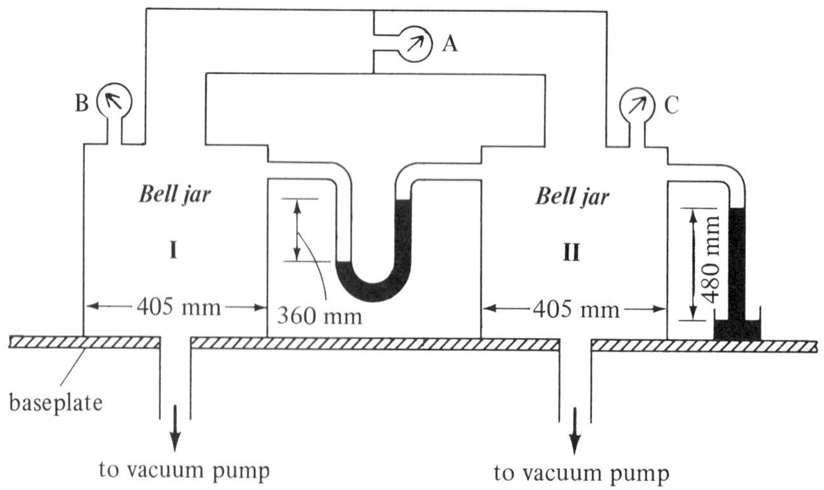

Figure 4P.1

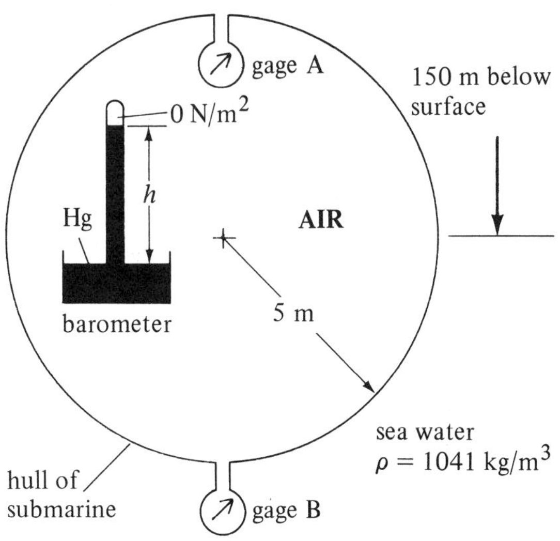

Figure 4P.3

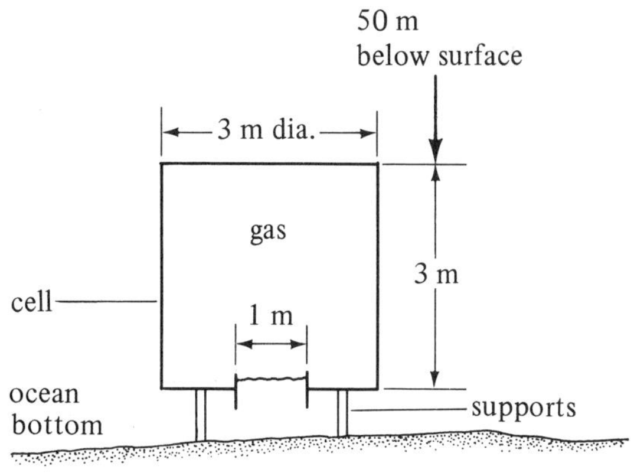

Figure 4P.4

opening. Heaters inside the cell maintain the temperature of the helium-oxygen gas mixture at 27 C.

(a) On a day when the atmospheric pressure at sea level is 101.3 kPa what mass of gas is required to fill the cell? Assume that the gas can be modeled as an ideal gas with $R = 315$ J/kgK.

(b) If the total barometric pressure variation is from 712 mm Hg to 788 mm, what must be the length of the cylindrical opening in order to prevent water from entering the cell proper using the mass of gas calculated in part (a). Assume sea level to remain fixed and that the temperature of the gas is a constant 27 C. The specific weight of mercury (relative to sea water at 15 C) is 13.25. The density of sea water at 15 C is 1041 kg/m³.

4.5 A vacuum tank is fitted with a mercury manometer. When the tank is pumped down, the manometer reads as shown in Figure 4P.5. Inside the vacuum tank is a chamber which is divided into two compartments as shown. The pressures inside these two compartments are constant and different from one another. Pressure gage A reads 360 kPa and pressure gage B reads 170 kPa.

(a) Determine the absolute pressures in the

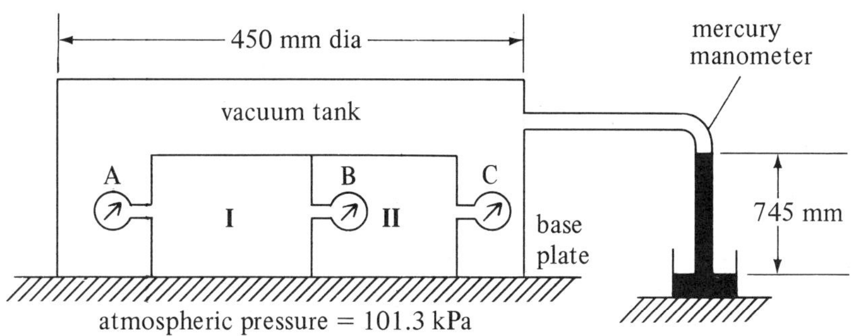

Figure 4P.5

two compartments and inside the vacuum tank.

(b) What is the reading of gage C?

(c) What is the force required to lift the vacuum tank off the base plate?

4.6 A vessel of 1.5 m^3 internal volume contains helium at 3 MPa. The vessel is connected through a valve to a balloon and the helium is used to slowly inflate it to a diameter of 4 m. The pressure inside the balloon after it is inflated may be assumed to be atmospheric. The pressure of the atmosphere is 1 bar and its temperature is 21 C. Assume that the helium and the atmosphere (air) can be modeled as ideal gases with gas constants and specific heats as given in Table 4.1. Assuming all of the helium is in thermal equilibrium with the atmosphere, what is the pressure of the remaining helium in the container after the balloon is filled?

4.7 A quantity of air is confined in the piston-cylinder apparatus as shown in Figure 4P.7. Initially the system is in equilibrium in the state shown. The system is then placed in thermal communication with a heat reservoir at T_2. When the system and the reservoir come to equilibrium, the pressure of the gas is P_2 = 200 kPa. Assume that the air can be modeled as an ideal gas.

(a) Determine the change of volume of the air.

(b) Determine the temperature T_2.

(c) Determine the heat transfer Q_{1-2} and the work transfer W_{1-2} for the air as a system.

4.8 The energy of a certain coupled thermodynamic system is a function of temperature only and is given by the expression E = 100 J + (1.0 J/K)T. When this system executes a certain process, the power (work per unit time) is a linear function of the time t as expressed by the relation P = (1.046 N m/s)t. A continuous temperature record of the system for this particular process indicates that at zero time the temperature is 100 C, and the change of the temperature is constant at 1 C/s. Find the heat transferred when the temperature changes from 100 C to 200 C.

4.9 As shown in Figure 4P.9, a gas is contained in a piston-cylinder apparatus which is also fitted with a damper. The following models can be used:

Gas: Ideal gas with c_v = 715.9 J/kgK
R = 287.0 J/kgK
P_1 = 70 kPa
T_1 = 330 K

Damper: Pure translational damper.

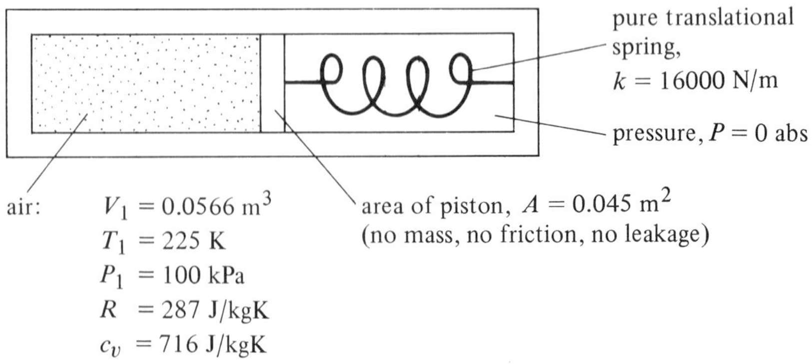

pure translational spring, k = 16000 N/m

pressure, P = 0 abs

air: $V_1 = 0.0566$ m^3
T_1 = 225 K
P_1 = 100 kPa
R = 287 J/kgK
c_v = 716 J/kgK

area of piston, A = 0.045 m^2
(no mass, no friction, no leakage)

Figure 4P.7

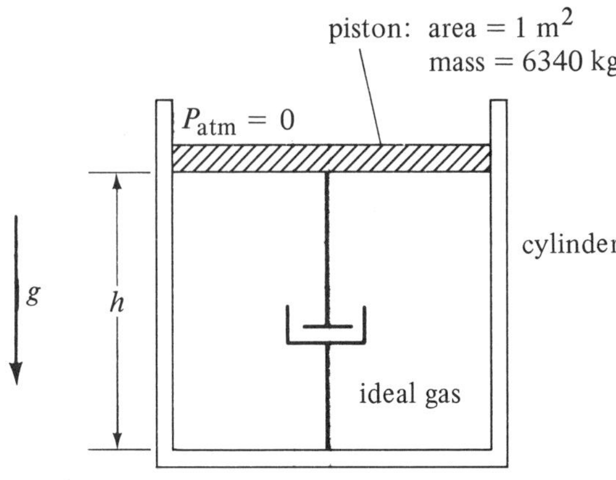

Figure 4P.9

Piston: Rigid, frictionless piston with mass of 6340 kg and a cross-sectional area of 1 m^2 and an initial height h_1 of 2 m.
Cylinder: Rigid
Gas-cylinder boundary: adiabatic
Gas-piston boundary: adiabatic
Gas-damper boundary: diathermal

The piston is initially held in mechanical equilibrium at the position $h_1 = 2$ m, by means of a pin. At some instant, the pin is removed and the piston comes to a new equilibrium position h_2.

(a) Determine the value of h_2 for this new equilibrium position.

(b) Determine the pressure and temperature of the gas for this new equilibrium position of the piston.

4.10 A manometer tube with a cross-sectional area of $A = 10$ mm^2 contains mercury (density $= 1.36 \times 10^4$ kg/m^3) as shown in Figure 4P.10. One leg of the manometer is open to the atmosphere while the other leg is sealed off and contains an ideal gas with $c_v = 3.14$ kJ/kgK and $R = 2.08$ kJ/kgK. The ideal gas is in thermal communication with the atmosphere which can be regarded as a heat reservoir with a temperature of 295 K. The pressure of the atmosphere is 101.3 kPa. The initial equilibrium state of the system is shown in Figure 4P.10. 1.50×10^{-4} m^3 of mercury (assumed incompressible) is added to the system.

(a) What is the length of the trapped gas column when thermal equilibrium is

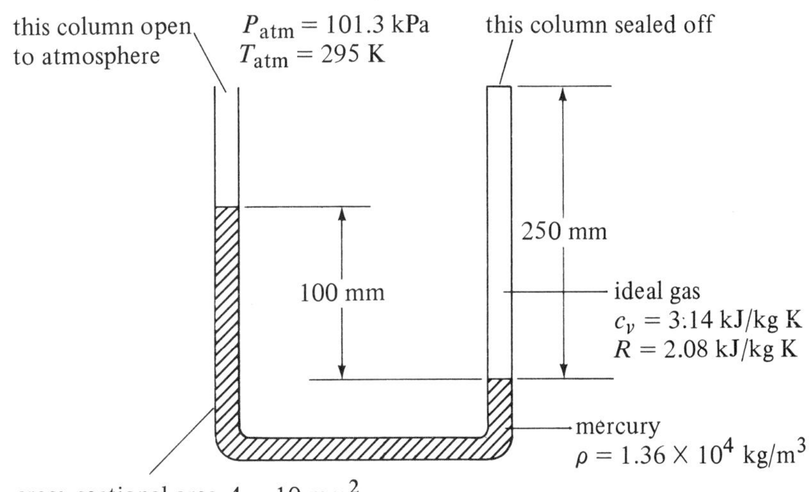

Figure 4P.10

finally established between the atmosphere and the gas?

(b) What is the difference in height for the two mercury columns?

(c) What is the final pressure of the gas trapped in the closed column?

4.11 An ideal gas is enclosed in a leak-proof piston-cylinder apparatus. The piston is displaced inward in such a manner that the pressure and volume are related to one another by the expression PV^n = constant where n has the constant value 1.4. The initial temperature of the ideal gas is 21 C.

(a) If the initial pressure is 100 kPa, the initial volume is 1 cubic meter and the final volume is 0.25 cubic meter, determine the necessary work transfer for the gas for the compression process assuming that the work is given by $\int P\,dV$ for the gas.

(b) Does the necessary work transfer for the gas depend upon the type of ideal gas involved provided the relation $PV^{1.4}$ = constant is valid?

(c) What is the work transfer for the gas per unit mass of gas if the gas is hydrogen? Helium? Assume that the ideal gas model is adequate.

(d) Assuming that the gas can be modeled as an ideal gas, determine the final temperature of the gas after the compression process. Does the type of ideal gas involved influence the final temperature as long as the relation $PV^{1.4}$ = constant is valid?

(e) Determine the heat transfer to the gas if the ideal gas model is (1) hydrogen, (2) air or (3) helium.

4.12 The *conservative* mechanical system shown in Figure 4P.12 has two means for work transfer interactions. One means is by translation, x, of the shaft while under tension, F. The other means is by a torque, τ, along the shaft rotating at an angular velocity, ω. The state of the system is completely specified when x and ω are specified. The constitutive relations for the system are

$$\tau = A\frac{d}{dt}[(B - x)^2\omega] \qquad 0 < x < B$$

$$F = A(B - x)\omega^2 \qquad 0 < x < B$$

The work transfer along the translating shaft is (thermodynamic sign convention)

$$-\delta W = F\,dx$$

The work transfer along the rotating shaft in terms of τ and ω is

$$-\delta W = \tau\,d\theta = \tau\frac{d\theta}{dt}\,dt = \tau\omega\,dt$$

$$-\delta W = A\omega\,d[(B - x)^2\omega]$$

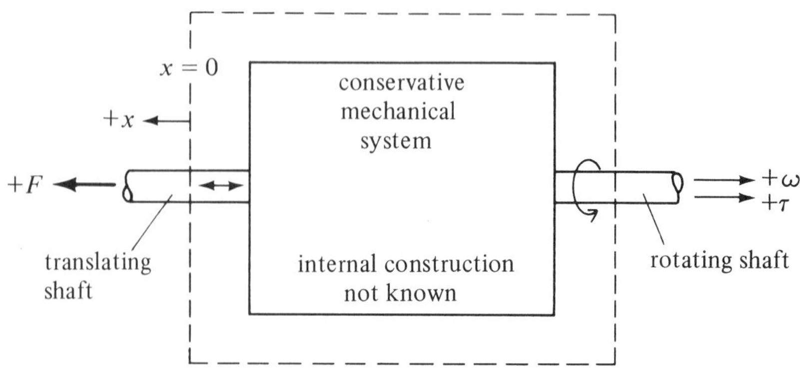

Figure 4P.12

(a) Are the two means of work transfer for this system coupled? Explain.
(b) Determine the energy of this system as a function of x and ω. The energy of the system is zero at $x = 0$, $\omega = 0$.
(c) Determine the two work transfers and the energy change for the system when the system changes from x_1 to x_2 at constant ω (when $\omega \neq 0$, and $x_2 > x_1$).

4.13 The solenoid shown in Figure 4P.13 has the following characteristics. The inductance of the coil depends upon the position of the plunger and is given by

$$L = \frac{5.2 \times 10^{-2} \text{ m henry}}{(x + 10^{-4} \text{ m})}$$

where x is the displacement of the plunger in mm from the full "in" position. The external tensile force to hold the plunger at a position x depends upon the current in the coil and is given by

$$F = \left(2.6 \times 10^{-2} \frac{\text{Nm}}{\text{amp}^2}\right) \frac{i^2}{(x + 10^{-4} \text{ m})^2}$$

These two relations are valid for $0 < x \leq 10$ mm and $F \leq 650$ N. Both electrical and mechanical dissipation are to be ignored in the model.

HINT: $\lambda = Li$ and $\dfrac{d\lambda}{dt} = \dfrac{d}{dt}(Li)$

(a) Should this solenoid be modeled as a coupled or an uncoupled system?
(b) How must the current i change with position x so that no electrical work transfer occurs when the plunger is moved? Determine the mechanical work transfer (in Joules) when the solenoid at $x = 10$ mm, $i = 1$ amp is changed with zero electrical work transfer, to $x = 0$.
(c) Determine the energy of the system (relative to $E_0 = 0$ at $i_0 = 1$ amp, $x_0 = 0$) as a function of i and x and again as a function of F and x.
(d) Construct a cycle for the model of the solenoid which will have $\oint \delta W_{\text{mechanical}} < 0$ and $\oint \delta W_{\text{electrical}} > 0$.
(e) Could this solenoid be used as an electric motor? How?

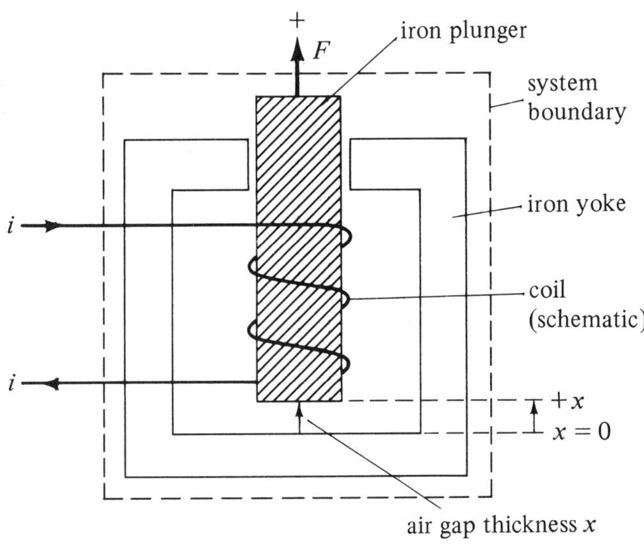

Figure 4P.13

CHAPTER 5

Equilibrium, Reversibility, and the Second Law of Thermodynamics

5.1 Equilibrium and Equilibrium States

The general principle which forms the foundation of thermodynamics is the tendency of all real systems to come to a particular state, the equilibrium state. This occurs when the system is isolated from the environment by restricting work transfer and heat transfer or when the conditions imposed by the environment do not change with time. When a system is isolated and in an equilibrium state, the system is said to be in internal equilibrium or simply, in equilibrium. When a system is in equilibrium while in communication with a second system or with the environment, the two systems are in mutual equilibrium. Mutual equilibrium requires internal equilibrium within each system. The internal equilibrium state is characterized by properties which are constant while the system is isolated from its environment. By definition, an isolated system cannot experience external heat transfer or work transfer; thus, in accordance with the first law of thermodynamics, the total stored energy is constant. Therefore, the isolated system must achieve a state of internal equilibrium by varying its properties while the total stored energy is constant. It is important to note that this variation of system properties occurs in a preferred direction. That is, during the approach to the equilibrium state, the system passes through any one state only once during the given process. Further, the succession of states along the path of the process converges to the equilibrium state as a limit. During the process by which internal equilibrium is attained, the thermodynamic properties usually increase or decrease in a monotonic fashion.

The case of a pure conservative mechanical system is somewhat different in that constant amplitude oscillations are possible so that the system can never achieve a true equilibrium state. However, when the equilibrium state of a mechanical system is our primary concern, the pure conservative mechanical system elements are not satisfactory models for the system. They must be combined with pure dissipative system elements and pure thermal system elements. It is only through the interactions among these system elements that the equilibrium state in a mechanical system can be assured. Because of the unidirectional nature of the interaction in a pure dissipative system element, the approach to equilibrium in a real mechanical system occurs in a specific direction even though the system may experience oscillations. In contrast to the pure conservative mechanical system, the amplitude of these oscillations decays with time because of the dissipation.

A number of the important features of the equilibrium state and the approach to equilibrium may be illustrated with the aid of the system shown in Figure 5.1. The mechanical aspect of the state of the system is fixed by the values of the properties x, x_i, and dx_i/dt. Here, x and x_i are measured relative to the fixed base and are each zero when the forces in both springs are zero. The thermal aspect of the system is fixed by the value of the temperature T of the pure thermal system element. Suppose now that x assumes some value different from zero and that the system is isolated from the environment. The system will be in an equilibrium state only if $x_i = (1/2)x$, $dx_i/dt = 0$, and $dT/dt = 0$. If the system is in any other state, that is, $x_i \neq (1/2)x$, $dx_i/dt \neq 0$ and $dT/dt \neq 0$, the values of the properties x_i, x, and T will continue to change until the equilibrium state is reached. As x_i changes, an internal work transfer will occur among the two pure translational springs, the pure translational mass, and the pure translational damper. An internal heat transfer will also occur between the pure translational damper and the pure thermal system element.

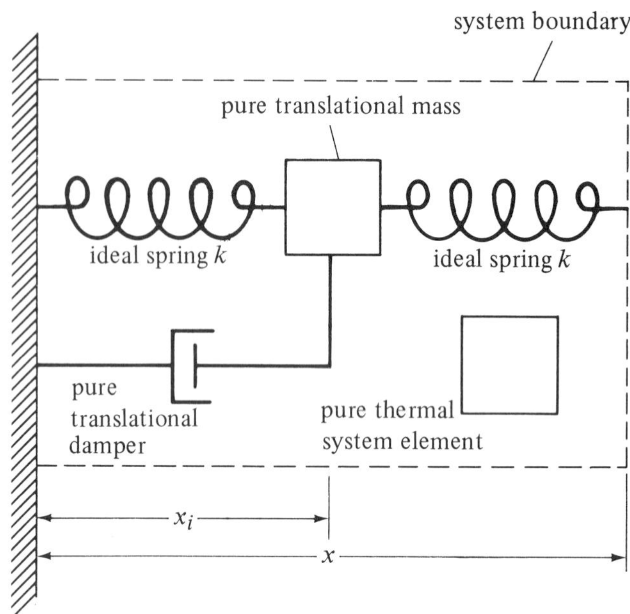

Figure 5.1 Mechanical system

In this simple system, it is evident that the approach to equilibrium is the result of the dissipative character of the pure translational damper. As a consequence of the work transfer to the damper, the total stored energy of the system is redistributed over the system elements. The sum of the elastic energy stored in the springs and the kinetic energy stored in the motion of the mass is continually decreased by a work transfer into the damper. Simultaneously, the stored thermal energy of the pure thermal system element is continually increased by a heat transfer from the damper. When the equilibrium state is reached, these internal interactions cease. For this simple system the equilibrium state is characterized by a minimum value of the sum of the elastic and kinetic energies consistent with the imposed value of x.

At equilibrium, the state of the system cannot change until the physical situation is changed. For example, the external conditions imposed by the environment may be changed thereby, causing a work transfer or heat transfer which in turn causes the state of the system to change spontaneously until the new equilibrium state, consistent with the new external conditions, is reached. As a second example, the state of an isolated system at equilibrium may also change spontaneously by means of internal interactions if the internal physical situation is changed. This change in the internal physical situation requires a change in the model to allow variations in a property which was previously considered fixed and not subject to change. For example, one of the springs in Figure 5.1 might fracture.

As a consequence of the dissipative aspects that exist in all real physical situations, the general mechanical requirement for equilibrium is that all forces, distributed or localized, must be in static balance so that the system does not have any accelerations. This condition applies to the complete system and to any subsystem contained within that system. Further, when a system is at equilibrium, no work transfers are occurring between any subsystems within the system. All parts of the system must be at rest relative to each other and relative to any part of the environment in communication with the system. In certain cases, the dissipative aspects of system behavior may be small enough so that it may be useful to model the situation as non-dissipative. For this model, the mode of motion which is frictionless will not necessarily be at rest when the system is otherwise in equilibrium.

As we stated previously in Chapter 3 during our discussion of temperature and heat transfer, the general thermal requirement for equilibrium is equality of temperature. Thus, in a system composed of several system elements, all subsystems must have the same temperature at thermal equilibrium. In a continuous system, the temperature must be uniform throughout for thermal equilibrium to prevail. Two systems in internal thermal equilibrium placed in thermal communication must have identical uniform temperatures in order to be in mutual thermal equilibrium. From the definition of heat transfer it is clear that a system in a state of internal thermal equilibrium cannot have any internal heat transfers among its subsystems, and two systems in a state of mutual thermal equilibrium cannot experience any heat transfer with one another at a finite rate.

All states which do not meet the general mechanical and thermal requirements for equilibrium are termed *non-equilibrium states*. These states have the characteristic of spontaneously changing their properties to approach the equilibrium state which is consistent with the conditions imposed by the environment. In the example of Figure 5.1, all the states with $x_i \neq (1/2)x$, $dx_i/dt \neq 0$ and $dT/dt \neq 0$ are the non-equilibrium states. The description of these non-equilibrium states requires a detailed description of the dynamics of the process and, hence, the specification of the values of more independent properties. On the other hand, the conditions of equilibrium relate some of the properties to one another so that the description of the equilibrium state is simplified. It is for this reason that the equilibrium state is so significant.

5.2 Quasi-Static States

The simple example discussed in the previous section illustrates the simplicity of equilibrium states as compared with non-equilibrium states. Equilibrium states are specified with a smaller number of independent properties than non-equilibrium states of the same system. The equilibrium states can be described without including the time as a variable. The science of classical thermodynamics is concerned with modeling situations that can be described in terms of the same set of properties that are required to describe equilibrium states. These properties are termed equilibrium properties. In a strict sense equilibrium properties are those properties which are defined only for equilibrium states (for example, temperature). A number of equilibrium properties (for example, volume or displacement) are also well defined for non-equilibrium states.

In thermodynamics the equilibrium properties may be used in two rather general classes of situations. In the first, the situation is modeled as equilibrium only at the initial and final states. The net heat transfer and the net work transfer for the entire change of state are then related to the end state equilibrium properties through the first law of thermodynamics without reference to the intermediate non-equilibrium states. A large variety of situations may be modeled in this way; however, the information which can be obtained is limited because the use of the first law of thermodynamics in this manner requires knowledge of the change of state *and* one of the interactions. This approach is useful for adiabatic, that is, work only situations and zero work situations. In the second class, the physical situation is near enough to equilibrium so that the situation may be modeled as being in equilibrium at every instant of time throughout the process. In other words, the equilibrium properties provide an adequate description of the situation even though the system is experiencing work transfer or heat transfer or both work transfer and heat transfer. Such a physical situation is termed *quasi-static* and the states of the system are said to be *quasi-static states*.

The quasi-static model gives a description of all the states of the system during the process and thus provides a more complete description of a more limited class of situations. As the term implies, the quasi-static model is suitable for processes which are slow relative to the time required for the system to reach equilibrium. The speed with which a system adjusts to the equilibrium state is studied in detail for mechanical and electrical systems in system dynamics and for thermal systems in heat transfer. The justification for the use of a quasi-static model in any physical situation requires either experimental experience with the system or an understanding of these other fields.

5.3 Reversible and Irreversible Processes

In thermodynamics the simple concepts of conservative processes or systems and dissipative processes or systems are broadened into more general classes termed reversible and irreversible, respectively. The models of the simple mechanical systems discussed earlier were termed conservative when the net work transfer for a cycle was always zero. The extension of this concept to any system is:

> A cycle executed by a system in communication with a single heat reservoir is reversible if the net work transfer for the cycle is zero. Thus,
>
> $$\oint \delta W = 0 \tag{5.1}$$
>
> and from the first law of thermodynamics
>
> $$\oint \delta Q = \oint dU + \oint \delta W = 0 \tag{5.2}$$

> A process is reversible if it can be a part of a reversible cycle.

The pure mechanical and electrical systems were termed conservative because they were capable of executing reversible processes only.

The pure translational damper was the simple model of the dissipative aspects of a mechanical system. The damper and the electrical resistance were termed dissipative because they were capable of processes which required that the net work transfer for a cycle was negative. The extension of this concept to any system is:

> A cycle executed by a system in communication with a single heat reservoir is irreversible if the net work transfer of the cycle is negative. A process is irreversible if the inclusion of the process in a cycle which is otherwise reversible makes the cycle irreversible.

The terms reversible and irreversible are generally applied to the process rather than the system because a system with dissipative elements may sometimes be adequately described by a reversible process when the situation is quasi-static. For example, if the elements of Figure 5.1 represent two springs, a mass, and a piston-in-cylinder dashpot, the displacement x may be changed so slowly that the force in the dashpot is negligible. The velocity of the mass is also so small that it has a negligible change in kinetic energy. Thus, the situation is quasi-static, and the boundary force is the same as the force in the spring. The work transfer is then equal to the change in the stored energy of the springs. In this case the process is adequately described by the equilibrium properties and a reversible process. The reversible process in a system with dissipative effects is the limit as the rate of change of state approaches zero so that the dissipative effects approach zero. In any real macroscopic physical situation, careful observation will reveal that all real processes are to some extent irreversible.

The model of a system can be termed reversible if it represents the idealization of the physical situation to such an extent that it is capable of only reversible processes. The pure conservative system elements which have been previously discussed are examples of this. These models are useful only when the processes are quasi-static, not only with respect to dissipative ef-

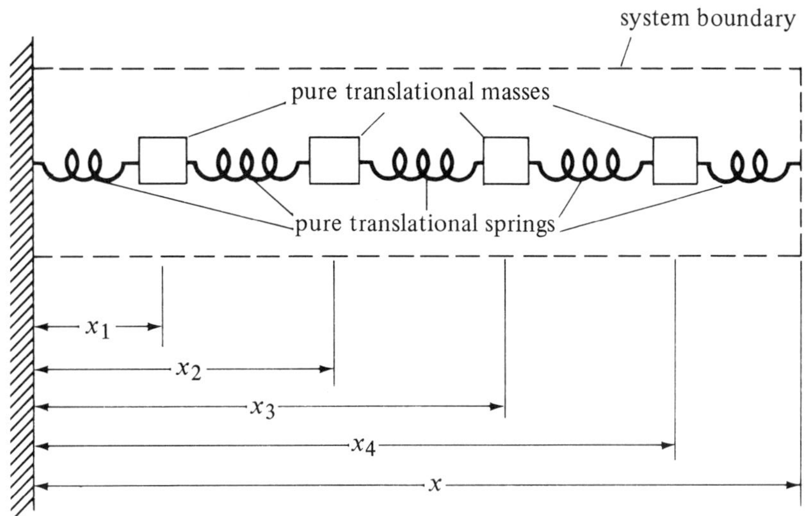

Figure 5.2 Distributed mechanical system

fects but also with regard to distributed effects. For example, the process in the system modeled as a pure spring must be slow enough so that the mass of the real spring is negligible.

Although a system composed of a number of pure conservative system elements is not capable of irreversible processes, it does exhibit a behavior rather analogous to an irreversible process. Figure 5.2 shows a system consisting of four pure translational masses and five pure translational springs.

The system is initially at equilibrium with all masses at rest and the same force for each spring. The right end is given a rapid displacement, x, which increases the existing tension in the springs. The work transfer into the system is larger in magnitude than the work transfer would have been if the displacement had been quasi-static. As a result, the new state is a non-equilibrium state with continuous internal work transfer between the components of the system. If the displacement x is now returned to its original value, the work transfer will not be equal and opposite to the work transfer for the first process. The result of the two processes is a net work transfer to the system with the equilibrium variable x having gone through a cycle. This aspect of the process resembles a dissipative irreversible process; however, the important difference is that by proper motion of the displacement x it would be possible to bring the system back to equilibrium and in the process, bring the system back to the initial energy by means of work transfer. In a dissipative process the work transfer in excess of the work transfer for the quasi-static process appears as a change in the thermal state of the system. At thermal equilibrium this energy is uniquely distributed among the microscopic variables of the system. Thermodynamics is based on the generalization that the information necessary for complete extraction (in the form of a positive work transfer) of the equilibrium energy associated with microscopic variables is not obtainable. In contrast the information necessary to extract the non-equilibrium energy from the four internal variables of the previous example is obtainable; however, it is easy to see how the difficulty would increase with the number of internal variables.

A number of physical situations are often described with simple models which are incapable of reversible processes. The most common example is Coulomb friction. Here the friction force is independent of the rate of change of state, except that the friction force changes direction when the rate of change of state alters direction. A second example is the perfectly plastic solid. In this model the force on the solid is independent of the deformation, except that the force changes direction when the deformation changes direction. For this class of models, the friction force cannot be eliminated even by reducing the rate of change of state to an arbitrarily small value.

5.4 Isothermal Work Transfer for an Ideal Gas

We now consider the behavior of gaseous systems subjected to various thermodynamic processes, and from these considerations we shall develop process models which can be applied to the ideal gas. On the basis of these processes in the ideal gas, we can make observations regarding the nature of thermodynamic processes. The generalization of these observations will be the second law of thermodynamics.

As a typical example of a gaseous system, consider a gas contained within the familiar piston-cylinder apparatus of Figure 5.3. As the piston is moved inside the cylinder in either direc-

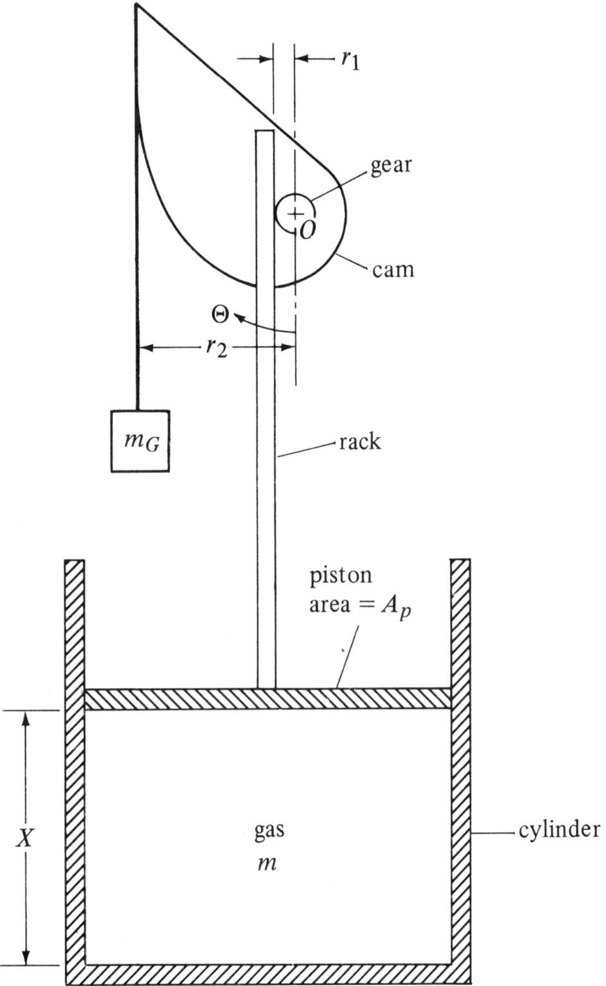

Figure 5.3 Piston-cylinder apparatus with frictionless force mechanism

tion, the state of the gas will change. The manner in which these changes of state take place depends on, among other things, the types of interactions that the gas system can experience and the rates at which these interactions take place. For example, suppose that the gas is in thermal communication with a large system which for purposes of analysis we can model as a heat reservoir. Moreover, let us move the piston by means of a work transfer at a rate such that any changes in temperature that result are adequately counteracted by heat transfer with the heat reservoir. Thus, the system will maintain an essentially constant temperature while experiencing work transfer. This situation implies that any temperature difference that might develop between the system and the heat reservoir will be quite small. Since it is an experimentally determined fact that the rate of heat transfer between two systems depends, in part, upon the temperature difference between them, we anticipate the rate of heat transfer to be small. Therefore, in order to maintain the temperature of the gas constant during the work transfer, it is necessary to carry out the work transfer process at a very slow rate. If these criteria are satisfied, we can model the process as an isothermal, that is, a constant temperature process.

As an example of how such a process might be effected in an actual physical situation, consider the apparatus shown in Figure 5.3. A gas, modeled as an ideal gas, is completely sealed within a cylinder by means of a piston which can be moved without friction. The cylinder is in thermal communication with a large system which can be suitably modeled as a heat reservoir. For our present purposes, the cylinder and piston can be adequately modeled as rigid, pure thermal system elements.

Let us now suppose that the piston, cylinder, gas, and heat reservoir are all in mutual equilibrium. This is by virtue of the action of an external force F applied to the piston by means of the frictionless mechanism shown in Figure 5.3. If the cam is properly designed, the entire apparatus can remain in equilibrium regardless of the position. That is, the gas exerts a pressure on the face of the piston which is exactly counteracted by a force due to the mass, m_G. There will be no net moment acting on the gear-cam subassembly. From the condition for equilibrium about the point O,

$$r_1 F_{\text{piston}} = m_G g r_2 \tag{5.3}$$

But for the rack and gear combination,

$$r_1 \Theta = X \tag{5.4}$$

Thus,

$$F_{\text{piston}} = \frac{m_G g r_2 \Theta}{X} \tag{5.5}$$

The force resulting from the gas pressure acting on the piston face is

$$F_{\text{piston}} = P_{\text{gas}} A_P \tag{5.6}$$

For an ideal gas

$$P_{\text{gas}} = \frac{mRT}{V_{\text{gas}}} = \frac{mRT}{XA_p}$$

Thus,

$$F_{\text{piston}} = \frac{mRT}{X} \tag{5.7}$$

Combining equations (5.5) and (5.7), we obtain

$$\frac{m_G g r_2 \Theta}{X} = \frac{mRT}{X}$$

or

$$r_2 \Theta = \frac{mRT}{m_G g} \tag{5.8}$$

Thus, for a fixed equilibrium temperature and a fixed geometry, the shape of the cam can be determined from equation (5.8).

Let us examine the results of upsetting this equilibrium by increasing the mass m_G by a small amount δm. When δm is first added, the force exerted by the cam mechanism will be greater than the force of the gas on the piston face. As a result, the piston will move down to establish mechanical equilibrium. By means of this downward motion of the piston, there is a work transfer from the m_G mass into the gas. This negative work transfer for the gas will increase the energy and the temperature of the gas until the rate of heat transfer from the gas to the reservoir is equal to the rate of work transfer. Thus, the piston will continue to move down so long as the additional mass δm is in place, The magnitude of the temperature increase can be determined from equation (5.8) to be

$$mR(T + \delta T) = r_2 \Theta (m_G + \delta m) g \tag{5.9}$$

The rate at which the process proceeds is controlled by the heat transfer rate from the gas at $T + \delta T$ to the reservoir at T. The process will stop and equilibrium will be reestablished at any state when δm is removed from m_G.

As the size of the additional mass δm is decreased, the gas temperature approaches the temperature of the reservoir, and the process proceeds at a lower rate. In the limit of an infinitesimal change in the mass, dm, the process is quasi-static, and the temperature of the gas becomes the temperature of the reservoir while the gas pressure becomes equilibrium pressure. The work transfer for the ideal gas system is then

$$W_{1-2} = \int_{V_1}^{V_2} P\,dV \tag{5.10}$$

for this quasi-static process. The pressure in equation (5.10) is the equilibrium pressure given by equation (4.36). The temperature in equation (4.36) is the equilibrium temperature, that is, the reservoir temperature. The work transfer for a quasi-static isothermal process in an ideal gas is thus

$$W_{1-2} = \int_{V_1}^{V_2} mRT\frac{dV}{V} = mRT \ln \frac{V_2}{V_1} \tag{5.11}$$

Since the temperature is the same throughout the process ($T_1 = T_2$), equation (4.41) requires that $U_2 = U_1$. The first law of thermodynamics then gives the required heat transfer

$$Q_{1-2} = W_{1-2} + (U_2 - U_1)$$

or

$$Q_{1-2} = W_{1-2} \tag{5.12}$$

With the system in equilibrium at state 2, the mass m_G is decreased by an infinitesimal amount dm. The system will then execute a quasi-static expansion which may be stopped at equilibrium state 1 by returning m_G to its original value. Thus, the system completes a cycle. The work transfer for this second process will be

$$W_{2-1} = \int_{V_2}^{V_1} mRT\frac{dV}{V} = mRT \ln \frac{V_1}{V_2} \tag{5.13}$$

by the same arguments as for the process from state 1 to state 2. The net work for the cycle will be

$$\oint \delta W = W_{1-2-1} = W_{1-2} + W_{2-1} = mRT\left(\ln\frac{V_2}{V_1} + \ln\frac{V_1}{V_2}\right) = 0 \quad \textbf{(5.14)}$$

Similarly the net heat transfer for the cycle will be

$$\oint \delta Q = Q_{1-2-1} = Q_{1-2} + Q_{2-1} = W_{1-2} + W_{2-1} = \oint \delta W = 0 \quad \textbf{(5.15)}$$

Thus, the quasi-static cycle is reversible because the net work transfer is zero for the cycle and the system has been in communication with a single heat reservoir. By definition, each of the processes in the reversible cycle must be a reversible process.

In the reversible isothermal compression of an ideal gas, energy is transferred into the gas by work transfer and is transferred out to the heat reservoir as a heat transfer. The stored energy of the gas remains unchanged; however, the gas does change state since the pressure increases and the volume decreases. In contrast to the similar process in a mechanical damper, this process is reversible and the work transfer is recovered when the system is returned to its original state. The interrelation between the work transfer and the heat transfer is the result of the coupled constitutive relation rather than the result of a dissipation. In contrast to an uncoupled mechanical system, the increased force on the piston confining the gas is not the result of an increase of the energy of the system.

5.5 Adiabatic Work Transfer for an Ideal Gas

A reasonably rapid change of the volume of a gas confined within a cylinder as in Figure 5.3 may be modeled as occurring without thermal communication between the gas and the cylinder walls. The basic physical fact that makes this a reasonable model is that the time required for pressure equalization in a gas is much shorter than the time required for temperature equalization. Hence, the time scale of a physical process may be long enough to permit pressure equalization but too short to permit temperature equalization. Such a process is modeled as an adiabatic work transfer to (or from) the gas. In many texts, this process is modeled by assuming that the confining walls are incapable of heat transfer, that is, zero thermal conductivity. A more realistic statement would be to assume that the gas was incapable of heat transfer. The thermal conductivities of metals range from 20 to 350 $Wm^{-1}K^{-1}$, while the thermal conductivities of gases range from 0.02 to 0.3 $Wm^{-1}K^{-1}$. In the real situation a steep temperature gradient occurs in the gas very close to the wall.

The student unfamiliar with the adiabatic process model is cautioned against applying the model in situations where it is an inadequate representation of the process. In particular, the student should avoid using this model for processes in very small systems. Since most systems of this type have a very high surface area to volume ratio, it is possible for heat transfer to dominate most other processes. Processes in systems of this type might be modeled more appropriately as isothermal processes. On the other hand, the system should not be so large that changes of pressure would require a sizable amount of time to propagate throughout the system. In such cases, the pressure would not be uniform; consequently, the process could not be represented by the quasi-static model.

For the present purposes, the model is an ideal gas which is not in thermal communication with the confining walls. In this case work transfer is the only interaction with the environment. Through the use of a device similar to the mechanism sketched in Figure 5.3, the force-displacement characteristics of the environment can be matched to the pressure-volume relation of the confined gas which cannot exchange heat with the confining walls. The first law of thermodynamics must be combined with the equation of state to determine the pressure-volume relation for the adiabatic gas. If the process is modeled as quasi-static, the equilibrium pressure and the equilibrium temperature (both uniform throughout the gas) are sufficient to describe the states of the system. For the adiabatic process ($Q_{1-2} = 0$), the first law of thermodynamics for the gas requires

$$-W_{1-2} = U_2 - U_1 \tag{5.16}$$

If the specific heat of the gas is constant, equation (4.41) may be used to evaluate the change in internal energy. Thus,

$$-W_{1-2} = mc_v(T_2 - T_1) \qquad \textbf{(5.17)}$$

It follows, then, that the energy of the system is increased by a work transfer into the system, and consequently the temperature increases. Since the process is modeled as quasi-static, the work transfer is again given by equation (5.10). Combining equations (5.10) and (5.17) we obtain

$$mc_v(T_2 - T_1) = -\int_{V_1}^{V_2} P\,dV \qquad \textbf{(5.18)}$$

For an infinitesimal change of state, equation (5.18) reduces to

$$mc_v\,dT = -P\,dV \qquad \textbf{(5.19)}$$

The equation of state, equation (4.37), may be expressed in differential form as

$$\frac{dP}{P} + \frac{dV}{V} = \frac{dT}{T} \qquad \textbf{(5.20)}$$

When equation (5.20) is used to eliminate dT and equation (4.37) is used to eliminate T, equation (5.19) becomes

$$\left(\frac{R + c_v}{c_v}\right)\frac{dV}{V} + \frac{dP}{P} = 0 \qquad \textbf{(5.21)}$$

Integrating equation (5.21), we obtain

$$PV^\gamma = P_1V_1^\gamma = P_2V_2^\gamma \qquad \textbf{(5.22)}$$

where γ is

$$\gamma = \frac{R + c_v}{c_v} \qquad \textbf{(5.23)}$$

Equation (5.22) is the pressure-volume relation for a *quasi-static adiabatic process* in an *ideal gas with a constant specific heat* c_v. Equation (5.22) can be written

$$PV^\gamma = \text{constant} \tag{5.24}$$

The work transfer for this process can be determined by substituting equation (5.22) into equation (5.10). Then

$$W_{1-2} = \int_{V_1}^{V_2} P_1 V_1^\gamma \frac{dV}{V^\gamma} = \frac{P_1 V_1^\gamma}{1-\gamma}[V_2^{1-\gamma} - V_1^{1-\gamma}] \tag{5.25}$$

which becomes

$$W_{1-2} = \frac{1}{1-\gamma}(P_2 V_2 - P_1 V_1) \tag{5.26}$$

when equation (5.22) is used. Equation (5.26) is the same as equation (5.17) with T substituted from the ideal gas equation of state, equation (4.37). The final temperature may be determined from equation (5.22) by substitution of the pressure from the ideal gas equation of state, equation (4.37). Then

$$P_1 V_1^\gamma = \frac{mRT_1}{V_1} V_1^\gamma = \frac{mRT_2}{V_2} V_2^\gamma$$

$$T_2 = T_1 \left[\frac{V_1}{V_2}\right]^{\gamma-1} \tag{5.27}$$

The student must bear in mind that equations (5.25) and (5.27) apply *only* to a quasi-static adiabatic process in an ideal gas with a constant specific heat c_v. Their use in any other physical situation or for any other system model will only lead to erroneous results.

For the ideal gas model just described it is left as an exercise for the student to show that for a quasi-static adiabatic process, the cam of equation (5.11) must be replaced by one for which

$$r_2\Theta = \frac{P_1 V_1^\gamma}{m_G g} V^{1-\gamma} \tag{5.28}$$

Here P_1 is the pressure for which the volume is $A_P X_1$. For the cam of equation (5.28), the adiabatic system will always be in a state of equilibrium regardless of the piston position. The nature of this equilibrium may be determined by a stepwise analysis similar to that of Section 5.4.

If the mass m_G on the mechanism is increased by δm, the piston will move down, and the gas will be compressed quasi-statically until the imbalance force due to δm is reduced by removing the excess mass δm. The gas will then follow equation (5.24) to the state (P_2, V_2) with the corresponding T_2 given by the equation of state (4.37). If the mass is now decreased by δm, the system piston will move up and the gas will expand quasi-statically following equation (5.24). When the gas has returned to the state (P_1, V_1), the temperature must be T_1 in order to satisfy the equation of state. From equation (5.26), it follows that W_{1-2-1} for the process 1-2-1 must be zero, and the cycle 1-2-1 is reversible. Thus, the quasi-static adiabatic processes of the cycle were reversible adiabatic processes. The reversible adiabatic process in the ideal gas is similar to a pure mechanical process in that a work transfer into the system increases the energy of the system and the force (pressure) at the system boundary. As a result of the coupling, the increase of the system's energy from the work transfer caused an increase in temperature even though there was no heat transfer.

The paths of the reversible isothermal process and the reversible adiabatic process can be graphically portrayed on the pressure-volume plane as shown in Figure 5.4. For processes such as these in which the only work transfer is the quasi-static displacement of the boundary, the work transfer is simply the integral $\int P\, dV$ between the two end states. This integral is represented by the area between the path of the process connecting the end states and the $P = 0$ axis. Hence, it is convenient to think of this area as a graphical representation of the quasi-static work transfer process with the following convention for the sign of the work transfer. If we traverse the path of the process on the pressure-volume plane in the direction of the process and find the area representing the integral $\int P\, dV$ to our right, the work transfer is positive. Conversely, if the area lies to our left, the work transfer is negative. Figure 5.4 (a) shows an example of the latter case in which the area lies to the left, hence the work transfer is negative for the reversible isothermal com-

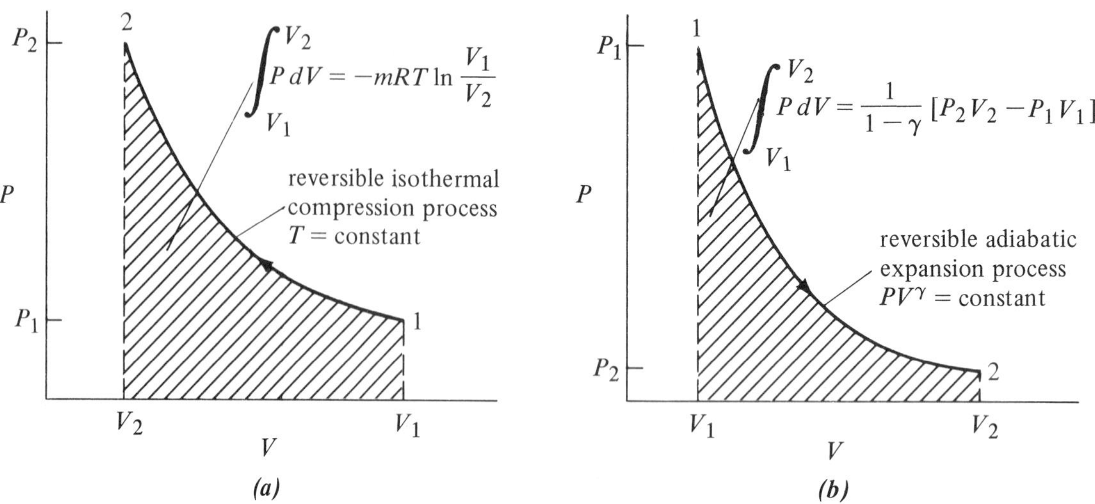

Figure 5.4 Graphical representation of reversible isothermal and reversible adiabatic processes for the ideal gas model

pression process from (P_1, V_1) to (P_2, V_2). Figure 5.4 (b) shows an example of the case in which the area lies to the right; hence the work transfer is positive for the reversible adiabatic expansion process from (P_1, V_1) to (P_2, V_2).

These isothermal and adiabatic processes illustrate two of the essential characteristics of a thermodynamically coupled system. First, a heat transfer is the result of a reversible isothermal work transfer, and second, a temperature change is the result of a reversible adiabatic work transfer. In the majority of cases the results are similar to those of the ideal gas. That is, a heat transfer out of the isothermal system and a temperature rise in the adiabatic system are both the result of a work transfer into the system.

5E.1 Sample Problem: As a first approximation, the compression process in an internal combustion engine can be modeled as a reversible adiabatic process while the air-fuel can be modeled as an ideal gas with R = 0.287 kJ/kgK, c_v = 0.716 kJ/kgK, and γ = 1.4. An engine cylinder with a volume of 617.8 cm^3 is filled with an air-fuel mixture at a pressure of 1.013 $\times$ 10^5 Pa and a temperature of 300 K. The air-fuel mixture is compressed to one-tenth of its original volume. Calculate the pressure and temperature of the air-fuel mixture at the end of compression.

Solution: From equation (5.22), the compression process can be described by

$$P_1 V_1^{\gamma} = P_2 V_2^{\gamma}$$

Thus

$$P_2 = P_1 \left(\frac{V_1}{V_2}\right)^{\gamma}$$

$$P_2 = 1.013 \times 10^5 (10)^{1.4}$$

$$P_2 = 2.545 \times 10^6 \text{ N/m}^2$$

The final temperature can be calculated with the aid of equation (5.27).

$$T_2 = T_1 \left(\frac{V_1}{V_2}\right)^{\gamma-1}$$

$$T_2 = 300(10)^{1.4-1}$$

$$T_2 = 753.6 \text{ K}$$

These are the thermodynamic conditions that prevail within the cylinder at the end of compression just prior to combustion.

5.6 An Irreversible Heat Transfer to an Ideal Gas at Constant Volume

Consider for the moment an ideal gas which has experienced a reversible adiabatic expansion from some initial state V_0 to some larger volume V_1. It is evident from equation (5.27) that in this final state, the gas temperature T_1 will be lower than the temperature of the gas in the initial state ($T_1 < T_0$). If we now place the gas in thermal communication with a heat reservoir at the original temperature T_0, the gas will experience a heat transfer and change to state 2 where $T_2 = T_0$, the original temperature. Suppose that we now permit the gas to experience this heat transfer while its volume remains constant. That is, let the gas return to T_0 from T_1 by a process for which the work transfer is zero.

For this special process of heat transfer, we would like to perform an analysis of a cycle similar to those of Sections 5.4

and 5.5. Because the work transfer for this process is zero, we cannot employ the mechanism of Figure 5.3. Therefore, a different approach must be employed. We must include the heat transfer process 1-2 in a cycle such as that of Figure 5.5. The cycle of Figure 5.5 consists of a reversible adiabatic expansion, 0-1, a constant volume heat transfer, 1-2, to bring the system back to the initial temperature, T_0, and finally a reversible isothermal compression to return the system to the original volume, V_0.

Process 0-1 shown in Figure 5.5 is a reversible adiabatic expansion from volume V_0 to volume V_1. The work transfer for this process is positive and the heat transfer is zero. The first law

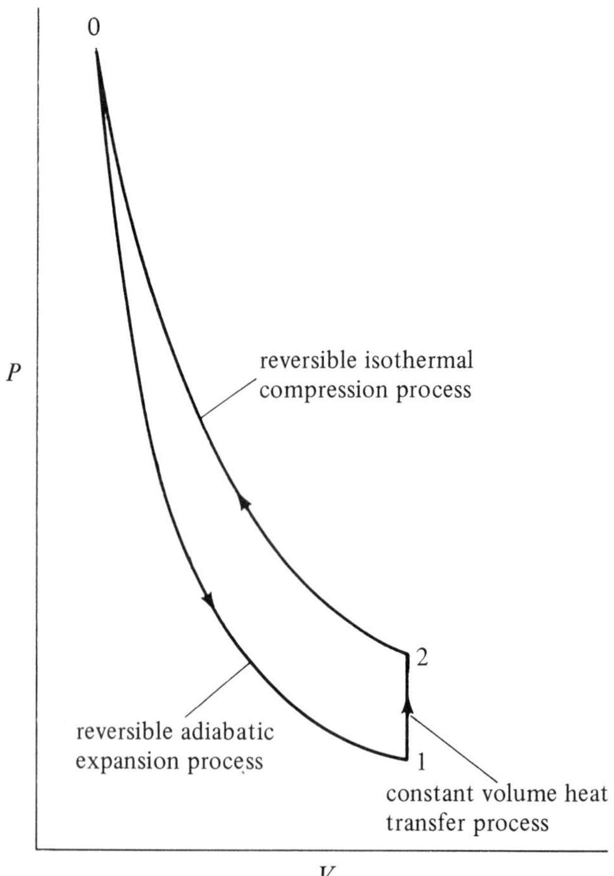

Figure 5.5 Ideal gas cycle

of thermodynamics thus requires a decrease in the internal energy U. As pointed out previously, the constitutive relation for the energy of the ideal gas then requires a decrease in the temperature as the gas expands from volume V_0 to volume V_1.

$$-W_{0-1} = U_1 - U_0 = mc_v(T_1 - T_0) \tag{5.29}$$

The volume of the ideal gas system is now maintained constant at V_1, and the system is allowed to come to mutual thermal equilibrium with the container wall and with the heat reservoir at temperature $T_2 = T_0$. The equilibrium is established by means of the heat transfer from the reservoir through the container to the gas. The work transfer for this process is zero because the boundary of the ideal gas system has not moved. As the result of the positive heat transfer, the first law of thermodynamics requires an increase in the internal energy which in turn requires an increase in the temperature. Because of this increase in temperature at constant volume, the equation of state for the ideal gas requires an increase in the pressure. Thus, state 1 in Figure 5.5 is at a lower pressure than state 2. The system is now recompressed to volume V_0 while in equilibrium with the heat reservoir. The negative work transfer and negative heat transfer for this reversible isothermal process are given by equations (5.11) and (5.12).

Since the system has been in thermal communication with only one heat reservoir during the cycle 0-1-2-0, the cycle may be tested for reversibility by simply evaluating the cyclic integral of the work transfer. This could be done by adding the work transfers for each of the processes. Thus,

$$\oint \delta W = W_{0-1} + W_{1-2} + W_{2-0} \tag{5.30}$$

From equations (5.11) and (5.26)

$$\oint \delta W = \frac{1}{1-\gamma}(P_1V_1 - P_0V_0) + 0 + mRT_0 \ln \frac{V_0}{V_2} \tag{5.31}$$

where the work transfer for the constant volume heat transfer process 1-2 is zero. It is not immediately evident from equation (5.31) that the right hand side is negative. This result can be seen more easily from the graphical representation of the work transfer on the (P, V) diagram of Figure 5.5. It is clear from equation (5.31) that the net work transfer for the cycle is represented by the area enclosed by the cycle. Since this area lies to the left as we traverse the path of the cycle in the same direction as the cycle, the area, and hence the net work transfer, is negative. Physically this must be so because the pressure at each volume during the recompression process 2-0 is higher than the pressure at the same volume during the expansion process 0-1.

The cycle consisting of the reversible adiabatic expansion, the constant volume heating, and the isothermal compression is an irreversible cycle. This is because the net work transfer for the cycle is less than zero and the system is in communication with a single heat reservoir. The adiabatic process and the isothermal process were shown to be reversible in Sections 5.4 and 5.5. Therefore, the constant volume process is the process which makes the cycle irreversible.

This heat transfer is "dissipative" in that the return of the system and its environment to their original state requires the decrease in the energy of a pure conservative mechanical system and the increase in the energy of a heat reservoir as the minimum external effect. On the other hand, the reversible processes were "non-dissipative" in that both the system and its environment could be returned to their original state with no permanent external effects.

In general, *any simple heat transfer process between two systems which are unequal in temperature will be irreversible.* However, as was seen in Section 5.4, the heat transfer process may approach the reversible process as the limit when the temperature difference approaches zero. As we shall see in the next chapter, the heat transfer between two systems which are unequal in temperature has the potential for producing a positive work transfer. Heat transfer by a simple thermal communication across the temperature difference is "dissipative" in that the potential for work transfer is not utilized.

5.7 A Reversible Heat Transfer to an Ideal Gas at Constant Volume

The methods of Sections 5.4 and 5.5 may be used to construct an apparatus for reversible heat transfer to an ideal gas held in a

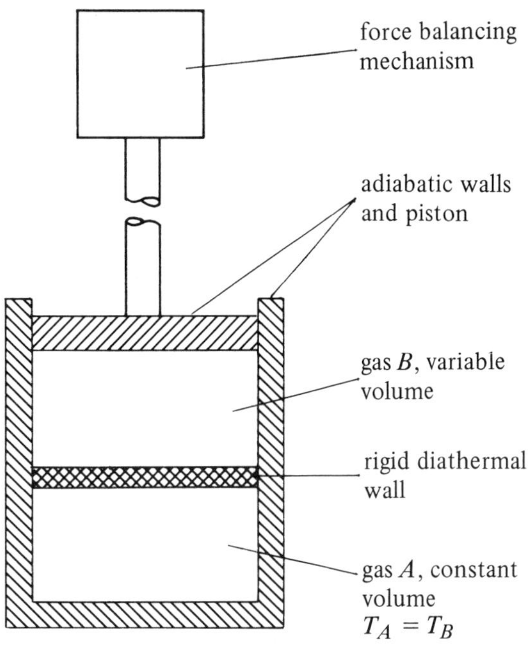

Figure 5.6 Apparatus for reversible heat transfer in an ideal gas at constant volume

rigid container as shown in Figure 5.6. The ideal gas A in the constant volume container is in thermal communication with a second ideal gas B confined in a piston-cylinder apparatus. The external walls and the piston are all adiabatic. The gas B is held in mechanical equilibrium by a force balancing mechanism similar to that of Figure 5.3. Gases A and B are kept in thermal equilibrium by heat transfer interactions through the diathermal wall.

If the force balancing mechanism has a force displacement characteristic which matches the equilibrium pressure-volume characteristics of gas B, then a composite system consisting of the mechanism, the piston-cylinder, the diathermal wall, and the gases A and B can be carried through a quasi-static adiabatic cycle with a vanishingly small external work transfer as the rate of change of state approaches zero. As with the processes of Sections 5.4 and 5.5, this adiabatic cycle is reversible in this limit. In the reversible limit, the mechanism transfers work to the gas B which transfers heat through the diathermal wall to gas A. At each state, the temperatures of gases A and B are uniform and equal. The temperatures of gases A and B increase together as work is transferred from the mechanism.

Since the processes for the composite system are reversible, the processes in all the subsystems must also be reversible. In particular we can consider the heat transfer between gas A and the diathermal wall as reversible. Thus, for the reversible heat transfer to an ideal gas at constant volume we have for the *gas* as a *system*

$$\delta W = 0, \qquad W_{1-2} = 0 \tag{5.32}$$

$$\delta Q = dU = c_v \, dT, \qquad Q_{1-2} = U_2 - U_1 = mc_v(T_2 - T_1) \tag{5.33}$$

In addition we could develop the detailed description of the process in the ideal gas B. Specifically, we could develop the pressure-volume path by integrating the differential form of the first law applied to gas B as a system. (See problem 5.6 .)

It is important to contrast the reversible constant volume process with the irreversible constant volume process. If the process is irreversible because of heat transfer at a finite rate, the resulting temperature gradient indicates a departure from thermal equilibrium. In this case equations (5.32) and (5.33) still apply if we consider the temperature in equation (5.33) as the average temperature rather than the actual nonuniform temperature. This distinction becomes especially important when the second law of thermodynamics is considered in Chapter 7.

If the constant volume process is irreversible because of a departure from mechanical equilibrium, the pressure will not be uniform throughout the gas. In this case, the work transfer will be negative regardless of the sign of the heat transfer. In this case the work transfer must be included in equation (5.33):

$$\delta Q - \delta W = dU = mc_v \, dT \tag{5.34}$$

$$Q_{1-2} - W_{1-2} = mc_v(T_2 - T_1) \tag{5.35}$$

Problem 5.5 is an example of such a constant volume process.

5.8 A Reversible Process in an Ideal Gas at Constant Pressure

The methods of Sections 5.4 and 5.5 may be used to construct an apparatus for a reversible process in an ideal gas at constant pressure as shown in Figure 5.7. The ideal gas A is confined within a piston-cylinder apparatus and is held at constant pressure by a constant force mechanism. Gas A is in thermal communication with a second ideal gas B confined in a second piston-cylinder apparatus. Gas B is held in mechanical equilibrium by a force balancing mechanism which matches the pres-

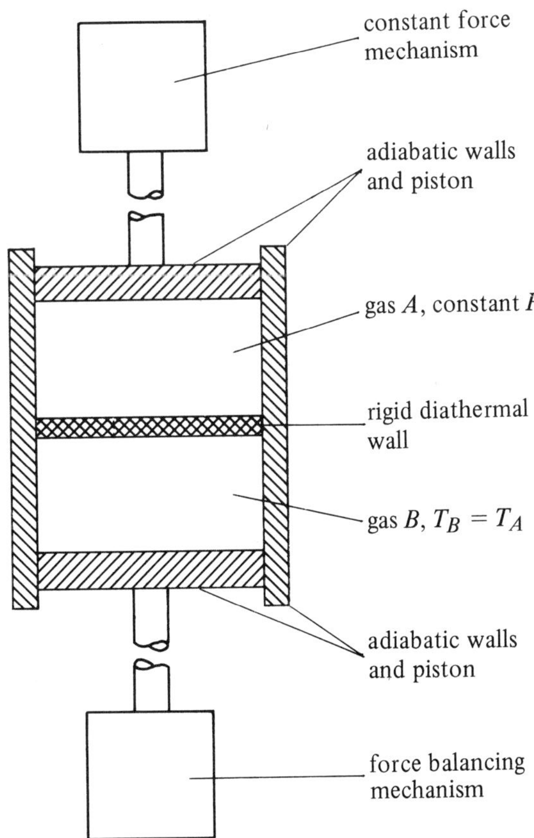

Figure 5.7 Apparatus for reversible constant pressure process in an ideal gas

sure-volume behavior of gas B when gas B is in thermal equilibrium with gas A through the diathermal wall.

Since the force balancing mechanism is matched, the entire apparatus will be in equilibrium with a specific temperature and a specific position of the upper and lower piston. Thus, the composite system consisting of the entire apparatus can be carried through a quasi-static adiabatic cycle with a vanishingly small external work transfer. In the limit, this adiabatic cycle is reversible so all subsystem processes of the cycle are reversible. In particular the process in gas A is a reversible process at constant pressure.

For this reversible constant pressure process in gas A as a system, we have

$$\delta W = P\,dV, \qquad W_{1-2} = P(V_2 - V_1) \tag{5.36}$$

$$\delta Q = mc_v\,dT + P\,dV \tag{5.37}$$

$$\delta Q = mc_v\,dT + P\left(\frac{mR}{P}\,dT\right) \tag{5.38}$$

$$\delta Q = m(c_v + R)\,dT, \qquad Q_{1-2} = m(c_v + R)(T_2 - T_1) \tag{5.39}$$

The expression $c_v + R$ is termed the specific heat at constant pressure[1] c_p. Thus,

$$c_p = c_v + R \tag{5.40}$$

It is important to contrast the reversible constant pressure process with the irreversible constant pressure process. If the irreversibility is the result of departure from thermal equilibrium, equation (5.39) still applies if the temperature is taken as the average temperature of the gas system rather than the actual nonuniform temperature. If the irreversibility is the result of a departure from mechanical equilibrium, the situation is more complex because the pressure is nonuniform. The work transfer is not simply the uniform pressure times the net change in the volume of the gas. The appropriate nonequilibrium pressure must be used at each point where the boundary is moving.

[1] The fundamental definition of the specific heat at constant pressure, c_p, is given in Chapter 9, equation (9.57), page 287.

Problems 5.12 and 5.15 are examples of the action of constant pressure pistons during processes for which the gas is not in mechanical equilibrium with a uniform pressure at each state along the process.

5.9 The Second Law of Thermodynamics

The cycles executed by the systems that we have examined have, in each case, experienced heat transfer only with a single heat reservoir and have had a net work transfer for the cycle that was less than or equal to zero. The generalization of this observation to all systems, no matter how complex, is called the *second law of thermodynamics.*

> Any cycle of any system in thermal communication with only a single heat reservoir must have a net work transfer for the cycle which is less than or, in the limit, equal to zero. That is,
>
> $$\oint \delta W \leq 0 \tag{5.41}$$
>
> In accordance with the first law of thermodynamics, it must also be the case that
>
> $$\oint \delta Q \leq 0 \tag{5.42}$$

This generalization may also be stated in the negative sense; it is not possible for a system to have a net positive work transfer for a cycle resulting from a net positive heat transfer with a single heat reservoir as the only heat interaction. Thus, a system operating in cycles cannot reduce the energy of a heat reservoir, that is, cool the reservoir by simply converting the energy into a work transfer (cf. Figure 5.8).

Note that the first law of thermodynamics establishes that the net work transfer for a cyclic system is equal to the net heat transfer. The second law of thermodynamics requires that both net interactions are negative or in the limit zero for a cyclic system in thermal communication with only a single heat reservoir. It is this unidirectionality that causes the cycle to be irreversible.

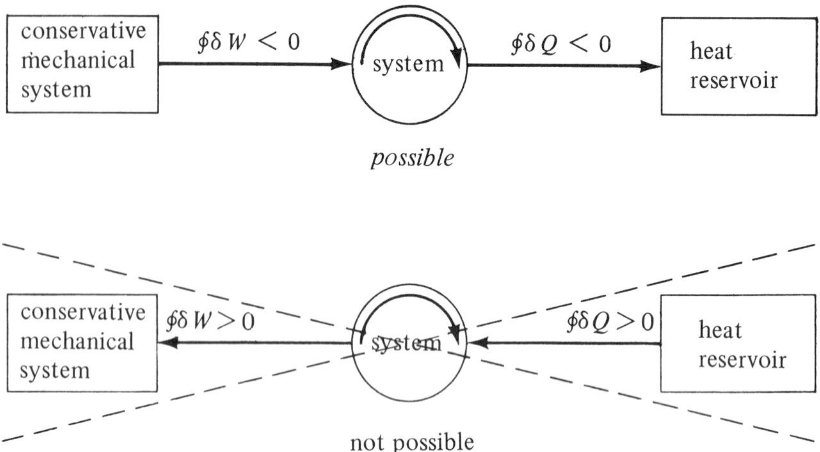

Figure 5.8 Pictorial representation of the second law of thermodynamics

A hypothetical machine which would violate the second law of thermodynamics is called a perpetual motion machine of the second kind. Such a machine could produce an unlimited positive work transfer while cooling an environment at a uniform temperature. A perpetual motion machine of the first kind would violate the first law of thermodynamics as well as the second law of thermodynamics.

5E.2 Sample Problem: An ideal gas system of 0.1 kg mass with $R = 2.078$ kJ/kgK and $c_v = 3.153$ kJ/kgK experiences a change of state from $T_1 = 300$ K and $P_1 = 10^5$ N/m^2 to $T_2 = 600$ K and $P_2 = 4 \times 10^5$ N/m^2. During the process by which the change of state is effected, the system experiences a heat transfer of $Q_{1-2} = 39.59$ kJ while in communication with a heat reservoir at 600 K. By application of the second law of thermodynamics, determine whether the process was reversible or irreversible.

Solution: To determine the reversibility of the process, we must combine the process in question with a series of *reversible* processes to form a cycle such that the system communicates with a single heat reservoir at 600 K while executing the cycle. For this cycle, the second law of thermodynamics requires that $\oint \delta W \leq 0$.

Consider the cycle consisting of the given process 1-2, a reversible isothermal compression, 2-3, while in communication with the heat reservoir at 600 K, and a reversible adiabatic expansion, 3-1. State 3 is determined by the temperature $T_3 = T_2 = 600$ K and the pressure P_3 which satisfies the reversible adiabatic relation

$$P_3 = P_1 \left(\frac{T_1}{T_3}\right)^{\gamma/(1-\gamma)}$$

[cf. equations (5.22) and (5.27)]. Since $\gamma = R/c_v + 1$ from equation (5.23),

$$\gamma = \frac{2.078}{3.153} + 1 = 1.66$$

Then

$$P_3 = 10^5 \left(\frac{300}{600}\right)^{1.66/-0.66} = 5.72 \times 10^5 \text{ N/m}^2$$

For the reversible isothermal process, 2-3,

$$W_{2-3} = \int P\,dV = mRT_2 \ln \frac{V_3}{V_2} = mRT_2 \ln \frac{P_2}{P_3}$$

$$W_{2-3} = (0.1)(2.078)(600) \ln \frac{4 \times 10^5}{5.72 \times 10^5} = -44.69 \text{ kJ}$$

For the reversible adiabatic process, 3-1, the first law of thermodynamics gives

$$-W_{3-1} = U_1 - U_3 = mc_v(T_1 - T_3)$$
$$-W_{3-1} = (0.1)(3.153)(300 - 600) = -94.59 \text{ kJ}$$

For the process in question, 1-2, the first law gives

$$Q_{1-2} - W_{1-2} = U_2 - U_1 = mc_v(T_2 - T_1)$$
$$W_{1-2} = Q_{1-2} - mc_v(T_2 - T_1)$$
$$W_{1-2} = 39.59 - (0.1)(3.153)(600 - 300) = -55 \text{ kJ}$$

Then

$$\oint \delta W = W_{1-2} + W_{2-3} + W_{3-1}$$

$$\oint \delta W = -55 - 44.69 + 94.59 = -5.10 \text{ kJ} < 0$$

Thus the process under study was irreversible according to the second law of thermodynamics as stated in equation (5.41).

Problems

5.1 Consider the system consisting of a large container of water as shown in Figure 5P.1. At some time, t_0, the plug in the bottom of the upper vessel is pulled out permitting the water to drain into the lower vessel. Formulate an appropriate model for the system and use it to determine whether the process consisting of draining the upper vessel and filling the lower vessel is reversible or irreversible. Prove your answer by use of the basic definitions of reversibility and irreversibility.

5.2 Consider the system consisting of the mass and springs as shown in Figure 5P.2. The two springs, a and b, with different spring constants k_a and k_b initially exert forces on the mass M that sum to zero net force on the mass. The force in each spring is a compression force. Attached to the mass M is a massless string and pan. Next to the pan is a series of shelves on each of which is a mass δm. The shelves are spaced vertically so that each δm can be placed on the pan at its own elevation. Thus, after placing a mass δm on the pan, the pan and masses are displaced a vertical distance equal to the shelf spacing. The large mass M is situated between two parallel guides and the gap between the guides and the mass M is filled with an oil film. The constitutive relation of the oil film is such that the shear force that the film exerts on the mass M is linearly related to the velocity of the mass M relative to the guides. The string and the springs are attached to the mass M in such a way that there is no couple exerted on the mass M.

(a) Show that as the size of the small masses δm becomes infinitesimally small (and the number of shelves becomes infinitely large), the change of state associated with the addition of a mass m becomes reversible. Prove your answer by use of the basic definitions of reversibility and irreversibility.

(b) Show that the process of part (a) takes place at a vanishingly small rate.

(c) Suppose that the oil film leaked out and the new interface between the mass M and the guides exhibited dry friction. If the process described in part (a) were now carried out at a vanishingly slow

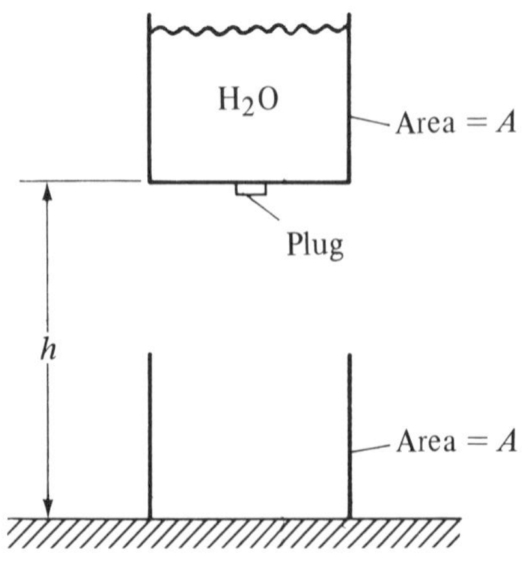

Figure 5P.1

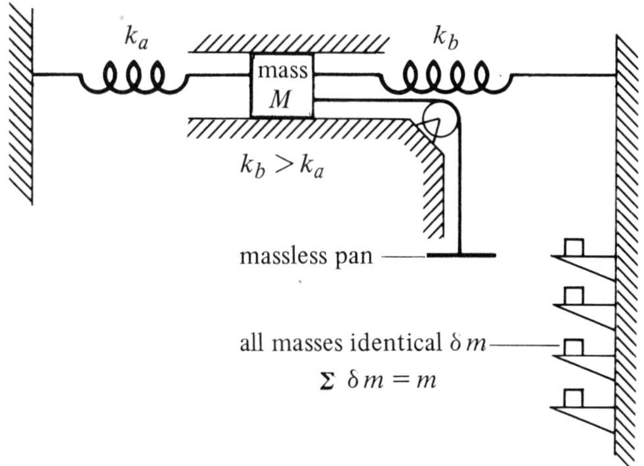

Figure 5P.2

rate with this new interface present, would the process be a reversible one? Why or why not?

(d) Suppose now that after the mass m has been added in infinitesimal increments to the systems with and without the oil film, the mass is removed abruptly by cutting the supporting string. Would the two systems attain the same equilibrium state? Is the process by which the equilibrium state is attained a reversible process? Prove your answer by the basic definition of a reversible process.

(e) From the answers to (a), (b), (c) and (d) above, can we conclude that *any* process executed by a mechanical system at a vanishingly small rate is reversible? That is, are all quasi-static processes necessarily reversible?

(f) For the system with the oil film, what is the maximum rate of change of state for which a reversible process is a satisfactory description?

5.3 Ten grams of an ideal gas with $R = 287$ J/kgK and $c_v = 716$ J/kgK are confined in a piston-cylinder apparatus which is surrounded by a vacuum as shown in Figure 5P.3. In-

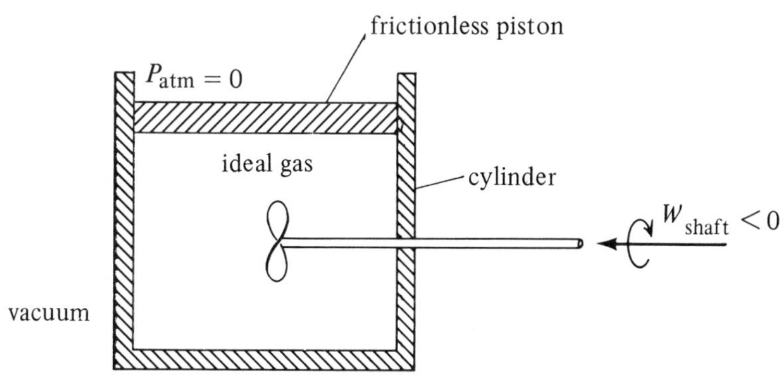

Figure 5P.3

serted in the gas is a paddle wheel which is attached to the shaft. The shaft and paddle wheel have a thermal conductivity of zero, so the surfaces are always adiabatic. Initially the system consisting of the ideal gas, piston-cylinder apparatus, and paddle wheel are in mechanical and thermal equilibrium (T_i = 21 C). The shaft is set rotating for a brief period of time and then stopped. After the system has reached its final state of mechanical and thermal equilibrium, it is found that the piston, whose mass is 100 kg, was displaced quasi-statically upwards a distance of 150 mm during the process. That is, the piston and the gas are always in mechanical equilibrium.

(a) Assuming the interface between the piston and cylinder surfaces and the ideal gas is adiabatic and the gas temperature increases from 21 C to some final equilibrium temperature during the process, determine the work transfer to the gas from the paddle wheel and shaft.

(b) Assuming the interface between the piston and cylinder surfaces and the ideal gas is diathermal, determine the heat transfer occurring between the piston and cylinder and the ideal gas for a piston displacement at 150 mm upwards. The heat capacity of the piston-cylinder apparatus is 45 kJ/K. The only heat transfer process for the piston-cylinder apparatus occurs between the gas and the piston-cylinder apparatus.

(c) Explain the physical difference between the shaft work transfer from the paddle wheel and the work transfer to the piston. Explain the difference between their effects on the system, for the two situations given in parts (a) and (b).

5-4 Assuming the atmosphere to be isothermal with a temperature T, show that if the ideal gas model is a suitable representation of the behavior of the atmosphere, then

$$P = P_0 e^{-(g/RT)y} \text{ and } \rho = \rho_0 e^{-(g/RT)y}$$

where y is the elevation above sea level, where the pressure is P_0, and the density is ρ_0.

5.5 A gas is confined within a piston-cylinder apparatus as shown in Figure 5P.5. The volume contained within the cylinder between the two piston faces is divided into two compartments by a partition with a small hole. The two movable pistons are clamped together so that when they move, the distance between their faces remains constant. An ex-

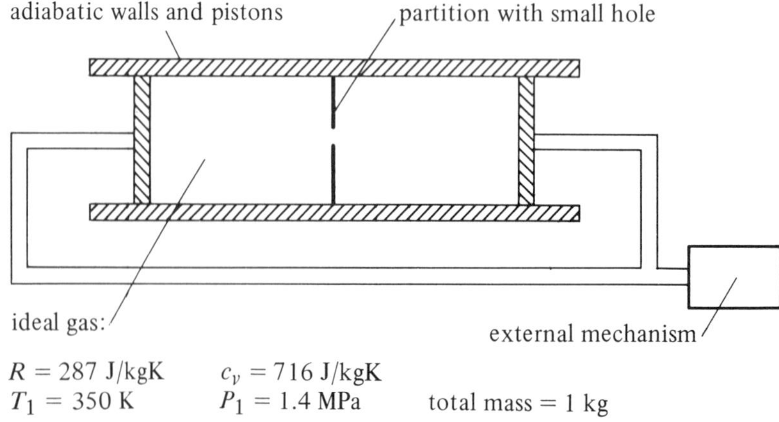

Figure 5P.5

ternal mechanism is now connected to the pistons and moves the pistons through some distance. Observation of the mechanism shows that a work transfer of -33.9 kJ was experienced by the gas. After the pistons came to rest, the system returns to equilibrium.

(a) Was the work transfer process quasi-static?

(b) What is the final temperature and pressure of the gas?

(c) Was the work transfer process reversible?

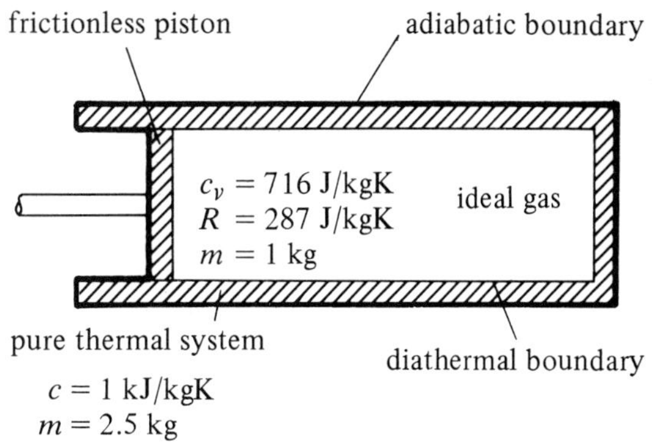

Figure 5P.6

5.6 An ideal gas is confined in a piston-cylinder apparatus by means of a frictionless piston as shown in Figure 5P.6. The gas is compressed quasi-statically so that thermal equilibrium is maintained between the gas and the piston-cylinder apparatus. The piston-cylinder apparatus is externally isolated so that there is no heat transfer between the piston-cylinder apparatus and the environment. The cylinder contains 1 kg of gas with a constant specific heat $c_v = 716$ J/kgK and a gas constant $R = 287$ J/kgK. The piston-cylinder apparatus has a mass of 2.5 kg and can be modeled as a pure thermal system with a specific heat of 1 kJ/kgK.

(a) If the initial state of the gas is an equilibrium state, determine the differential relationship between the pressure and volume of the gas for a quasi-static compression or expansion.

(b) Determine the integral form of the differential relation of part (a) above.

(c) If the initial state of the gas is $P_1 = 1.8$ MPa and $T_1 = 55$ K, determine the final temperature of the complete system (consisting of gas and piston-cylinder apparatus) for a compression process by which the volume is reduced to one-half its original value, $V_2 = 0.5V_1$.

(d) What is the work transfer experienced by the gas for this process?

(e) What is the heat transfer experienced by the piston-cylinder apparatus for this process?

(f) What is the change in internal energy of the gas?

(g) Show that the process is reversible.

5.7 A schematic representation of one design for an air rifle is shown in Figure 5P.7. A gas charge in the reservoir volume V accelerates the BB shot down the barrel when the valve is opened quickly. The initial pressure of the gas is 900 kPa and the temperature is 21 C. Determine the minimum volume V and barrel length L required to give the shot a velocity of 100 m/sec when the shot leaves the barrel. Assume the gas is air, $c_v = 716$ J/kgK, $R = 287$ J/kgK, and that atmospheric pressure is 100 kPa. The spherical steel shot is 3 mm in diameter and $\rho_{steel} = 7770$ kg/m³. State clearly the model you are using for your calculations.

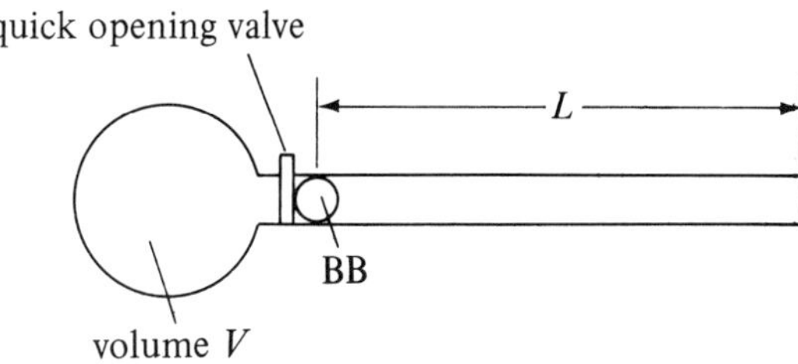

Figure 5P.7

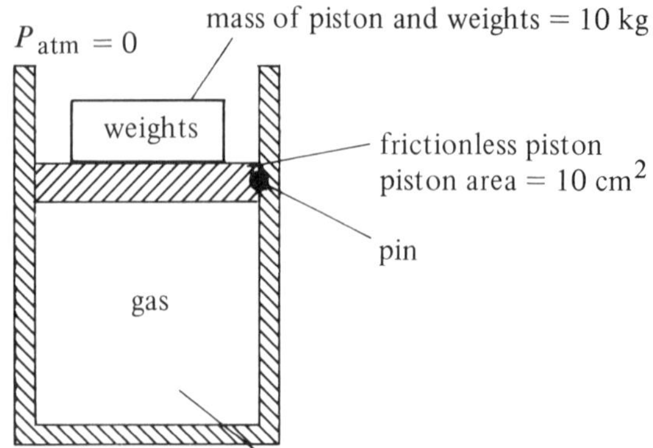

Figure 5P.8

5.8 A heavy piston encloses an ideal gas within a vertical cylinder as shown in Figure 5P.8. The piston moves in the cylinder without leakage or friction. Initially the ideal gas is in mutual equilibrium with the rigid cylinder walls, the piston and a heat reservoir. Initially the piston must be held in place with a pin. The pin is removed and the piston moves until finally a new mutual equilibrium is established between the ideal gas, the piston-cylinder and the heat reservoir. For a system consisting of the ideal gas only:

- **(a)** Determine the final pressure and temperature of the gas.
- **(b)** Determine the heat transfer, the work transfer and the change in the energy U for the change from the initial state to the final state.
- **(c)** Is the process executed by the ideal gas reversible or irreversible? Prove your answer.

Assume that the gas can be modeled as an ideal gas with R = 287 J/kgK and c_v = 716 J/kgK.

5.9 A rigid, insulated, that is, adiabatic, chamber is divided by a movable piston into two compartments of equal volumes as shown in Figure 5P.9. The piston, which moves without friction, is initially held in place with a pin. One compartment is filled with air at 6×10^5 Pa and 25 C while the other compartment is evacuated. Assume that the piston can be modeled as a solid body of zero mass. Also, assume that the air can be modeled as an ideal gas with R = 287 J/kgK and c_v = 716 J/kgK.

- **(a)** Determine the final temperature of the air after the pin is removed and the system comes to equilibrium.
- **(b)** What is the final pressure for this process?
- **(c)** What is the work transfer for the gas?
- **(d)** Prove by thermodynamic arguments whether the process was reversible or irreversible.

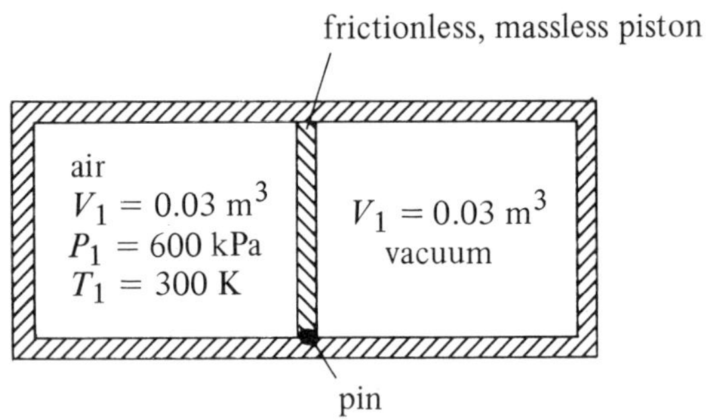

Figure 5P.9

5.10 Consider a closed system of an ideal gas with constant specific heat c_v and gas constant R. Let the gas be in an initial state with a pressure P_1 and a volume V_1. Let the gas expand adiabatically to a new volume V_2. If the expansion is reversible, the final pressure and temperature after the expansion can be determined from the relation $P_1 V_1^\gamma = P_2 V_2^\gamma$. Let the values of the pressure and temperature so determined be P_{2r} and T_{2r}, respectively. If the expansion had been irreversible, the final pressure and temperature would have assumed the values P_{2i} and T_{2i}, respectively. Clearly, these values depend upon the work transfer out of the system during the expansion.

(a) Show that it is always true that $P_{2r} \neq P_{2i}$ and $T_{2r} \neq T_{2i}$.

(b) Decide by proper thermodynamic arguments whether

$$P_{2i} < P_{2r}, \quad P_{2i} > P_{2r}, \quad \text{or } P_{2i} \lessgtr P_{2r}$$

(c) Decide by proper thermodynamic arguments whether

$$T_{2i} < T_{2r}, \quad T_{2i} > T_{2r}, \quad \text{or } T_{2i} \gtrless T_{2r}$$

5.11 As shown in Figure 5P.11, a gas is contained in a piston-cylinder apparatus which is also fitted with a damper. The following models can be used:

Gas: Ideal gas with $c_v = 716$ J/kgK
$R = 287$ J/kgK
$P_1 = 7 \times 10^4$ Pa
$T_1 = 350$ K

Damper: Pure translational damper

Piston: Rigid, frictionless piston with mass of 1300 kg and a cross-sectional area of 0.1 m^2 and an initial height h_1 of 1 m

Cylinder: Rigid

Gas-cylinder boundary: adiabatic

Gas-piston boundary: adiabatic

Gas-damper boundary: diathermal

The piston is initially held in mechanical equilibrium at the position $h_1 = 1$ m, by means of a pin. At some instant, the pin is removed and the piston comes to a new equilibrium position h_2.

(a) Determine the value of h_2 for this new equilibrium position.

(b) Determine the pressure and temperature of the gas for this new equilibrium position of the piston.

(c) Was the process h_1-h_2 a reversible or irreversible one? Use the proper thermodynamic arguments to substantiate your answer.

Figure 5P.11

5.12 A well-insulated tank A is connected to a well-insulated cylinder B through a valve and insulated line as shown in Figure 5P.12. The cylinder B is closed by a movable, frictionless, adiabatic piston of cross-sectional area 0.01 m^2. The mass of the piston is such that a pressure of 140 kPa is required to support it. $P_{atm} = 101.3$ kPa. Initially tank A contains 1 kg of air at 1.4 MPa and 40 C. Assume the air can be modeled as an ideal gas with $R = 287$ J/kgK and that cylinder B is initially empty. The volume in the connecting tube is negligible. The valve in the tube is opened thereby allowing air to flow from A to B.

(a) Determine the final height of the piston B.

(b) Assuming all the gas is at the same final temperature determine the final temperature of the gas.

(c) Determine the work transfer to the piston.

(d) Determine by proper test whether the process is reversible or irreversible.

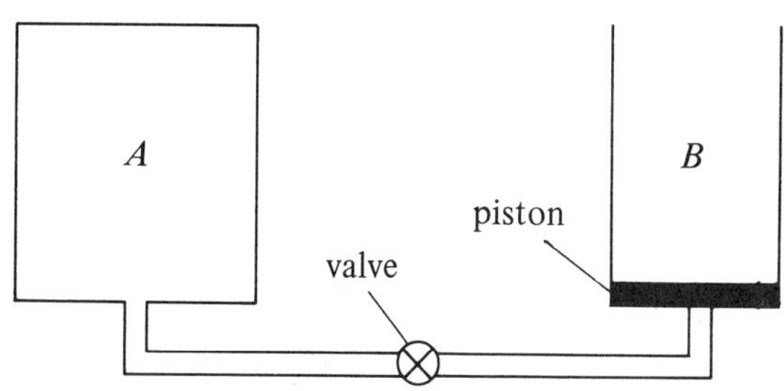

Figure 5P.12

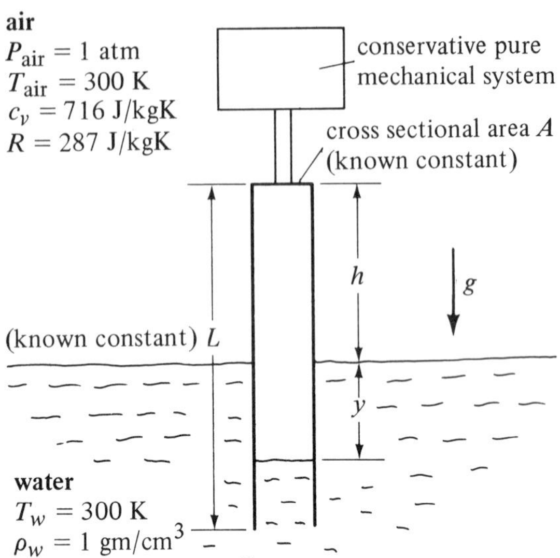

Figure 5P.13

5.13 A vertical cylinder which is closed at the top end is lowered by a pure mechanical system to the surface of a large lake as shown in Figure 5P.13. When the lower end of the cylinder touches the surface, atmospheric air is trapped inside. The internal air pressure is increased by lowering the cylinder further while the gas is in thermal and mechanical equilibrium with the air and water in the environment. The cylinder has walls of negligible thickness and negligible mass. Model the air as an ideal gas with a constant specific heat c_v.

(a) Derive an expression for the height h in terms of the pressure P, and the constants of the problem.

(b) Determine the work transfer and heat transfer for the internal air as a system for the process from P_{atm} to a final pressure P_2.

(c) Determine the work transfer for the pure mechanical system as the cylinder is lowered from the surface to the level required to reach the final pressure P_2.

5.14 A rigid, insulated tank with a volume of 2 m³ is initially evacuated and isolated from the environment by means of a valve. At some instant, the valve is opened and the tank fills with air from the atmosphere. Atmospheric pressure is 101.3 kPa and atmospheric temperature is 21 C. If the air can be modeled as an ideal gas with R = 287 J/kgK and c_v = 716 J/kgK, determine the final pressure and temperature of the air in the tank when the tank is full.

HINT: Approach this problem by considering a system consisting of the atmospheric air which will finally fill the tank. Since no mass crosses the system boundary, such a system can be regarded as a closed system. Carefully examine the system boundary for work transfer during the process.

5.15 Two vertical, well-insulated cylinders, A and B, closed on top by pistons are connected through a tube with a valve as shown in Figure 5P.15. The pistons are loaded with weights such that pressures of 1.4 MPa and 140 kPa, respectively, are necessary to support the pistons. The cross-sectional areas of the pistons are 0.25 m² and 2.5 m², respectively. Initially cylinder A contains 16 kg of air which can be modeled as an ideal gas with R = 287 J/kgK and c_v = 716 J/kgK while cylinder B is empty.

The initial temperature of the gas is 40 C. The valve in the tube connecting the two cyl-

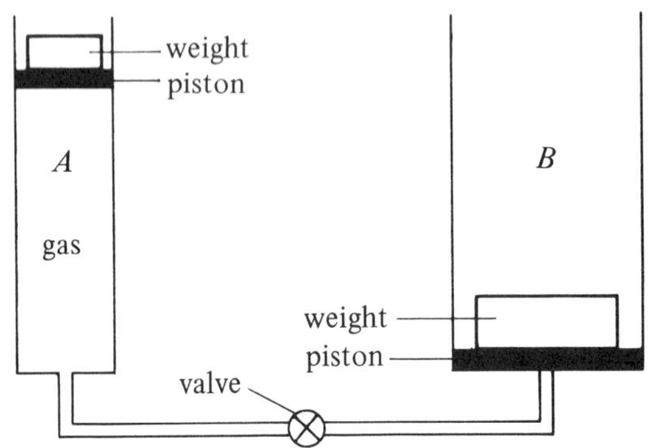

Figure 5P.15

inders is opened thereby allowing air to flow from A to B. The volume of the connecting tube is negligible.

(a) Determine the final height of the piston and weights in cylinder B.

(b) Determine the final temperature of the gas.

(c) Determine the work transfer experienced by the gas.

(d) Determine by the proper test whether the process is reversible or irreversible.

5.16 A column of mercury 10 m high rests on a frictionless, massless piston as shown in Figure 5P.16. Trapped behind the piston is a pocket of air. The pressure of the air in this state is sufficient to support the mass of the mercury. It is now proposed to raise the piston and the mercury until the piston is at the top of the tube. The process is to be carried out quasi-statically by means of a heat transfer to the trapped air. As the piston rises in the tube, the mercury displaced in the process spills over the sides of the tube.

(a) On a set of coordinates in which the pressure is the ordinate and the volume of the gas is the abscissa, show the locus of states of the gas for this process.

(b) Show the locus of states of the gas on a temperature-volume set of coordinates.

(c) What is the work transfer experienced by the air in pushing all of the mercury out of the tube?

(d) What is the change of internal energy of the air for this process?

(e) What is the necessary heat transfer to the air in order to displace all of the mercury?

(f) Can all of the mercury be displaced by means of a heat transfer interaction with a single heat reservoir?

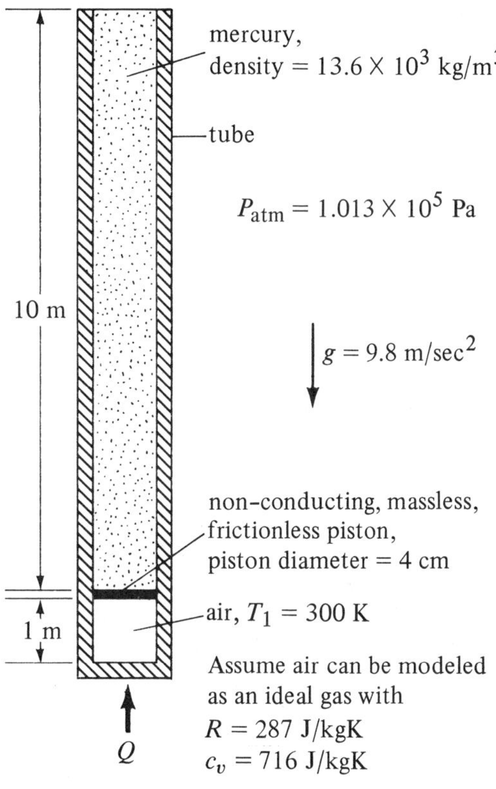

Figure 5P.16

CHAPTER 6

Systems in Thermal Communication with More than One Heat Reservoir

6.1 Introduction

We began our development of thermodynamics by considering pure mechanical systems. By virtue of their dissipative aspects, these systems were able to communicate with a single heat reservoir. For this simple class of systems, the second law of thermodynamics requires that the net work transfer for a cycle be into the system, $\oint \delta W < 0$. As the dissipative aspects of the system behavior decrease in significance, the net work transfer for the cycle approaches zero. In the limit of vanishingly small dissipation, the net work transfer for the cycle is zero, $\oint \delta W = 0$. The pure conservative mechanical system developed previously is in fact the model for this limiting case. For thermodynamic systems with coupled constitutive relations, these same restrictions on the net work transfer for a cycle apply, provided the system is in communication with only one heat reservoir. The restriction to a single heat reservoir is necessary because a system which can experience heat transfer with two reservoirs at different temperatures can have a positive net work transfer for a cycle. The temperature difference between the heat reservoirs introduces the potential for producing a net positive work transfer in a cyclic process.

We now illustrate that even though the potential for positive work transfer exists, its magnitude is also restricted by the second law of thermodynamics. In the cases involving a single heat reservoir, the second law restrictions were expressed in terms of the limits on the work transfer for a cycle. In more complex cases involving more than one heat reservoir, the second law requirements are simpler when expressed as restrictions on the heat transfer.

These requirements will be developed for the special case of the Carnot cycle employing an ideal gas as the working substance. This result will then be shown to be generally applicable to any cycle of any system exchanging heat with the same two reservoirs. The requirements will then be extended to cover a cycle exchanging heat with any number of reservoirs. In the next chapter the property entropy will be defined so that the second law requirements may be specified for processes as well as cycles.

6.2 Carnot Cycle

The Carnot cycle is conceptually the simplest cycle which will operate reversibly and transfer heat with heat reservoirs at two different temperatures. The cycle is composed of reversible isothermal processes that occur while the system is in thermal equilibrium with either of the heat reservoirs and reversible adiabatic processes that occur while the temperature of the system is changed from the temperature of one reservoir to that of the other. Any system having coupled constitutive relations can be employed for the Carnot cycle; however, for the sake of simplicity in the present treatment, we shall use the ideal gas with constant specific heat, c_v. The mechanisms for carrying out the reversible adiabatic processes with the ideal gas have been illustrated in Sections 5.4 and 5.5.

The reversibility of the processes that form the ideal gas Carnot cycle allows us to relate the heat transfers in the cycle to the ideal gas temperatures of the reservoirs, one at T_H and the other at T_L. The second law of thermodynamics allows us to generalize this relation to any reversible cycle of any system. In order to establish this relation consider the states of the ideal gas system during the Carnot cycle as shown in Figure 6.1. In the initial state, state 1, the system is in equilibrium with the heat reservoir at the temperature T_H. For the reversible adiabatic

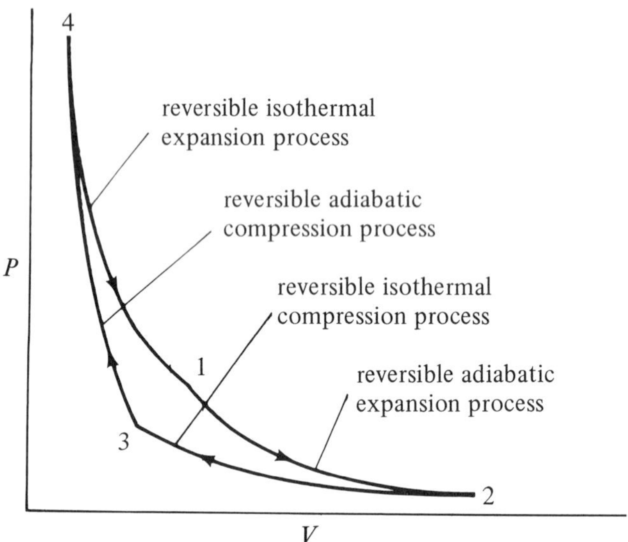

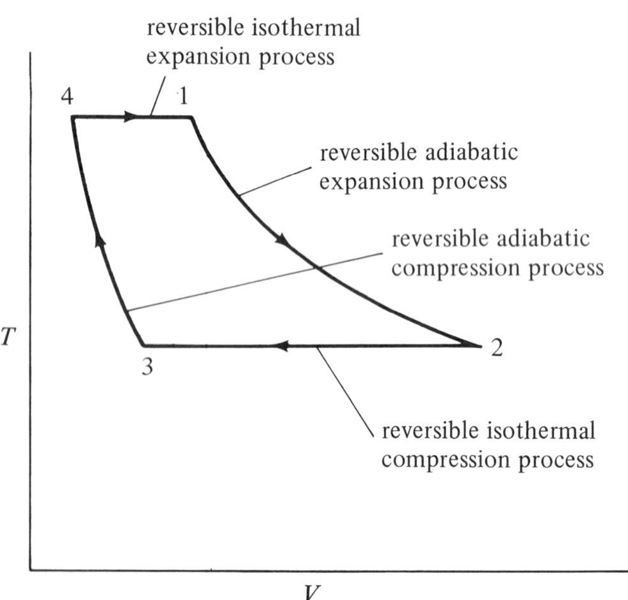

Figure 6.1 Carnot cycle

process 1–2 the system expands from volume V_1 to volume V_2 while thermally isolated from the environment. During this process, the system does positive work; thus, its internal energy decreases as required by the first law. For an infinitesimal process, the first law may be written

$$\delta Q - \delta W = m\,du \tag{6.1}$$

or since δQ is zero for an adiabatic process,

$$-\delta W = m\,du \tag{6.2}$$

Substitution of the constitutive relation for the ideal gas then gives the decrease in the temperature

$$-\delta W = mc_v\,dT \tag{6.3}$$

For this process the only equilibrium work transfer for the ideal gas is due to the pressure on the moving boundary (for example, the piston in Figure 5.3). Since the process is reversible, the work transfer can also be expressed as

$$\delta W = P\,dV \tag{6.4}$$

Substituting the pressure as given by the constitutive relation, we obtain

$$\delta W = mRT\frac{dV}{V} \tag{6.5}$$

When equations (6.3) and (6.5) are combined, the differential equation for the volume-temperature relation for the reversible adiabatic process becomes

$$-c_v\frac{dT}{T} = R\frac{dV}{V} \tag{6.6}$$

Upon integrating equation (6.6)[1] for the adiabatic process from state 1 to state 2, we obtain

$$\ln \frac{T_2}{T_1} = -\frac{R}{c_v} \ln \frac{V_2}{V_1}$$

$$T_1 = T_H \text{ and } T_2 = T_L \tag{6.7}$$

$$\ln \frac{T_L}{T_H} = -\frac{R}{c_v} \ln \frac{V_2}{V_1}$$

Thus, the volume V_2 corresponding to the temperature T_L of the system is determined. The system is now brought into thermal communication with the reservoir at T_L. However, because the temperatures of the system and the reservoir are equal, no heat transfer occurs.

The heat transfer during the reversible isothermal process 2-3 is the result of the work transfer into the system during the compression from V_2 to V_3. The work transfer and the heat transfer are given by equation (5.11).

$$W_{2-3} = Q_{2-3} = mRT_L \ln \frac{V_3}{V_2} \tag{6.8}$$

Once in the state 3, the system is again thermally isolated while being compressed reversibly from volume V_3 to volume V_4, simultaneously increasing the temperature from T_L back to T_H. Equation (6.7) for the reversible adiabatic expansion, process 1-2, applies to this process when states 1 and 2 are replaced by states 3 and 4, respectively.

$$\ln \frac{T_4}{T_3} = -\frac{R}{c_v} \ln \frac{V_4}{V_3} = \ln \frac{T_H}{T_L} \tag{6.9}$$

The system is now brought into thermal communication with the reservoir at T_H and is subsequently expanded reversibly and

[1] Integration of equation (6.6) to equation (6.7) is similar to the integration of equation (5.21) to equation (5.22).

isothermally back to the original state (T_1, V_1). The heat transfer for this process is

$$Q_{4-1} = mRT_H \ln \frac{V_1}{V_4} \tag{6.10}$$

The two isothermal processes require

$$T_1 = T_4 = T_H \text{ and } T_2 = T_3 = T_L \tag{6.11}$$

Consequently, equations (6.11), (6.9) and (6.7) may be combined to give

$$\frac{V_2}{V_1} = \frac{V_3}{V_4} \text{ or } \frac{V_2}{V_3} = \frac{V_1}{V_4} \tag{6.12}$$

Equation (6.12) shows that the volume ratio for the reversible isothermal expansion, process 4-1, is the reciprocal of the volume ratio for the reversible isothermal compression, process 2-3. Hence,

$$\ln \frac{V_1}{V_4} = -\ln \frac{V_3}{V_2} \tag{6.13}$$

Substituting equations (6.8) and (6.10) into equation (6.13), we obtain the relation between the heat transfers and the ideal gas temperatures, namely

$$\frac{Q_{4-1}}{T_H} = -\frac{Q_{2-3}}{T_L} \tag{6.14}$$

or after rearrangement

$$\frac{Q_{4-1}}{T_H} + \frac{Q_{2-3}}{T_L} = 0 \tag{6.15}$$

From equations (6.8), (6.10), and (6.12) it is clear that equation (6.15) is not influenced by the volume ratio of the cycle, V_3/V_2. That is, equation (6.15) holds regardless of the size of the cycle.

In order to further generalize equation (6.15), we note that in each of the terms the heat transfer is divided by the temperature of the reservoir in communication with the system while this heat transfer interaction took place. For this reason, it is convenient to use the symbols Q_H and Q_L to indicate the heat transfer to the system at T_H and T_L, respectively. Thus, equation (6.15) becomes

$$\frac{Q_H}{T_H} + \frac{Q_L}{T_L} = 0 \tag{6.16}$$

6.3 Systems in Thermal Communication with Two Heat Reservoirs

We shall now show that the second law of thermodynamics requires that equation (6.16) must apply to *any* reversible cycle of *any* system (regardless of constitutive relation) which experiences heat transfer with only two heat reservoirs. To this end, consider a Carnot cycle which transfers heat with the same two heat reservoirs as the general system during its general reversible cycle. The arrangement is shown schematically in Figure 6.2. The heat transfers to the general system are $(Q_H)_S$ and $(Q_L)_S$, and the heat transfers to the Carnot cycle are $(Q_H)_C$ and $(Q_L)_C$. The technique for the proof is to arrange the Carnot cycle so that the reservoir at T_H undergoes a cycle. Then, the basic statement of the second law of thermodynamics (Section 5.9) may be applied to the reversible cycle of the composite system enclosed in the dashed boundary shown in Figure 6.2.

Even though the temperature of the reservoir remains constant, the stored thermal energy of the reservoir is changed by these heat transfers. Thus in order for a heat reservoir to execute a cycle, there must be no net change in the stored thermal energy and no net heat transfer.

$$\oint \delta Q = \oint dU = 0 \tag{6.17}$$

Then for the heat reservoir at T_H in Figure 6.2 to execute a cycle,

$$(Q_H)_C + (Q_H)_S = 0$$
$$(Q_H)_C = -(Q_H)_S \tag{6.18}$$

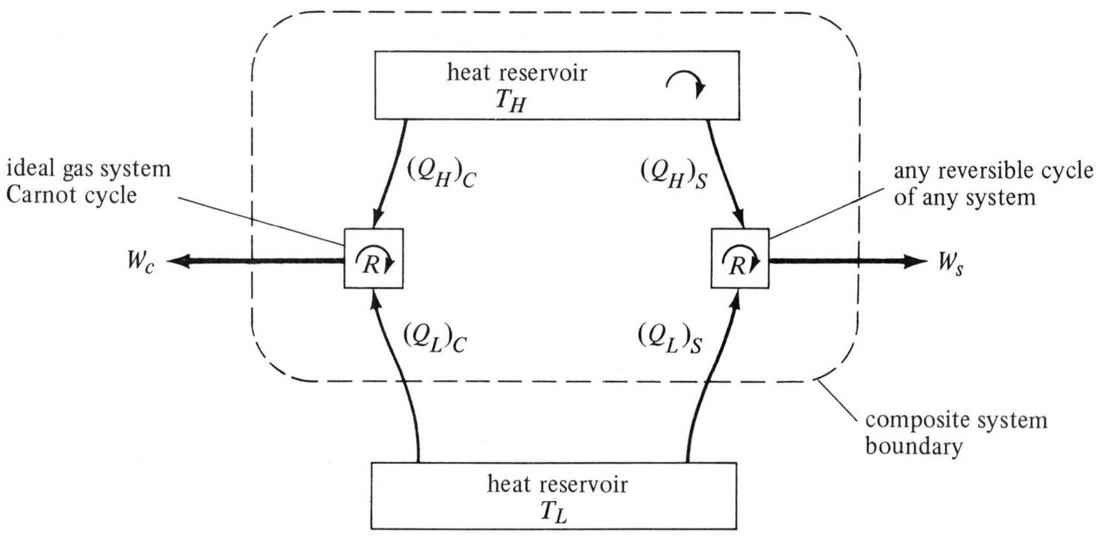

Figure 6.2 General system in thermal communication with two heat reservoirs

The Carnot cycle in Figure 6.2 is sized so that it will complete a cycle when the system completes a cycle. This is so the heat transfer interaction with the high temperature reservoir is equal in magnitude and opposite in sign to the heat transfer interaction between the system and the high temperature reservoir (as required by equation (6.18)). After each cycle of the general system, all parts of the composite system have completed a reversible cycle; hence, the composite system has completed a reversible cycle. The composite system experiences external heat transfer only with the single heat reservoir at T_L. Under these circumstances, the basic statement of the second law of thermodynamics requires that the net heat transfer for the reversible cycle of the composite system be zero. Thus, for the composite system

$$\oint \delta Q = (Q_L)_C + (Q_L)_S = 0 \tag{6.19}$$

For the Carnot cycle, equation (6.16) applies.

$$\frac{(Q_H)_C}{T_H} + \frac{(Q_L)_C}{T_L} = 0 \tag{6.20}$$

When equations (6.18) and (6.19) are substituted into equation (6.20), we obtain

$$\frac{(Q_H)_S}{T_H} + \frac{(Q_L)_S}{T_L} = 0 \tag{6.21}$$

Thus, we have shown that this general relation (equation (6.21)) between the heat transfers and the ideal gas temperatures of the corresponding heat reservoir must apply to any system for any reversible cycle which transfers heat only at T_H and T_L.

We can now show that equation (6.21), which applies to reversible cycles, is the upper limit for all irreversible cycles. The arguments are the same as before except that now the general system experiences a general irreversible cycle. The second law of thermodynamics requires that the heat transfer for the composite system of Figure 6.2 be negative when the cycle is irreversible. Thus,

$$\oint \delta Q = (Q_L)_C + (Q_L)_S < 0 \tag{6.22}$$

Since the reservoir at T_H still executes a cycle, equation (6.18), which is a consequence of the first law only, holds for the irreversible cycle also. Equation (6.20) still applies to the Carnot cycle since it is by definition always reversible. Then, combining equations (6.18) and (6.20), we obtain

$$(Q_L)_C = -(Q_H)_C \frac{T_L}{T_H} = +(Q_H)_S \frac{T_L}{T_H} \tag{6.23}$$

Upon substitution of equation (6.23), equation (6.22) becomes

$$\frac{T_L}{T_H}(Q_H)_S + (Q_L)_S < 0 \tag{6.24}$$

Since the ideal gas temperature is non-negative, we may divide by T_L without modifying the inequality.

$$\frac{(Q_H)_S}{T_H} + \frac{(Q_L)_S}{T_L} < 0 \tag{6.25}$$

It then follows, from equations (6.21) and (6.25) that the requirement of the second law of thermodynamics for any cycle which experiences heat transfer at two temperatures is

$$\frac{Q_H}{T_H} + \frac{Q_L}{T_L} \leq 0 \qquad \textbf{(6.26)}$$

where in the limit of the reversible cycle, the equality applies.

6E.1 Sample Problem: In a simple model of a power plant, a cyclic device experiences a heat transfer interaction with an energy source at high temperature and a heat transfer interaction with an energy sink at low temperature. As a consequence of the temperature difference between the heat reservoirs, a net positive work transfer is produced by the cycle. The magnitude of the heat transfer interaction with the energy sink is an issue of concern in the energy conversion industry. Since the environment is usually used as the energy sink, there can be adverse ecological consequences on a local scale from excessive heat transfer with the environment. Because of their low operating temperatures, nuclear energy sources are particularly troublesome in this regard. Compare the magnitudes of the thermal pollution produced by an oil fired power plant and a nuclear plant given that both plants produce 10^3 MW of power and both plants use the environment at a temperature of 300 K as an energy sink. The temperature of the oil fired energy source is 800 K while the temperature of the nuclear energy source is 550 K.

Solution: From the second law of thermodynamics for a system in communication with two heat reservoirs, equation (6.26), we have

$$\frac{Q_H}{T_H} + \frac{Q_L}{T_L} \leq 0$$

For the sake of comparison, we will assume that the cycle is reversible; therefore, equality holds, From the first law of thermodynamics, we have

$$\oint \delta Q = \oint \delta W \qquad \text{or} \qquad Q_H + Q_L = W$$

Substituting this result into the second law, we get

$$\frac{W - Q_L}{T_H} + \frac{Q_L}{T_L} = 0$$

or

$$Q_L = \left(\frac{T_L}{T_H - T_L}\right) W$$

Then

$$(Q_L)_{\text{oil}} = \left(\frac{300}{800 - 300}\right) 10^3 = 0.60 \times 10^3 \text{ MW}$$

$$(Q_L)_{\text{nuclear}} = \left(\frac{300}{550 - 300}\right) 10^3 = 1.20 \times 10^3 \text{ MW}$$

Thus, the thermal pollution emanating from the nuclear plant is double that of the oil fired plant.

6.4 Thermodynamic Temperature Scale

Equation (6.21) is a general relation between the heat transfers for a reversible cycle and the temperature at which the heat transfers take place. This equation may be used as the least restrictive method of defining temperature ratios. Equation (6.21) can be written as

$$\frac{T_H}{T_L} = \frac{Q_H}{-Q_L} \tag{6.27}$$

Thus, the ratio of the temperatures of the two heat reservoirs may be defined in terms of the ratio of the heat transfers for any reversible cycle of any system. Any temperature scale which is defined in terms of these heat transfers is termed a thermodynamic temperature scale, and its definition is independent of the constitutive relation of the system. The thermodynamic Kelvin scale of temperature and the thermodynamic Rankine scale of temperature are identical with the ideal gas scales because of the special nature of the constitutive relation of the ideal gas.

6.5 Limitations Imposed by the Second Law of Thermodynamics on Systems in Communication with Two Heat Reservoirs

Some of the more important applications of thermodynamics involve situations which are modeled with a system known as a *heat engine*. A heat engine is *a closed system that operates in cycles with heat transfers and work transfers as the only interactions at the boundary*. Heat engines assume many forms, but the simplest is a closed system that communicates with only two heat reservoirs (with different thermodynamic temperatures) and produces a positive net work transfer while experiencing a positive heat transfer with the high temperature heat reservoir.

According to the first law of thermodynamics, the net work transfer is equal to the net heat transfer, equation (2.1).

$$\oint \delta W = \oint \delta Q = Q_H + Q_L \tag{6.28}$$

Unfortunately, the net positive work transfer from equation (6.28) cannot be made to approach the heat transfer from the high temperature heat reservoir simply by adjusting the negative heat transfer with the low temperature heat reservoir. Once the thermodynamic temperatures of the heat reservoirs have been established, the second law of thermodynamics limits the possible heat transfers between these heat reservoirs and a heat engine. Consequently, the performance of the heat engine is also limited.

One convenient measure of the performance of a heat engine is the *heat engine efficiency*, η (also known as the energy conversion efficiency and the thermal efficiency). *This efficiency is defined as the positive net work transfer,* $\oint \delta W$, *per unit of positive heat transfer with the high temperature heat reservoir,* Q_H. Thus,

$$\eta \equiv \frac{\oint \delta W}{Q_H} \tag{6.29}$$

The efficiency η is a positive quantity with a value less than unity. This definition is motivated by practical considerations of power production. The positive net work transfer is the desired output and the positive heat transfer from the high temperature heat reservoir is the necessary input supplied by the consumption of expensive fuel. The local atmosphere (normally air or water) serves as a "free" low temperature heat reservoir to receive the negative heat transfer of the engine. In some cases, especially power plants operating in space, this low temperature reservoir is not free since a large radiating surface must be provided to radiate the energy to deep space.

Combining equations (6.28) and (6.29), we obtain

$$\eta = 1 + \frac{Q_L}{Q_H} \tag{6.30}$$

The second law of thermodynamics for a reversible cycle of a heat engine of this type, equation (6.26), requires that

$$\frac{Q_L}{Q_H} = -\frac{T_L}{T_H} \tag{6.31}$$

Thus, the heat engine efficiency for a reversible cycle becomes

$$\eta_{\text{rev}} = 1 - \frac{T_L}{T_H} \tag{6.32}$$

where the subscript "rev" has been added to emphasize that equation (6.32) applies only to a reversible heat engine experiencing heat transfer with two heat reservoirs. The positive net work transfer for such a heat engine can be evaluated by combining equations (6.29) and (6.32).

$$\left(\oint \delta W\right)_{\text{rev}} = Q_H \eta_{\text{rev}} = Q_H \left(1 - \frac{T_L}{T_H}\right) \tag{6.33}$$

Thus, for a reversible cycle involving a given set of heat reservoirs with temperatures T_H and T_L, there is only one possible value for the positive net work transfer that corresponds to any one value for the positive heat transfer with the high temperature heat reservoir. This unique relationship is a consequence of the combined influence of the first and second laws of thermodynamics.

In the case of an irreversible cycle of a heat engine in communication with two heat reservoirs, the relationship between the positive heat transfer with the high temperature heat reservoir and the positive net work transfer is not so unique as equation (6.33). In this case, the second law of thermodynamics, equation (6.26), requires that

$$\frac{Q_L}{T_L} < -\frac{Q_H}{T_H} \tag{6.34}$$

For a heat engine with a positive net work transfer, Q_H and T_L are themselves both positive quantities. Then, both sides of equation (6.34) can be multiplied by the fraction T_L/Q_H without changing the inequality. Hence, for the irreversible cycle

$$\frac{Q_L}{Q_H} < -\frac{T_L}{T_H} \tag{6.35}$$

If we add $+1$ to both sides of equation (6.35) and make use of equation (6.30), we obtain

$$1 + \frac{Q_L}{Q_H} < 1 - \frac{T_L}{T_H} \tag{6.36}$$

or

$$\eta_{\text{irrev}} < \eta_{\text{rev}} \tag{6.37}$$

Thus, for the irreversible cycle, equation (6.33) becomes

$$\left(\oint \delta W\right)_{\text{irrev}} = Q_H \eta_{\text{irrev}} < Q_H\left(1 - \frac{T_L}{T_H}\right) = \left(\oint \delta W\right)_{\text{rev}} \tag{6.38}$$

Thus, the relationship between the positive heat transfer with the high temperature heat reservoir and the positive net work transfer is not a unique one but is in the form of limit. The output of a real (irreversible) heat engine is less than the output of an ideal reversible heat engine. The real and ideal cycles are compared for the same two heat reservoirs and for the same Q_H. Clearly, in order to satisfy the first law of thermodynamics, the heat transfer with the low temperature heat reservoir will be smaller *in the algebraic sense* for the irreversible cycle than for the reversible cycle.

A common refrigerator is an apparatus that is normally modeled as a heat engine that communicates with only two heat reservoirs and involves a negative net work transfer while experiencing a positive heat transfer with the low temperature heat reservoir. For a heat engine of this type, equation (6.28) still applies; however, the performance parameter defined in equation (6.29) is no longer physically meaningful even though the equation itself is mathematically valid. A more physically significant performance parameter is the *coefficient of performance* (COP). This is defined as the refrigeration effect (positive heat transfer from the low temperature heat reservoir), Q_L, per unit of net negative work transfer, $-\oint \delta W$. Thus,

$$\text{COP} = \frac{Q_L}{-\oint \delta W} \tag{6.39}$$

When the first law is used to eliminate $\oint \delta W$

$$\text{COP} = \frac{1}{-(Q_H/Q_L) - 1} \tag{6.40}$$

The COP is a positive quantity that can assume values greater than or less than unity depending upon the physical situation. This definition, like that of equation (6.29), is also motivated by practical considerations. On the right side of equation (6.39) the numerator represents the desired output, the refrigeration effect, while the denominator represents the necessary input, the expensive negative net work transfer. For a reversible cycle of a refrigerator, equation (6.31) is still valid. Thus, equation (6.40) can be written

$$\text{COP}_{\text{rev}} = \frac{1}{(T_H/T_L) - 1} \tag{6.41}$$

for the reversible case. As in the case of the reversible heat engine, the desired output of the reversible refrigerator is uniquely related to the necessary input once the thermodynamic temperatures of the heat reservoirs have been established.

$$(Q_L)_{\text{rev}} = \left(-\oint \delta W\right) \text{COP}_{\text{rev}} = \left(-\oint \delta W\right)\left[\frac{1}{(T_H/T_L) - 1}\right] \tag{6.42}$$

As we might expect from our experience with heat engines, the uniqueness of equation (6.42) is lost in the case of an irreversible refrigerator. Equation (6.34) is still a valid form for the second law of thermodynamics. Both sides of that equation can be multiplied by the fraction T_H/Q_L without affecting the inequality since both T_H and Q_L are positive quantities for a refrigerator in communication with only two heat reservoirs. Hence, for the irreversible refrigerator, equation (6.34) becomes

$$\frac{T_H}{T_L} < -\frac{Q_H}{Q_L} \tag{6.43}$$

If we add -1 to both sides of equation (6.43), we obtain

$$\frac{T_H}{T_L} - 1 < -\frac{Q_H}{Q_L} - 1 \tag{6.44}$$

Since $T_H > T_L$, both sides of equation (6.44) must be positive; therefore, equation (6.44) can also be written in the form

$$\frac{1}{(T_H/T_L) - 1} > \frac{1}{-(Q_H/Q_L) - 1} \tag{6.45}$$

or

$$\text{COP}_{\text{rev}} > \text{COP}_{\text{irrev}} \tag{6.46}$$

Thus, for the irreversible cycle, equation (6.42) becomes

$$(Q_L)_{\text{irrev}} = \left(-\oint \delta W\right) \text{COP}_{\text{irrev}} < \left(-\oint \delta W\right)\left[\frac{1}{(T_H/T_L) - 1}\right] = (Q_L)_{\text{rev}} \tag{6.47}$$

Equation (6.47) shows that the refrigeration effect (positive heat transfer with low temperature heat reservoir) of a real (and irreversible) cycle of a refrigerator in communication with two heat reservoirs must be less than the refrigeration effect for the limiting reversible cycle. For this comparison both cycles experience the same negative net work transfer and both experience heat transfer with the same two heat reservoirs. To satisfy the first law of thermodynamics, the heat transfer with the high temperature reservoir will be smaller *in the algebraic sense* for the reversible cycle than for the irreversible cycle.

6.6 Systems in Communication with Any Number of Heat Reservoirs

The arguments of Section 6.3 can be generalized to a system in thermal communication with any number of heat reservoirs as shown in Figure 6.3. Each reservoir executes a cycle by means of counteracting heat transfers with the system and with one of the Carnot cycles. The Carnot cycles are sized so that they complete a cycle when the system completes a cycle. In addition, one of the heat transfer interactions for each of the Carnot cycles is with the common heat reservoir at a temperature T_0.

The composite system consisting of heat reservoirs, Carnot cycles, and system is therefore arranged to undergo a cycle when the system undergoes a cycle. The basic statement of the second law of thermodynamics requires that the net heat transfer for the cycle of the composite system be zero if the cycle is reversible or negative if the cycle is irreversible. Thus, for the composite system

$$Q = \sum_{i=1}^{n} (Q_{i0})_C \leq 0 \tag{6.48}$$

Each of the Carnot cycles must satisfy equation (6.16). Thus,

$$-(Q_{i0})_C = (Q_i)_C \frac{T_0}{T_i} \tag{6.49}$$

and equation (6.48) becomes

$$\sum_{i=1}^{n} [-(Q_i)_C] \frac{T_0}{T_i} \leq 0 \tag{6.50}$$

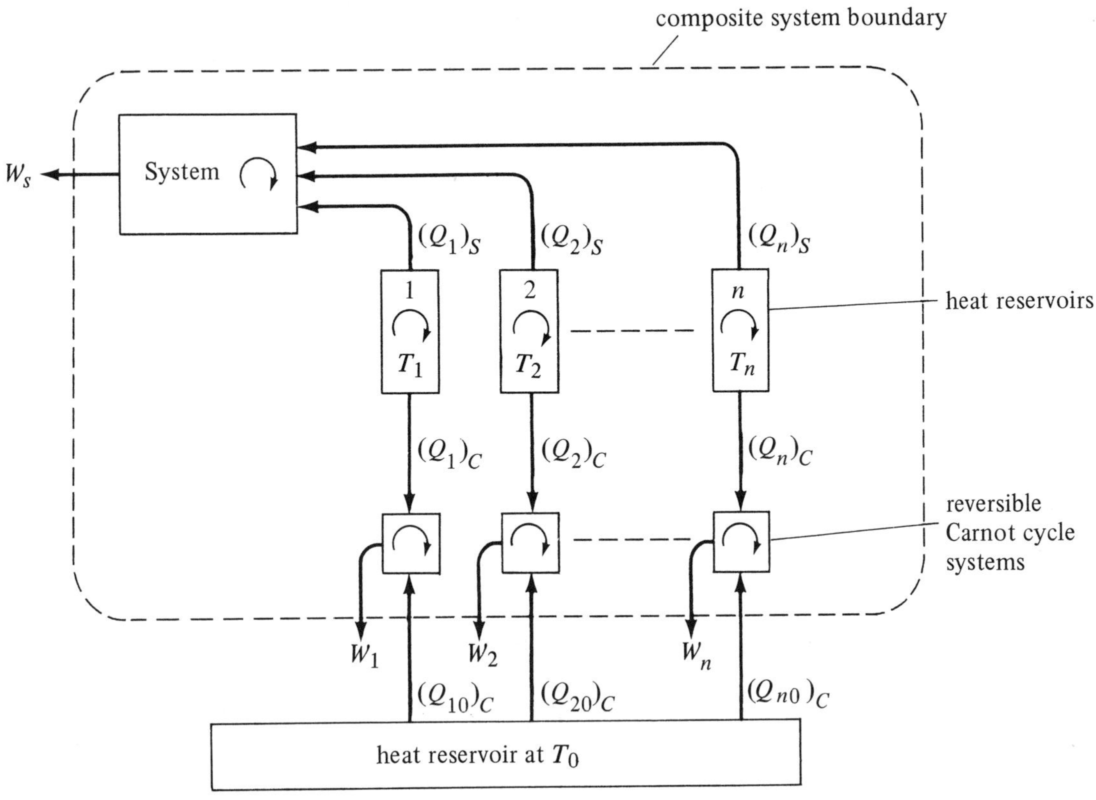

Figure 6.3 System in communication with several heat reservoirs

If we apply the first law of thermodynamics to the individual heat reservoirs during the cycle, we find that

$$\oint dU_i = \oint \delta Q_i = 0 \tag{6.51}$$

which requires that the net heat transfer to the reservoir during the cycle is zero. Therefore,[2]

$$-(Q_i)_C = (Q_i)_S \tag{6.52}$$

[2] The student should prove this equality by considering the ith reservoir as a separate system, and noting that the heat transfer to the reservoir is a negative heat transfer for the environment of the reservoir.

Substituting equation (6.52) into equation (6.50), we obtain

$$\sum_{i=1}^{n} (Q_i) \frac{T_0}{T_i} \leq 0 \tag{6.53}$$

Since T_0 is finite, equation (6.53) may be divided by T_0. Then for a cycle of any system, the second law of thermodynamics becomes

$$\sum_{i=1}^{n} \frac{Q_i}{T_i} \leq 0 \tag{6.54}$$

where we have dropped the subscript s and used the subscript i to indicate that Q_i is the heat transfer experienced *by the system* while in communication with the heat reservoir with temperature T_i. Note that in equation (6.54), equality holds only in the case of a reversible cycle.

We may now consider the limiting case of an infinite number of heat reservoirs in thermal communication with a finite system which is undergoing a reversible cycle. In this case, the number of heat reservoirs in Figure 6.3 approaches infinity, and in general the heat transfer Q_i with any one heat reservoir becomes an infinitesimal quantity δQ_i. Then for a reversible cycle, the summation in equation (6.54) becomes the integral

$$\oint \left(\frac{\delta Q_i}{T_i}\right)_{\text{rev}} = 0 \tag{6.55}$$

where the subscript i is retained to indicate that δQ_i is the infinitesimal heat transfer experienced by the system while in thermal communication with the heat reservoir with temperature T_i. The subscript "rev" is added to the cyclic integral to indicate that equation (6.55) applies only to reversible cycles. Note that since the cycle is reversible, the heat transfer interaction that occurs between the system and each reservoir is also reversible; therefore, the temperature of the system is identical to the temperature of the heat reservoir during the heat transfer process in which δQ_i was the heat transfer.

For an irreversible cycle, equation (6.55) becomes

$$\oint \left(\frac{\delta Q_i}{T_i}\right)_{\text{irrev}} < 0 \tag{6.56}$$

Note that in this case, T_i is the temperature of the heat reservoir in thermal communication with the system while the heat transfer δQ_i occurred. Since the cycle is irreversible, T_i is not necessarily the temperature of the system. Note that in both equations (6.55) and (6.56) a finite number of the heat transfers δQ_i may themselves be finite during the cycle.

6E.2 Sample Problem: A heat engine operating between two heat reservoirs, one at a temperature of 600 K and a second at a temperature of 300 K, is used to provide part of the power necessary to operate a refrigerator between a heat reservoir, at 300 K and a refrigeration load at 250 K. The additional power required to operate the refrigerator, 10^2 W, is supplied from an external source. Calculate the maximum refrigeration load, that is, the maximum heat transfer from the load at 250 K, if the heat transfer from the 600 K heat reservoir to the heat engine is 10^3 W.

Solution: Schematically, the system of heat engine and refrigerator would appear as shown in Figure 6E.2. For the

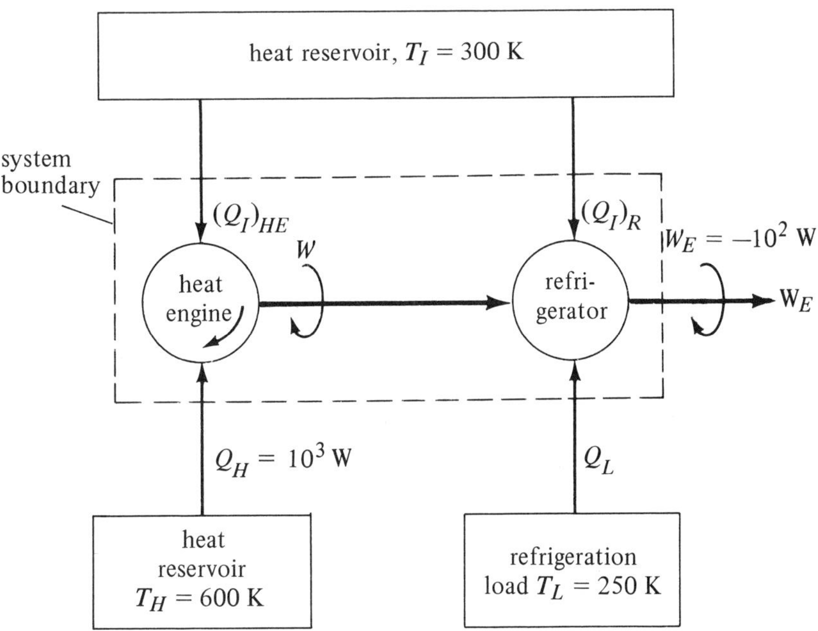

Figure 6E.2 Heat engine operating a refrigerator

system boundary shown, the first law of thermodynamics becomes

$$Q_H + (Q_I)_{HE} + (Q_I)_R + Q_L = W_E$$

The second law of thermodynamics gives

$$\frac{Q_H}{T_H} + \frac{(Q_I)_{HE}}{T_I} + \frac{(Q_I)_R}{T_I} + \frac{Q_L}{T_L} \leq 0$$

The maximum value of Q_L is given by equality in the second law expression. Then from the first law,

$$\begin{aligned}(Q_I)_{HE} + (Q_I)_R &= W_E - Q_H - Q_L = -0.1 \times 10^3 - 1 \times 10^3 - Q_L \\ &= -1.1 \times 10^3 - Q_L\end{aligned}$$

Substituting this result into the second law expression, we obtain

$$\frac{10^3}{600} - \frac{1.1 \times 10^3}{300} - \frac{Q_L}{300} + \frac{Q_L}{250} = 0$$

or

$$Q_L = 3 \times 10^3 \text{ W}$$

If any irreversibilities occur (as indeed they would in the case of a real system), the refrigeration load would be reduced below this value.

Problems

6.1 A home freezer operates with an electric motor rated at 250 watts. During a hot summer day ($T_{\text{atm}} = 33$ C) it is observed that the motor is on half the time while the freezer executes an integral number of cycles.

(a) Estimate the heat transfer into the freezer per hour, if its interior temperature is maintained at $T_0 = -17$ C.

(b) What is the heat transfer from the freezer to the atmosphere during the hour?

6.2 In order to provide power to an undersea community, it has been proposed to build a heat engine operating in a cycle that uses the warm surface water of the Gulf Stream as the high temperature heat reservoir, and the cold deep water as the low temperature heat reservoir. If the surface water is at 30 C and the deep water is at 4 C, determine the heat transfer rate (kW) from the high temperature heat reservoir necessary to provide 1 megawatt (10^6 watts) of power with a reversible engine.

6.3 A closed thermodynamic system employs the cycle shown in Figure 6P.3.

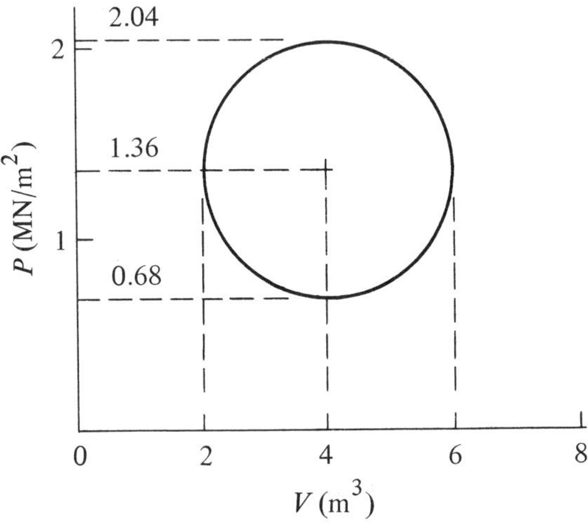

Figure 6P.3

(a) What is the net quasi-static work transfer for this cycle? If the system traverses the cycle in the counterclockwise sense, is the work transfer positive or negative? If the system traverses the cycle in the clockwise sense, is the work transfer positive or negative?

(b) If the system performs as a heat engine and the heat transfer to the low temperature heat reservoir is 50 MJ, determine the thermal efficiency of the cycle.

(c) If the system performs as a refrigerator and the heat transfer to the high temperature heat reservoir is 50 MJ, determine the coefficient of performance of the system (COP).

6.4 For the Boston Garden hockey rink, the heat transfer from the air above the ice to the ice surface is at a rate of 400 kW. The ice surface temperature is maintained by a refrigerator that uses the air outside the building as the high temperature heat reservoir and the lowest layer of ice as the low temperature heat reservoir.

(a) When the lowest layer of ice is at a temperature of -9 C and air outside the building is at a temperature of 15 C, estimate the minimum cost of maintaining the ice at this fixed temperature for 24 hours. For this level of power consumption, electric power costs \$0.02/kWh.

(b) Would you expect the operating costs of the refrigerator to increase or decrease as the temperature of the outside environment decreases? Assume everything else remains unchanged.

(c) In actual practice, a substantial temperature gradient in the ice is required in order to establish heat transfer from the ice surface to the lowest layer where the cooling pipes of the refrigeration system are located. For the conditions described above, the ice surface temperature is -3 C while the temperature of the lowest layer is -9 C. Show from considerations of the COP that it would be substantially cheaper to operate the refrigerator if the refrigeration load were removed at the actual ice surface temperature rather than the temperature of the lowest layer.

6.5 During winter, the interior of a building is to be maintained at a temperature T_i. Because the exterior walls of the building are imperfectly insulated, the interior of the building experiences a negative heat transfer at a rate $\dot{Q}$ when the temperature of the atmosphere is T_{atm}, $T_{atm} < T_i$.

(a) What is the electric power required to maintain an interior temperature T_i, if the power is dissipated in electric resistors inside the building?

(b) What would be the minimum electric power required if a heat pump were employed? As shown schematically in Figure 6P.5, a heat pump is a system which operates in cycles, has a net work transfer into the system, a heat transfer

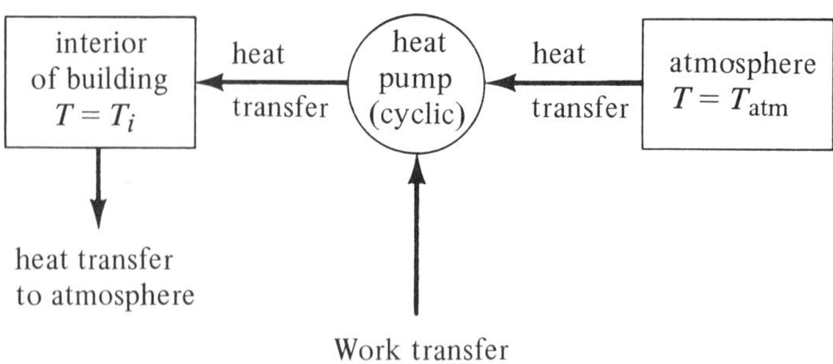

Figure 6P.5

to the interior of the building, and a heat transfer from the atmosphere during each cycle.

6.6 Recently ecologists have expressed considerable concern over the thermal pollution produced by electrical power plants. However, there is another source of thermal pollution. In the summer, the increase in electrical power consumption is considerable. The bulk of this increase is used to operate air conditioning systems. For example, in New York City, statistics compiled by the Consolidated Edison Co. show that the daily increase in electrical power consumption during the summer months is 2×10^9 watts. Let us assume that all of this increase is used to operate air conditioning systems. Let us also assume that we can model these air conditioning systems as refrigerators that operate in cycles while maintaining the temperature of the air conditioned spaces at a steady temperature of 21 C. Model the environment (atmosphere) as a heat reservoir at 33 C.

(a) Estimate the rate of thermal pollution (watt) of the environment (atmosphere) due to these air conditioning systems.

(b) Does this estimate represent a lower bound or upper bound on the rate of thermal pollution? Explain your answer.

6.7 Currently the Stirling engine is receiving considerable attention in the engineering literature. The ideal Stirling engine employs a closed ideal gas system. The cycle of the working fluid consists of two reversible isothermal processes (state 1 to state 2 and state 3 to state 4) and two reversible constant volume processes, (state 2 to state 3 and state 4 to state 1). The heat transfer for the isothermal process 1-2 is from the heat reservoir at T_H and the heat transfer for the isothermal process 3-4 is to the heat reservoir at T_L. The gas is cooled at constant volume (process 2-3) by a reversible heat transfer to the elements of the thermal regenerator. The thermal regenerator stores this energy and returns it to the gas when the gas is heated at constant volume (process 4-1). The thermal regenerator consists of a series of elements each at an infinitesimally different temperature so that the element of the regenerator in contact with the gas is always at the temperature of the gas.

(a) Show that for a perfect regenerator, i.e. a regenerator with no net change of energy, the specific heat of the ideal gas at constant volume must be independent of pressure. (That is, the ideal cycle requires an ideal gas.)

(b) Show that the energy conversion efficiency (thermal efficiency; heat engine efficiency) is identical to that of a Carnot cycle.

6.8 The Servel gas refrigerator may be considered a cyclic device. It experiences positive heat transfers Q_H at a high temperature T_H and

Q_L at a low temperature T_L and a heat transfer Q_I with the environment at T_I. The system has no work transfer in or out. The performance of this machine may be given by the ratio (Q_L/Q_H), the refrigeration effect divided by the positive heat transfer at the high temperature.

(a) Determine the limits imposed on this ratio by the second law of thermodynamics.

(b) Find the numerical value of this limit for $T_H = 450$ K, $T_L = 270$ K and $T_I = 290$ K.

6.9 A Carnot engine is used as a source of mechanical power for a Carnot refrigerator. The engine experiences a positive heat transfer Q_{HE} with a heat reservoir at the temperature T_{HE} and a heat transfer Q_{LE} with a heat reservoir at a temperature T_{LE}. On the other hand, the refrigerator experiences a positive heat transfer Q_{LR} with an isothermal refrigeration load at a temperature T_{LR} and a heat transfer Q_{HR} with a heat reservoir at a temperature T_{HR}.

(a) Formulate an expression for the ratio Q_{LR}/Q_{HE} in terms of the various temperatures of the heat reservoirs.

(b) On the basis of this expression, what statements can you make regarding the possible range of values for this ratio of heat transfers?

(c) Show that if $T_{LE} = T_{HR}$, this system becomes identical to the reversible Servel refrigerator. (See problem 6.8.)

6.10 Consider a reversible heat engine which executes an integral number of cycles while having heat transfer interactions with three heat reservoirs, HR_1, HR_2, and HR_3 at 400K, 300 K, and 200 K, respectively, as shown in Figure 6P.10. For one cycle, the engine experiences a positive heat transfer of 1600 J with HR_1 and a positive work transfer of 250 J with the environment.

(a) Find the heat transfer for the other heat reservoirs.

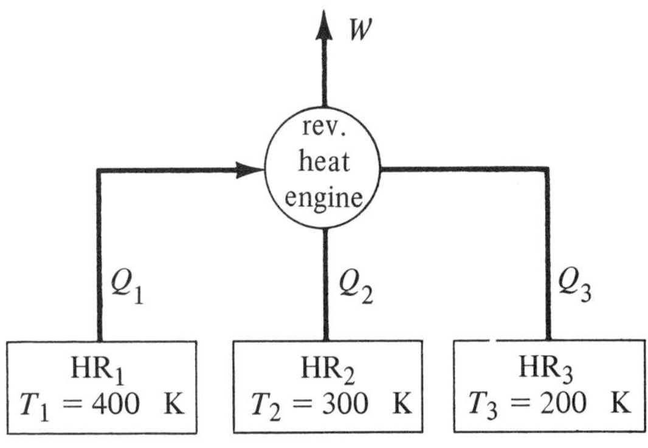

Figure 6P.10

(b) Evaluate $\int \delta Q/T$ for each reservoir and for the engine.

6.11 A heat engine is in thermal communication with two heat reservoirs HR_1 and HR_2. The temperatures of these heat reservoirs are 600 K and 300 K, respectively. The thermal communication between the two heat reservoirs and the engine is not perfect in the sense that a finite temperature difference is required to establish heat transfer between each of the heat reservoirs and the heat engine. For heat reservoir HR_1, the temperature difference is 50 K and for heat reservoir HR_2, the temperature difference is 25 K.

(a) If the heat engine experiences a positive heat transfer of 1000 kJ/cycle with the heat reservoir HR_1, estimate the reduction in the work transfer/cycle associated with the imperfect thermal communication.

(b) As an engineer, it is your job to improve the performance of this heat engine. One of the obvious ways to accomplish this feat is to improve the thermal communication between the heat reservoirs and the heat engine. Which of the two heat interactions would yield the greater return for a given reduction in temperature difference? Explain your answer.

6.12 A convenient and very effective means of heat transfer from a space vehicle or satellite is by thermal radiation to the depths of space which is used as a heat reservoir with an extremely low temperature, namely, 0 K. For this situation the rate at which a body radiates to this heat reservoir is proportional to the surface area of the radiator and the fourth power of its thermodynamic temperature. However, in order to minimize the weight of a space vehicle or satellite, it is necessary to minimize the area of the radiator since its weight is directly proportional to its surface area. Suppose that a space vehicle or satellite contains a reversible Carnot engine which experiences a positive heat transfer Q_H (not fixed) from an energy source (heat reservoir) whose temperature is fixed at T_H. This engine delivers a fixed amount of power W and ultimately uses the depths of space as an energy sink (heat reservoir). Show that the least radiator weight for this system is obtained when the temperature of the radiator, T_R, is such that $T_R = 0.75T_H$.

CHAPTER 7

Entropy

7.1 Introduction

In Chapter 6 the general statement of the second law of thermodynamics was extended to any cyclic process of any system, and is expressed by equations (6.55) and (6.56). In this chapter we shall use $\oint (\delta Q/T)_{\text{rev}} = 0$ to define the change in the property entropy, and we shall consider $\int \delta Q/T$ as a transfer of entropy across the system boundary. The concept of $\int \delta Q/T$ as an entropy transfer across the boundary of a closed system is motivated by cause-effect considerations of *reversible* interactions, for example between two systems A and B. As we will develop in this chapter, a decrease in the entropy of system A requires an equal magnitude increase in the entropy of system B. The entropy which disappeared in A has appeared in B. The logical cause of such compensating changes in entropy is a transfer of entropy from A to B in the same way we think of the heat transfer causing a decrease in the internal energy of A and an equal magnitude increase in B.

The property entropy will enable us to express the second law of thermodynamics in a form that is directly applicable to any process. We shall then illustrate how the second law of thermodynamics imposes limitations on the heat transfer, the work transfer, and the entropy transfer during a change of state of a system.

7.2 Definition of Entropy Change

In Chapter 6 we showed that any reversible cycle of any system must satisfy the condition

$$\oint \left(\frac{\delta Q}{T}\right)_{\text{rev}} = 0 \tag{7.1}$$

Equation (7.1) guarantees the existence of a quantity that is a function of the state of the system such that the change in the quantity between two states will be independent of the path followed by the system in making the change of state. This can be shown by considering the two cycles shown in Figure 7.1. The first cycle consists of a reversible process from state 1 to state 2 along path B and a reversible process from state 2 to state 1 along path A. The second cycle consists of a reversible process from state 1 to state 2 along any path N and a reversible process from state 2 to state 1 along path A as in the first cycle. Equation (7.1) requires for the first cycle

$$\int_1^2 \left(\frac{\delta Q}{T}\right)_B + \int_2^1 \left(\frac{\delta Q}{T}\right)_A = 0 \tag{7.2}$$

For the second cycle

$$\int_1^2 \left(\frac{\delta Q}{T}\right)_N + \int_2^1 \left(\frac{\delta Q}{T}\right)_A = 0 \tag{7.3}$$

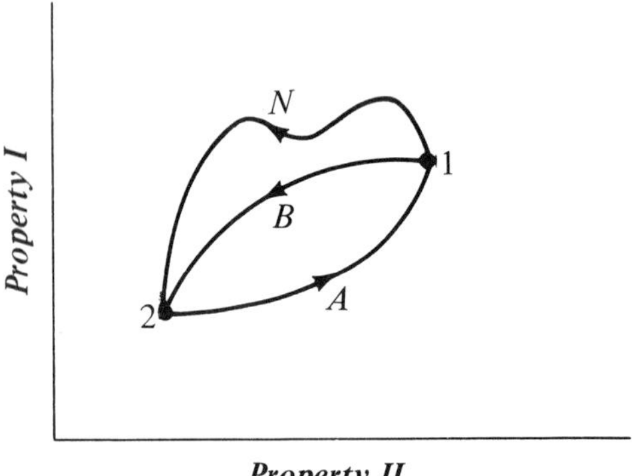

Figure 7.1 Paths of reversible cycles

When equation (7.2) is subtracted from equation (6.3), we have

$$\int_1^2 \left(\frac{\delta Q}{T}\right)_N = \int_1^2 \left(\frac{\delta Q}{T}\right)_B \tag{7.4}$$

Thus, the integral of $\delta Q/T$ for a *reversible process along any path* has the same value as the reversible process along path B. Equation (7.4) is the basis for defining the change in the property called entropy and for defining the boundary interaction which results in the transfer of entropy from the environment to a system. The entropy transfer from the environment to the system is defined as $\int (\delta Q/T)$. As a result of the interaction, the state of the system changes from state 1 with entropy S_1 to state 2 with entropy S_2. The *entropy change* is defined to be equal to the reversible entropy transfer. Thus,

$$S_2 - S_1 = \int_1^2 \left(\frac{\delta Q}{T}\right)_{\text{rev}} \tag{7.5}$$

Since the entropy transfer required to change the state of the system reversibly from one fixed state to another fixed state is independent of the path, we have omitted in equation (7.5) the subscript indicating the path as in equation (7.4). The subscript "rev" is added only as a reminder that the change in entropy is equal to the entropy transfer only when the integration of the entropy transfer is carried out for a reversible process. For an infinitesimal change of state, equation (7.5) reduces to

$$dS = \left(\frac{\delta Q}{T}\right)_{\text{rev}} \tag{7.6}$$

Equation (7.6) represents the conservation of entropy in reversible processes. Conservation of entropy is illustrated in Figure 7.2 for two systems a and b which interact with a reversible heat transfer. For the heat transfer

$$\delta Q_b = -\delta Q_a \tag{7.7}$$

For the reversible heat transfer (Section 5.6)

$$T_b = T_a \tag{7.8}$$

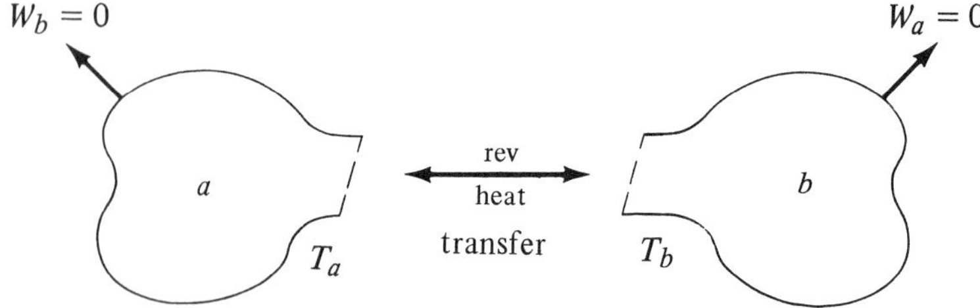

Figure 7.2 Reversible heat transfers between two systems

Thus the entropy transfer out of a is equal to the entropy transfer into b

$$\frac{\delta Q_b}{T_b} = -\frac{\delta Q_a}{T_a} \tag{7.9}$$

at each infinitesimal step of the reversible process. For a finite change from state 1 to state 2 equation (7.5) requires

$$(S_2 - S_1)_b = \int_1^2 \frac{\delta Q_b}{T_b} = -(S_2 - S_1)_a \tag{7.10}$$

Equation (7.10) shows that the entropy increase in b is the entropy transfer from a to b and is the same as the entropy decrease in a. The conservation of entropy for reversible processes follows from the second law in the same manner as the conservation of energy follows from the first law. The first law for the reversible heat transfer of Figure 7.2 is

$$(U_2 - U_1)_b = \int_1^2 \delta Q_b = -(U_2 - U_1)_a \tag{7.11}$$

The energy increase in b is the energy transfer from a to b and is the same as the energy decrease in a. Equations (7.10) and (7.11) are consistent with the application of the first and second laws to the composite system consisting of a and b.

The first law for the combined system gives

$$(U_2 - U_1)_a + (U_2 - U_1)_b = 0 \tag{7.12}$$

since for the combined system Q and W are zero. The second law (equation (7.5)) gives

$$(S_2 - S_1)_a + (S_2 - S_1)_b = 0 \tag{7.13}$$

since for the combined system the entropy transfer $\int \delta Q/T$ is zero.

Equation (7.5) shows that the second law of thermodynamics defines only entropy changes. Thus, entropy values must be calculated relative to an arbitrarily selected zero point for the entropy. To compute the entropy change between two given states, one need only evaluate the integral of the entropy transfer $\delta Q/T$ (equation (7.5)) for reversible processes along any path between these two states. This flexibility allows one to choose the path to make the integration simple. Since the processes involved are reversible, the system and its environment are in thermal equilibrium. Therefore, there is a single uniform temperature throughout both the system and the environment, and the identification of the temperature to be used in the entropy transfer of equation (7.5) is a simple task.

Normally the integration is simplest if the states are joined by a two step reversible process. For example, the entropy difference, $S_2 - S_1$, between the two states of an ideal gas system shown on the pressure-volume diagram of Figure 7.3, may be computed by integrating the entropy transfer $\delta Q/T$ for a reversible adiabatic process from state 1 to state a and a reversible isothermal process from state a to state 2. Since the entropy is a property, the difference in entropy between states 1 and 2 is simply the sum of the entropy changes for the two reversible processes.

$$S_2 - S_1 = (S_a - S_1) + (S_2 - S_a) \tag{7.14}$$

The entropy change $(S_a - S_1)$ for the reversible adiabatic process is

$$S_a - S_1 = \int_1^a \left(\frac{\delta Q}{T}\right)_{\text{rev}} = 0 \tag{7.15}$$

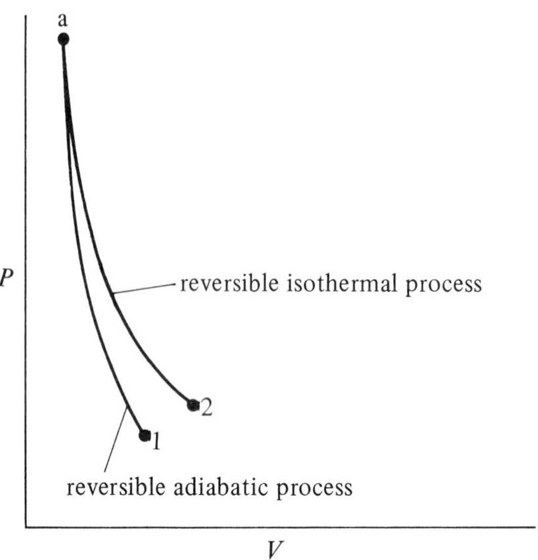

Figure 7.3 Reversible processes for the evaluation of the change of entropy

since the heat transfer and the associated entropy transfer are zero for an adiabatic process. From equation (7.15) we conclude that a reversible adiabatic process is a constant entropy (isentropic) process. An adiabatic process which is irreversible can produce a change in entropy within the system boundary even though there is no entropy transfer across the boundary. (We shall consider this latter case in greater detail in Section 7.3.)

The entropy change for the reversible isothermal process is given by

$$S_2 - S_a = \int_a^2 \left(\frac{\delta Q}{T}\right)_{\text{rev}} \tag{7.16}$$

Before we can evaluate the integral of the entropy transfer in equation (7.16), we must first determine the heat transfer for an infinitesimal reversible isothermal process in an ideal gas. For such a process we have from the first law of thermodynamics

$$\delta Q - \delta W = dU = mc_v\, dT = 0 \tag{7.17}$$

Then since the only reversible work transfer is $P\,dV$,

$$\delta Q = \delta W = P\,dV \tag{7.18}$$

From the ideal gas equation of state, equation (4.36),

$$P = \frac{mRT}{V} \tag{7.19}$$

Then with the substitution of equations (7.18) and (7.19) into equation (7.16), the entropy change $(S_2 - S_a)$ for a reversible isothermal process becomes

$$S_2 - S_a = \int_a^2 \left(\frac{\delta Q}{T}\right)_{\text{rev}} = \int_{V_a}^{V_2} mR\frac{dV}{V} = mR \ln \frac{V_2}{V_a} \tag{7.20}$$

The temperature and volume for the reversible adiabatic process are related according to equation (5.27). Thus,

$$V_a = V_1 \left(\frac{T_1}{T_2}\right)^{1/(\gamma-1)} \tag{7.21}$$

$$S_2 - S_a = mR \ln \frac{V_2}{V_1}\left(\frac{T_2}{T_1}\right)^{1/(\gamma-1)}$$

$$S_2 - S_a = mR \ln \frac{V_2}{V_1} + \frac{mR}{\gamma - 1} \ln \frac{T_2}{T_1} \tag{7.22}$$

We can substitute for γ in equation (7.22) the value given by equation (5.23). Then

$$S_2 - S_a = mR \ln \frac{V_2}{V_1} + mc_v \ln \frac{T_2}{T_1} \tag{7.23}$$

Thus, the difference in entropy between any two states 1 and 2 in an ideal gas with constant specific heat c_v is given by the combination of equations (7.14), (7.15), and (7.23).

$$S_2 - S_1 = mR \ln \frac{V_2}{V_1} + mc_v \ln \frac{T_2}{T_1} \tag{7.24}$$

In effect we have established the entropy of every state *in the ideal gas* relative to the entropy of some arbitrary datum state by means of a combination of reversible processes which enables us to reach any state from the datum state. Equation (7.24) is especially convenient when V and T are the independent properties. When the volume is not an independent property, we can use the ideal gas equation of state to express the volume ratio in terms of the pressure and temperature ratios. Equation (7.24) then can be used to express the entropy change of an ideal gas in terms of P and T.

$$S_2 - S_1 = m(R + c_v) \ln \frac{T_2}{T_1} - mR \ln \frac{P_2}{P_1} \tag{7.25}$$

In a similar manner the entropy change of an ideal gas can be expressed in terms of V and P.

$$S_2 - S_1 = mc_v \ln \frac{P_2}{P_1} + m(R + c_v) \ln \frac{V_2}{V_1} \tag{7.26}$$

In more advanced treatments the third law of thermodynamics establishes in a formal manner the entropy of a particular datum state at $T = 0$. Thus, it is now possible to assign an absolute value to the entropy. The appearance of the mass in equation (7.24) indicates that the entropy is an extensive property. If two systems have the same intensive state but the mass of one system is twice that of the other, the entropy of the larger system is twice the entropy of the smaller system. For a given change in intensive properties, the entropy change is proportional to the mass of the system.

7.3 The Entropy Change for an Irreversible Process

We can now combine equation (6.56) and equation (7.5) to determine the relationship between the entropy change and the integral of the entropy transfer $\delta Q/T$ for an irreversible process. This relationship can be developed by considering the cycle shown in Figure 7.4.

The cycle consists of an irreversible process I along an arbitrary path from state 1 to state 2 and a reversible process R along

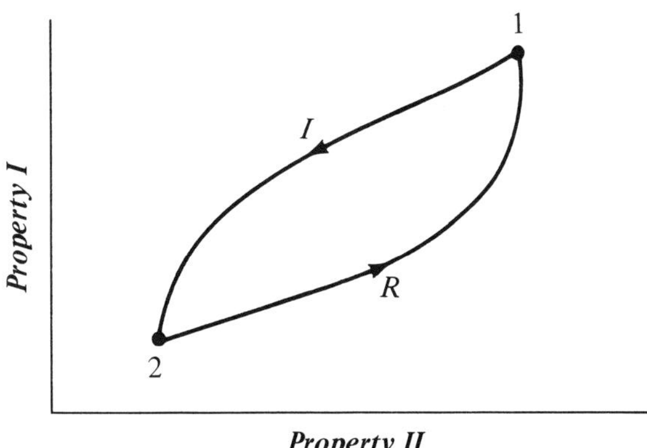

Figure 7.4 Path of an irreversible cycle

an arbitrary path from state 2 to state 1. Equation (6.56) requires that the net entropy transfer for an irreversible cycle is negative; thus,

$$\oint \left(\frac{\delta Q}{T}\right)_{\text{irrev}} = \int_1^2 \left(\frac{\delta Q}{T}\right)_I + \int_2^1 \left(\frac{\delta Q}{T}\right)_R < 0 \qquad \textbf{(7.27)}$$

The second entropy transfer integral is by equation (7.5) the entropy change from state 2 to state 1. Thus, equation (7.27) becomes

$$\int_1^2 \left(\frac{\delta Q}{T}\right)_I + S_1 - S_2 < 0 \qquad \textbf{(7.28)}$$

Equation (7.28) is not changed by adding $S_2 - S_1$ to both sides; therefore,

$$S_2 - S_1 > \int_1^2 \left(\frac{\delta Q}{T}\right)_{\text{irrev}} \qquad \textbf{(7.29)}$$

Here we have replaced the subscript I with "irrev" to emphasize the irreversible nature of the process between states 1 and 2. For an infinitesimal change of state, equation (7.29) becomes

$$dS > \left(\frac{\delta Q}{T}\right)_{\text{irrev}} \qquad \textbf{(7.30)}$$

Equations (7.29) and (7.30) show that in an irreversible process the gain in entropy of the system is greater than the entropy transfer across the boundary of the system during the process. The entropy generated within the system by the irreversibility of the process may be defined as the excess of the entropy change of the system over the entropy transfer across the system boundary. Thus,

$$(S_{\text{generated}})_{1-2} \equiv (S_2 - S_1) - \int_1^2 \left(\frac{\delta Q}{T}\right)_{\text{irrev}} > 0 \tag{7.31}$$

As the irreversible process approaches reversibility, the inequalities of equations (7.29) and (7.30) approach the equalities of equations (7.5) and (7.6) and the entropy generated, equation (7.31), approaches zero. Therefore, the second law of thermodynamics requires entropy to be generated during irreversible processes or to be conserved in the limit for reversible processes. Note in equation (7.31) that even in those cases for which the entropy transfer is negative, that is, entropy is being transferred across the boundary and out of the system, the entropy within the system does not decrease by the same amount. The magnitude of the entropy decrease is less than the magnitude of the entropy transfer out of the system. This is because irreversibilities internal to the system generate entropy during the process and prevent the entropy within the system from decreasing as much as would be expected from the entropy transferred out of the system. Conversely, any process for which the magnitude of the decrease of entropy within a system exceeds the magnitude of the entropy transfer out of the system (negative entropy generation) would be a violation of the second law of thermodynamics.

We can now see an analogy between the role of entropy and entropy transfer and the role of thermal energy and heat transfer in uncoupled systems. For dissipative processes in an uncoupled system, the increase in thermal energy exceeds the heat transfer [equation (3.46)]. For irreversible processes in a coupled system, the increase in entropy exceeds the entropy transfer [equation (7.29)]. In the uncoupled case, the excess thermal energy is the result of mechanical dissipation. In the coupled case, the excess entropy is the result of any irreversibility.

In equation (7.29) the evaluation of the integral of the entropy transfer $\delta Q/T$ for irreversible processes which proceed through states which are far from equilibrium requires special care. In such cases the temperature in the integral can be identified with the heat reservoir that is in communication with the system while δQ is transferred, or the temperature of the reservoir or reservoirs which could have replaced the actual environment of the system. When the system undergoing the irreversible process is non-uniform in temperature yet still near enough to equilibrium for satisfactory identification of the local quasi-static temperature, the temperature in equation (7.29) or (7.30) is the quasi-static temperature of the boundary at the point where the δQ is transferred. In each case, the relevant heat transfer in equation (7.29) or (7.30) is the actual heat transfer for the irreversible process in question. In some complex interactions which proceed through states far from equilibrium, it may not be possible to identify a boundary temperature for the evaluation of the entropy transfer $\delta Q/T$.

To evaluate the entropy change appearing in equation (7.29), we make use of the fact that the entropy of a system is a property and its value for a given state is independent of the process by which the state is attained. Therefore, to compute the entropy change for an irreversible process, we first identify the equilibrium end states of the irreversible process and then proceed from the initial state to the final state through a reversible process or series of reversible processes along some path which is easily defined. We then evaluate equation (7.5) along this path; however, in doing so, we must be certain to use *not* the entropy transfer associated with the actual irreversible process but rather the entropy transfer associated with the reversible processes connecting the equilibrium end states. Thus, in effect we have evaluated both sides of equation (7.29) independently.

7E.1 Sample Problem: A system consisting of a pure dissipative element and a pure thermal element with a heat capacity of 50 kJ/K experiences an irreversible work transfer interaction which takes the system from state 1 of $T_1 = 300$ K to state 2 of $T_2 = 310$ K. There is no heat transfer during this irreversible process. Calculate the change in entropy of the system for this process.

Solution: To calculate the change in entropy, we must find a series of reversible processes that take the system from state 1 to state 2 and then evaluate the integral $\int \delta Q/T$ for this series of reversible processes. In the case of the pure thermal system, the situation is straightforward since there is only one property, temperature, that characterizes the state of the system. A reversible process which can effect the desired change of state is one in which the heat transfer occurs at a rate slow enough so that at any instant of time, every element of the system has exactly the same temperature, i.e., the process is quasi-static with uniform temperature. Under these conditions, as the temperature changes by an amount dT, and the necessary heat transfer is $C\,dT$. Then, from equation (7.5),

$$S_2 - S_1 = \int_1^2 \left(\frac{\delta Q}{T}\right)_{\text{rev}} = \int_{T_1}^{T_2} \frac{C\,dT}{T} = C \ln \frac{T_2}{T_1}$$

For this system,

$$S_2 - S_1 = 50 \ln \frac{310}{300} = 1.6395 \text{ kJ/K}$$

Notice that for the actual process, $\int (\delta Q/T)_{\text{actual}} = 0$ since the process was adiabatic. Thus entropy was generated within the system by virtue of the work transfer interaction. Although we are unable to identify the details of the irreversibility without further information about the system, we are able to calculate the magnitude of the entropy generated by the irreversibility by connecting the same initial and final states with a reversible process and evaluating the integral $\int (\delta Q/T)_{\text{rev}}$.

7.4 Entropy as a Test for Reversibility

Equations (7.5) and (7.29) are the basis for a positive test for the reversibility of a given process between two fixed equilibrium states in a system. The elements of the test are as follows:

1. If the change in entropy of the system is equal to the integral of the entropy transfer $\delta Q/T$ and δQ is the heat transfer for the actual process under test, the process must be completely reversible.

2. If the change in entropy of the system is greater than the integral of the entropy transfer $\delta Q/T$ and δQ is the heat transfer for the actual process under test, the process is irreversible.
3. If the change in entropy is less than the integral of the entropy transfer $\delta Q/T$ and δQ is the heat transfer for the actual process under test, the process is a violation of the second law of thermodynamics.

These statements may also be expressed in mathematical form as follows:

reversible process $$dS = \frac{\delta Q}{T} \quad \text{or} \quad S_2 - S_1 = \int_1^2 \frac{\delta Q}{T} \tag{7.32}$$

irreversible process $$dS > \frac{\delta Q}{T} \quad \text{or} \quad S_2 - S_1 > \int_1^2 \frac{\delta Q}{T} \tag{7.33}$$

violation of second law of thermodynamics $$dS < \frac{\delta Q}{T} \quad \text{or} \quad S_2 - S_1 < \int_1^2 \frac{\delta Q}{T} \tag{7.34}$$

The application of equations (7.32), (7.33), and (7.34) is especially simple in the case of completely adiabatic processes since the integral of the entropy transfer $\delta Q/T$ is always zero. In this case constant entropy means a reversible process, an increase in entropy means an irreversible process, and a decrease in entropy would mean a violation of the second law of thermodynamics. Conversely, a process at constant entropy is reversible if adiabatic, and must involve a negative entropy transfer and heat transfer if it is irreversible.

One can usually select an adiabatic system for analysis by including all things significantly influenced by the process. For any process in such a system all heat transfer is internal; therefore, the external heat transfer and entropy transfer are zero. Then as is evident from equations (7.32) and (7.33),

$$dS \geq 0 \quad \text{or} \quad S_2 - S_1 \geq 0 \tag{7.35}$$

Since a system of this type is isolated with respect to heat transfer and entropy transfer, equation (7.34) is often stated semantically as: "The entropy of *an isolated* (adiabatic) system can never decrease". Equation (7.35) is often known as the principle of entropy increase. This type of system, however, will not allow us to locate the source of an irreversibility during a particular process since all interactions are internal to the system. Quite often the universe is regarded as an adiabatic system. This rather unjustifiable consideration leads to the widely discussed concept of an entropy death for the universe. At this state, the universe would have reached a state of maximum entropy corresponding to complete thermal equilibrium.

The use of the foregoing test for reversibility may be simply illustrated by considering a pure thermal system initially at T_1 which is allowed to reach equilibrium with a heat reservoir at T_0 by means of a simple conduction heat transfer. For this situation (shown schematically in Figure 7.5) the reversibility of the process is most easily established by applying the test for reversibility, equation (7.35). Thus, we first establish an adiabatic system and then calculate the change in entropy for this system. The composite system consisting of the pure thermal system and the heat reservoir is sufficient. As shown in Figure 7.5, there is no external heat transfer for this system and the heat transfer between the pure thermal system and the heat reservoir is the only significant internal interaction. Therefore, any possible irreversibility that we might detect must originate with this internal interaction.

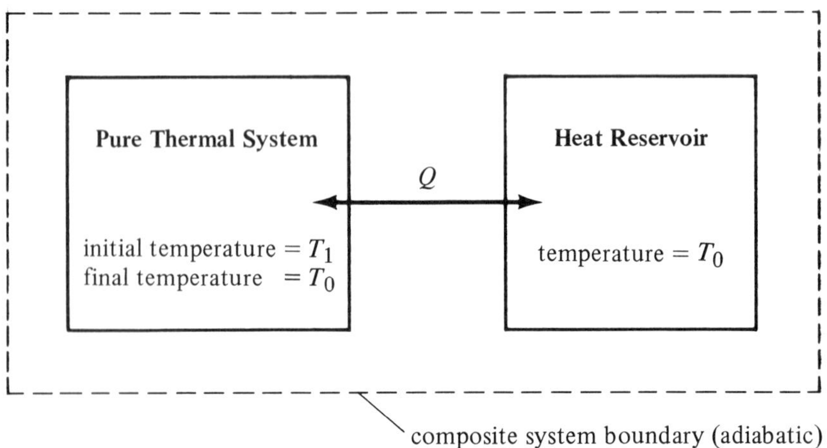

Figure 7.5 Heat transfer by simple conduction between a pure thermal system and a heat reservoir

Since the entropy is an extensive property, the change in entropy of the composite system involves two contributions—one due to the change in entropy of the heat reservoir, the other due to the change in entropy of the pure thermal system. Thus,

$$(S_2 - S_1)_{\substack{\text{composite}\\\text{system}}} = (S_2 - S_1)_{\substack{\text{heat}\\\text{reservoir}}} + (S_2 - S_1)_{\substack{\text{pure}\\\text{thermal}\\\text{system}}} \tag{7.36}$$

To compute the change in entropy of the heat reservoir, we apply equation (7.5); however, we must first establish the end states of the interaction and then find a reversible process that will produce the same final state from the same initial state. Although the heat reservoir does not experience a change in temperature as a result of the heat transfer, it does experience a definite change of state. From the first law of thermodynamics for the heat reservoir,

$$(\delta Q)_{\substack{\text{heat}\\\text{reservoir}}} = (dU)_{\substack{\text{heat}\\\text{reservoir}}} \tag{7.37}$$

Thus, the change of state of the heat reservoir is manifested as a change in its stored thermal energy. The heat transfer that produces this change of state in the present case is determined most readily from the application of the first law of thermodynamics to the composite system. Since there is no net external heat transfer for the composite system, we have

$$(\delta Q)_{\substack{\text{composite}\\\text{system}}} = (\delta Q)_{\substack{\text{heat}\\\text{reservoir}}} + (\delta Q)_{\substack{\text{pure thermal}\\\text{system}}} = 0 \tag{7.38}$$

Thus,

$$(\delta Q)_{\substack{\text{heat}\\\text{reservoir}}} = -(\delta Q)_{\substack{\text{pure thermal}\\\text{system}}} \tag{7.39}$$

From the combination of the first law of thermodynamics and the constitutive relation for the pure thermal system, we have

$$(\delta Q)_{\substack{\text{pure thermal}\\\text{system}}} = (C\,dT)_{\substack{\text{pure thermal}\\\text{system}}} \tag{7.40}$$

Here C is the heat capacity of the pure thermal system. Then combining equations (7.37), (7.39), and (7.40), we obtain

$$(\delta Q)_{\substack{\text{heat}\\\text{reservoir}}} = -(C\,dT)_{\substack{\text{pure thermal}\\\text{system}}} = (dU)_{\substack{\text{heat}\\\text{reservoir}}} \tag{7.41}$$

Equation (7.37) shows that a given heat transfer always produces the same change of state in a heat reservoir regardless of whether the heat transfer interaction was reversible or irreversible. Thus in the case of a heat reservoir, the subscript "rev" appearing in equation (7.5) is unnecessary, and the actual heat transfer can be used without any concern for its reversibility. Since the temperature of the heat reservoir is constant, the change of entropy of the heat reservoir becomes

$$(S_2 - S_1)_{\substack{\text{heat}\\\text{reservoir}}} = \int_1^2 \left(\frac{\delta Q}{T}\right)_{\substack{\text{heat}\\\text{reservoir}}} = \frac{1}{T_0}\int_1^2 (\delta Q)_{\substack{\text{heat}\\\text{reservoir}}} \tag{7.42}$$

or

$$(S_2 - S_1)_{\substack{\text{heat}\\\text{reservoir}}} = \frac{1}{T_0}\int_1^2 -(C\,dT)_{\substack{\text{pure thermal}\\\text{system}}}$$

$$= \frac{C(T_1 - T_2)}{T_0} = \frac{C(T_1 - T_0)}{T_0} \tag{7.43}$$

Note that the final temperature, T_2, of the pure thermal system is identical to the temperature of the heat reservoir, T_0.

In the case of the pure thermal system, the calculation of the entropy change is slightly more complicated. Because the first law of thermodynamics for the pure thermal system is identical to that of the heat reservoir (heat reservoirs are a special class of pure thermal systems), a given heat transfer always produces the same change of state in a pure thermal system regardless of whether the heat transfer was reversible or irreversible *provided the temperature of the pure thermal system is uniform.* Then, substituting equation (7.40) into equation (7.5), we obtain

$$(S_2 - S_1)_{\substack{\text{pure thermal}\\\text{system}}} = \int_1^2 \left(\frac{\delta Q}{T}\right)_{\substack{\text{pure thermal}\\\text{system}}} = \int_1^2 \frac{C\,dT}{T} \tag{7.44}$$

which upon integration becomes

$$(S_2 - S_1)_{\substack{\text{pure thermal}\\\text{system}}} = C\ln\frac{T_2}{T_1} = C\ln\frac{T_0}{T_1} \tag{7.45}$$

Substituting equations (7.43) and (7.45) into (7.36), we obtain

$$(S_2 - S_1)_{\substack{\text{composite}\\\text{system}}} = \frac{C(T_1 - T_0)}{T_0} - C \ln \frac{T_1}{T_0} \tag{7.46}$$

If $T_1 - T_0 = \Delta T$,

$$(S_2 - S_1)_{\substack{\text{composite}\\\text{system}}} = C\frac{\Delta T}{T_0} - C \ln \left(1 + \frac{\Delta T}{T_0}\right) \tag{7.47}$$

Figure 7.6 clearly shows that the entropy increase of the reservoir always exceeds the entropy decrease of the system (for all possible values of ΔT). Since the composite system is adiabatic, this net gain of entropy shows that this process is irreversible

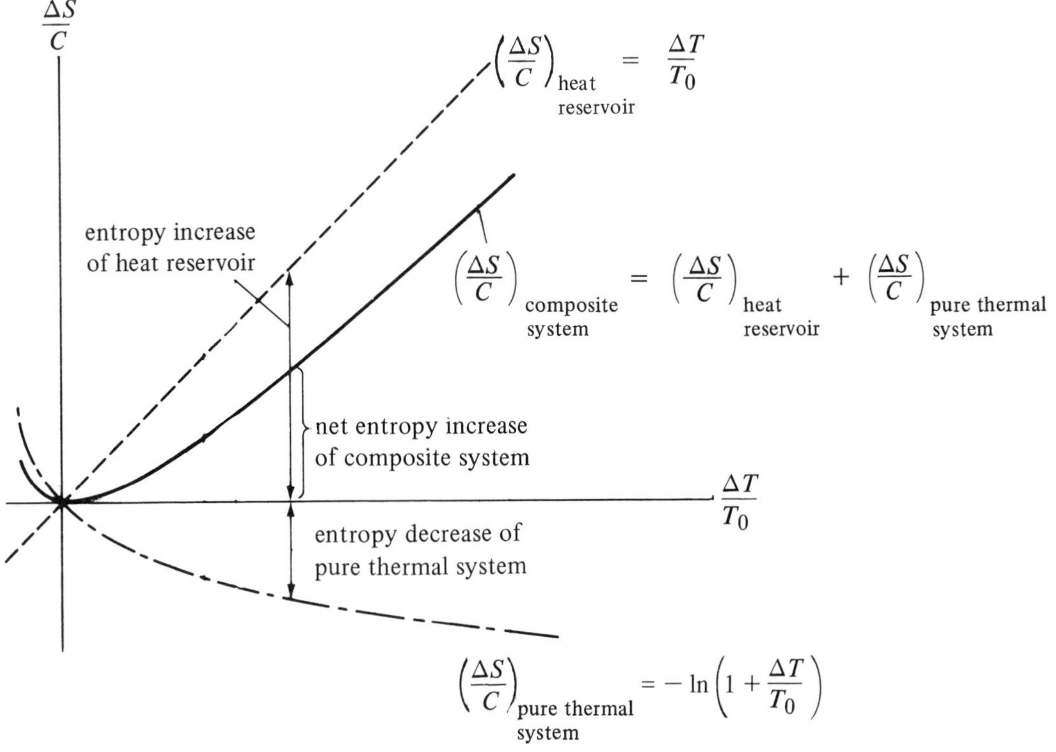

Figure 7.6 Entropy change for simple conduction heat transfer between a pure thermal system and a heat reservoir

regardless of the sign of ΔT. For $\Delta T \ll T_0$, the logarithm appearing in equation (7.47) can be expanded.

$$\ln\left(1 + \frac{\Delta T}{T_0}\right) \approx \frac{\Delta T}{T_0} - \frac{1}{2}\left(\frac{\Delta T}{T_0}\right)^2 + \cdots \tag{7.48}$$

Substitution of this result into equation (7.47) shows that as ΔT approaches zero, $(S_2 - S_1)_{\text{composite system}}$ approaches zero as $(C/2)(\Delta T/T_0)^2$. Since $(S_2 - S_1)_{\text{composite system}}$ approaches zero faster than ΔT, the total entropy change for the process becomes zero in the limit of zero temperature difference, and the process becomes reversible. By similar reasoning, we can conclude that the only possible processes between two pure thermal systems are heat transfers which decrease the temperature difference between the two systems.

7E.2 Sample Problem: As shown in Figure 7E.2, a rigid, insulated (adiabatic), chamber is divided into two compartments of equal volume ($V = 0.03\ \text{m}^3$) by means of a frictionless, massless piston which is initially held in place with a pin. One compartment is evacuated while the other compartment is filled with air at 6×10^5 Pa and 300 K. (The air can be modeled as an ideal gas with $R = 0.287$ kJ/kgK and $c_v = 0.716$ kJ/kgK.) The pin is removed and the gas comes to equilibrium. Show from entropy considerations whether the process is reversible or irreversible.

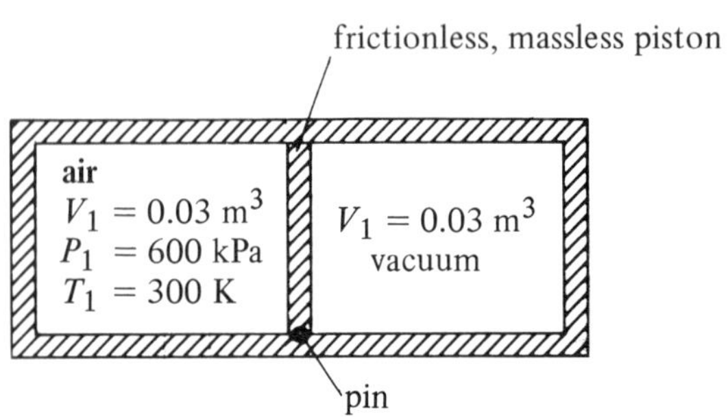

Figure 7E.2 Free expansion of an ideal gas

Solution: We can determine whether the process is reversible or irreversible by comparing the change in entropy of the gas with $\int (\delta Q/T)_{actual}$; thus we must first determine the change of state and the interactions by which the change of state is effected.

Since the container is adiabatic, the gas experiences no heat transfer. When the pin is removed, the piston moves due to the force exerted by the gas. However, since the vacuum exerts no force on the opposite face of the piston, the gas does no work. That is, even though the boundary of the gas is displaced during the process by which equilibrium is attained, there is no work transfer because there is no force exerted by the environment on that moving boundary. Thus, $Q_{1-2} = 0$ and $W_{1-2} = 0$. It follows, then, from the first law of thermodynamics that

$$U_2 - U_1 = 0$$

$$mc_v(T_2 - T_1) = 0$$

$$T_2 = T_1 = 300 \text{ K}$$

Since

$$V_2 = 2V_1 = 2(0.03) = 0.06 \text{ m}^3$$

the constitutive relation for the gas gives

$$P_2 = 0.5P_1 = 3 \times 10^5 \text{ Pa}$$

Also from the constitutive relation,

$$m = \frac{PV}{RT} = \frac{6 \times 10^5(0.03)}{287(300)} = 0.2091 \text{ kg}$$

Then from equation (7.24)

$$S_2 - S_1 = 0.2091(0.287) \ln \frac{2V_1}{V_1} + 0.2091(0.716) \ln \frac{T_1}{T_1}$$

$$S_2 - S_1 = 0.0416 \text{ kJ/K}$$

Since the process was adiabatic,

$$\int_1^2 \left(\frac{\delta Q}{T}\right)_{\text{actual}} = 0$$

Then

$$S_2 - S_1 > \int_1^2 \left(\frac{\delta Q}{T}\right)_{\text{actual}}$$

and the process is irreversible by equation (7.33).

7.5 Methods of Reversible Heat Transfer

In the preceding illustration of the use of entropy as a test for reversibility, we noted that because of the especially simple nature of a pure thermal system it was not necessary to distinguish between reversible and irreversible heat transfer processes for calculating the change in entropy (provided the temperature remained uniform). Certainly this simplification is not possible in the case of a coupled thermodynamic system. In such cases, the distinction between reversible and irreversible processes is an important one indeed. It is now necessary to effect reversible heat transfer interactions, at least in principle if not in actual practice, so that the change in entropy can be properly calculated through the application of equation (7.5).

For the sake of simplicity, we consider first the case of reversible heat transfer for a pure thermal system. Based on the results of Section 7.4, we conclude that one method of reversible heat transfer is the successive communication of the pure thermal system with an infinite series of heat reservoirs such that each successive heat reservoir differs in temperature from its predecessor by an infinitesimal amount. This limiting process is reversible because when the system is communicated with the series of reservoirs in the reverse order, the system and all reservoirs (except the first and last) are returned to the initial state. The changes of state in the end reservoirs are vanishingly small since the number of reservoirs in the series is infinite.

A second method of reversible heat transfer is the thermal communication of the pure thermal system with an ideal gas system that is initially in thermal equilibrium with it (Section 5.7). A subsequent quasi-static work transfer to the ideal gas system would result in a reversible heat transfer to the pure thermal system because of the coupled thermodynamic behavior of the ideal gas. The process would be quasi-static only if the temperature of the gas and the temperature of the thermal system were both equal and uniform.

A third method of reversible heat transfer is the thermal communication of the pure thermal system with a single heat reservoir by means of an intermediate system with a very low thermal conductance and a negligible heat capacity. In this case, all significant temperature gradients during the process are within the intermediate conductor system. The interaction for the composite system consisting of the pure thermal system, conductor, and heat reservoir is irreversible; however, the irreversibility is confined to the conductor system. One such situation is shown in Figure 7.7. Since the temperature of the reservoir is uniform and equal to the temperature at the boundary with the conductor system, the entropy transfer at the boundary of the reservoir is equal to the entropy change of the reservoir and the process within the reservoir is reversible. In the same way, the temperature of the pure thermal system is uniform (but not constant) and equal to the temperature of the boundary with the conductor; therefore, the entropy transfer at the boundary is the same as the change in entropy of the pure thermal system so the process within the pure thermal system is reversible. For the conductor system, the heat transfer into the conductor crosses the boundary at a high temperature, flows by conduction to the low temperature boundary where it leaves the system at a lower temperature. Since the conductor has a negligible heat capacity, the heat transfer in has the same magnitude as the heat transfer out, which requires that the magnitude of the entropy transfer $\int (\delta Q/T)_{HR}$ in at the high temperature is smaller than the magnitude of the entropy transfer $\int (\delta Q/T)_{PTS}$ out at the lower temperature. However, since the heat transfer in at the high temperature is positive while the heat transfer out at the low temperature is negative, the net entropy transfer for the conductor system is negative while the conductor system experiences a negligible change of state.

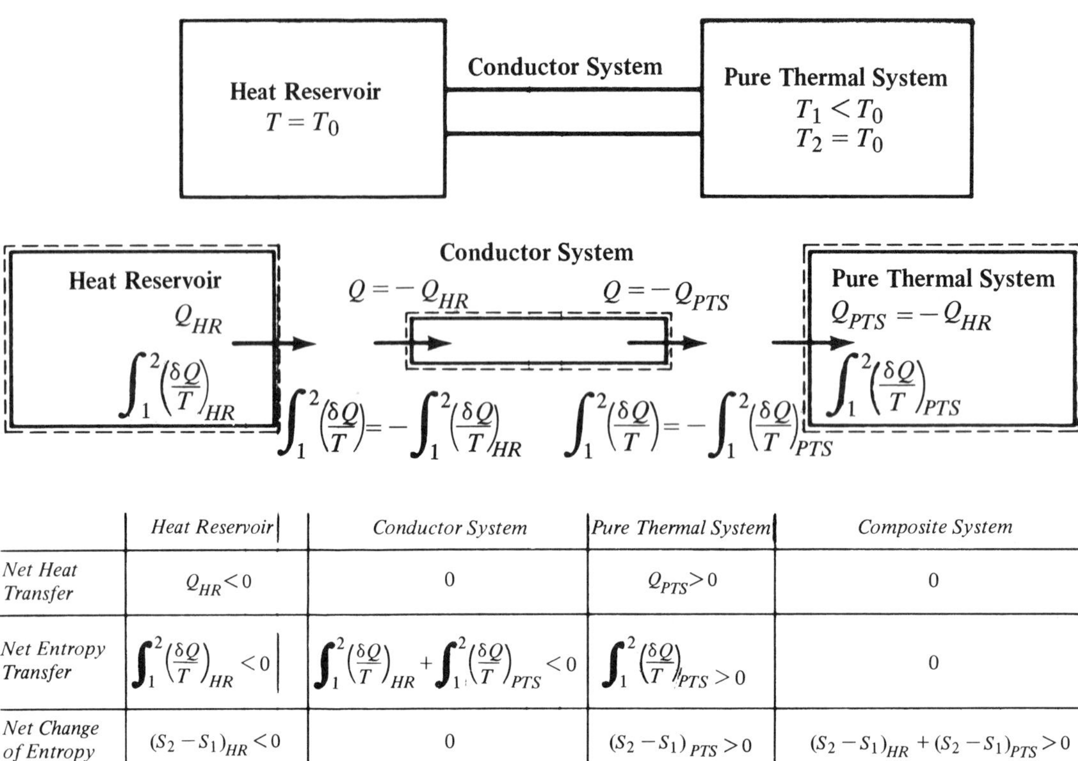

	Heat Reservoir	*Conductor System*	*Pure Thermal System*	*Composite System*
Net Heat Transfer	$Q_{HR} < 0$	0	$Q_{PTS} > 0$	0
Net Entropy Transfer	$\int_1^2 \left(\frac{\delta Q}{T}\right)_{HR} < 0$	$\int_1^2 \left(\frac{\delta Q}{T}\right)_{HR} + \int_1^2 \left(\frac{\delta Q}{T}\right)_{PTS} < 0$	$\int_1^2 \left(\frac{\delta Q}{T}\right)_{PTS} > 0$	0
Net Change of Entropy	$(S_2 - S_1)_{HR} < 0$	0	$(S_2 - S_1)_{PTS} > 0$	$(S_2 - S_1)_{HR} + (S_2 - S_1)_{PTS} > 0$

Figure 7.7 Interaction between heat reservoir and pure thermal system via a conductor

The irreversibility associated with heat transfer across a temperature discontinuity can be shown by letting the length of the conductor system of Figure 7.7 go to zero while the thermal conductance remains finite (thermal conductivity decreasing proportional to length). In this limit of zero length the heat transfer remains continuous, but the temperature changes discontinuously across the conductor system. Thus the entropy transfer is discontinuous to account for the irreversibility from the heat transfer across the temperature discontinuity.

A fourth method of reversible heat transfer is the communication of the pure thermal system with a single heat reservoir by means of an intermediate reversible Carnot cycle system. The Carnot cycle must be arranged so that it will experience only a differential heat transfer δQ during each cycle. Thus during any one cycle, the temperature of the pure thermal system is essentially constant but will be dT lower in temperature during the next cycle of the Carnot cycle system. In this case, the Carnot cycle can operate reversibly in spite of the variation of the temperature of the heat source. This is because an infinite number of cycles occur while the temperature of the thermal system is decreased by a finite amount.

Now it is apparent that many processes may be used to accomplish reversible heat transfer (in principle) for the pure thermal system. We can also use any one of these to effect a reversible heat transfer for a coupled thermodynamic system. For example, consider the case of an ideal gas with constant specific heat c_v. For an infinitesimal reversible process with $P\,dV$ as the only work transfer, the first law of thermodynamics becomes

$$\delta Q - \delta W = dU \tag{7.49}$$

$$\delta Q - P\,dV = mc_v\,dT \tag{7.50}$$

For a reversible constant volume process, equation (7.50) reduces to

$$(\delta Q)_{\substack{\text{constant}\\\text{volume}}} = mc_v\,dT \tag{7.51}$$

Thus, the change in entropy for this reversible, constant volume process in an ideal gas is given by

$$S_2 - S_1 = \int_1^2 \left(\frac{\delta Q}{T}\right)_{\text{rev}} = \int_{T_1}^{T_2} \frac{mc_v\,dT}{T} = mc_v \ln \frac{T_2}{T_1} \tag{7.52}$$

which is consistent with equation (7.24) for a reversible process with $V_2 = V_1$.

For a reversible constant pressure process in the ideal gas, equation (7.50) becomes

$$\delta Q = mc_v\,dT + P\,dV \tag{7.53}$$

From the ideal gas equation of state, equation (4.36),

$$V = \frac{mRT}{P} \tag{7.54}$$

Since the pressure is constant,

$$dV = \frac{mR}{P}\,dT \tag{7.55}$$

Substituting equation (7.55) into equation (7.53), we obtain

$$(\delta Q)_{\substack{\text{constant}\\\text{pressure}}} = mc_v\,dT + mR\,dT \tag{7.56}$$

$$(\delta Q)_{\substack{\text{constant}\\\text{pressure}}} = (c_v + R)m\,dT \tag{7.57}$$

We note that equation (7.57) can be cast in a form similar to equation (7.51) by defining a specific heat at constant pressure[1] for the ideal gas to be

$$c_P = c_v + R \tag{7.58}$$

Then equation (7.57) becomes

$$(\delta Q)_{\substack{\text{constant}\\\text{pressure}}} = mc_P\,dT \tag{7.59}$$

The change in entropy for the reversible constant pressure process in the ideal gas is

$$S_2 - S_1 = \int_1^2 \left(\frac{\delta Q}{T}\right)_{\text{rev}} = \int_{T_1}^{T_2} \frac{mc_P\,dT}{T} = mc_P \ln\frac{T_2}{T_1} \tag{7.60}$$

which is consistent with equation (7.25), developed by changing variables in equation (7.24).

The physical interpretation of equations (7.25) and (7.26) are now evident. Equation (7.25) is the entropy change for an ideal gas calculated by means of a reversible constant pressure process followed by a reversible isothermal process. Equation (7.26) is the entropy change of an ideal gas calculated by means of a reversible constant volume process followed by a reversible constant pressure process. Clearly, the change in entropy for a

[1] The general definition of specific heat c_P is given by equation (9.57) in Chapter 9, page 287.

fixed pair of states does not depend upon the combination of reversible processes used.

7.6 Definitions of Heat Transfer and Work Transfer

Now that we have developed the second law of thermodynamics and described the associated physical phenomena, we can give final definitions for heat transfer and work transfer:

> Heat transfer is an interaction between two systems which transfers both energy and entropy across the boundary between the two systems.
>
> Work transfer is an interaction between two systems which transfers energy but no entropy across the boundary between the two systems.

In the case of a reversible interaction, there is no ambiguity or problem in distinguishing heat transfer from work transfer. The change of energy dU and the change of entropy dS of the two systems are evaluated. The heat transfer is then given by

$$\delta Q = T\,dS \tag{7.61}$$

and the work transfer by

$$\delta W = T\,dS - dU \tag{7.62}$$

Here T is the uniform thermodynamic temperature which is identical for the two systems. The pure conservative mechanical system serves as our standard of measure of energy changes and the pure thermal system (especially the heat reservoir) serves as our standard of measure of entropy changes. For a finite interaction, we simply integrate equations (7.61) and (7.62) along the path of the interaction to determine the net heat transfer and net work transfer.

In the case of an irreversible interaction, we can still apply these definitions simply as in a reversible interaction provided we can model the states at the boundary between the systems as quasi-static. In particular, we must be able to identify a quasi-static temperature for the boundary. If the boundary states are quasi-static, we may replace one of the irreversibly interacting systems with a reversibly interacting system, while leaving the process in the other of the irreversibly interacting systems unchanged. We then evaluate the heat transfer and the work transfer with the aid of the reversible changes in the substitute system as we did for the completely reversible interactions, equations (7.61) and (7.62).

When two interacting systems are far from equilibrium, it may not be possible to model the states at the boundary as quasi-static. For example, when two bodies at very different temperatures are interacting by thermal radiation across an intervening vacuum space, it is not possible to find a quasi-static temperature for a boundary through the vacuum. Although both energy and entropy are being transferred across the boundary, we are unable to evaluate the entropy transfer at the boundary with the quasi-static concepts developed in this chapter. The study of entropy and entropy flux in systems which are not quasi-static is called irreversible thermodynamics, an area of advanced research in thermodynamics which is beyond our present scope[2,3]. In any case the entropy gain of the low temperature system exceeds the entropy loss of the high temperature system because of the entropy generated by the irreversibility. With our quasi-static methods we cannot determine the location of the irreversibility.

7.7 Limitations Imposed by the Second Law of Thermodynamics on the Heat Transfer During a Change of State

When taken together, equations (7.32) through (7.34) constitute a valid statement of the second law of thermodynamics for a change of state. In effect, these equations establish an upper limit on the integral of the entropy transfer $\delta Q/T$ for a given change of state. Further, with the addition of certain restrictions, these equations can be used to establish a limit on the heat

[2] DeGroot, S. R. and Mazur, P., *Non-Equilibrium Thermodynamics,* North-Holland Pub. Co., Amsterdam, 1962.

[3] Prigogine, I., *Introduction to Thermodynamics of Irreversible Processes,* 3rd ed., Wiley–Interscience, N.Y., 1968.

transfer that a system can experience during a given change of state.

For example, for a reversible process, equation (7.32) can be written in the differential form

$$(\delta Q)_{\text{rev}} = T\,dS \tag{7.63}$$

Integrating equation (7.63) for a reversible change of state from state 1 to state 2, we obtain the total heat transfer for that change of state.

$$(Q_{1-2})_{\text{rev}} = \int_1^2 T\,dS \tag{7.64}$$

Clearly in carrying out the integration of equation (7.64) we must first establish the path connecting states 1 and 2. Graphically the integral in equation (7.64) is represented by the area between the path of the process and the abscissa on the temperature-entropy diagram as is shown for path A in Figure 7.8. Since this area represents the heat transfer in a reversible process, the temperature-entropy diagram is useful for graphically illustrating the reversible processes. If some other reversible process proceeds from state 1 to state 2 along path B in Figure 7.8, then the

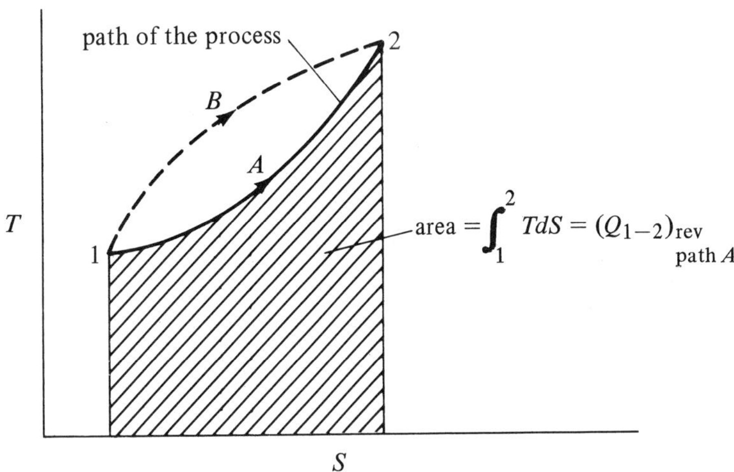

Figure 7.8 Graphical representation of the heat transfer for a reversible process

area between this path and the abscissa, and the heat transfer Q_{1-2}, will be larger for this new reversible process along path B than for the first reversible process along path A. The sign convention for this graphical representation is such that the area, and hence the reversible heat transfer, is positive if the area lies to the right as we proceed from the initial state to the final state along the path of the process. When the area lies to the left as we proceed, its sign is negative.

In the case of an irreversible process between two end states, the second law of thermodynamics does not determine the heat transfer itself but rather the upper limit for the heat transfer. In fact this upper limit can be established only when the path of the irreversible process is known. Specifically, equation (7.33) can be written

$$(\delta Q)_{irrev} < T\,dS \tag{7.65}$$

If the path of the irreversible process by which the system changes state is known, equation (7.65) can be integrated.

$$(Q_{1-2})_{irrev} < \int_1^2 T\,dS \tag{7.66}$$

Equation (7.66) can be used only for irreversible processes that proceed through states with identifiable temperature and entropy. Also, it follows from equation (7.66) that the heat transfer for an irreversible process cannot be represented graphically by the area between the path of the process and the abscissa on a temperature-entropy diagram.

For a reversible process connecting the same two end states along the same path as the irreversible process of equation (7.66), we may substitute equation (7.64) into equation (7.66). Thus,

$$(Q_{1-2})_{irrev} < (Q_{1-2})_{rev} \tag{7.67}$$

Equation (7.67) can be used only to compare the heat transfer for reversible and irreversible processes along the same path.

There are many interesting and physically important applications of equations (7.64) and (7.66), but several merit special attention. For example, consider the case of a system that is experiencing an isothermal change of state. The path for such a process, be it reversible or irreversible, is one of constant temperature. Then, the integrals in equations (7.64) and (7.66) become especially simple and the heat transfer must be of such a magnitude that

$$(Q_{1-2})_{\text{isothermal}} \leq T(S_2 - S_1) \tag{7.68}$$

where equality holds only in the case of a reversible isothermal process.

As another, somewhat different example of the limitations imposed by the second law of thermodynamics on heat transfer, consider the case of a system that experiences heat transfer with a single heat reservoir. If we establish a system boundary that includes both the system and the heat reservoir, this boundary is adiabatic. Therefore, we can apply the second law of thermodynamics in the form of equation (7.35).

$$(S_2 - S_1)_{\substack{\text{composite}\\\text{system}}} \geq 0 \tag{7.69}$$

Since the entropy is an extensive property,

$$(S_2 - S_1)_{\substack{\text{composite}\\\text{system}}} = (S_2 - S_1)_{\text{system}} + (S_2 - S_1)_{\substack{\text{heat}\\\text{reservoir}}} \tag{7.70}$$

Then combining equations (7.69) and (7.70), we obtain

$$(S_2 - S_1)_{\text{system}} \geq -(S_2 - S_1)_{\substack{\text{heat}\\\text{reservoir}}} \tag{7.71}$$

As we have already seen in Section 7.4, equation (7.42),

$$(S_2 - S_1)_{\substack{\text{heat}\\\text{reservoir}}} = \frac{(Q_{1-2})_{\substack{\text{heat}\\\text{reservoir}}}}{T_{\substack{\text{heat}\\\text{reservoir}}}} \tag{7.72}$$

Since the composite system is adiabatic, we have

$$(Q_{1-2})_{\substack{\text{composite}\\\text{system}}} = (Q_{1-2})_{\text{system}} + (Q_{1-2})_{\substack{\text{heat}\\\text{reservoir}}} = 0 \tag{7.73}$$

Thus,

$$(Q_{1-2})_{\text{system}} = -(Q_{1-2})_{\substack{\text{heat}\\\text{reservoir}}} \tag{7.74}$$

Combining equations (7.71), (7.72), and (7.74), we obtain

$$(S_2 - S_1)_{\text{system}} \geq \frac{(Q_{1-2})_{\text{system}}}{T_{\substack{\text{heat}\\\text{reservoir}}}} \tag{7.75}$$

Hence,

$$(Q_{1-2})_{\text{system}} \leq \left(T_{\substack{\text{heat}\\\text{reservoir}}}\right)(S_2 - S_1)_{\text{system}} \tag{7.76}$$

where equality holds only in the case of a reversible process.

Finally, consider the case of a system undergoing a cycle. If the cycle is reversible, equation (7.64) becomes

$$\oint (\delta Q)_{\text{rev}} = \oint T\,dS \tag{7.77}$$

Thus the net heat transfer for a reversible cycle is represented by the area enclosed by the path of the cycle on the T–S plane. For example, a reversible Carnot cycle is shown in Figure 7.9. According to the sign convention, $\oint (\delta Q)_{\text{rev}}$ is positive (heat engine) if the cycle proceeds clockwise on the T–S plane. On the other

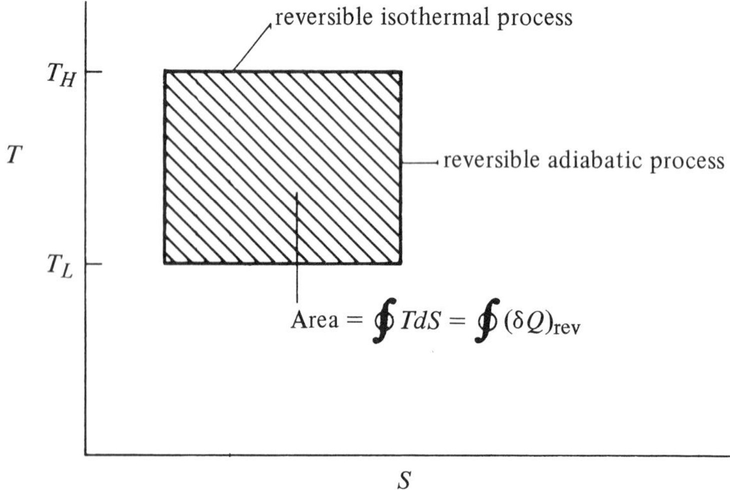

Figure 7.9 Reversible Carnot cycle

hand, $\oint (\delta Q)_{rev}$ is negative (refrigerator) if the cycle proceeds counter-clockwise on the $T-S$ plane. In the case of an irreversible cycle, equation (7.66) becomes

$$\oint (\delta Q)_{irrev} < \oint T\, dS \tag{7.78}$$

provided the cycle proceeds through states with identifiable temperature and entropy. Obviously there is no correlation between the heat transfer for an irreversible cycle and the area enclosed by the path on the $T-S$ plane.

7.8 Entropy Transfer For Cyclic Systems

We can now apply the concepts of entropy to a heat engine operating between two heat reservoirs as shown in Figure 7.10. The second law, equation (6.26), requires that the net entropy

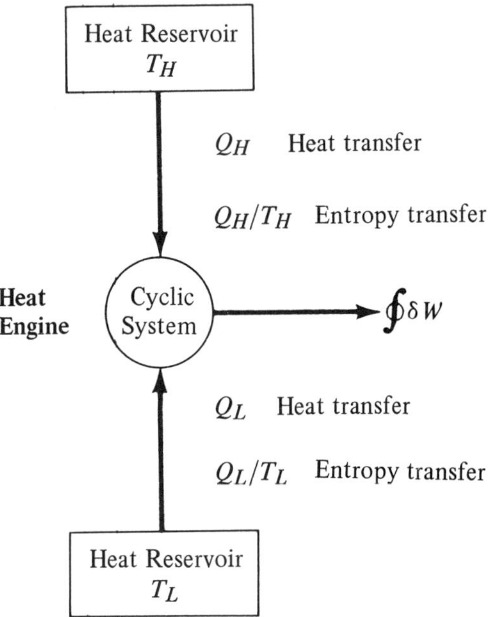

Figure 7.10 Heat transfer and entropy transfer for a heat engine

transfer for the cycle is negative or in the reversible limit zero. Consider the case of the reversible Carnot cycle (Figure 7.9) as a sequence of reversible processes. Entropy is transferred from the high temperature heat reservoir (Q_H positive), decreasing the entropy of the reservoir and temporarily increasing the entropy of the system executing the cycle. This entropy stored within the system is processed to a lower temperature during the cycle and is then transferred out to the low temperature reservoir (Q_L negative), increasing the entropy of the reservoir and returning the entropy of the system to its original value in completing the cycle. In effect a given quantity of entropy has been transferred reversibly (unchanged in quantity) from the high temperature reservoir to the low temperature reservoir. The work transfer from the engine must result because the energy associated with the entropy which was transferred was decreased as the entropy was processed by the cycle from T_H to T_L.

On the other hand if the cycle of the system were irreversible (rather than a reversible Carnot cycle), then equation (6.26) requires that entropy transferred to the T_L reservoir exceed the entropy transferred from the T_H reservoir. The excess entropy rejected to T_L is the entropy generated by the irreversibility, equation (7.31), $\oint dS = 0$. The energy associated with the rejection of additional entropy to the T_L reservoir is at the expense of the net work transfer from the cycle. In effect the entropy generated by the irreversibilities drained off energy that would have been net work transfer in a reversible cycle.

When the cycle of the system of Figure 7.9 operates as a reversible refrigerator, entropy is transferred from the low temperature reservoir (Q_L positive), processed by the cycle of the system and is rejected to the high temperature reservoir (Q_H negative). Since the energy associated with the entropy transfer at T_H is higher than at T_L; the net work transfer for the cycle must be negative. A work transfer into the system is required to transfer the entropy from T_L up to T_H. If the refrigeration cycle is irreversible, the entropy generated appears as extra entropy to be transferred to the T_H reservoir. The energy associated with the extra entropy requires extra work transfer into the irreversible cycle.

We can describe any cyclic thermodynamic energy conversion system as an entropy processing operation. If the system operates reversibly the entropy is conserved during the process. All entropy transferred in is exactly matched by the entropy transferred out. The net work transfer is positive if the entropy is processed from a high energy (high T) to a lower energy (lower T) as in power production. The net work transfer is negative if the entropy is processed from a low energy (low T) to a higher energy (higher T) as in refrigeration systems. Any entropy generated by irreversibilities must be discarded. The energy associated with this extra discarded entropy reduces the net work transfer (reduces work production or increases work input).

7.9 Limitations Imposed by the Second Law of Thermodynamics on the Work Transfer During a Change of State

The limit that the second law of thermodynamics places on the heat transfer during a given process may be combined with the first law of thermodynamics to establish a limit for the work transfer during a given process. As with the heat transfer, the upper limit on the work transfer between two states is not established until the path followed during the change of state is established or other equivalent restrictions are imposed.

Consider first the case of a reversible process. From the first law of thermodynamics, equation (2.2), we have

$$(W_{1-2})_{rev} = (Q_{1-2})_{rev} - (U_2 - U_1) \tag{7.79}$$

Substituting equation (7.64) into equation (7.79), we obtain

$$(W_{1-2})_{rev} = \int_1^2 T\,dS - (U_2 - U_1) \tag{7.80}$$

Thus once the path of a reversible process between two given states is specified, equation (7.64) establishes the heat transfer and the first law of thermodynamics establishes the work transfer. In other words the specification of the path of a process and the fact that it is reversible are sufficient to completely establish all thermodynamic information about the process, interactions included.

In the case of an irreversible process, however, a great deal more information is necessary. From the first law of thermodynamics, equation (2.2), we have

$$(W_{1-2})_{\text{irrev}} = (Q_{1-2})_{\text{irrev}} - (U_2 - U_1) \tag{7.81}$$

Substituting equation (7.66) into equation (7.81), we obtain

$$(W_{1-2})_{\text{irrev}} < \int_1^2 T\,dS - (U_2 - U_1) \tag{7.82}$$

Thus in the case of an irreversible process, specification of the path connecting two given end states is not sufficient to establish the interactions with the environment. Some other means must be used to determine the heat transfer and the work transfer. Given only the path and the end states, the first and second laws of thermodynamics only establish the upper limits for these interactions in the case of an irreversible process.

The limits on the work transfer can also be established for the special cases considered in Section 7.7. For the isothermal process, the heat transfer is given by equation (7.68) which can also be written in the form

$$(Q_{1-2})_{\text{isothermal}} \leq (TS)_2 - (TS)_1 \tag{7.83}$$

since the temperature is constant. The first law of thermodynamics, equation (2.2), can be written in the form

$$(W_{1-2})_{\text{isothermal}} = (Q_{1-2})_{\text{isothermal}} - (U_2 - U_1) \tag{7.84}$$

Combining equations (7.83) and (7.84), we obtain

$$(W_{1-2})_{\text{isothermal}} \leq -[(U - TS)_2 - (U - TS)_1] \tag{7.85}$$

The combination of properties $U - TS$ is itself a property called the *Helmholtz free energy* of the system and is denoted by the symbol F. Then equation (7.85) can be written

$$(W_{1-2})_{\text{isothermal}} \leq -(F_2 - F_1) \tag{7.86}$$

The physical interpretation of equation (7.86) is that in a reversible isothermal process, the work transfer from the system will be given by the decrease in the property F. For this restricted class of processes (isothermal) the decrease in F is the amount of energy which is "free" to be extracted as a work transfer. In any real (irreversible) isothermal process, the work transfer from the system will be less than this limiting reversible work transfer.

Equation (7.86) is most useful for systems which are considerably more complex than the systems which we have considered. It is especially useful for systems with coupling between the thermal, mechanical, electrical, and chemical properties, for example an electrolytic cell. In this case the work transfer in equation (7.86) includes electrical work transfer as well as mechanical work transfer.

An even more restricted class of processes than those just considered is the isothermal process at constant pressure. Obviously this class of processes can occur only in complex systems in which more properties than just the pressure and temperature are required to fix the state of the system. An electrolytic cell would be such a system. For this class of processes, equation (7.86) must apply; however, the uniformity of pressure allows us to evaluate that part of the work transfer $(W_{1-2})_{\text{isothermal}}$ which results from the motion of the system boundary against the constant pressure. The work transfer is thus

$$(W_{1-2})_{\text{isothermal}} = (W_{1-2})_{\substack{\text{isothermal} \\ \text{non } P\,dV}} + P(V_2 - V_1) \tag{7.87}$$

When we combine equations (7.86) and (7.87), we have

$$(W_{1-2})_{\substack{\text{isothermal} \\ \text{non } P\,dV}} \leq -[(U + PV - TS)_2 - (U + PV - TS)_1] \tag{7.88}$$

The combination of properties $(U + PV - TS)$ is itself a property, called the *Gibbs free energy*, of the system and is denoted by the symbol G. Then equation (7.88) can be written

$$(W_{1-2})_{\substack{\text{isothermal} \\ \text{non } P\,dV}} \leq -(G_2 - G_1) \tag{7.89}$$

The physical interpretation of equation (7.89) is that in a reversible isothermal process which is also a constant pressure process, the work transfer not associated with the boundary motion will be given by the decrease in the property G. For example, in an electrolytic cell operating at constant T and P, the amount of energy which is "free" to produce an electrical work transfer will be given by the decrease in the Gibbs free energy for the process. In the real (irreversible) case, the electrical work transfer will be less than this reversible work transfer.

Consider once again the case of a system that can experience heat transfer with a single heat reservoir. This situation is physically important because quite often engineering systems are in thermal communication with only the atmosphere which is often modeled as a heat reservoir. In equation (7.76) we have already established the limit on the heat transfer for this case. Substituting this result into the first law of thermodynamics, equation (2.2), we obtain

$$W_{1-2} \leq -[(U - T_{HR}S)_2 - (U - T_{HR}S)_1] \qquad \textbf{(7.90)}$$

Here T_{HR} is the thermodynamic temperature of the heat reservoir and W_{1-2} is the total work transfer from the system as it changes from state 1 to state 2 by any process involving a heat transfer with a reservoir at T_{HR}. Obviously, when the system is continuously in thermal equilibrium with the heat reservoir, equations (7.90) and (7.85) are identical since the temperature of the system and the reservoir are identical.

If in addition to being in thermal communication with the heat reservoir, the system is immersed in an atmosphere at a uniform pressure P_{atm}, then part of the work transfer W_{1-2} in equation (7.90) will be associated with the motion of the boundary of the system against the pressure of the atmosphere. Since the work transfer to the atmosphere cannot usually be used to any advantage, we may think of the total work transfer as the sum of a useful work transfer and the work transfer associated with the displacement of the atmosphere. Thus,

$$W_{1-2} = (W_{1-2})_{\text{useful}} + P_{\text{atm}}(V_2 - V_1) \qquad \textbf{(7.91)}$$

Substituting equation (7.91) into equation (7.90), we obtain

$$(W_{1-2})_{\text{useful}} \leq -[(U + P_{\text{atm}} V - T_{\text{atm}} S)_2 - (U + P_{\text{atm}} V - T_{\text{atm}} S)_1] \quad \textbf{(7.92)}$$

Since the atmosphere will usually also serve as the heat reservoir, atmospheric temperature has replaced the temperature of the heat reservoir. The combination of properties $(U + P_{\text{atm}} V - T_{\text{atm}} S)$ is itself a property (of the system-atmosphere combination) which we denote by the symbol Φ. Then equation (7.92) can be written

$$(W_{1-2})_{\text{useful}} \leq -(\Phi_2 - \Phi_1) \quad \textbf{(7.93)}$$

For the system-atmosphere combination, equation (7.93) establishes the limit on the useful work transfer for a given change of state as determined by the change in the property Φ. There is, however, a limit on the values assumed by the property Φ. For some particular state, necessarily one in which the pressure of the system is P_{atm} and the temperature of the system is T_{atm}, the property Φ assumes its minimum value Φ_{min}. Then, the change of state determined by $\Phi - \Phi_{\text{min}}$ represents the absolute maximum for the useful work transfer available from a given state of the system-atmosphere combination. Accordingly, Keenan[4] has defined a new property, called the *availability* and denoted by the symbol Λ, which represents the maximum useful work transfer available from a given state of a system-atmosphere combination. Thus,

$$\Lambda = \Phi - \Phi_{\text{min}} \quad \textbf{(7.94)}$$

where for any given system-atmosphere combination Φ differs from Λ by a constant.

Equation (7.94) can then be combined with equation (7.93) so that the useful work transfer for a particular change of state of the system-atmosphere is given by

$$(W_{1-2})_{\text{useful}} \leq (\Lambda_2 - \Lambda_1) \quad \textbf{(7.95)}$$

[4] Keenan, J. H., *Thermodynamics,* John Wiley & Sons, New York, 1941, Chapter 17.

The physical interpretation of equation (7.95) is that the useful work transfer from a reversible process of a system-atmosphere combination is given by the decrease in the availability of the system-atmosphere combination. An irreversible process between the same two end states for the system, but different end state for the atmosphere, would have a smaller useful work transfer.

The difference between the reversible and irreversible useful work transfer for a given change of state is termed the *irreversibility, I,* of the irreversible process experienced by the system-atmosphere combination. The reversible and the irreversible processes are schematically illustrated in Figure 7.11. Thus

$$I = (W_{1-2})_{\text{useful, rev}} - (W_{1-2})_{\text{useful, irrev}} \tag{7.96}$$

The first law for the system plus the atmosphere requires

$$-(W_{1-2})_{\text{useful}} = (U_2 - U_1)_{\text{atm}} + (U_2 - U_1)_{\text{system}} \tag{7.97}$$

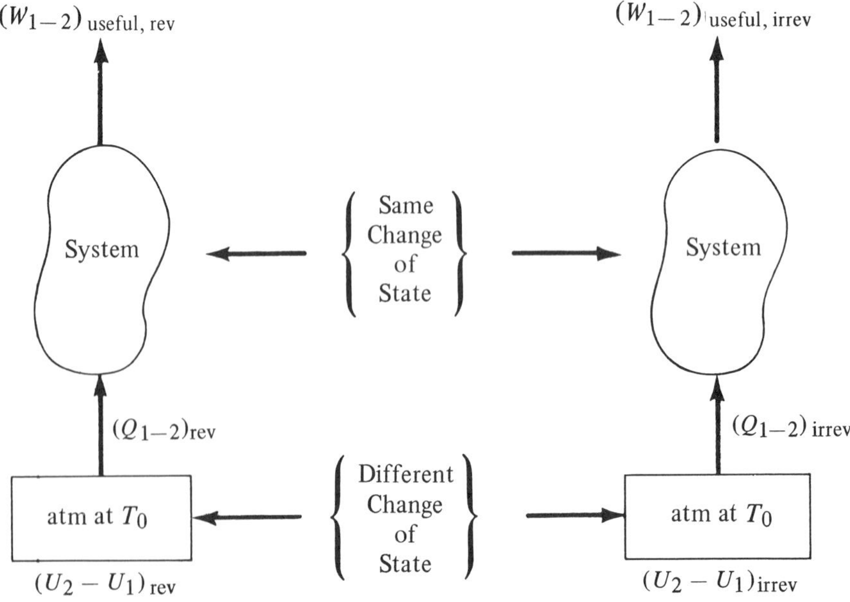

Figure 7.11 Reversible and irreversible processes for a system plus an atmosphere

Since the system has the same change of state in the reversible and the irreversible process, then

$$(W_{1-2})_{\text{useful, rev}} - (W_{1-2})_{\text{useful, irrev}} = (U_2 - U_1)_{\text{atm, irrev}} - (U_2 - U_1)_{\text{atm, rev}} \quad \textbf{(7.98)}$$

For the constant temperature atmosphere

$$(U_2 - U_1)_{\text{atm}} = T_{\text{atm}}\,(S_2 - S_1)_{\text{atm}} \quad \textbf{(7.99)}$$

Thus

$$I = T_{\text{atm}}[(S_2 - S_1)_{\text{atm, irrev}} - (S_2 - S_1)_{\text{atm, rev}}] \quad \textbf{(7.100)}$$

For the reversible process

$$(S_2 - S_1)_{\text{atm, rev}} = -(S_2 - S_1)_{\text{system}} \quad \textbf{(7.101)}$$

Finally we have

$$I = (W_{1-2})_{\text{useful, rev}} - (W_{1-2})_{\text{useful, irrev}} = T_{\text{atm}}\, S_{\text{gen, irrev}}$$

since S_{gen} in the irreversible process is the sum of the entropy change of the system plus the entropy change of the atmosphere.

For the same change of state in the system, all entropy generated by the irreversible processes in the system must be transferred to the atmosphere since only the atmosphere is allowed to have a different state 2 as a result of all irreversibilities. If we divide the system into subsystems and apply equation (7.31) to each subsystem, we can find the origin of the generated entropy and thus localize the loss of useful work with minimum calculation effort. The use of equation (7.92) or equation (7.96) to localize the irreversibility requires additional calculations.

The non zero entropy generation given by equation (7.31) is direct evidence of an irreversible process. However, the irreversibility cannot be related to the loss of useful work until a specification is given for the final location (specifically the temperature) of the entropy which has been generated. In a heat engine the dissipation of a unit of mechanical energy during a high temperature process does not generate as much entropy as the same mechanical process at ambient temperature. The high temperature process produces a smaller loss of useful work because useful work can be realized during the series of processes that transfer the generated entropy from the high temperature to the entropy sink at ambient temperature. On the other hand, no useful work can be realized in transferring entropy generated at ambient temperature to the entropy sink at ambient temperature. In contrast, a unit of mechanical energy dissipated during a very low temperature process in a cryogenic refrigerator generates much more entropy than the same dissipation at ambient temperature. A large quantity of useful work must be used to transfer (or pump) the generated entropy from the low temperature to the entropy sink at ambient temperature.

Each of these functions (F, G, and Λ) which place upper limits on the work transfer are useful for determining the changes of state (within the restricted class) that can occur spontaneously, that is, without a negative work transfer to the system. For example, if the reversible work transfer for a given change of state (within the restricted class) is negative, that change of state cannot occur spontaneously. On the other hand, if the reversible work transfer for the given change of state is positive, the change of state can occur spontaneously with zero work transfer since in accordance with the second law of thermodynamics this zero work transfer is less than the positive reversible work transfer. It follows, then, that the spontaneous process is irreversible.

7E.3 Sample Problem: In a particular automotive engine design, the cylinder contains gas at a pressure of 5×10^5 N/m^2 and a temperature of 1600 K at the end of the power stroke just prior to the opening of the exhaust valve. If the atmosphere is at a pressure of 10^5 N/m^2 and a temperature of 300 K, calculate the maximum useful work transfer that could be extracted from each kg of this "waste" gas. Assume that the gas can be modeled as an ideal gas with $R = 0.287$ kJ/kgK and $c_v = 0.716$ kJ/kgK.

Solution: We need to determine the maximum useful work that can be extracted from this gas as it attains thermal and mechanical equilibrium with the atmosphere. Thus, for the gas $P_1 = 5 \times 10^5$ N/m^2 and $T_1 = 1600$ K; $P_2 = 10^5$ N/m^2 and $T_2 = 300$ K. From the second law of thermodynamics,

$$Q_{1-2} \leq T_{atm}(S_2 - S_1)$$

where from equation (7.25)

$$S_2 - S_1 = m(R + c_v) \ln \frac{T_2}{T_1} - mR \ln \frac{P_2}{P_1}$$

or

$$S_2 - S_1 = 1(0.287 + 0.716) \ln \frac{300}{1600} - 1(0.287) \ln \frac{10^5}{5 \times 10^5}$$

$$= -1.2171 \text{ kJ/K}$$

or

$$Q_{1-2} \leq 300(-1.2171)$$

$$Q_{1-2} \leq -365.13 \text{ kJ}$$

The total work transfer for the gas is the sum of the useful work and the work done against the atmosphere.

$$W_{1-2} = (W_{1-2})_{useful} + P_{atm}(V_2 - V_1)$$

where from the constitutive relation $V = mRT/P$.

$$V_1 = \frac{mRT_1}{P_1} = \frac{(1)(0.287)(1600)}{5 \times 10^5} = 9.184 \times 10^{-4} \text{ m}^3$$

$$V_2 = \frac{(1)(0.287)(300)}{10^5} = 8.610 \times 10^{-4} \text{ m}^3$$

Then

$$W_{1-2} = (W_{1-2})_{useful} + 10^5(8.610 \times 10^{-4} - 9.184 \times 10^{-4})$$

$$W_{1-2} = (W_{1-2})_{useful} - 5.74 \text{ J}$$

Thus from the first law of thermodynamics

$$W_{1-2} = Q_{1-2} - (U_2 - U_1) = Q_{1-2} - mc_v(T_2 - T_1)$$

or

$$(W_{1-2})_{useful} \leq 5.74 \times 10^{-3} - 365.13 - (1)(0.716)(300 - 1600)$$

$$(W_{1-2})_{useful} \leq 565.68 \text{ kJ}$$

Thus the maximum useful work transfer is 565.68 kJ. Clearly, a considerable amount of potential work transfer is being lost in the conventional exhaust process of automotive engines. It is for this reason that turbo charging has been given so much attention in the automotive industry in recent years.

Problems

7.1 Consider a system consisting of two identical balls of putty at 21 C, each ball having a mass of 1 kg. Initially one ball of putty is at rest and the other is traveling at a velocity of 3 m/s. The moving ball strikes the stationary ball producing an inelastic collision so that the two balls stick together and travel on at a reduced velocity. The specific heat of the putty is 2.1 kJ/kgK.

(a) With the assumption of no energy transfer to the surroundings, what is the state of the two balls after the collision?

(b) What reversible process will produce the same change of state?

(c) For the system, what is the change in entropy as a result of the collision process?

(d) On the basis of this entropy change, what can you infer about the reversibility of the collision process?

(e) What is the entropy transfer, $\int \delta Q/T$, to the system (two balls) during the collision process of part (a)? What is the entropy generation within the two balls during the collision?

7.2 A mass m of water at temperature T_{1A} is mixed with an equal mass of water at temperature T_{1B} in an insulated container. Assume that the water can be modeled as a pure thermal system with a constant specific heat c.

(a) What is the state of the water after mixing?

(b) What is the change in entropy of all the water in terms of m, c, T_{1A}, T_{1B}, and T_2? Does this change denote an increase or decrease in entropy?

(c) If $m = 5$ kg, $T_{1A} = 345$ K, $T_{1B} = 288$ K and $c = 4.18$ kJ/kgK, what is the entropy change?

(d) What is the entropy transfer, $\int \delta Q/T$, to the system (2 m) during the mixing process? What is the entropy generated within the system during the mixing?

(e) The water is now divided up into two equal masses and their temperatures are changed back to T_{1A} and T_{1B}, respectively, by means of a reversible process. What is the change in entropy for this process?

7.3 The water shown in Figure 7P.3 drains from the upper tank into an identical tank 15 m below. The initial temperature of both tanks and the water is $T_1 = 375$ K. The heat capacity of the water is 8.374 kJ/K and the heat capacity of each tank is 0.9 kJ/K. At equilib-

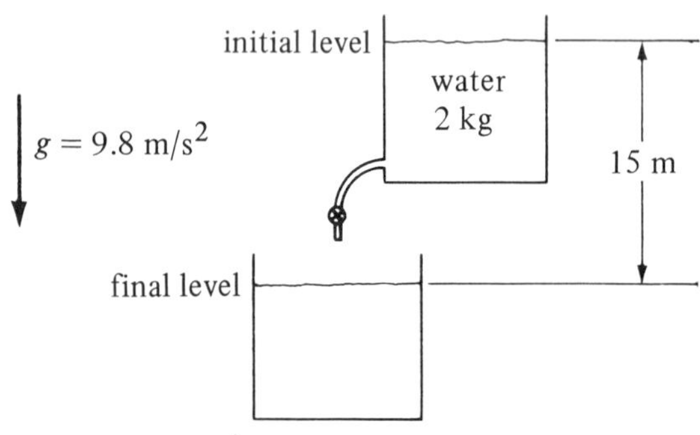

Figure 7P.3

rium the tanks and the water may be modeled as a pure thermal system plus a pure gravitational mass.

(a) What is the state of the water and the lower tank after the water drains into the tank and comes to equilibrium with the tank? Assume no heat transfer to the environment.

(b) Determine the entropy change of the water plus the lower tank when the water has come to rest in the lower tank. Assume no heat transfer to the environment.

(c) Determine the entropy change if the initial temperature had been $T_1 = 300$ K with all other elements remaining the same.

(d) What conclusions can you reach regarding the nature of the reversibility of the process?

7.4 One kg of air at 288 K and 7×10^5 N/m^2 is blown down into the atmosphere at 288 K. In the end, the remaining air in the tank and the atmosphere are all at the same temperature. The system is the one kg of air.

(a) What is the final state?

(b) Describe a reversible process connecting the initial and final states.

(c) What is the total entropy change for the system for the process of part (b)?

(d) What is the entropy transfer, $\int \delta Q/T$, and the entropy generation for the blow down process?

(e) What is the entropy transfer, $\int \delta Q/T$, and the entropy generation for the reversible process of part (b)?

7.5 One kg of water at 40 C is brought into thermal communication with a heat reservoir at 95 C and the water and heat reservoir are allowed to attain thermal equilibrium. Assume that the water is incompressible and can be modeled as a pure thermal system with specific heat of $c = 4.187$ kJ/kgK.

(a) What is the change in entropy of the water for the process by which the water attains thermal equilibrium with the heat reservoir?

(b) What is the change in entropy of the heat reservoir?

(c) What is the change in entropy, the entropy transfer, $\int \delta Q/T$, and the entropy generation of the composite system (water plus heat reservoir)?

(d) If the water had been heated from 40 C to 95 C by first bringing it into thermal equilibrium with a reservoir at 70 C and then with a reservoir at 95 C, what would have been the change in entropy of the composite system consisting of the water and the two heat reservoirs?

(e) Explain how water might be heated from 40 C to 95 C with essentially no change in entropy of the composite system consisting of the water and any heat reservoirs involved. Would this new heating process be reversible?

7.6 An electric current of 10 amp is maintained for 1 sec in a 25 ohm electrical resistor while the temperature of the resistor is kept constant at 0 C by an ice-water bath. Assume that the resistor can be modeled as a pure dissipative system element and a pure thermal system element with a mass of 0.01 kg and a constant specific heat c of 850 J/kgK.

(a) Evaluate the entropy transfer, $\int \delta Q/T$, for the resistor and for the ice-water bath assuming that the bath is thermally isolated from the environment.

(b) What is the change of entropy of the resistor?

(c) What is the change of entropy of the bath?

(d) If the resistor had been insulated, what would have been the change of entropy of the resistor?

7.7 Sketch the paths followed on the T–s diagram by each of the following reversible processes in an ideal gas. The important aspects of the paths are the slope dT/ds and the curvature of the path.

(a) Constant temperature
(b) Constant volume
(c) Constant pressure
(d) Constant entropy
(e) Constant internal energy

7.8 One kg of an ideal gas with a constant specific heat c_v and a gas constant R is initially at a temperature T_1 and a pressure P_1.

(a) Determine the maximum work transfer for this gas in expanding adiabatically to zero pressure.
(b) Determine the limit of the work transfer if this same gas at T_1 and P_1 were expanded isothermally to zero pressure.

7.9 A closed system containing an ideal gas undergoes an adiabatic work transfer. One of the end states of this interaction is given by (P_A, V_A) while the other end state is given by (P_B, V_B). However, it is not known which state is the initial state and which state is the final state.
Given:

$$
\begin{aligned}
P_A &= 6.894 \times 10^5 \text{ N/m}^2 \\
V_A &= 8.49 \times 10^{-2} \text{ m}^3 \\
P_B &= 1.013 \times 10^5 \text{ N/m}^2 \\
V_B &= 2.831 \times 10^{-1} \text{ m}^3 \\
c_v &= 3.15 \times 10^3 \text{ J/kgK} \\
R &= 2071.2 \text{ J/kgK}
\end{aligned}
$$

(a) Determine by the proper thermodynamic arguments the orientation of end states for the process. That is, which state is the initial state and which state is the final state?
(b) Evaluate the work transfer experienced by the gas.

7.10 A spherical steel tank 3 m in diameter contains air at $P = 1.4 \times 10^6$ N/m^2 and $T =$ 21 C. A blow out disk which is 150 mm in diameter bursts and the pressure within the tank decreases to atmospheric pressure in a few seconds. Establish a lower bound on the temperature of the gas which remains in the tank when the pressure first reached atmospheric pressure. For air $c_v = 715.9$ J/kgK, $R = 287$ J/kgK.

7.11 Any given thermodynamic process or class of processes may be termed (1) reversible, (2) irreversible or (3) a violation of the second law of thermodynamics. Classify each of the following processes or group of processes (actual or hypothetical) into one of the above three classes. Prove your answer in each case.

(a) An isothermal process in an ideal gas which doubles the volume of the system. The work transfer is

$$\frac{W_{1-2}}{mRT} = 2$$

(b) A constant entropy process with a negative heat transfer.
(c) An adiabatic process with constant entropy.
(d) A completely adiabatic cycle.
(e) An adiabatic and isothermal process in an ideal gas.
(f) An adiabatic ($\delta Q = 0$) and zero work ($\delta W = 0$) process which increases the temperature of an uncoupled system. By uncoupled it is implied that changes in the temperature of the system do not change any of the mechanical or electrical properties of the system.
(g) A cycle of an uncoupled system with $\oint \delta W > 0$.
(h) A system which can be modeled as a pure thermal system plus a pure gravitational mass decreases in height and increases in temperature during an adiabatic process.

7.12 One tenth kg of air in a closed system is compressed from a state in which its pressure is 10^5 N/m^2 and its temperature is 5 C to a final pressure of 2×10^5 N/m^2. During the process, the air experiences a heat transfer of -1.7 kJ with the environment at a temperature of 5 C. Also, the negative work transfer for the air during this process is 2.6 kJ. Assume that the air can be modeled as an ideal gas with $R = 287$ J/kgK and $c_v = 715.9$ J/kgK.

(a) Determine the final temperature of the air.

(b) Determine the change in entropy of the air for this process.

(c) Use this entropy change to determine whether the process was reversible, irreversible, or represents an impossible situation which violates the second law.

7.13 A tank of compressed air with a volume of 1.0 m^3 is to be used to power a portable tool. The air can experience heat transfer only with the atmosphere at 300 K. The pressure of the atmosphere is 10^5 N/m^2. Determine the maximum work transfer from the air to the tool if the initial state of the air is 10^7 N/m^2 and 600 K. Assume that the air can be treated as an ideal gas with $R = 287$ J/kgK and $c_v = 715.9$ J/kgK.

7.14 A piston and cylinder apparatus contains 1 kg of an ideal gas with $c_v = 653.1$ J/kgK. The following sequence of processes is experienced by the gas.

Process 1: Reversible positive heat transfer of 40 kJ at a constant temperature of 600 K.

Process 2: Reversible adiabatic expansion from 600 K to 400 K.

Process 3: Reversible negative heat transfer of 40 kJ at a constant temperature of 400 K.

Process 4: Reversible adiabatic compression from 400 K to 600 K.

(a) Show the sequence of processes on P-v and T-s diagrams.

(b) Determine the work transfer for each process in the sequence.

(c) Determine the net work transfer for the sequence.

(d) Determine the heat transfer for each process in the sequence.

(e) Determine the net heat transfer for the sequence.

(f) Determine the entropy transfer, $\int \delta Q/T$, and the change in entropy for each process and for the sequence as a whole.

(g) Is the first law of thermodynamics satisfied for the sequence? Explain.

(h) Is the second law of thermodynamics satisfied for the sequence? Explain.

7.15 An ideal gas with constant specific heat $c_v = 5R/2$ executes the reversible cycle ABC shown in Figure 7P.15. Complete the following tables.

State	P (N/m^2)	V (m^3)	T (K)
State A	1.4×10^6	0.1	50
State B	2.8×10^6		
State C	1.4×10^6		

Process	W (J)	Q (J)	ΔU (J)	ΔS (J/K)
Path AB				
Path BC				
Path CA				
Cycle ABC				

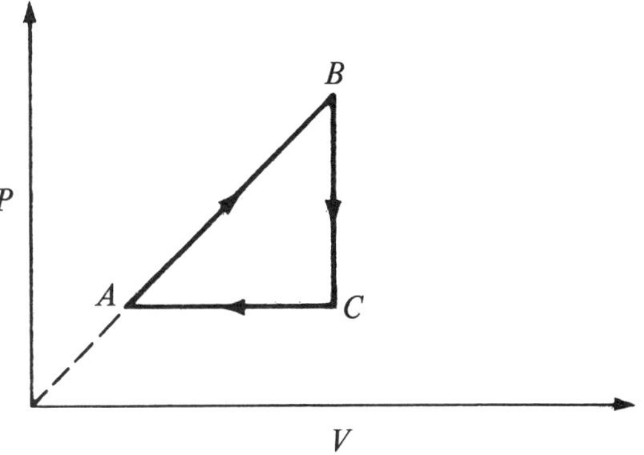

Figure 7P.15

7.16 An electrolytic cell containing an aqueous solution of hydrochloric acid has an open circuit voltage of 1.262 volts at 25 C. When the cell is charged reversibly while a temperature of 25 C is maintained by a constant temperature bath, the heat transfer for the cell is minus one joule per coulomb of charge which flows through the system.

(a) What is the change in the entropy of the cell per coulomb of charge?

(b) What is the change in the entropy of the system consisting of the cell and the constant temperature bath?

(c) The cell is now stored in a warehouse at 25 C, and internal leakage causes an internal current so that the cell loses its charge and returns to its initial uncharged state. What is the change in entropy of the cell per coulomb transferred in this manner? Compare this change in entropy with the entropy transfer, $\int \delta Q/T$, for this process.

7.17 An ideal gas in the piston-cylinder A (Figure 7P.17) expands reversibly and *adiabatically* from $V_1 = 1\ \text{m}^3$, $T_1 = 300$ K, $P_1 = 1.013 \times 10^5\ \text{N/m}^2$ to a final temperature $T_2 = 150$ K and in the process does 125 kJ of work on the piston. The gas has a constant specific heat c_v. The same amount of the same ideal gas in piston-cylinder B expands irreversibly and *adiabatically* from $V_1 = 1\ \text{m}^3$, $T_1 = 300$ K and in the process does 62.5 kJ of work in expanding to the same final volume as the first process.

(a) Find the final volume of the gas in cylinder A after the first reversible expansion.

(b) What is the entropy transfer to the gas in cylinder A during this process?

(c) Find the pressure and temperature in cylinder B after the irreversible adiabatic expansion.

(d) Describe a reversible process or series of processes with the same end states as the irreversible adiabatic process in cylinder B.

(e) What is the entropy transfer to the gas in cylinder B during the process of part (d)?

(f) What is the change of entropy of the gas in cylinder A during the process of part (a) and of the gas in cylinder B during the process of part (c)?

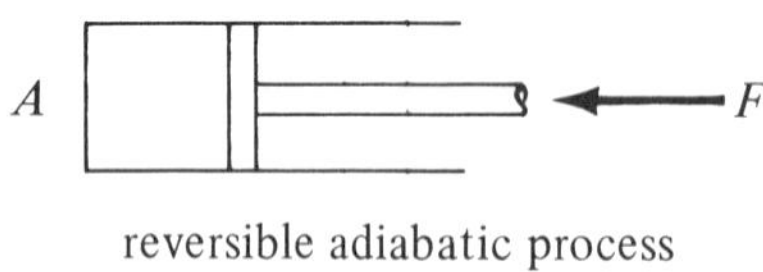

reversible adiabatic process

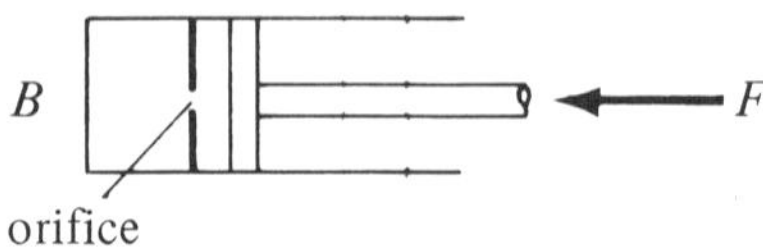

irreversible adiabatic process

Figure 7P.17

7.18 The rigid container shown in Figure 7P.18 contains 1 kg of an ideal gas with a constant specific heat of $c_v = 3.14$ kJ/kgK. The heat capacity of the container is 36 kJ/K. The gas and the container are initially in equilibrium at a temperature of 300 K and a pressure of $10^5\ \text{N/m}^2$. The gas experiences a negative work transfer by means of the stirring action of the paddlewheel.

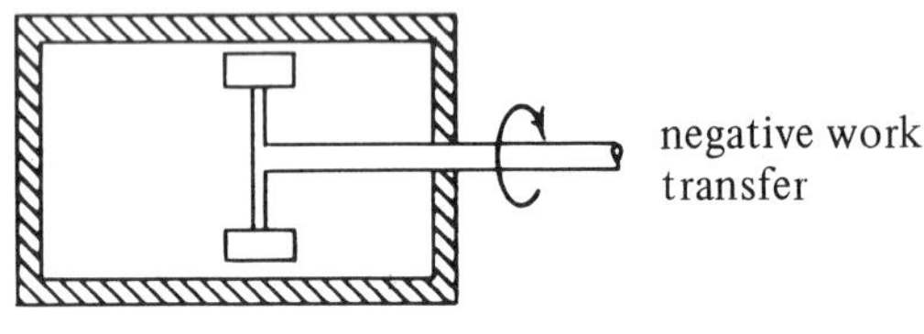

Figure 7P.18

(a) After the gas has experienced a negative work transfer of 500 kJ while in thermal communication with the container, the gas and the container attain thermal equilibrium. Assuming no heat transfer from the outer walls of the container, determine change in entropy experienced by the gas only.

(b) Determine the final temperature and pressure of the gas.

(c) Determine the change in entropy experienced by the gas and the container as a system.

(d) In a second process, the gas experiences a negative work transfer of 500 kJ while thermally isolated from the container. After the work transfer is completed, the gas then attains thermal equilibrium with the container. Again assuming no heat transfer from the outer walls of the container, determine the entropy changes for the gas and the gas-container systems.

(e) By considering entropy changes, show that the gas and container cannot return to the initial state by a positive work transfer to the paddle wheel as the only external interaction.

7.19 The pressure of a certain gas (a photon gas) is a function of temperature only and is related to the energy and volume by $P(T) = U/3V$. A system consisting of this gas is confined in a piston-cylinder apparatus and is subjected to a Carnot cycle between two pressures P_1 and P_2.

(a) Show the Carnot cycle on T-S and P-V diagrams.

(b) Formulate expressions for the work transfers and heat transfers experienced by the gas during the various processes.

(c) Find the energy conversion efficiency of the cycle in terms of pressures.

(d) Evaluate the functional relationship between pressure and temperature.

7.20 As shown in Figure 7P.20, a pure thermal system with mass m_1 and specific heat c_1, and a pure thermal system with mass m_2 and specific heat c_2, have heat transfer interac-

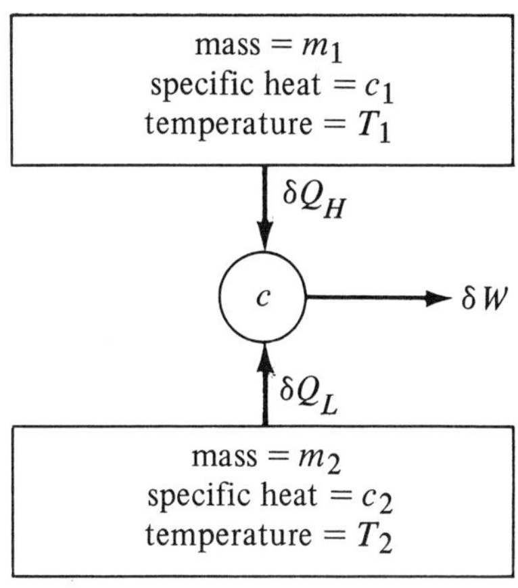

Figure 7P.20

tions with a Carnot cycle c. The Carnot cycle operates with infinitesimal cycles so the interactions in any one cycle are infinitesimal. Initially the temperatures are T_{10} and T_{20}. The Carnot cycle operates with completely reversible interactions until the temperatures equalize to T_e (an unknown temperature).

(a) What is the entropy change for system 1, the entropy change for system 2, and the entropy change for the Carnot cycle?

(b) Use part (a) results to show that T_e is

$$T_e = (T_{10})^a(T_{20})^b$$

Here $a = C_1/(C_1 + C_2)$ and $b = C_2/(C_1 + C_2)$, with $C_1 = m_1c_1$ and $C_2 = m_2c_2$.

(c) Show that the work transfer from the Carnot cycle must be

$$W = C_1T_{10} + C_2T_{20} - (C_1 + C_2)T_e$$

(d) What work transfer is required to return system 1 and system 2 to T_{10} and T_{20} from T_e?

(e) If the two systems are equilibrated to $T_{e\ \mathrm{irrev}}$ by means of an irreversible cycle rather than the Carnot cycle, is $T_{e\ \mathrm{irrev}}$ larger or smaller than T_e of part (b)? Is the work transfer larger or smaller than W of part (c)?

7.21 An ideal gas is confined in an ideal piston-cylinder. The gas is compressed quasi-statically so that thermal equilibrium is maintained between the gas and the cylinder. No external heat transfer takes place between the piston-cylinder and the environment. The cylinder contains one kg of gas with a constant specific heat, $c_v = 3.14$ kJ/kgK and with a gas constant $R = 2.077$ kJ/kgK. The piston and cylinder may be modeled as a pure thermal system with a heat capacity, $C = 9$ kJ/K.

(a) If the initial state of the gas is an equilibrium state at P_1, V_1, and T_1, determine the entropy change, $S - S_1$, of the system (piston, cylinder, and the gas) in terms of T, T_1, V, and V_1. Rewrite this answer in terms of P, P_1, V, and V_1.

(b) Determine the pressure-volume relation for the gas of this system (piston, cylinder, and gas) during the externally adiabatic quasi-static compression or expansion process.

(c) Determine the entropy change, $S - S_1$, for the gas alone as a system.

7.22 The perfectly isolated system shown in Figure 7P.22 consists of a cylinder divided into two compartments by means of a frictionless, massless, thermally insulated piston of area 0.001 m². One compartment of the system contains an ideal gas with $c_v = 3.153$ kJ/kgK and $R = 2.078$ kJ/kgK. The other compartment is perfectly evacuated with the exception of a pure translational spring with a spring constant of 900 N/m. The free length of the spring is 0.3 m. The initial state of the gas is $T_1 = 40$ C, $P_1 = 1.4 \times 10^5$ N/m², and $V_1 = 10^{-4}$ m³. The piston is held in equilibrium in the initial state by means of a pin. At some instant of time, the pin is removed thereby allowing the gas and spring to come to a state of mutual mechanical equilibrium.

(a) For the system consisting of the gas and the spring determine whether the work transfers, heat transfers, and changes in stored energy are positive, negative, or zero for the process by which equilibrium is attained.

(b) Determine the pressure and temperature of the gas in the final equilibrium state.

(c) What is the change in entropy experienced by the gas for the process by which equilibrium is attained? Use this change in entropy to determine whether this process is reversible or irreversible.

HINT: Be very careful of the signs that you assign to the various quantities involved.

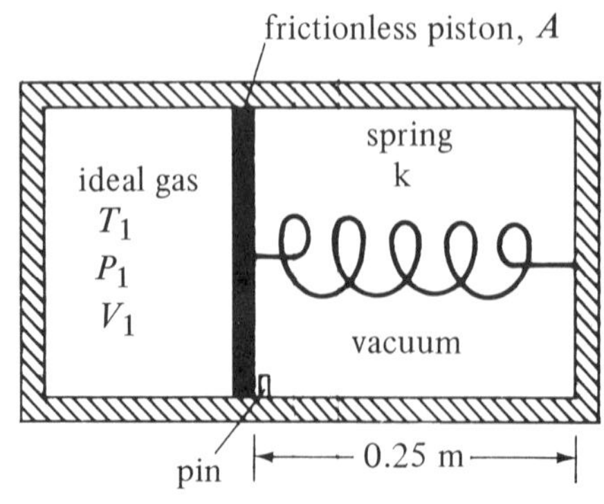

Figure 7P.22

7.23 An ideal gas ($c_v = 715.9$ J/kgK and $R = 287$ J/kgK) is initially in tank A shown in Figure 7P.23. Tank B is initially evacuated. The valve in the pipe between the tanks is opened and the pressure is allowed to equalize. When the pressure equalizes, the gas in

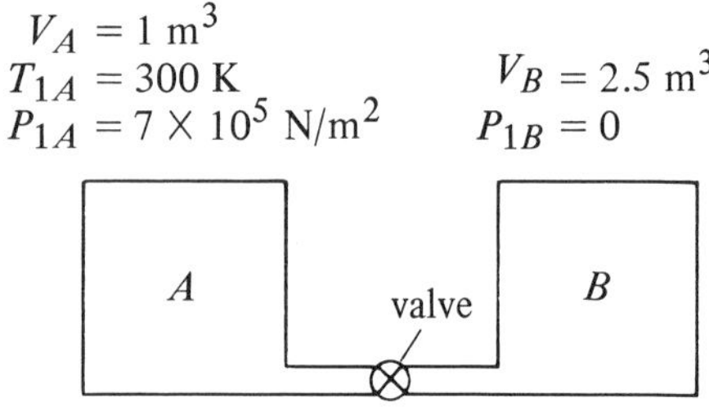

Figure 7P.23

each tank comes to equilibrium internally but is thermally isolated from the gas in the other tank. Assume no heat transfer between either tank and the gas. Determine the final uniform pressure and the final uniform temperature of the gas within each of the tanks. (The two final temperatures are not the same.)

Hint 1: Select the gas in tank A at state 2 as one system and the remainder of the gas as a second system.

Hint 2: Write the first law for both systems taken together. Express all unknown products mT in terms of P and V and solve for the final pressure.

CHAPTER 8

Thermodynamic Properties of Pure Substances

8.1 Introduction

In developing the first and second laws of thermodynamics in the preceding chapters, we have used simple models to represent the thermodynamic behavior of a system. The constitutive relations for the equilibrium properties of these models embody the essential elements of thermodynamic behavior and serve to illustrate the use of the laws of thermodynamics. Unfortunately, because of their simplicity, these special models are incapable of representing certain significant aspects of thermodynamic behavior. Accordingly, in this chapter, we consider the constitutive relations for the more complex simple system model.

By definition a *simple system* is a thermodynamic system without significant effects from the following phenomena:

(1) Surface forces from capillarity and surface tension.
(2) Forces due to gravitational, electric, and magnetic fields.
(3) Shear stresses resulting from the distortion of solids.
(4) Bulk motion.

In keeping with our simplified approach, the simple system model is still an idealization of thermodynamic behavior. In the model, the state of stress is hydrostatic everywhere and all body forces due to external force fields are taken to be insignificant. The state of stress is hydrostatic if the force per unit area acting on a test plane is independent of the orientation of the plane at each point within the system. A solid under uniform three-dimensional compression or tension will not have shear stresses. This classification may seem highly restrictive since these phenomena are present in so many physical situations; however, experience shows that the simple system represents the significant coupled thermodynamic behavior of many systems. The most significant characteristic of the simple system model is that the pressure is uniform and hydrostatic at equilibrium and all reversible work transfer is given by $W = \int P\, dV$. Here P is the uniform pressure of the system and V is the volume of the system.

When no chemical reactions are involved, the simple system model becomes the pure substance model. The definition of a *pure substance* is:

> A pure substance is a simple system which is uniform and invariable in chemical composition.

Further, the internal energy, U, of the pure substance can be changed *reversibly* only by heat transfer or by work transfer associated with the uniform pressure of the system and the displacement of the system boundary, that is, the reversible work transfer is given by the integral $\int P\, dV$. This restriction on the *reversible* work transfer requires that bulk motion, gravity, surface tension, solid shear stresses, electric fields, and magnetic fields all have an insignificant influence on the thermodynamic equilibrium properties of the pure substance. With all of these influences eliminated, the requirement for mechanical equilibrium is that the pressure is uniform throughout the system and that the force on the boundary of the system is directly related to this uniform pressure. In addition, the requirement for thermal equilibrium is that the temperature is uniform throughout the system.

In some cases the influence of bulk velocity and gravity are included; however, the energies associated with these effects are considered to be uncoupled from the internal energy, U, in the manner described in Chapter 4. Each of these effects, which are ignored for the pure substance, represent reversible methods of work transfer in addition to $\int P\,dV$, and thus would add to the complexity of the interactions between the system and the environment.

The pure thermal system element and the ideal gas, which we have used previously, are both special cases of the pure substance. In particular, the pure thermal system element is in effect the degenerate case of the pure substance in which the volume is constant and no reversible work transfer is possible. In the more general case, the pure substance describes the properties of a chemically pure material in the gaseous, liquid, and solid forms, or any combination of the three. In addition, the pure substance describes the properties of a mixture of gases if no liquefaction or solidification takes place. For example, air may be modeled as a pure substance in the gaseous state; however, if some of the air condenses into liquid, the liquid will be richer in oxygen than the remaining gas. This condition then violates the requirement that a pure substance be uniform in composition. The pure substance model also describes the properties of a solution if no solidification or vaporization takes place. A solution of salt in water may be modeled as a pure substance, but if freezing takes place, the solid will be of lower salt concentration than the remaining liquid, again violating the uniform composition requirement. On the other hand, pure H_2O may be modeled as a pure substance in the solid, liquid, or gaseous form or any combination of the three.

8.2 Independent Properties of the Pure Substance Model

In establishing the independent properties of the pure substance model, we make use of a fundamental principle, the *state principle*. This principle, although deduced by empirical means, is based upon the observation of so many thermodynamic systems that it can be regarded as a fundamental law of thermodynamics. *The state principle,* stated rather simply, is:

> Any two independent properties are sufficient to establish a stable equilibrium thermodynamic state of a simple system. For each identifiable departure from the requirements for a simple system, one additional independent property is necessary.

Note that the state principle is restricted to *stable* equilibrium thermodynamic states rather than to equilibrium states in general. Stable equilibrium states are those equilibrium states for which a finite change of state can occur only if there is some corresponding change of state in the environment. Since most equilibrium thermodynamic states are of this type, this restriction poses no serious limitations from a thermodynamic point of view. The significance of the state principle itself is contained in the fact that it establishes in a formal way the minimum number of properties necessary to describe the state of a thermodynamic system.

A convenient, but not necessarily unique, interpretation of this principle is that there is one independent property associated with each reversible mode of energy interaction. The only reversible mode of work transfer for the simple system is given (in differential form) by $P\,dV$. In this case, the volume V can be regarded as the independent property describing the work transfer. Similarly, we have seen in Section 7.7 that the reversible heat transfer for any system is given (in differential form) by $T\,dS$. For the heat transfer interaction, the entropy assumes the role of the independent property.

Thus for the pure substance, the simple system of interest here, the first law of thermodynamics in differential form becomes

$$dU = T\,dS - P\,dV \tag{8.1}$$

The first law of thermodynamics guarantees that the internal energy is a property; consequently, equation (8.1) has an integrated form

$$U = U(S,V) \tag{8.2}$$

Equation (8.2) expresses the internal energy of a pure substance in terms of the two independent properties entropy and volume. According to the state principle, only two independent properties are necessary to describe the equilibrium state of a pure substance; equations (8.1) and (8.2) are just such a description. Therefore, equations (8.1) and (8.2) relate equilibrium states and are valid regardless of whether the interactions experienced by the pure substance during the change of state are reversible or irreversible. Another way of interpreting this result is to note that any one equilibrium state of a pure substance can be reached (in principle) from any other equilibrium state by a series of reversible processes. In this manner, we can establish for every equilibrium state the value of any property (relative to some datum state). Equation (8.1) is an interrelation among the system properties that must be satisfied for every infinitesimal change of state. From the change of state itself, we cannot establish whether reversible or irreversible interactions were involved; however, equation (8.1) is true regardless of the nature of the interactions. The important point is that the product $T\,dS$ represents the heat transfer and the product $P\,dV$ represents the work transfer only for reversible interactions. In the case of irreversible interactions, the product $T\,dS$ exceeds the actual heat transfer by the same amount that $P\,dV$ exceeds the actual work transfer.

Although we have just established the entropy and volume as a set of independent properties for a pure substance, we are by no means restricted to this set. That is, any other two properties of the pure substance could be selected as the independent set provided, of course, that the two properties really are independent. For example, in the case of the ideal gas, the temperature and the pressure form a set of independent properties, but the temperature and the internal energy do not. This is because the internal energy is a function of only temperature for the ideal gas. As we shall see later, the temperature and the pressure are not a set of independent properties for the states of a pure substance in which the vapor and liquid forms coexist in equilibrium.

Although equation (8.1) has been discussed in the context of a pure substance, the equation is equally applicable to any simple system. In fact, equation (8.1) occupies a rather special position in thermodynamics, and for this reason, some authors refer to it as the *canonical relation*. The significance of equation (8.1) lies in the fact that there is a great deal of information that can be obtained from it. For example, if we mechanically restrain a simple system so that its volume is fixed, it cannot experience any reversible work transfer. If we further isolate it so that it cannot experience any heat transfer or irreversible work transfer, either internal or external, its entropy is fixed.

Under these circumstances, equation (8.1) can be shown to indicate that the internal energy is an extremum. In more advanced treatments, stability considerations reveal that this extremum is a minimum. The very foundation of thermodynamics is that, under these imposed restrictions, the system will achieve a unique state as determined by the entropy and volume, and that the internal energy of this state is smaller than all other states of the system with the same entropy and volume. By definition, this state is the stable equilibrium state and all other states with the same entropy and volume are nonequilibrium states. Thus, equation (8.1) is not necessarily satisfied for nonequilibrium states.

Equation (8.1) serves to distinguish between *extensive* properties and *intensive* properties. The extensive properties energy, entropy, and volume are directly proportional to the mass present in the system. For example, if a pure substance system at equilibrium is subdivided into two subsystems the value of each extensive property of the subsystem will be less than the value for the total system. The value of an extensive property for a composite system is the sum of the values for that property for the subsystems.

The intensive properties, temperature and pressure, in equation (8.1) are the properties with values that are not changed by subdivision of the system. The intensive properties can be defined at a point within a system and have values which are uniform for the equilibrium states of a pure substance. For nonequilibrium states the intensive properties may be non uniform or may be undefined.

Another noteworthy feature of the canonical relation is that the thermodynamic properties appear as pairs which are said to be canonically conjugate. For example, the temperature and the entropy are canonically conjugate properties that completely describe a reversible heat transfer interaction while the pressure and the volume are canonically conjugate properties that completely describe a reversible work transfer interaction. Note that in each case the pair of properties consists of one intensive property such as the temperature or pressure, and one extensive property such as the entropy or volume. We can generalize these observations to the case of closed systems which cannot be classed as simple. For each independent, reversible interaction capable of altering the internal energy of the system, there is a pair of properties, one intensive and one extensive, that completely describes the interaction. The set of independent extensive properties associated with these interactions represents the minimum number of properties necessary to specify the stable equilibrium state of a given system. Since all of the properties affect the internal energy, they are coupled. The simplest form for the constitutive relations are then equations that express each of the intensive properties in terms of all of the independent extensive properties. For each aspect of system behavior that is uncoupled, there is not only an independent property that is used to describe the interaction but also a separate stored energy form. Note that the conditions of equilibrium require that the intensive property associated with each possible reversible interaction be uniform. When two systems (or subsystems) are not in a state of mutual equilibrium, the difference in their intensive properties serves as the driving potential for the interaction which will bring the systems to a state of mutual equilibrium characterized by equal intensive properties.

Equation (8.1) can also be expressed in terms of specific properties, which are the extensive properties per unit mass of the system. If we divide equation (8.1) by the mass of the system we have

$$d\left(\frac{U}{m}\right) = T\,d\left(\frac{S}{m}\right) - P\,d\left(\frac{V}{m}\right) \tag{8.3}$$

or

$$du = T\,ds - P\,dv \tag{8.4}$$

Here u is the specific internal energy, U/m, s is the specific entropy, S/m, and v is the specific volume V/m. As we will discuss in detail later, the specific properties may be multivalued in a pure substance in an equilibrium state, when more than one of the liquid, gaseous or solid forms are present.

8.3 Equilibrium Liquid-Vapor States

One of the primary reasons for formulating the pure substance model is to be able to model the thermodynamic behavior of a substance that can experience changes between the solid, liquid, and gaseous forms. None of the other system models that we have considered previously is capable of experiencing such a transformation.

The transformation between gaseous and liquid forms has a high degree of thermodynamic coupling and is thus quite important in thermo-mechanical energy conversion. It is natural to begin the discussion of the pure substance with the consideration of liquid-vapor equilibrium states.

As a consequence of the state principle, any two independent properties are sufficient to fix the state of a pure substance. Thus, an expression relating any other property to the two independent properties selected for the characterization of the state constitutes a valid constitutive relation, or equation of state. It is often convenient to graphically represent the equilibrium states of a pure substance in terms of this constitutive relation. Accordingly, we most commonly use a two-dimensional representation in which one independent property appears as the abscissa, the dependent property appears as the ordinate, and the other independent property appears as a parameter indexing the family of curves. The constitutive relations for pure substances are usually determined by actual measurements of the properties. The pressure-volume-temperature relation is very common since these properties are most easily measured.

There is no one universal constitutive relation which will model all of the pure substances. However, with a few notable exceptions, the equilibrium thermodynamic behavior of most pure substances is qualitatively the same. Thus, to illustrate the general thermodynamic behavior of a pure substance (in a qualitative fashion, of course) it is sufficient to perform a hypothetical experiment on a hypothetical pure substance. To this end, consider the following situation in which a unit of mass of a pure substance, for example a gas, is enclosed in a piston-cylinder apparatus and the whole configuration is in thermal equilibrium with a heat reservoir at the temperature T_1 (cf. Figure 8.1).

We now proceed to compress the pure substance by decreasing the volume. If we perform this compression at a sufficiently slow rate, the temperature of the substance will remain fixed at T_1 since the system is in thermal communication with the heat reservoir. As we might expect, the pressure of the sub-

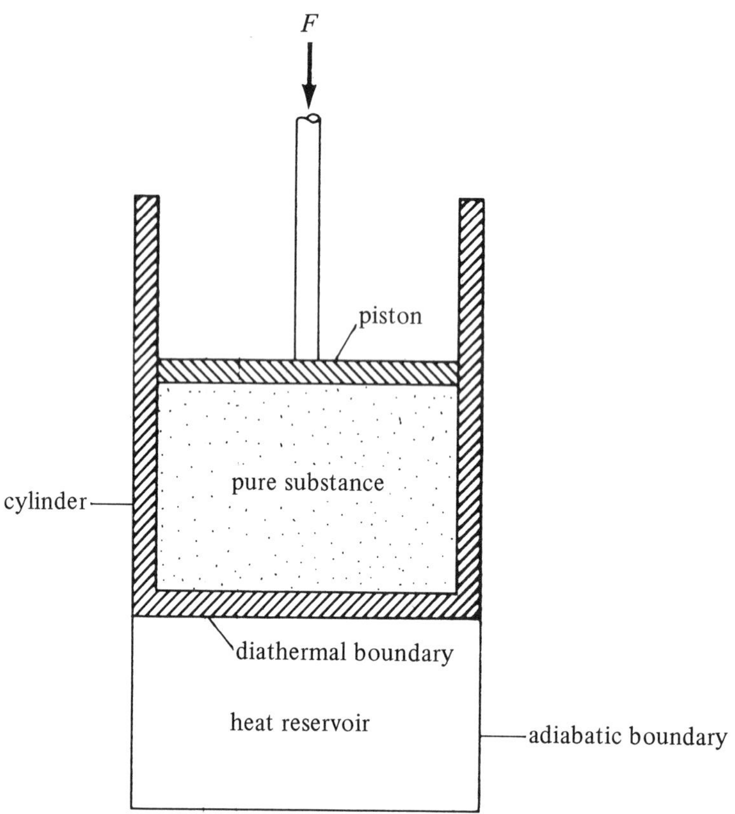

Figure 8.1 Piston-cylinder apparatus for isothermal processes of the pure substance model

stance will increase during this compression process. To be more specific, if we decrease the volume from v_a to v_b (Figure 8.2) while the temperature remains constant at T_1, the pressure increases accordingly from P_a to P_b. In fact, if we were to monitor the pressure constantly during this process *a-b*, we would find that it would trace out the path *a-b* shown on Figure 8.2. Clearly, because of the manner in which we are performing the experiment, the states which lie on the line *a-b* are equilibrium states.

If we were to continue to decrease the volume in the manner described above, we would continue to see a corresponding increase in pressure until the state *g* is reached. Suddenly, when this state is reached, the pressure ceases to increase with a further decrease in volume. In fact, if we were to continue to de-

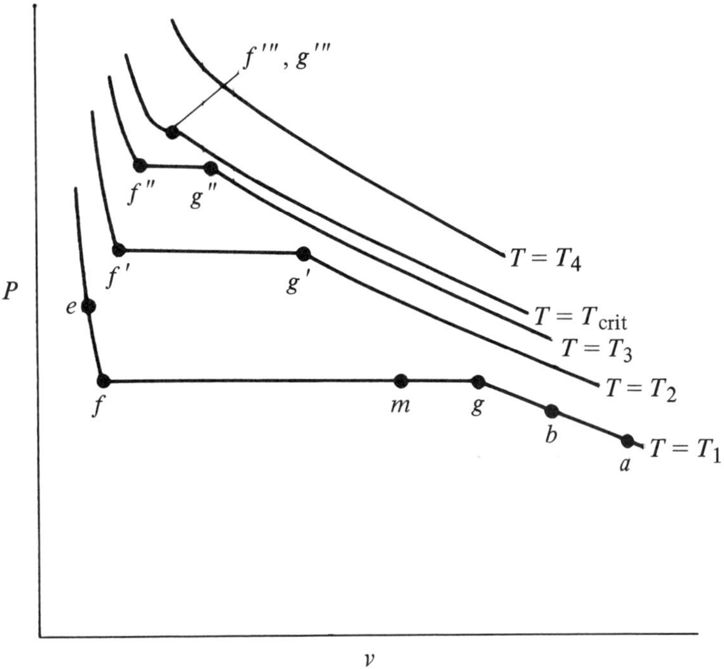

Figure 8.2 Isotherms for the pure substance model

crease the volume to v_m for example, the pressure in state m would still be identical to that in state g. Obviously, something drastic happened when the system reached state g. Closer inspection of the apparatus reveals that when the system reached state g, the substance was present only in the gaseous form. For all volumes less than v_g the liquid form of the pure substance was present. Further, as the system progressed from state g to state m, the amount of the liquid form increased while the amount of gas, or vapor as it is often called, decreased.

If we examine the liquid form more closely, we find that the state of the liquid is homogeneous, that is, uniform throughout its extent. Similar inspection of the vapor form reveals that it, too, is homogeneous in state. The conditions of thermal and mechanical equilibrium require that the temperature and the pressure for these two forms be identical. However, in spite of the homogeneity that exists in the two forms, the only properties that are the same for these two forms of the pure substance are the temperature and pressure. All of the specific properties, such as entropy, energy, and volume, have different values for the two forms, but each of these properties is uniform throughout the extent of the particular form. Thus, the liquid and the vapor forms have the same intensive state, that is, temperature and pressure, but different thermodynamic states. In order to better describe this situation, J. Willard Gibbs introduced the concept of *phase* which he defined in the following manner:

> A *phase* is the collection of all macroscopic parts of the system with the same homogeneous thermodynamic state. That is, all parts of a system which have identical and uniform values for each of the specific properties as well as identical and uniform temperature and pressure are said to constitute one phase.

Thus, in state m described above, the liquid form and the vapor form constitute separate, distinct phases, namely, the liquid phase and the vapor phase. State m is said to be a *two-phase state* or *mixed state* in which the two phases are in equilibrium.

As we continue to decrease the volume of the pure substance from state m in the manner described previously, we find that both the pressure and temperature remain constant. All during this process, the amount of liquid present continues to increase while the amount of vapor present continues to decrease. Suddenly, when we reach state f, we find that the vapor has vanished. Only the liquid phase is present. If we now continue to decrease the volume of the pure substance to v_e, for example, we find that the pressure again increases. Then, it is apparent that unlike single phase states the pressure and temperature are *not* independent properties when two phases of a pure substance exist together in equilibrium. Thus, a two-phase state can be specified only by temperature *or* pressure *and* some other property such as volume.

In proceeding from state g to state f, the pure substance is said to have experienced a *change in phase* or *phase transition* since only the vapor phase is present in state g while only the liquid phase is present in state f. A definite change in the specific properties is associated with this phase change which took place both at constant temperature and at constant pressure. The locus of equilibrium states traced out by the system while in thermal communication with the heat reservoir at the temperature T_1 is represented on the pressure-volume plane as a continuous line known as an *isotherm*. Notice from Figure 8.2 that at the two states g and f, the *isotherm*, $T = T_1$, undergoes a discontinuous change in slope. That is, between states g and f, $(\partial P/\partial v)_T = 0$ while everywhere else $(\partial P/\partial v)_T < 0$. A discontinuous change of slope of this type is characteristic of phase changes which occur at constant temperature and pressure.

If we were to perform the same experiment while the system was in thermal communication with a heat reservoir at a temperature T_2, $T_2 > T_1$, we would observe the same general behavior. That is, the pressure would continue to increase with a decrease in volume until some state g' is reached. For this state, the slope of the isotherm $T = T_2$ changes discontinuously to zero as the liquid phase appears. The pressure and temperature remain constant as the volume is decreased until the state f' is reached. For this state the slope of the isotherm again changes discontinuously, this time from zero to something less than zero, as the vapor phase vanishes. From this state on, the pressure increases as the volume is decreased. Once again, the states between g' and f' are two phase states with liquid and vapor phases. As the volume is decreased from v at g' to v at f', the relative amount of the liquid phase present increases from zero at g' to all liquid at f'. Associated with this phase change at constant pressure and temperature is a definite change in the specific properties such as the specific volume. Of course the pressure and temperature are not independent properties during the phase change $g' \rightarrow f'$. The locus of equilibrium states for the system while in communication with the heat reservoir at T_2 is shown on Figure 8.2 as the isotherm $T = T_2$.

If we were to perform this same experiment on this apparatus while it was in thermal communication with a heat reservoir at a temperature T_3, where $T_3 > T_2 > T_1$, the system would trace out the isotherm $T = T_3$ as shown in Figure 8.2. Notice that this isotherm possesses the same general characteristics as the previous two isotherms, $T = T_1$ and $T = T_2$. However, in comparing the three isotherms $T = T_1$, $T = T_2$, and $T = T_3$, we note the occurrence of a rather curious phenomenon. The change in specific volume associated with the phase change at constant pressure and temperature decreases as the temperature at which the experiment is executed increases. Changes in all other specific properties behave in a similar manner. Apparently, the liquid and vapor phases are becoming more and more similar in nature as the temperature is increased. The question then arises as to whether there exists some critical temperature $T = T_c$ for which there is no difference in specific properties between the liquid and vapor phases. For such a temperature there must exist a state for which there is no distinction between the liquid and vapor phases. Such a critical temperature does indeed exist, and the *critical state,* is the state at the critical temperature for which the liquid and vapor phases are identical. At this state the isotherm $T = T_c$ undergoes an inflection with $(\partial P/\partial v)_T = 0$ and $(\partial^2 P/\partial v^2)_T = 0$ (cf. Figure 8.2). No other state possesses this feature. Thus, the critical state is a unique state and every pure substance is observed to possess only one.

For experiments carried out at higher temperatures, we find that the isotherms, for example $T = T_4$, $T_4 > T_c$, possess no discontinuity of slope. Everywhere $(\partial P/\partial v)_T < 0$ if $T > T_c$. Along such isotherms, sometimes called *supercritical isotherms,* the pressure increases continuously as the volume is decreased. As is shown on Figure 8.2, no liquid-vapor phase change occurs.

If we again look at the collection of isotherms that we have generated in our hypothetical experiment, we note that we can form the locus of states for which the slope $(\partial P/\partial v)_T$ changes discontinuously. These states, because of this change in slope, are very special states that we call *saturation states.* The saturation states are those states of the pure substance for which $(\partial P/\partial v)_T = 0$ *and* only one phase is present. The locus of saturated states for a typical pure substance is shown in Figure 8.3. Any saturated state for which $v < v_c$ is known as a *saturated liquid state.* On the other hand, any saturated state for which $v > v_c$ is known as a *saturated vapor state.* A change in phase from a saturated liquid state to a saturated vapor state is commonly known as *boiling, evaporation* or *vaporization.* Conversely, a change in phase from a saturated vapor state to a saturated liquid state is commonly known as *condensation* or *liquefaction.* The locus of saturated liquid states and the locus of saturated vapor states meet at the critical state since in this state the two phases, liquid and vapor, are identical. By way of convention we denote the specific properties of the saturated vapor state with the subscript g (gas) and the specific properties of the saturated liquid state by the subscript f (liquid). Thus, the specific volume of the saturated vapor state would be v_g while that of the saturated liquid state would be v_f.

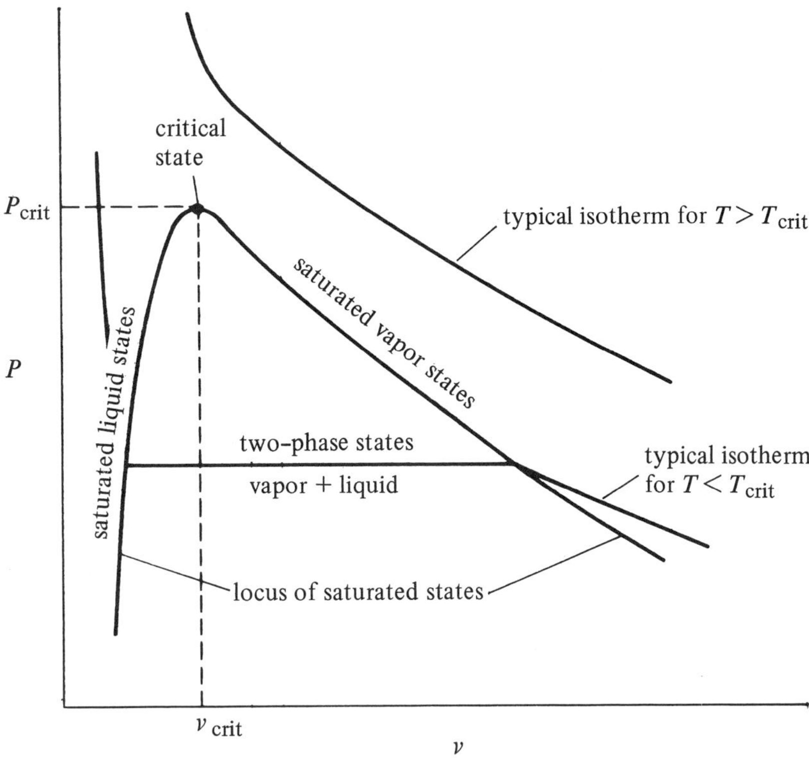

Figure 8.3 Pressure-volume diagram of vapor-liquid two-phase states for pure substance model

For a given temperature, there is at most one and only one set of liquid-vapor saturation states. Further, there is a unique pressure, known as the *saturation pressure* or *vapor pressure,* that corresponds to a given temperature for which a saturated liquid state and a saturated vapor state can coexist. Conversely, to any one pressure at which the two types of saturated states can coexist there corresponds a single temperature known as the *saturation temperature, dew point temperature,* or *boiling point temperature.* There is a definite relationship between the saturation temperature and saturation pressure.

In Figure 8.3 there are other regions of interest in addition to the region of two-phase states. For example, at a given temperature, those states for which $v > v_g$ are known as *superheated vapor* states. Note that for these states the temperature is greater than the saturation temperature at the pressure of the state—hence the name superheated vapor. The difference in these two temperatures is often called the *degree of superheat*. All superheated vapor states are single phase states since they involve the vapor phase only.

Another region of interest in Figure 8.3 involves those states at a given temperature for which $v < v_f$. Note that for these single phase (liquid) states the pressure is greater than the saturation pressure that corresponds to the temperature of the state. For this reason states of this group are often called *compressed liquid states*. For these same states we note that the temperature of the state is lower than the saturation temperature which corresponds to the pressure of the state. Accordingly, these states are sometimes referred to as *subcooled liquid states*.

There is one other region of interest in Figure 8.3. This region involves states which are most appropriately termed supercritical states, but because of their special nature, we shall delay their discussion until later.

8.4 Critical State

We have already introduced the existence of a special state, known as the critical state, for which the saturated liquid and vapor phases become identical. We have also pointed out that the isotherm for this state, i.e. the critical isotherm, possesses an inflection point for which both the slope and the curvature of the isotherm vanish. That is, $(\partial P/\partial v)_{T=T_{\text{crit}}} = 0$ and $(\partial^2 P/\partial v^2)_{T=T_{\text{crit}}} = 0$. The critical state is specified by means of its pressure, the critical pressure, and its temperature, the critical temperature. Since the common substances with critical temperatures near room temperature have rather high critical pressures (on the order of 60 MPa), the critical state is not within our everyday experience. However, because the critical state exhibits such unusual behavior, it is worthwhile to study the nature of this state and its neighbors even though such states are not encountered at moderate temperatures and pressures.

The physical behavior of a pure substance in states near the critical state can be illustrated by considering several simple thermodynamic processes. If a substance at state a, with spe-

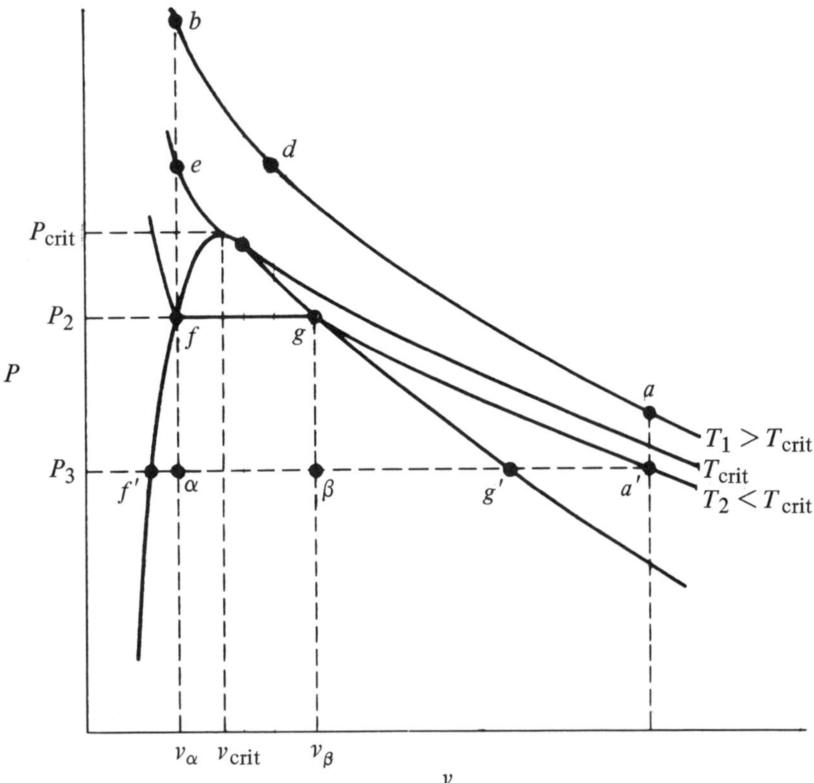

Figure 8.4 Thermodynamic processes for the pure substance model

cific volume v_a ($v_a > v_{crit}$) as shown in Figure 8.4, is compressed at constant temperature T_1 ($T_1 > T_{crit}$) to a specific volume v_b ($v_b < v_{crit}$) the state of the system will remain homogeneous throughout the process. One could then say that the substance had been compressed to state *b* without any condensation. On the other hand, suppose the substance in state *a* had been first cooled to state *a'*, where $v_a = v_{a'}$ and $T_{a'} = T_2 < T_{crit}$. If the substance in this state is now compressed at constant temperature, the substance will begin to condense when state *g* (saturated vapor) is reached, and will continue to condense at constant pressure until state *f* (saturated liquid) is reached. The system at state *f* may now be heated at constant volume to state *b* without passing through any two phase states. Thus, in the first process (*a* to *b*) the substance at *a* was still vapor when it reached state b. In the second series of processes, the substance at state a' condensed to liquid before reaching the same state b. Is the substance at state b liquid or vapor? The only completely satisfactory way out of this paradox is to abandon the simple idea that every state of a given material can be uniquely classified as a solid, liquid, or vapor. In fact, at the critical point the system is entirely liquid and vapor at the same time since the distinction between liquid and vapor has vanished. It follows that any liquid can be changed to a vapor without a change of phase (i.e. passing through two-phase states) by a series of processes which take place at temperatures and pressures in excess of their critical values. In a strict sense the terms liquid and vapor can be applied only to the phases of the two-phase states. The location of a boundary between liquid and vapor in the supercritical region would be completely arbitrary.

Historically the term vapor was applied to that form of a substance which could be compressed to two-phase states while at room temperature, that is, to substances with critical temperatures above room temperature. The term permanent gases was originally applied to hydrogen, oxygen, and carbon monoxide because the behavior of these gases is such that their two-phase states cannot be reached by compression at room temperature. The critical temperature of these substances is below room temperature. Now it is common, but not universal, to use the term vapor for states with $v > v_g$ and $T < T_{crit}$, the term liquid for states with $v < v_f$ and $P < P_{crit}$, and the term supercritical for states with either $P > P_{crit}$ or $T > T_{crit}$.

An additional feature of the supercritical states is illustrated by the constant pressure process d to e shown in Figure 8.4. Since the pressure is greater than the critical pressure, the constant pressure cooling from state d to e cannot reach the two-phase region and no condensation can occur. Conversely the system at state e cannot be made to evaporate by a constant pressure heating.

Subcritical states also exhibit unusual behavior. Consider first a substance in the state α shown in Figure 8.4. The pressure in this state, P_3, is less than the saturation pressure corresponding to the saturated liquid state with the same specific volume as this state, $v < v_{crit}$. Clearly this state is a mixture of liquid and vapor phases. The properties of the liquid present are the properties characteristic of the saturated liquid at the pressure P_3 (state f'), and the properties of the vapor present are the properties characteristic of the saturated vapor at the pressure P_3 (state g'). If this system is enclosed in a rigid container and heated, the pressure and the temperature will increase while the total volume remains constant. The temperature in each state is the saturation temperature corresponding to the pressure of the two phases. Because of the increase in temperature, the liquid phase will expand and increase in specific volume along the saturated liquid line from $v_{f'}$ to v_f. In the vapor phase, the effect of the increase in pressure overshadows the effect of the increase in temperature, and the specific volume decreases from $v_{g'}$ to v_g along the saturated vapor line. Since the amount of liquid present is large relative to the amount of vapor present, the expansion of the liquid is larger than is necessary to compress the vapor; consequently, the vapor condenses. By the time the pressure has reached P_2, all of the vapor has condensed and only liquid remains. Thus, we have condensed the vapor in the system by heating.

If the system at state β has an initial specific volume v such that $v > v_{\text{crit}}$, the expansion of the liquid when heated is not sufficient to compress the vapor, and thus some of the liquid must evaporate in order to increase the density of the vapor. This evaporation process continues until finally only vapor is present at state g. If the initial specific volume is such that $v = v_{\text{crit}}$, the two effects balance and the system remains in a two-phase state up to the critical point. When the experiment is carried out in a container with glass viewing ports, the approach to the critical state is easily followed by shaking the container and observing the sloshing motion of the interface between the liquid and vapor phases. As the density of the vapor approaches the density of the liquid, the sloshing becomes very slow. At the critical point, the interface separating the liquid and vapor just fades away. At the critical point, the fluid usually changes color and may become reasonably opaque.

The critical constants of a great many pure substances have been compiled in the International Critical Tables. Every pure substance has a critical state; however, some substances decompose chemically before the critical state can be reached. On the other hand, there are some materials with critical states at temperatures and pressures which are above the available experimental range. Consequently, these critical states have never been observed experimentally.

8.5 Equilibrium Liquid-Solid and Vapor-Solid States

In Sections 8.3 and 8.4 we have described the nature of various equilibrium states of the pure substance model. The states described there were by no means inclusive since a solid phase is possible for the pure substance model. Thus, two other types of two-phase states are possible—liquid-solid two-phase states and vapor-solid two-phase states. Since these two-phase states occur at extremes in temperatures and pressures, it is difficult to represent them on the P-v plane with liquid-vapor two-phase states. Accordingly, we expand that portion of the P-v plane in the vicinity of these two-phase states. Figure 8.5 is an example of such an expansion.

In general, these states exhibit a behavior similar to the liquid-vapor two-phase states. That is, if we were to extend the isotherms as in Figure 8.5 to include the solid phase, we would find much the same behavior that we observed in the liquid-vapor region. The slope of an isotherm is less than zero everywhere except in the two-phase regions where $(\partial P/\partial v)_T = 0$. There exist saturated solid, liquid, and vapor states for which there is only one phase and $(\partial P/\partial v)_T = 0$. For all two-phase states, pressure and temperature are related so that an additional property is required to specify the state. Definite changes in the values of the specific properties are associated with a change in phase at constant temperature and pressure. From Figure 8.5 it is apparent that if the isothermal compression process producing the change in phase is carried out at sufficiently low temperature, the vapor phase will condense directly to the solid phase without ever becoming liquid. In distinct contrast to liquid-vapor two-phase states, there is no critical state within our physical experience for which the liquid and solid become indistinguishable or for which the vapor and solid become indistinguishable.

A phase change from saturated liquid to saturated solid is commonly referred to as *freezing* while the reverse change of

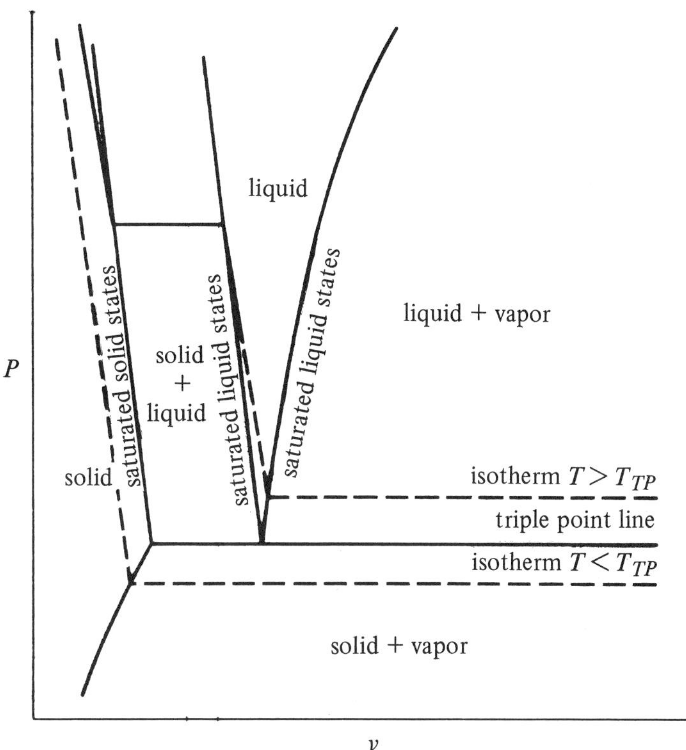

Figure 8.5 Equilibrium states in the solid-liquid region of the *P-v* plane of substance that contracts upon freezing

phase is known as *melting*. (Note that the saturated liquid phase in equilibrium with the saturated solid phase is completely different from the saturated liquid phase in equilibrium with the saturated vapor phase.) A phase change from saturated vapor to saturated solid is commonly known as *frosting* while the reverse change of phase is referred to as *sublimation*.

The properties that establish these states are also given special names. At a given pressure, the temperature for which the solid and liquid phases can coexist in equilibrium is known as the *melting point* or *freezing point*. The pressure at which these two phases coexist for a given temperature is called simply the *saturation pressure*. Solid and vapor phases at a given pressure can coexist in equilibrium at a temperature known as the *sublimation point* or *frost point*. For a given temperature, the pressure at which the solid and vapor phases are in equilibrium is also known as the *saturation pressure*.

In dealing with the solid phase, we note that there are two phenomena peculiar to this phase only. The first of these is a consequence of the variety of possible molecular orientations in the solid phase. As a result, some pure substances are capable of forming more than one solid phase, usually of different crystal structure. Thermodynamically, the two-phase state formed by two solid phases in equilibrium is much the same as the two-phase states that we have already discussed. Two-phase states of this type, as well as liquid-solid two-phase states, are of major importance in metallurgical thermodynamics.

The second peculiarity of the solid phase is a consequence of the nature of intermolecular forces at close molecular spacings typical of the solid phase. These forces can be of two distinctly different types which can cause the substance to expand or contract upon freezing. That is, for a freezing process which takes place at constant pressure and temperature, the change in specific volume associated with the liquid-to-solid phase change may be either positive or negative according to whether the substance expands or contracts. All other specific properties, however, decrease during the freezing process. Thus, in the neighborhood of the liquid-solid equilibrium states, the P-v plane would have the appearance of Figure 8.5 for a substance that contracts upon freezing or the appearance of Figure 8.6 for a substance that expands upon freezing. Figure 8.6 is difficult to visualize because it is a two-dimensional projection of a folded surface. For example, it is impossible to reach the saturated solid state from the saturated liquid state by isothermal compression along the isotherm shown in Figure 8.6. Although such behavior would at first sight seem anomalous, one of the most common substances known, water, behaves in this manner. In fact, it is because of this behavior that ice floats in liquid water.

8.6 Equilibrium Solid-Liquid-Vapor States

The final combination of phases in the pure substance model is a state of mutual equilibrium among the solid phase, the liquid phase, and the vapor phase. These three-phase equilibrium

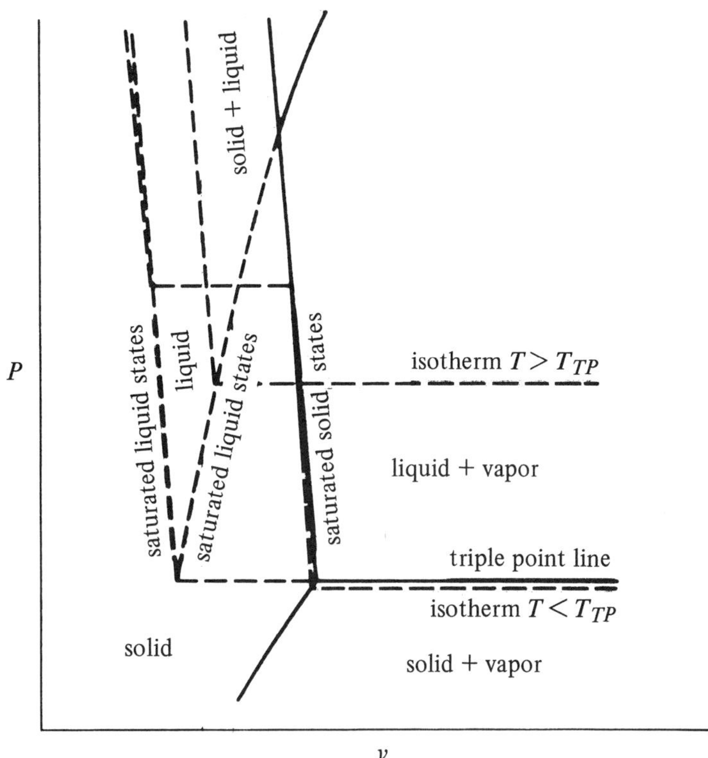

Figure 8.6 Equilibrium states in the liquid-solid region of the *P-v* plane of a substance that expands upon freezing

states are unique in that they occur only at one temperature and one pressure for a given pure substance. This intensive state is known as the *triple point*. Thus, for the pure substance model a three-phase equilibrium state is even more restrictive than a two-phase equilibrium because the temperature and pressure are not only related but they are fixed. Neither temperature nor pressure can be arbitrarily established if the three phases are to remain in equilibrium. For an isothermal compression process carried out at the triple point temperature, the vapor phase can condense to either the solid or the liquid.

Although three-phase states are unique with regard to their intensive state, they still possess considerable freedom with regard to their extensive states. At the triple point, each of the specific properties (e.g. the specific volume) has a unique value; however, the relative amounts of the three phases present can be varied over a relatively large range so that there is essentially an infinite number of three-phase states. For this reason, the locus of three-phase states is a line in the P-v plane (cf. Figures 8.5 and 8.6). This line joins the saturated solid, saturated liquid, and saturated vapor states. Note that at the triple point there is a single saturated liquid state.

With the exception of helium, all real substances possess a triple point. Helium, however, cannot exist as a solid under its own vapor pressure and can only be solidified, even at temperatures near 0 K, at pressures in excess of 25 atmospheres. At any lower pressure, helium remains liquid to the lowest temperatures attainable.

8.7 Gibbs Phase Rule

In comparing the mutual equilibrium that prevails in three-phase states, two-phase states, and single-phase states for a given pure substance, we make an interesting observation. Three-phase states can occur only at a single intensive state. That is, only at a unique combination of temperature and pressure will we find the solid, liquid, and vapor phases in equilibrium for a given substance. On the other hand, two-phase states can occur at a variety of temperatures and pressures, and we are free to arbitrarily specify either the temperature *or* the pressure. Of course, once we have specified one of these properties, the other is automatically determined since there is a unique relationship between pressure and temperature for two-phase states. Finally, we note that single-phase states can occur at any combination of pressure and temperature. There is no restriction placed on the intensive properties of a single-phase state. Temperature and pressure are independent properties for these states.

An alternate way of making this observation is to note from Figure 8.7 that three-phase states are confined to a single point in the P-T plane, two-phase states can occur only along certain lines in the P-T plane, but single-phase states can occur anywhere in the P-T plane. Note from Figure 8.7 the two possible configurations of the intensive states of a pure substance. One for the substance that contracts upon freezing, Figure 8.7(a), and the other for the substance that expands upon freezing,

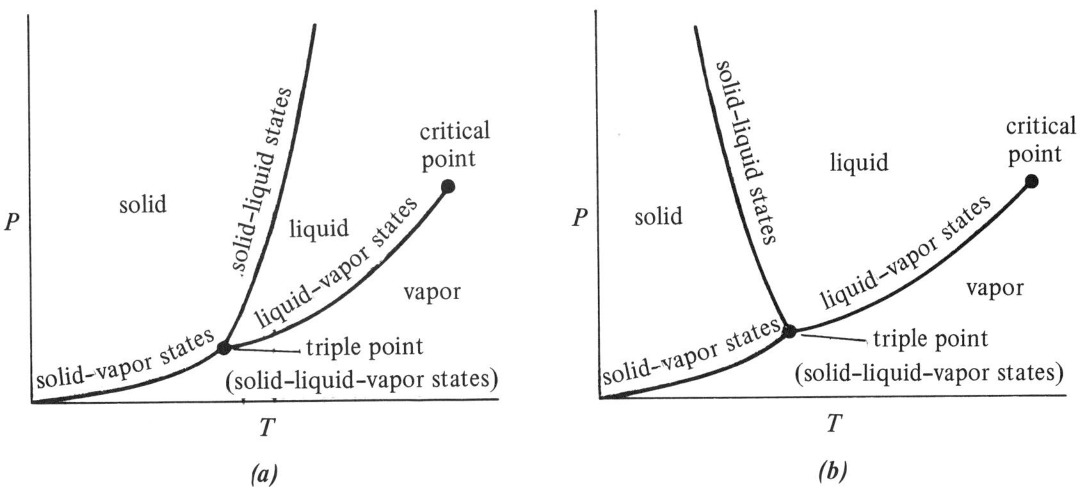

Figure 8.7(a) Intensive states of the pure substance model that contracts upon freezing (b) Intensive states of the pure substance model that expands upon freezing

Figure 8.7(b). The two differ in the sign of the slope of the line that represents the locus of solid-liquid two-phase states. The significance of the term triple point becomes immediately apparent from Figure 8.7.

J. Willard Gibbs generalized these observations of the number of independent intensive properties in the form of an empirical expression which is now known as the *Gibbs phase rule*.

$$f = 2 + c - \zeta \tag{8.5}$$

Here f is the number of thermodynamic degrees of freedom, that is, the number of intensive properties which can be arbitrarily specified, c is the number of chemically independent components, and ζ is the number of phases present. For a pure substance, c must be 1 since only one component is present. Then Gibbs phase rule reduces to

$$F = 3 - \zeta \tag{8.6}$$

Thus, if only one phase is present, $\zeta = 1$ and $f = 2$. Then the pressure and temperature can be fixed arbitrarily, and the system is said to be *divariant*. If there are two phases present, $\zeta = 2$ and $f = 1$. Then only the temperature *or* the pressure, but not both, can be fixed arbitrarily. Once one of these properties is specified, the other is automatically determined in order to preserve the equilibrium between the two phases. The system is then said to be *monovariant*. Finally, if three phases are present, $\zeta = 3$ and $f = 0$. Thus, no intensive property can be arbitrarily fixed. The intensive state is automatically specified by virtue of the equilibrium among the three phases. The system is then said to be *invariant*. Note that the three phases involved need not be solid, liquid, and vapor. There can be two or even three solid phases involved, and the Gibbs phase rule will still be valid.

8.8 Energy Interactions During Changes of Phase

Change of phase processes at constant pressure are important practical processes for the pure substance. Since the two phases are in equilibrium with each other during such a change of phase, it follows from our previous discussions in this chapter that if the pressure remains constant, then the temperature must also remain constant. Under these circumstances, the evaluation of the energy interactions becomes particularly simple. For example, consider the case of a phase change from saturated liquid to saturated vapor due to a heat transfer process in which the temperature and pressure of the pure substance are always constant. If the process is reversible, the only work transfer (for a unit mass of the substance) is $\int P\, dv$, but since the pressure is constant, the work transfer is

$$(W_{f-g})_{\text{rev}} = mP(v_g - v_f) \tag{8.7}$$

Here the subscript f refers to the saturated liquid and the subscript g refers to the saturated vapor. The first law of thermodynamics requires

$$Q_{f-g} - W_{f-g} = m(u_g - u_f) \tag{8.8}$$

When we substitute equation (8.7) for the work transfer, the required heat transfer for the system is

$$(Q_{f-g})_{\text{rev}} = m[(u + Pv)_g - (u + Pv)_f] = m[h_g - h_f] \qquad \textbf{(8.9)}$$

In equation (8.9) we have used the property *enthalpy* defined as

$$H = U + PV = mh = m(u + Pv) \qquad \textbf{(8.10)}$$

(Although the enthalpy has a special significance in the present case, we shall see later that it has an even greater significance in the case of the open system.) If the system were initially all saturated liquid and finally all saturated vapor, then

$$(Q_{f-g})_{\text{rev}} = m(h_g - h_f) = mh_{fg} \qquad \textbf{(8.11)}$$

The quantity h_{fg} is called the *latent heat* of vaporization and involves the same historic misnomer as the specific heat. The term latent heat originated in caloric theory which attempted to describe all thermal processes in terms of the simple concepts of stored heat. We now know these simple concepts to be adequate only for pure thermal systems. The term latent heat was used in contrast to the term *sensible heat* in an attempt to justify the absence of the temperature increase expected in the common case. In spite of its name, the latent heat h_{fg} is generally useful for liquid-vapor phase changes at constant pressure. For example, if a saturated liquid confined in a constant-pressure cylinder is evaporated by an adiabatic irreversible process such as stirring the fluid with a paddle wheel, the work transfer *to* the fluid is given by the latent heat h_{fg}.

The heat transfer required for a reversible evaporation at constant temperature and pressure may also be determined directly from the second law of thermodynamics.

$$(Q_{f-g})_{\text{rev}} = m \int_f^g T\, ds \qquad \textbf{(8.12)}$$

Since the temperature is constant, the integration of equation (8.12) is straightforward. Thus,

$$(Q_{f-g})_{\text{rev}} = mT(s_g - s_f) = mTs_{fg} \qquad \textbf{(8.13)}$$

When we equate this result to the heat transfer computed with the first law, equation (8.11), we obtain the important relation

$$Ts_{fg} = h_{fg} \tag{8.14}$$

which must be satisfied by the specific properties of two phases in equilibrium.

Although we have calculated the heat transfer and work transfer for the case of a liquid-vapor phase change, the analysis is also valid for solid-liquid and solid-vapor phase changes. In each case, it is simply a matter of using the properties for the appropriate saturated states.

8.9 Phase Equilibrium

The results of equation (8.14) can be used to establish an additional requirement for equilibrium between two phases in the pure substance model. Equation (8.14) can be rewritten in the form

$$h_g - Ts_g = h_f - Ts_f \tag{8.15}$$

Making use of the definition of the Gibbs free energy equation (7.88), we can write equation (8.15) in the form

$$g_g = g_f \tag{8.16}$$

Equation (8.16) is a general result and holds for any two phases of a pure substance in equilibrium. Thus for such a system, the specific Gibbs free energy is the same for both phases. This result is consistent with the second law limit on the work transfer (equation (7.89)) for the evaporation of a saturated liquid at constant temperature and pressure since the reversible work transfer other than $\int P\,dv$ is zero.

We have now developed an additional requirement for thermodynamic equilibrium of a pure substance in a multiphase state. This requirement is that the specific Gibbs free energy be uniform throughout the system if the system is to be at equilibrium with respect to mass passing from one phase to another.

Thus, the specific Gibbs free energy, $h - Ts$, plays the same role with respect to mass transfer between phases as the temperature plays with respect to heat transfer and the pressure plays with respect to work transfer by means of mechanical motion. Accordingly, equation (8.1) can be modified to allow for interactions involving mass transfer. For a pure substance,

$$dU = T\,dS - P\,dV + g\,dm \tag{8.17}$$

Here m is the mass of the system and g is the specific Gibbs free energy. For a closed system, the mass is constant, and equation (8.17) reduces to equation (8.1). The open system is another matter altogether since mass can be transferred across the open system boundary.

In more complex cases in which the system is made up of a number of different substances and chemical reactions are possible, a property called the chemical potential must be uniform throughout the system for equilibrium with respect to mass transfer between phases. Thus, in the simple case of the pure substance the chemical potential reduces to the Gibbs free energy.

8.10 Thermodynamic Surfaces

In the preceding section we have assumed that all the equilibrium properties of the pure substance are known. Because of the complex thermodynamic behavior of the pure substance we cannot evaluate these properties from simple algebraic constitutive relations of the type we have used previously. Instead, we must rely upon the values of properties determined from experimental measurements. We have already seen that it is possible to represent the thermodynamic properties of the pure substance model in terms of curves plotted on an appropriate set of coordinates. Although such a representation is often very convenient, it suffers from two disadvantages. First, for clarity, the data must be presented as a family of curves spaced at reasonable intervals while the actual physical data vary continuously. Interpolation must be used to obtain data for engineering calculations from the two-dimensional graphical representation. The accuracy of this method is at best limited. Second, the two-dimensional graphical representation is sometimes difficult to interpret. This situation is particularly true in the case of the pure substance that expands on freezing (cf. Figure 8.6). Thus, it is worthwhile to consider a representation that does not involve these disadvantages.

We recognize that according to the state principle, we can specify the state of a pure substance by establishing the values of two independent properties. Any other property can then be related to these two through the appropriate set of constitutive relations. These constitutive relations can be reduced to a single expression for the desired property as a function of the two independent properties. For a unit mass of the pure substance this single expression can be represented as a surface in a three-

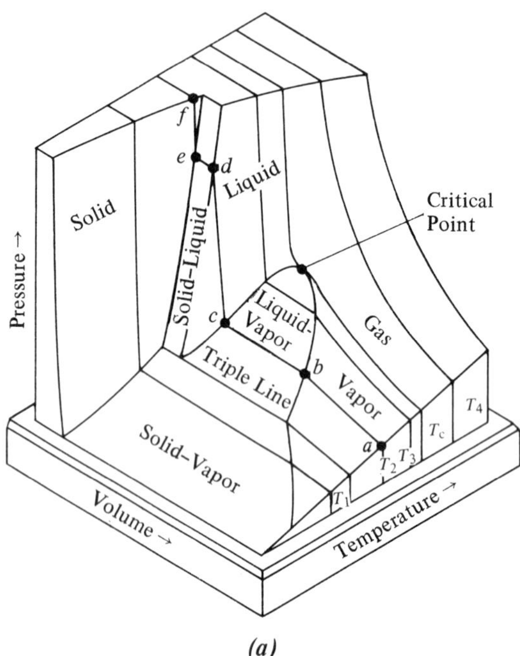

(a)

Figure 8.8(a) *P-v-T* surface for a substance that contracts upon freezing. Figure reprinted from Lee and Sears, *Thermodynamics: An Introductory Text for Engineering Students,* © 1963, 2/3, Addison-Wesley Publishing Company, Inc., page 38, figure 2-7. Reprinted with permission.

dimensional coordinate space defined by the two independent properties and the property of interest. Such a surface is called a *thermodynamic surface,* and for a unit mass of a given pure substance, every equilibrium state of the substance must lie somewhere on the surface.

The most common, and perhaps most useful, thermodynamic surface is the three-dimensional representation of the P-v-T data for the pure substance. These three properties are the easiest properties to measure. Figure 8.8 shows schematically typical P-v-T surfaces for two classes of pure substances, those that contract upon freezing and those that expand upon freezing. In this representation the interrelation between temperature and pressure for two-phase states becomes immediately apparent. Because of this interrelationship the loci of all two-phase states are surfaces whose generators are normal to the pressure-

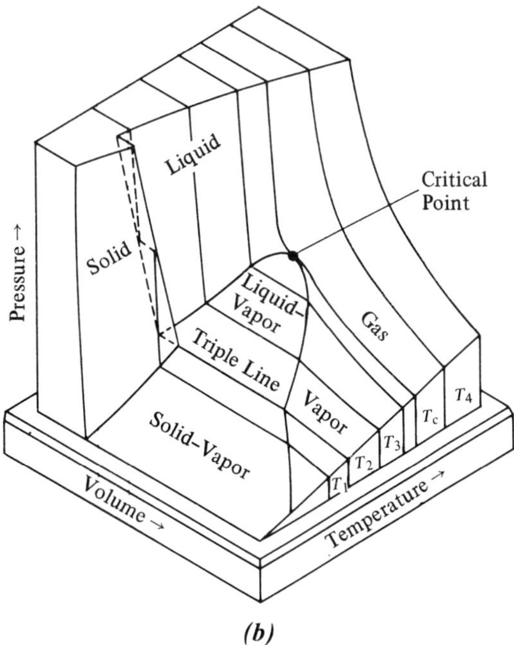

Figure 8.8(b) P-v-T surface for a substance that expands upon freezing. Figure reprinted from Lee and Sears, *Thermodynamics: An Introductory Text for Engineering Students,* © 1963, 2/3, Addison-Wesley Publishing Company, Inc., page 39, figure 2-8. Reprinted with permission.

temperature plane and parallel to the volume axis. A single curved surface of this type is often called a ruled surface. There are three such regions of phases. All three of these ruled surfaces intersect at a particular value of the pressure and temperature, the triple point. The locus of the intersection is actually a line called the triple line and every three-phase state for a unit mass of the pure substance must lie somewhere on this line. Although the three ruled surfaces have the same temperature and pressure at their line of intersection, they each have a different slope. These slopes, however, do not vary with the average specific volume of the three-phase state.

On the P-v-T surface the saturated states lie on the lines of intersection between the ruled surfaces representing the two-phase states and the double curved surfaces representing the single-phase states. In Figure 8.8 the relative magnitudes of the volume changes associated with constant pressure phase changes are greatly exaggerated. The volume change for the constant pressure liquid-solid phase change is extremely small (too small to show unless exaggerated) relative to the volume change for either the vapor-liquid or vapor-solid phase change which are both of the same order of magnitude.

Note that there are two distinct differences between the P-v-T surfaces shown in Figures 8.8 (a) and 8.8 (b). The first difference is the location of the liquid-solid two-phase surface with respect to the location of the liquid-vapor two-phase surface. In Figure 8.8 (a) we see that at the triple point, the volume change for the vapor-liquid phase change is smaller than the volume change for the vapor-solid phase change. Thus, at this pressure the specific volume of the solid is smaller than the specific volume of the liquid, and the substance contracts upon freezing. On the other hand, in Figure 8.8 (b) we note that at the triple point the volume change for the vapor-liquid phase change is greater than the volume change for the vapor-solid phase change. Hence, at this pressure the specific volume of the solid is greater than the specific volume of the liquid, and the substance expands upon freezing. This difference between Figures 8.8 (a) and 8.8 (b) becomes even more obvious when we project the P-v-T surface onto the already familiar P-v plane as in Figures 8.5 and 8.6.

The second difference between Figures 8.8 (a) and 8.8 (b) is in the sign of the slope of the ruled surface representing the liquid-solid two-phase states. In Figure 8.8 (a) it is clear that an increase in pressure results in an increase in the freezing point. This situation is also typical of the liquid-vapor and solid-vapor two-phase states for which an increase in pressure causes an increase in the saturation temperature. For a substance that expands upon freezing, the liquid-solid two-phase states exhibit the anomalous behavior that an increase in pressure results in a decrease in the freezing point, a distinct contrast to other two-phase states. The projection of the P-v-T surface onto the P-T plane as in Figure 8.7 reveals this difference in even more striking fashion.

By now it is obvious that the P-v-T surface for a pure substance provides a great deal of information about the equilibrium thermodynamic behavior of the substance. However, we must remember that the actual P-v-T surface of a given pure substance may look radically different from the two schematic surfaces that we have just considered. For example, one of the most common of all pure substances, water, exhibits unusual behavior of its P-v-T surface at high pressures. When water freezes at pressures which are within our normal range of experience (e.g. several atmospheres), it expands. Thus the solid form, ice, will float on the surface of the liquid. At extremely high pressures (thousands of atmospheres), the solid form becomes more dense than the liquid and hence will not float. This type of behavior together with the seven different solid phases of ice is shown in Figure 8.9. The important point to remember is that regardless of the anomalies that might occur, *every* equilibrium state of a unit mass of a pure substance must lie somewhere on the P-v-T surface.

The P-v-T surface is just one example of the many possible thermodynamic surfaces that we can construct. Each one serves some special purpose; however, these surfaces are awkward to use. For this reason, we usually rely on projections of these surfaces onto one of the coordinate planes for a graphical representation of the states of the pure substance model. In spite of their disadvantages, these projections are quite often useful because the liquid-vapor two-phase states are of most interest.

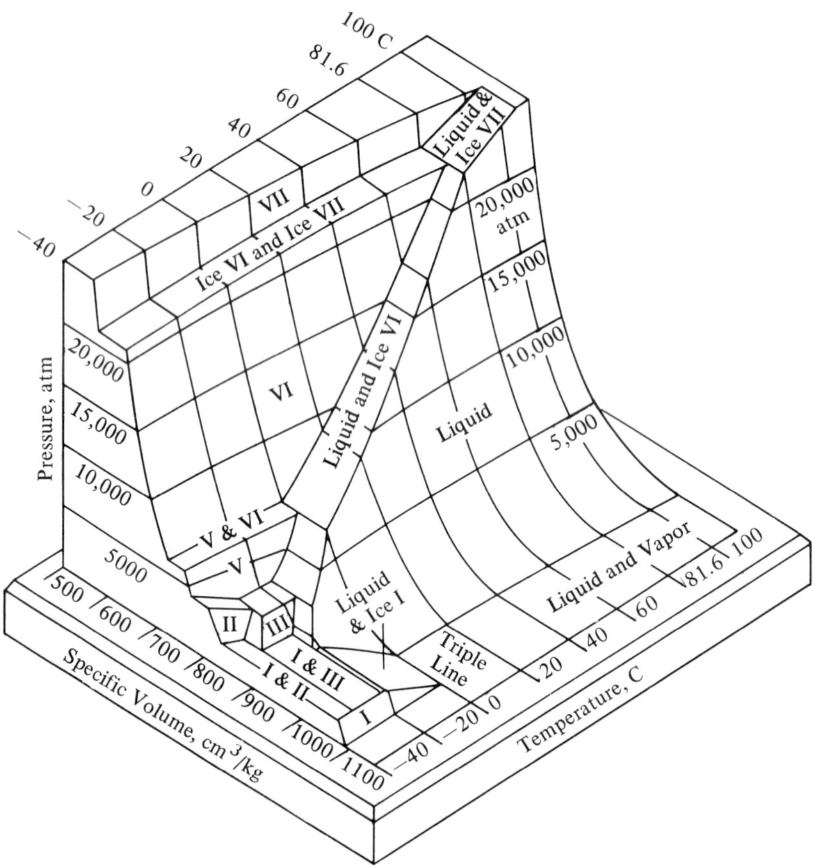

Figure 8.9 *P-v-T* surface of water at high pressure. Figure reprinted Lee and Sears, *Thermodynamics: An Introductory Text for Engineering Students,* © 1963, 2/3, Addison-Wesley Publishing Company, Inc., page 42, figure 2-11.

8.11 Tabulation of the Thermodynamic Properties

In the previous section we saw that all possible thermodynamic states of a pure substance can be represented by a surface in a three-dimensional coordinate space. In principle, a collection of such surfaces could be used to record all of the thermodynamic properties; however, this is certainly not a very practical method. A more convenient record of the properties is the projections of these surfaces onto appropriate two-dimensional planes such as the P-v plane (cf. Figure 8.3) or the T-s plane (cf. Figure 8.10). These two projections are useful since the interactions for quasi-static processes are easily depicted on these planes. On the P-v plane, the quasi-static work transfers are $\int P\,dv$ while on the T-s plane reversible heat transfers are $\int T\,ds$.

Since any property diagram of any reasonable size is limited in accuracy to a few percent, tables of the values of the thermodynamic properties are more useful for accurate engineering calculations. Typically, these tables are a listing of the values of the specific volume, specific entropy, specific enthalpy, and specific internal energy at convenient intervals of temperature and pres-

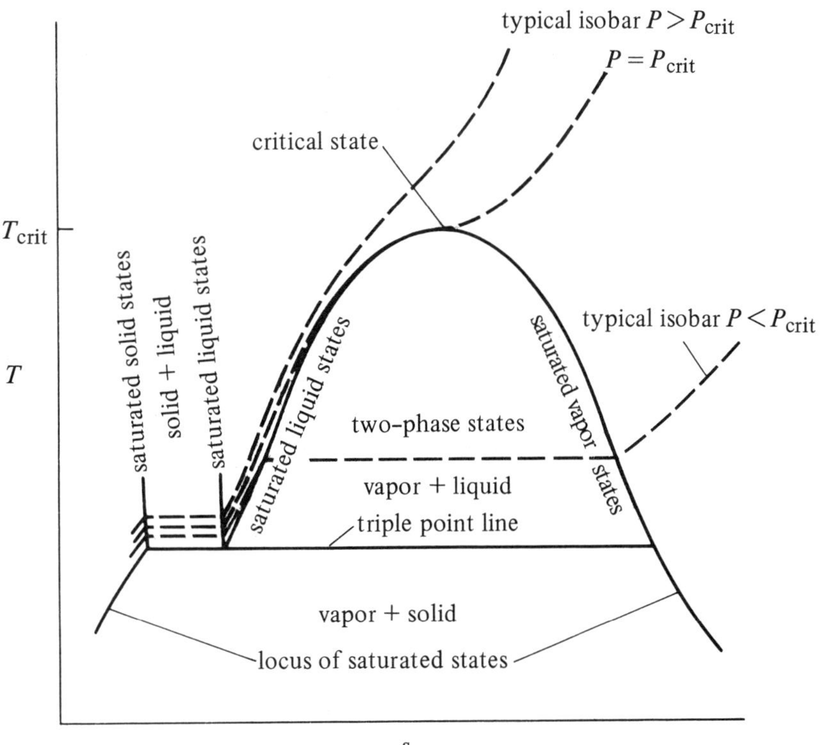

Figure 8.10 Temperature-entropy diagram of the pure substance model

sure. The listing is usually in the form of a matrix with the rows and columns corresponding to either a fixed temperature or a fixed pressure. Table 8.1 is an example extracted from the tabulated properties of H_2O given in Appendix A. To determine the values of the specific properties for a known state, we simply select the appropriate column corresponding to the pressure of the system and search down this column until we reach the appropriate row corresponding to the temperature of the system. The values of u, v, s, and h can be read directly from the entry in the tables. To calculate the total value of these properties we simply multiply the specific value by the mass of the system. The intervals of temperature and pressure used in full and complete tables are selected to permit linear interpolation with reasonable accuracy. Abstracted tables for several pure substances are given in Appendix A.

This scheme of tabulation is sufficient for those states in which only one phase of the pure substance is present. However, for two-phase states the situation is a bit more complicated since the pressure and temperature are no longer independent of one another. Fortunately, each of the two phases involved is in a single-phase saturated state so that for a given temperature or pressure it is possible to tabulate the specific properties for each of the two coexistent phases. A table of thermodynamic properties of these saturated states is called a *saturation table*. Either the temperature or the pressure is selected as the independent property and is indexed at convenient intervals to permit linear interpolation. The value of the other intensive property (pressure or temperature) is usually included in the list of tabulated properties. Table 8.2 is an example, extracted from Appendix A, of a saturation table for equilibrium liquid-vapor two-phase states of H_2O with the temperature as the independent property. Abstracted tables for the saturated states for several pure substances are given in Appendix A.

Table 8.1 also lists the two-phase states for the pressure values listed at the top of the table. The saturation temperature corresponding to the pressure is listed in parentheses after the pressure. The values of the specific properties of saturated liquid (sub f states) at each pressure are in row one of the table (indicated by Sat liq in the left column). The values for saturated vapor (sub g states) are given in row three (indicated by Sat vap). The difference between saturated vapor properties and saturated liquid properties (sub fg) are given in row two (indicated by Evap). At each pressure the states above the horizontal line are compressed liquid states and the states below are superheated vapor states.

We can now use these tabulated specific properties of the two saturated phases to evaluate the extensive properties of a two-phase state simply by computing the values of these extensive properties for the separate phases and then summing the values for all the phases present. Since the state of each phase is saturated, the value of an extensive property for a given phase is obtained by multiplying the tabulated value of the specific property for the saturated state of the phase by the mass of the phase. The volume for a two-phase state is the sum of the volume of the liquid phase, $V_f = m_f v_f$, and the volume of the vapor phase, $V_g = m_g v_g$, as in equations (8.18) and (8.19).

$$V = V_f + V_g \tag{8.18}$$

or

$$V = m_f v_f + m_g v_g \tag{8.19}$$

In equation (8.19) the subscript f denotes the saturated liquid phase and the subscript g the saturated vapor phase. The relations for the other extensive properties are of the same form.

$$U = m_f u_f + m_g u_g \tag{8.20}$$
$$S = m_f s_f + m_g s_g \tag{8.21}$$
$$H = m_f h_f + m_g h_g \tag{8.22}$$

Equations (8.19) through (8.22) show that the extensive properties of a system in a two-phase state depend upon the relative amounts of the two phases. Thus, a system in a two-phase state can have as one of its independent properties a property which expresses the relative mass in the two phases. This property is called the *quality,* denoted by the symbol x, and is defined for liquid-vapor states as *the fraction of the mass which is in the vapor phase*. Thus,

$$x = \frac{m_g}{m_f + m_g} = \frac{m_g}{m} \tag{8.23}$$

TABLE 8.1
Properties of H_2O

$P(t_{sat})$ Temp C	1×10^3 N/m² (6.98 C)				10×10^3 N/m² (45.81 C)				20×10^3 N/m² (60.06 C)	
	v m³/kg	u kJ/kg	h kJ/kg	s kJ/kgK	v m³/kg	u kJ/kg	h kJ/kg	s kJ/kgK	v m³/kg	u kJ/kg
Sat liq	1.0002−3	29.30	29.30	0.1059	1.0102−3	191.82	191.83	0.6493	1.0172−3	251.38
Evap	129.21	2355.70	2484.90	8.8697	14.673	2246.08	2392.87	7.5009	7.6473	2205.32
Sat vap	129.21	2385.0	2514.2	8.9756	14.674	2437.9	2584.7	8.1502	7.649	2456.7
0	1.0002−3	−0.03	−0.03	−0.0001	1.0002−3	−0.03	−0.02	−0.0001	1.0002−3	−0.03
10	130.60	2389.2	2519.8	8.9956	1.0010−3	41.91	41.92	0.1483	1.0010−3	41.96
20	135.23	2403.2	2538.5	9.0603	1.0018−3	83.85	83.86	0.2966	1.0018−3	83.95
30	139.85	2417.3	2557.1	9.1228	1.0048−3	125.70	125.71	0.4345	1.0048−3	125.75
40	144.47	2431.3	2575.8	9.1835	1.0098−3	167.56	167.57	0.5725	1.0078−3	167.56
50	149.09	2445.4	2594.5	9.2423	14.869	2443.9	2592.6	8.1749	1.0125−3	209.34
60	153.71	2459.5	2613.2	9.2994	15.336	2458.2	2611.5	8.2327	1.0172−3	251.12
70	158.33	2473.7	2632.0	9.3550	15.801	2472.5	2630.5	8.2887	7.883	2471.1
80	162.95	2487.9	2650.8	9.4090	16.267	2486.8	2649.5	8.3432	8.117	2485.6
90	167.57	2502.1	2669.7	9.4616	16.731	2501.1	2668.5	8.3963	8.351	2500.1
100	172.19	2516.4	2688.6	9.5129	17.196	2515.5	2687.5	8.4479	8.585	2514.6
120	181.42	2545.0	2726.5	9.6119	18.123	2544.4	2725.6	8.5474	9.051	2543.6
140	190.65	2573.9	2764.5	9.7063	19.050	2573.3	2763.8	8.6423	9.516	2572.7
160	199.89	2602.9	2802.8	9.7968	19.975	2602.5	2802.2	8.7331	9.980	2602.0
180	209.12	2632.2	2841.3	9.8836	20.900	2631.8	2840.8	8.8201	10.444	2631.4
200	218.35	2661.6	2880.0	9.9671	21.825	2661.3	2879.5	8.9038	10.907	2660.9
220	227.58	2691.3	2918.8	10.0476	22.749	2691.0	2918.5	8.9844	11.370	2690.7
240	236.82	2721.2	2958.0	10.1254	23.674	2720.9	2957.7	9.0623	11.832	2720.7
260	246.05	2751.3	2997.3	10.2006	24.598	2751.1	2997.0	9.1376	12.295	2750.8
280	255.28	2781.6	3036.9	10.2735	25.521	2781.5	3036.7	9.2105	12.757	2781.3
300	264.51	2812.2	3076.8	10.3443	26.445	2812.1	3076.5	9.2813	13.219	2811.9
320	273.74	2843.1	3116.8	10.4130	27.369	2842.9	3116.6	9.3501	13.681	2842.8
340	282.97	2874.2	3157.2	10.4799	28.292	2874.1	3157.0	9.4170	14.143	2873.9
360	292.20	2905.6	3197.8	10.5450	29.216	2905.4	3197.6	9.4821	14.605	2905.3
380	301.43	2937.2	3238.6	10.6085	30.139	2937.0	3238.4	9.5457	15.067	2936.9
400	310.66	2969.0	3279.7	10.6705	31.063	2968.9	3279.6	9.6077	15.529	2968.8
420	319.89	3001.2	3321.1	10.7310	31.986	3001.1	3320.9	9.6682	15.991	3001.0
440	329.12	3033.6	3362.7	10.7902	32.909	3033.5	3362.6	9.7275	16.453	3033.4
460	338.35	3066.2	3404.6	10.8482	33.832	3066.1	3404.5	9.7854	16.915	3066.1
480	347.58	3099.2	3446.7	10.9049	34.756	3099.1	3446.6	9.8422	17.376	3099.0
500	356.81	3132.4	3489.2	10.9605	35.679	3132.3	3489.1	9.8978	17.838	3132.2
550	379.89	3216.6	3596.5	11.0950	37.987	3216.5	3596.4	10.0323	18.992	3216.4
600	402.97	3302.5	3705.5	11.2235	40.295	3302.5	3705.4	10.1608	20.147	3302.4
650	426.04	3390.2	3816.2	11.3469	42.603	3390.1	3816.2	10.2842	21.301	3390.1
700	449.12	3479.6	3928.7	11.4655	44.911	3479.6	3928.7	10.4028	22.455	3479.5
750	472.19	3570.8	4043.0	11.5800	47.218	3570.8	4043.0	10.5174	23.609	3570.8
800	495.27	3663.8	4159.1	11.6908	49.526	3663.8	4159.0	10.6281	24.763	3663.8
850	518.34	3758.6	4276.9	11.7981	51.834	3758.6	4276.9	10.7354	25.916	3758.5
900	541.42	3855.1	4396.5	11.9023	54.141	3855.0	4396.4	10.8396	27.070	3855.0
950	564.49	3953.2	4517.7	12.0035	56.449	3953.2	4517.7	10.9408	28.224	3953.2
1000	587.57	4053.0	4640.6	12.1019	58.757	4053.0	4640.6	11.0393	29.378	4053.0
1100	633.72	4257.5	4891.2	12.2914	63.372	4257.5	4891.2	11.2287	31.686	4257.5
1200	679.87	4467.9	5147.8	12.4718	67.987	4467.9	5147.8	11.4091	33.994	4467.9
1300	726.02	4683.7	5409.7	12.6438	72.602	4683.7	5409.7	11.5811	36.301	4683.7

TABLE 8.1 (*cont'd.*)
Properties of H_2O

$20 \times 10^3\ N/m^2$ (60.06 C)		$30 \times 10^3\ N/m^2$ (69.10 C)				$50 \times 10^3\ N/m^2$ (81.33 C)				
h kJ/kg	s kJ/kgK	v m³/kg	u kJ/kg	h kJ/kg	s kJ/kgK	v m³/kg	u kJ/kg	h kJ/kg	s kJ/kgK	$P(t_{sat})$ Temp C
251.40	0.8320	1.0223−3	289.20	289.23	0.9349	1.0300−3	340.44	340.49	1.0910	Sat liq
2358.30	7.0765	5.2280	2179.20	2336.07	6.8337	3.2390	2143.46	2305.41	6.5029	Evap
2609.7	7.9085	5.229	2468.4	2625.3	7.7686	3.240	2483.9	2645.9	7.5939	Sat vap
−0.01	−0.0001	1.0002−3	−0.03	0.00	−0.0001	1.0002−3	−0.03	0.02	−0.0001	0
41.98	0.1483	1.0010−3	41.96	41.99	0.1483	1.0010−3	41.96	42.01	0.1483	10
83.97	0.2966	1.0018−3	83.95	83.98	0.2966	1.0018−3	83.95	84.00	0.2966	20
125.77	0.4345	1.0048−3	125.75	125.78	0.4345	1.0048−3	125.75	125.80	0.4345	30
167.58	0.5725	1.0078−3	167.56	167.59	0.5725	1.0078−3	167.56	167.60	0.5725	40
209.35	0.7018	1.0125−3	209.34	209.36	0.7018	1.0125−3	209.34	209.38	0.7018	50
251.13	0.8312	1.0712−3	251.12	251.14	0.8312	1.0172−3	251.12	251.16	0.8312	60
2628.8	7.9649	5.243	2469.7	2627.0	7.7737	1.0231−3	292.99	293.02	0.9532	70
2647.9	8.0199	5.401	2484.4	2646.4	7.8293	1.0291−3	334.87	334.89	1.0753	80
2667.1	8.0734	5.558	2499.0	2665.7	7.8833	3.323	2496.8	2662.9	7.6414	90
2686.2	8.1255	5.715	2513.6	2685.0	7.9357	3.418	2511.6	2682.5	7.6947	100
2724.6	8.2256	6.027	2542.8	2723.6	8.0365	3.608	2541.3	2721.6	7.7968	120
2763.0	8.3209	6.338	2572.1	2762.2	8.1323	3.796	2570.8	2760.6	7.8935	140
2801.6	8.4120	6.648	2601.4	2800.9	8.2237	3.983	2600.4	2799.6	7.9856	160
2840.2	8.4993	6.958	2630.9	2839.7	8.3112	4.170	2630.1	2838.6	8.0736	180
2879.1	8.5831	7.267	2660.6	2878.6	8.3952	4.356	2659.9	2877.7	8.1580	200
2918.1	8.6639	7.577	2690.4	2917.7	8.4762	4.542	2689.8	2916.9	8.2392	220
2957.3	8.7418	7.885	2720.4	2957.0	8.5542	4.728	2719.9	2956.2	8.3174	240
2996.7	8.8172	8.194	2750.6	2996.4	8.6297	4.913	2750.2	2995.8	8.3931	260
3036.4	8.8903	8.502	2781.0	3036.1	8.7028	5.099	2780.6	3035.6	8.4663	280
3076.3	8.9611	8.811	2811.7	3076.0	8.7736	5.284	2811.3	3075.5	8.5373	300
3116.4	9.0299	9.119	2842.6	3116.2	8.8425	5.469	2842.3	3115.7	8.6062	320
3156.8	9.0968	9.427	2873.8	3156.6	8.9095	5.654	2873.5	3156.2	8.6732	340
3197.4	9.1620	9.735	2905.2	3197.2	8.9747	5.839	2904.9	3196.8	8.7385	360
3238.3	9.2256	10.043	2936.8	3238.1	9.0383	6.024	2936.5	3237.8	8.8021	380
3279.4	9.2876	10.351	2968.7	3279.2	9.1003	6.209	2968.5	3278.9	8.8642	400
3320.8	9.3482	10.659	3000.9	3320.6	9.1609	6.394	3000.6	3320.4	8.9249	420
3362.4	9.4074	10.967	3033.3	3362.3	9.2202	6.579	3033.1	3362.0	8.9841	440
3404.3	9.4654	11.275	3066.0	3404.2	9.2781	6.764	3065.8	3404.0	9.0421	460
3446.5	9.5222	11.583	3098.9	3446.4	9.3349	6.949	3098.7	3446.2	9.0989	480
3489.0	9.5778	11.891	3132.1	3488.9	9.3906	7.134	3132.0	3488.7	9.1546	500
3596.3	9.7123	12.661	3216.4	3596.2	9.5251	7.596	3216.2	3596.0	9.2891	550
3705.3	9.8409	13.430	3302.3	3705.3	9.6537	8.057	3302.2	3705.1	9.4178	600
3816.1	9.9642	14.200	3390.0	3816.0	9.7770	8.519	3389.9	3815.9	9.5412	650
3928.6	10.0829	14.969	3479.5	3928.6	9.8957	8.981	3479.4	3928.5	9.6599	700
4043.0	10.1975	15.739	3570.7	4042.9	10.0103	9.443	3570.6	4042.8	9.7744	750
4159.1	10.3082	16.508	3663.7	4158.9	10.1210	9.904	3663.6	4158.9	9.8852	800
4276.9	10.4155	17.277	3758.5	4276.8	10.2284	10.366	3758.4	4276.7	9.9926	850
4396.4	10.5197	18.047	3855.0	4396.4	10.3325	10.828	3854.9	4396.3	10.0967	900
4517.7	10.6209	18.816	3953.1	4517.6	10.4337	11.289	3953.1	4517.6	10.1979	950
4640.6	10.7193	19.585	4053.0	4640.5	10.5322	11.751	4052.9	4640.5	10.2964	1000
4891.2	10.9088	21.124	4257.4	4891.1	10.7217	12.674	4257.4	4891.1	10.4859	1100
5147.8	11.0892	22.662	4467.9	5147.7	10.9020	13.597	4467.8	5147.7	10.6662	1200
5409.7	11.2612	24.201	4683.6	5409.7	11.0740	14.521	4683.6	5409.6	10.8382	1300

TABLE 8.2
Properties of Saturated H_2O

t C	T K	P N/m^2	v_f m^3/kg	v_{fg} m^3/kg	v_g m^3/kg
0.01	273.16	611.3	1.0002−3	206.14	206.14
5	278.15	872.1	1.0001−3	147.12	147.12
10	283.15	1.2276+3	1.0004−3	106.38	106.38
15	288.15	1.7051+3	1.0009−3	77.925	77.926
20	293.15	2.339+3	1.0018−3	57.790	57.791
25	298.15	3.169+3	1.0029−3	43.359	43.360
30	303.15	4.246+3	1.0043−3	32.893	32.894
35	308.15	5.628+3	1.0060−3	25.215	25.216
40	313.15	7.384 3	1.0078−3	19.522	19.523
45	318.15	9.593+3	1.0099−3	15.257	15.258
50	323.15	12.349+3	1.0121−3	12.031	12.032
55	328.15	15.758+3	1.0146−3	9.567	9.568
60	333.15	19.940+3	1.0172−3	7.670	7.671
65	338.15	25.03+3	1.0199−3	6.196	6.197
70	343.15	31.19+3	1.0228−3	5.041	5.042
75	348.15	38.58+3	1.0259−3	4.130	4.131
80	353.15	47.39+3	1.0291−3	3.406	3.407
85	358.15	57.83+3	1.0325−3	2.827	2.828
90	363.15	70.14+3	1.0360−3	2.360	2.361
95	368.15	84.55+3	1.0397−3	1.9809	1.9819
100	373.15	101.35+3	1.0435−3	1.6719	1.6729
105	378.15	120.82+3	1.0475−3	1.4184	1.4194
110	383.15	143.27+3	1.0516−3	1.2091	1.2102
115	388.15	169.06+3	1.0559−3	1.0355	1.0366
120	393.15	198.53+3	1.0603−3	0.8908	0.8919
125	398.15	232.1+3	1.0649−3	0.7695	0.7706
130	403.15	270.1+3	1.0697−3	0.6674	0.6685
135	408.15	313.0+3	1.0746−3	0.5811	0.5822
140	413.15	361.3+3	1.0797−3	0.5078	0.5089
145	418.15	415.4+3	1.0850−3	0.4452	0.4463
150	423.15	475.8+3	1.0905−3	0.3917	0.3928
155	428.15	543.1+3	1.0961−3	0.3457	0.3468
160	433.15	617.8+3	1.1020−3	0.3060	0.3071
165	438.15	700.5+3	1.1080−3	0.2761	0.2727
170	443.15	791.7+3	1.1143−3	0.2417	0.2428
175	448.15	892.0+3	1.1207−3	0.2157	0.2168
180	453.15	1.0021+6	1.1274−3	0.19292	0.19405
185	458.15	1.1227+6	1.1343−3	0.17296	0.17409
190	463.15	1.2544+6	1.1414−3	0.15399	0.15654
195	468.15	1.3978+6	1.1488−3	0.13990	0.14105

Data from *Steam Tables* by Joseph H. Keenan, Frederick G. Keyes, Philip G. Hill, and Joan G. Moore. Copyright © 1969 by John Wiley & Sons, Inc. Reproduced by permission of John Wiley & Sons, Inc.

TABLE 8.2 (*cont'd.*)
Properties of Saturated H_2O

u_f kJ/kg	u_{fg} kJ/kg	u_g kJ/kg	h_f kJ/kg	h_{fg} kJ/kg	h_g kJ/kg	s_f kJ/kgK	s_{fg} kJ/kgK	s_g kJ/kgK
0	2375.3	2375.3	0.01	2501.3	2501.4	0	9.1562	9.1562
20.97	2361.3	2382.3	20.98	2489.6	2510.6	0.0761	8.9496	9.0257
43.00	2347.2	2389.2	42.01	2477.7	2519.8	0.1510	8.7498	8.9008
62.99	2333.1	2396.1	62.99	2465.9	2528.9	0.2245	8.5569	8.7814
83.95	2319.0	2402.9	83.96	2454.1	2538.1	0.2966	8.3706	8.6672
104.88	2304.9	2409.8	104.89	2442.3	2547.2	0.3674	8.1905	8.5580
125.78	2290.8	2416.6	125.79	2430.5	2556.3	0.4369	8.0164	8.4533
146.67	2276.7	2423.4	146.68	2418.6	2565.3	0.5053	7.8478	8.3531
167.56	2262.6	2430.1	167.57	2406.7	2574.3	0.5725	7.6845	8.2570
188.44	2248.4	2436.8	188.45	2394.8	2583.2	0.6387	7.5261	8.1648
209.32	2234.2	2443.5	209.33	2382.7	2592.1	0.7238	7.3725	8.0763
230.21	2219.9	2450.1	230.23	2370.7	2600.9	0.7679	7.2234	7.9913
251.11	2205.5	2456.6	251.13	2358.5	2609.6	0.8312	7.0784	7.9096
272.02	2191.1	2463.1	272.06	2346.2	2618.3	0.8935	6.9375	7.8310
292.95	2176.6	2469.6	292.98	2333.8	2626.8	0.9549	6.8004	7.7553
313.90	2162.0	2475.9	313.93	2321.4	2635.3	1.0155	6.6669	7.6824
334.86	2147.4	2482.2	334.91	2308.8	2643.7	1.0753	6.5369	7.6122
355.84	2132.6	2488.4	355.90	2296.0	2651.9	1.1343	6.4102	7.5445
376.85	2117.7	2494.5	376.92	2283.2	2660.1	1.1925	6.2866	7.4791
397.88	2102.7	2500.6	397.96	2270.2	2668.1	1.2500	6.1659	7.4159
418.94	2087.6	2506.5	419.04	2257.0	2676.1	1.3069	6.0480	7.3549
440.02	2072.3	2512.4	440.15	2243.7	2683.8	1.3630	5.9328	7.2958
461.14	2057.0	2518.1	461.30	2230.2	2691.5	1.4185	5.8202	7.2387
482.30	2041.4	2523.7	482.48	2216.5	2699.0	1.4734	5.7100	7.1833
503.50	2025.8	2529.3	503.71	2202.6	2706.3	1.5276	5.6020	7.1296
524.74	2009.9	2534.6	524.99	2188.5	2713.5	1.5813	5.4962	7.0775
546.02	1993.9	2539.9	546.31	2174.2	2720.5	1.6344	5.3925	7.0269
567.35	1977.7	2545.0	567.69	2159.6	2727.3	1.6870	5.2907	6.9777
588.74	1961.3	2550.0	589.13	2144.7	2733.9	1.7391	5.1908	6.9299
610.18	1944.7	2554.9	610.63	2129.6	2740.3	1.7907	5.0926	6.8833
631.68	1927.9	2559.5	632.20	2114.3	2746.5	1.8418	4.9960	6.8379
653.24	1910.8	2564.1	653.84	2098.6	2752.4	1.8925	4.9010	6.7935
674.87	1893.5	2568.4	675.55	2082.6	2758.1	1.9427	4.8075	6.7502
696.56	1876.0	2572.5	697.34	2066.2	2763.5	1.9925	4.7153	6.7078
718.33	1858.1	2576.5	719.21	2049.5	2768.7	2.0419	4.6244	6.6663
740.17	1840.0	2580.2	741.17	2032.4	2773.6	2.0909	4.5347	6.6256
762.09	1821.6	2583.7	763.22	2015.0	2778.2	2.1396	4.4461	6.5857
784.10	1802.9	2587.0	785.37	1997.1	2782.4	2.1879	4.3586	6.5465
806.19	1783.8	2590.0	807.62	1978.8	2786.4	2.2359	4.2720	6.5079
828.37	1764.4	2592.8	829.98	1960.0	2790.0	2.2835	4.1863	6.4698

The quality is often expressed in percent. As an alternate description of a liquid-vapor two-phase state, we often employ the property called the *moisture content,* denoted by the symbol y, which is defined as *the fraction of the mass in the liquid phase*. Then

$$y = \frac{m_f}{m_f + m_g} = \frac{m_f}{m} \tag{8.24}$$

From equations (8.23) and (8.24) it follows that

$$y = 1 - x \tag{8.25}$$

Thus, when a system enters the two-phase region of the P-v-T surface, we lose the temperature or the pressure as an independent property but gain the quality or the moisture content as an independent property.

The average value for a specific property of a two-phase state is defined as the corresponding total extensive property divided by the total mass of the system. For example, the average specific volume of a two-phase state is

$$v = \frac{V}{m} \tag{8.26}$$

It must be remembered that this is not a true specific property since a random sample from the two-phase state might contain only a single phase, and therefore will not have this volume per unit mass. Substituting equation (8.19) into equation (8.26), we obtain

$$v = \frac{m_f v_f + m_g v_g}{m_f + m_g} \tag{8.27}$$

Therefore,

$$v = \frac{m_f}{m_f + m_g} v_f + \frac{m_g}{m_f + m_g} v_g \tag{8.28}$$

Substituting equations (8.23) and (8.24) into equation (8.27), we obtain

$$v = yv_f + xv_g \tag{8.29}$$

or

$$v = (1 - x)v_f + xv_g \tag{8.30}$$

Equation (8.30) can be simplified to

$$v = v_f + x(v_g - v_f) \tag{8.31}$$

or

$$v = v_f + xv_{fg} \tag{8.32}$$

Here v_{fg} is a symbol representing the difference in the specific volume for the saturated liquid and vapor phases, $v_{fg} = v_g - v_f$. This specific volume difference is a property fixed by the saturation temperature or pressure and is often tabulated for convenience in calculation. In a similar fashion, the other specific properties can be calculated.

$$u = u_f + xu_{fg} \tag{8.33}$$
$$h = h_f + xh_{fg} \tag{8.34}$$
$$s = s_f + xs_{fg} \tag{8.35}$$

In many practical situations it is necessary to compute the properties of a two-phase state from a knowledge of the temperature or pressure and the value of an extensive property and the mass of the system. We begin the calculation by first dividing the extensive property by the mass, thereby converting the extensive property into an average specific property. We next establish that the system is indeed in a two-phase state. This information is obtained by recalling that the specification of the temperature or the pressure immediately fixes all saturated specific properties of the two phases. Then, if the average specific property that we have just calculated has a value between the two tabulated saturation values for that property at the given temperature or pressure, the system is in a two-phase state. However, if the value of the specific property had been less than the value of that specific property for the saturated liquid state at the system temperature or pressure, the system would have been in a compressed liquid state. Conversely, if the value of the specific property had been greater than the value of that specific property for the saturated vapor state, the system would have been in a superheated vapor state.

Once we have established the fact that the system is in a two-phase state, we next compute the quality of the two-phase state by substitution of the known specific property into the appropriate equation of equations (8.32) through (8.35). The value for the quality together with the tabulated values for the saturated properties then permits us to evaluate all other average specific properties for the two-phase state. These specific properties can then be converted into extensive properties simply by multiplying by the mass.

8E.1 Sample Problem: A system with a total volume of 3 m^3 contains 2 kg of H_2O at a pressure of 7×10^4 N/m^2. What is the entropy of this steam?

Solution: Following the procedure just described, we first calculate the average specific volume.

$$v = \frac{3}{2} = 1.5 \text{ m}^3/\text{kg}$$

Entering the table of properties of H_2O at a pressure 7×10^4 N/m^2 we find at the top of the column for this pressure the specific volume of saturated liquid and of saturated vapor.

$$v_f = 1.0360\text{E}{-3} \text{ m}^3/\text{kg} \quad v_g = 2.365 \text{ m}^3/\text{kg}$$

Since $1.0360\text{E}{-3} < 1.5 < 2.365$ the system is in a two-phase state. We now compute the quality. From equation (8.32)

$$x = \frac{v - v_f}{v_{fg}} = \frac{v - v_f}{v_g - v_f}$$

Substituting the appropriate values, we obtain

$$x = \frac{1.5 - 0.001}{2.365 - 0.001} = 0.634$$

Then according to equation (8.35) the specific entropy becomes

$$s = s_f + xs_{fg} = 1.1919 + (0.634)(6.2878)$$

or

$$s = 1.1919 + 3.9865 = 5.1784 \text{ kJ/kgK}$$

The total entropy of the steam is then

$$S = ms = 2(5.1784) = 10.3568 \text{ kJ/K}$$

8.12 Metastable States

All of the equilibrium properties of the pure substance model that we have been discussing thus far are in fact the properties of *stable equilibrium* states. A system can change from one stable equilibrium state to another only if there is a corresponding, finite, permanent change of state of the environment of the system. *Metastable* equilibrium states are states of a system that can be changed at a finite rate to some other stable equilibrium state by means of a finite, but temporary, change of state of the environment in excess of a certain minimum value. In other words, when a system is in a metastable equilibrium state, a fluctuation of sufficient magnitude in the state of the environment can cause the system to experience a change of state at a finite rate. On the other hand, had the initial state of the system been a stable equilibrium state, no finite change of state could have occurred.

For example, when very clean deaerated water is heated at constant pressure in a new unscratched glass container, the water can be heated significantly above the boiling point corresponding to the pressure without the appearance of any bubbles of the vapor phase. This liquid is termed a superheated liquid and is in a metastable state because the introduction of a small quantity of the vapor phase will cause the rapid (explosive) change to a stable two-phase state consisting of a liquid phase and a vapor phase in mutual equilibrium. The superheated liquid is maintained in the metastable state by the action of surface tension which is ignored in considerations of the pure substance. The extra energy needed to form the liquid surface of the first vapor bubble prevents bubble formation. As a result of metastability of the liquid-vapor phase change, a boiling liquid will normally be at a temperature somewhat higher than the saturation temperature corresponding to the pressure.

Metastable states are also associated with the freezing of a liquid. A clean liquid may be supercooled considerably below the freezing point without the formation of any solid phase. Here again the system is maintained in the metastable state by the surface energy required to form the first small crystal of solid. The introduction of a small amount of solid will produce a rapid solidification in a supercooled liquid.

A metastable state associated with a phase change is a single-phase state where at stable equilibrium two phases are present. The energy associated with the formation of the first particle of the second phase is not available so the substance remains in the single phase. Quite often the necessary energy is provided by an external perturbation such as a vibration, and the substance violently reverts to the stable two-phase state.

In some situations metastable states can be used to advantage. In fact many heat treating and other metallurgical processes involve freezing the material in a metastable state by cooling the substance at a rate sufficient to prevent the formation of the stable equilibrium state. Another example of the useful exploitation of metastable states is the bubble chamber which is used to detect the presence of high energy nuclear and subnuclear particles. The particles pass through a liquid, usually liquid hydrogen, which has been intentionally placed in a metastable superheated liquid state. The interaction between the particles and the liquid releases sufficient energy to produce a series of bubbles along the path of the particles.

On the other hand, metastable states can sometimes be troublesome. For example, the steam (vapor) flowing at high speed through the nozzles and blades of a steam turbine may be supercooled well below the saturation temperature corresponding to the local pressure. Suddenly, the stream of vapor is perturbed sufficiently to produce a liquid nucleus, and a rapid condensation process ensues, perhaps resulting in damage to the turbine blades.

8.13 Applications of the Pure Substance Model

In the present chapter we have described the thermodynamic properties of the pure substance model from a phenomenological point of view. This model is very general for a simple system without chemical reactions. Because of this general nature, the pure substance model is useful in a very broad range of physical situations and thus occupies a prominent position in the science of thermodynamics.

The general nature of the model requires more care in the application of the first and second laws of thermodynamics than was necessary for the ideal gas model. To be more specific, the possibility of a phase change in the pure substance model necessitates the evaluation of properties from tables or charts. As we saw in Section 8.9, the proper use of these tables and charts requires that we first establish the number and nature of the phases present in a given state. Having done this, we can then apply the first and second laws of thermodynamics by using the appropriate table or chart to evaluate the relevant properties.

The calculations for a change of state in the pure substance model are best illustrated by numerical examples.

8E.2 Sample Problem: A closed, rigid container with an internal volume of 1 m^3 contains 2kg of steam at a pressure of 7×10^4 N/m^2. The system is heated quasi-statically until the final absolute pressure is 3×10^5 N/m^2.

(a) What are the initial and final temperatures of the system?

(b) What is the quality of the steam in the initial and final states?

(c) What is the work transfer experienced by the steam during the change of state?

(d) What is the heat transfer experienced by the steam during the change of state?

Solution:

(a) From the table of the properties of H_2O at pressure of 7×10^4 N/m^2.

$$v_f = 1.0360\text{E}-3 \text{ m}^3/\text{kg and } v_g = 2.365 \text{ m}^3/\text{kg}$$

The specific volume of the steam in the container is

$$v_1 = 1 \text{ m}^3/2 \text{ kg} = 0.5 \text{ m}^3/\text{kg}$$

Since at this pressure $v_f < v_1 < v_g$, the steam in the initial state is a two-phase mixture of vapor and liquid. Therefore, the temperature of the steam in this state is the saturation temperature at 7×10^4 N/m^2. From the top of the pressure column in the table this temperature is $T_1 = 89.95$ C (the number in parentheses after the pressure). For the final state with an absolute pressure of 3×10^5 N/m^2, the table gives $v_f = 1.0732\text{E}-3$ m^3/kg and $v_g = 0.6058$ m^3/kg. Since the tank is closed, the total mass of steam (liquid and vapor) does not change. Also, since he tank is rigid, its volume does not change. Thus, $v_2 = 0.5$ m^3/kg. Again, since $v_f < v_2 < v_g$ at this new pressure, the final state is a two-phase state with a temperature equal to the saturation temperature at 3×10^5 N/m^2. From the table, $T_2 = 133.55$ C. The path of the process would appear as shown in the P-v and T-s diagrams in Figure 8E.2.

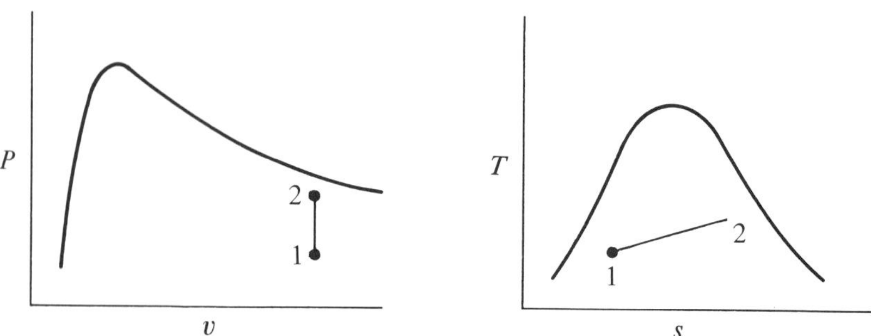

Figure 8E.2 P-v and T-s diagrams

(b) The quality in either the initial or final state is determined from the average specific volume which is 0.5 m^3/kg for both states. Then from equation (8.31)

$$x = \frac{v - v_f}{v_g - v_f}$$

The value for the initial state is

$$x_1 = \frac{0.5 - 0.001036}{2.365 - 0.001036} = 0.2111$$

For the final state

$$x_2 = \frac{0.5 - 0.0010732}{0.6058 - 0.0010732} = 0.8250$$

(c) Since we are modeling the steam as a pure substance, the only possible work transfer in this quasi-static case is associated with the displacement of the boundary of the system. Since the container is rigid and since the two phases fill the container, there is no work transfer for this quasi-static process. That is,

$$W_{1-2} = \int P\,dv = 0$$

(d) From the first law of thermodynamics,

$$Q_{1-2} = U_2 - U_1 + W_{1-2}$$

which in this case reduces to

$$Q_{1-2} = U_2 - U_1$$

since $W_{1-2} = 0$. From equation (8.33),

$$u = u_f + xu_{fg}$$

Then for the initial state the table gives $u_f = 376.63$ kJ/kg and $u_{fg} = 2117.9$ kJ/kg and

$$u_1 = 376.63 + (0.2111)(2117.9) = 376.63 + 447.09$$
$$u_1 = 823.71 \text{ kJ/kg}$$

For the final state, the table gives $u_f = 561.15$ kJ/kg and $u_{fg} =$ 1982.5 kJ/kg. Then

$$u_2 = 561.15 + (0.825)(1982.5) = 561.15 + 1635.56$$
$$u_2 = 2196.71 \text{ kJ/kg}$$

Then the quasi-static heat transfer for this constant volume change of state is

$$Q_{1-2} = (2.0)(2196.71 - 823.71) = 2746 \text{ kJ}$$

8E.3 Sample Problem: A closed system contains 0.10 kg of steam at a temperature of 180 C and at a pressure of 10^5 N/m^2. The steam is expanded by means of a process, modeled as adiabatic and reversible, to a final pressure of 2×10^4 N/m^2.

(a) What is the temperature of the steam in the final state?

(b) What is the heat transfer for the steam during the expansion process?

(c) What is the work transfer for the steam during the expansion process?

Solution:

(a) Since the process connecting the end states is both reversible and adiabatic, the second law of thermodynamics, equation (7.35), shows that the entropy of the steam is constant during this process. From the table of the properties of H_2O the initial state is clearly a superheated state with $s_1 =$ 7.7489 kJ/kgK. For the final state at 2×10^4 N/m^2, the table also shows that $s_f = 0.8329$ kJ/kgK, $s_g = 7.9085$ kJ/kgK, and $s_{fg} = 7.0765$ kJ/kgK. Thus, since $s_f < s_1 = s_2 < s_g$ at the pressure 2×10^4 N/m^2, the final state must be a two-phase state. From equation (8.35)

$$x_2 = \frac{s_2 - s_f}{s_{fg}}$$

Substituting the appropriate values for s_2, s_f, and s_{fg}, we have

$$x_2 = \frac{7.7489 - 0.8320}{7.0765} = 0.9774$$

Clearly the temperature in this final state is the saturation temperature at 2×10^4 N/m². From the table, $T_2 = 60.60$ C. The path of the process would appear as shown in the *P-v* and *T-s* diagrams in Figure 8E.3.

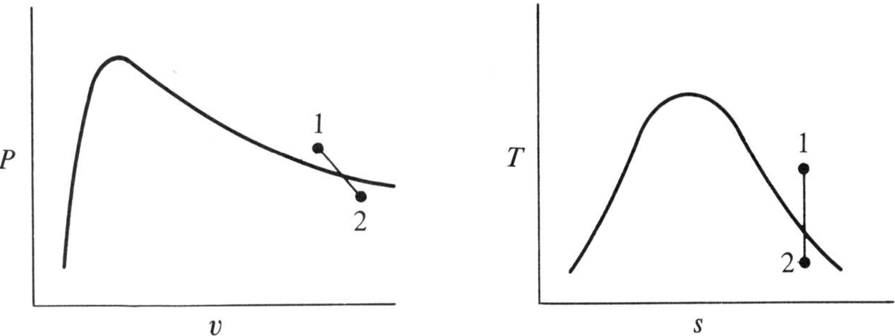

Figure 8E.3 *P-v* and *T-s* diagrams

(b) Since the process is being modeled as an adiabatic process, there is no heat transfer during the expansion.

$$Q_{1-2} = 0$$

(c) To determine the work transfer, we use the first law of thermodynamics.

$$W_{1-2} = Q_{1-2} - (U_2 - U_1)$$

Since in this case $Q_{1-2} = 0$

$$W_{1-2} = -(U_2 - U_1) = -m(u_2 - u_1)$$

From equation (8.33) and the data of the table, $u_2 = 251.38 + (0.9774)(2205.31) = 2406.86$ kJ/kg. For the specific internal energy of the initial state, the table gives

$$u_1 = 2627.9 \text{ kJ/kg}$$

Thus,

$$W_{1-2} = (0.1)(2627.9 - 2406.86) = 22.10 \text{ kJ}$$

8E.4 Sample Problem: A piston-cylinder apparatus is fitted with a frictionless piston capable of maintaining a constant absolute pressure of 10^5 N/m² on 0.1 kg of steam. The steam, which is all saturated liquid in the initial state, is heated quasi-statically until all of the liquid is converted to saturated vapor.

(a) What is the temperature of the steam in the initial and final states?

(b) What is the reversible heat transfer necessary to achieve this change of state?

(c) What is the reversible work transfer associated with this change of state?

Solution:

(a) Since the process involves a change of state from saturated liquid to saturated vapor at constant pressure and since the pressure and temperature are directly related for two-phase equilibrium states of a pure substance, the temperature must remain constant throughout the process. The temperature is the saturation temperature corresponding to the pressure of the system, 10^5 N/m² absolute. From the table of the properties of H_2O the saturation temperature corresponding to this pressure is $T_1 = T_2 = 99.63$ C. Thus the path of this process would appear as shown in the *P-v* and *T-s* diagrams in Figure 8E.4.

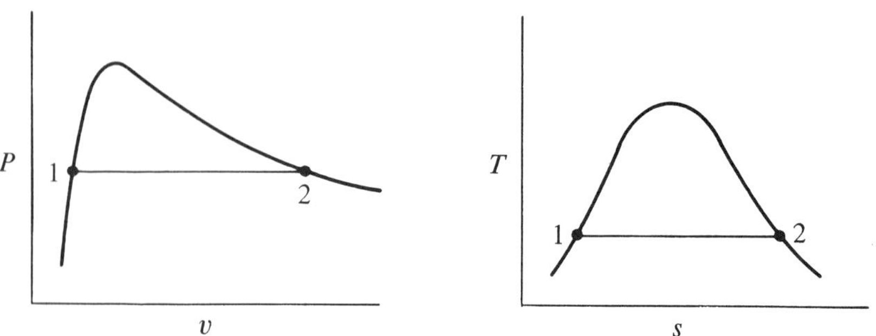

Figure 8E.4 *P-v* and *T-s* diagrams

(b) Since the process from state 1 to state 2 is quasi-static and isothermal, we can according to the second law of thermodynamics calculate the heat transfer for this process from the relation

$$Q_{1-2} = mT(s_2 - s_1)$$

Since state 1 is a saturated liquid state and state 2 is a saturated vapor state, the expression for the heat transfer reduces to

$$Q_{1-2} = mT(s_g - s_f) = mTs_{fg}$$

From the table at 10^5 N/m², $s_{fg} = 6.0568$ kJ/kgK. Then since $T = 99.63 + 273.15 = 372.78$ K

$$Q_{1-2} = (0.1)(372.78)(6.0568) = 225.8 \text{ kJ}$$

Alternatively, we could have calculated the heat transfer from the first law of thermodynamics. For this quasi-static constant pressure process the first law can be written

$$Q_{1-2} - mP(v_2 - v_1) = m(u_2 - u_1)$$

or

$$Q_{1-2} = m(u_2 - u_1) + m(P_2v_2 - P_1v_1) = m(h_2 - h_1)$$

Again since the end states are saturated states, this expression reduces to

$$Q_{1-2} = (0.1)(2258.0) = 225.8 \text{ kJ}$$

(c) The work transfer for this quasi-static constant pressure process is

$$W_{1-2} = mP(v_2 - v_1)$$

Since the end states are saturated states,

$$W_{1-2} = mP(v_g - v_f)$$

From the table, $v_g = 1.6940$ m^3/kg and $v_f = 0.0010$ m^3/kg. Then

$$W_{1-2} = (0.1)(10^5)(1.6940 - 0.0010) = 1.693 \times 10^4 \text{ J} = 16.93 \text{ kJ}$$

Note that the above results are in distinct contrast to what we would have expected from an ideal gas which was subjected to an isothermal process. For the ideal gas the constant temperature requires that the internal energy be constant. Then according to the first law of thermodynamics, the work transfer and heat transfer are equal. Clearly this is not the case in the present example for which the heat transfer is a whole order of magnitude greater than the work transfer. Obviously the internal energy in the present case is not constant. In fact the change in internal energy is equal to mu_{fg}

$$m(u_2 - u_1) = (0.1)(2088.7) = 208.87 \text{ kJ}$$

Problems

8.1 A boiling water nuclear reactor contains m kg of saturated liquid water at 260 C. In case the reactor pressure vessel fails, a secondary containment structure must be provided to avoid the spread of radioactive water.

(a) How large must the secondary containment chamber be if the maximum design pressure is 2×10^5 N/m^2 and the space between the reactor and the containment vessel is initially at zero pressure?

(b) In the case of a sudden release of the H_2O, it would be reasonable to assume that pressure equilibrium would be established in a time too short for significant heat transfer to the H_2O. Determine the final temperature of the H_2O.

8.2 A sealed tube of liquid and vapor H_2O is heated at constant volume.

(a) Find the proportions *by volume* of liquid and vapor H_2O sealed in the tube at atmospheric pressure that will pass through the critical state when heated.

(b) What is the necessary heat transfer to the contents of the tube between atmospheric pressure and the critical state if the enclosed volume is 8×10^{-6} m^3?

8.3 A rigid vessel (constant volume bomb) is charged with carbon dioxide at 20 C. If the initial charge contains the correct proportions of liquid and vapor, the carbon dioxide will pass through the critical state when heated with the fill line closed off. In the initial state, what are the proper proportions by volume of liquid and vapor CO_2 that will produce the desired change of state when the mixture is heated?

8.4 One kg of Freon-12 at an initial pressure of 686.5×10^3 N/m^2 and 30 C undergoes an adiabatic free expansion (unrestrained expansion into a vacuum) in which the final volume is 20 times the original volume. Find the changes in the temperature and in the entropy for this process by each of the following methods:

(a) Assuming Freon is an ideal gas ($R = 0.0687$ kJ/kgK).

(b) Using the tabulated pure substance properties.

8.5 One kg of nitrogen at a temperature of 300 K and a pressure of 10^5 N/m^2 is compressed reversibly and isothermally to a pressure of 2×10^7 N/m^2.

(a) Using the ideal gas model, compute the work transfer and heat transfer for the compression process.

(b) Using the tabulated properties of nitrogen based on the pure substance model, compute the work transfer and heat transfer for the process.

(c) What error is involved in using the ideal gas model at these temperatures and pressures?

(d) Repeat parts (a) through (c) above for a reversible isothermal compression process between the same two limiting pressures but at a temperature of 90 K.

(e) From these results what conclusions can you draw regarding the use of the ideal gas model?

8.6 A new method of stress-relieving metal castings has been developed. The method, known as "cryoquenching", consists of plunging the casting into liquid nitrogen at a pressure of 10^5 N/m^2 and allowing the casting to attain thermal equilibrium with the liquid before it is removed.

In a particular application of this technique, a steel casting of 10 kg is to be cooled at a constant pressure of 10^5 N/m^2 from a temperature of 300 K to the saturation temperature of nitrogen. Only 5 kg of nitrogen are available, not all of which are liquid. Assume that the casting can be modeled as a pure thermal system with a specific heat of 0.419 kJ/kgK.

(a) What is the minimum fraction of the mass of nitrogen that must be liquid in order to attain an equilibrium temperature equal to the saturation temperature of nitrogen at a pressure of 10^5 N/m^2?

(b) From an explicit calculation of the entropy change of some suitably defined system, determine whether the cryoquenching process is reversible or irreversible.

8.7 A Dewar storage vessel has a volume of 25 liters (1 liter $= 10^{-3}$ m^3) and contains liquid nitrogen at 10^5 N/m^2. At some particular instant of time when the Dewar contains 22 liters of liquid and 3 liters of vapor, the vapor vent tube is accidentally sealed off. Heat transfer calculations show that the heat transfer from the environment to the liquid nitrogen across the vacuum space is 1.67 J/s. Stress analysis shows that the containment vessel will rupture when the internal pressure reaches 5×10^5 N/m^2.

(a) How long will it take to rupture the vessel?

(b) Estimate how much mass will be lost if the vapor vent tube is opened when the pressure reaches 5×10^5 N/m^2.

8.8 As shown in Figure 8P.8, 5 kg of water at 30 C are contained in a verticle cylinder by a frictionless piston with a mass such that the pressure on the water is 1.4×10^6 N/m^2. The

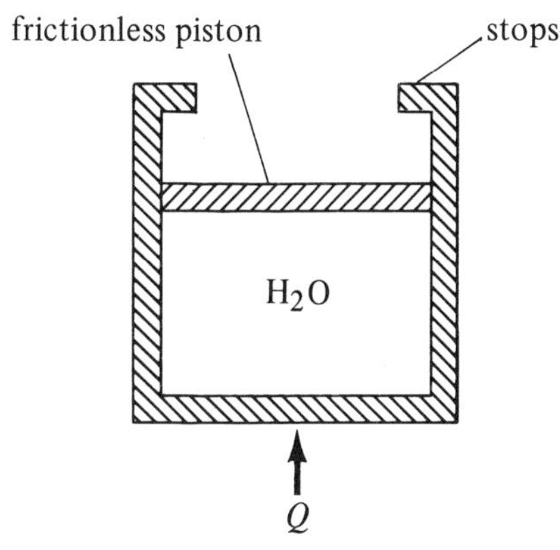

Figure 8P.8

water experiences a quasi-static heat transfer, causing the piston to rise until it reaches the stops at which point the volume inside the cylinder is 0.5 m^3. The heat transfer process continued until the water exists as a saturated vapor.

- **(a)** Show the heat transfer process on a *T-s* diagram.
- **(b)** What is the final pressure of the vapor in the cylinder?
- **(c)** Determine the heat transfer to the H_2O for this process.
- **(d)** Evaluate the work transfer to the piston.

8.9 The piston-cylinder apparatus shown in Figure 8P.9 is fitted with a leakproof, frictionless piston loaded with enough weights to maintain a pressure of 980.7×10^3 N/m^2. Initially the piston rests upon stops so that the volume trapped in the cylinder is 0.01 m^3. This volume is initially filled with 1 kg of Freon-12 at a pressure 39.22×10^3 N/m^2. The Freon now experiences a quasi-static heat transfer until its temperature reaches 170 C.

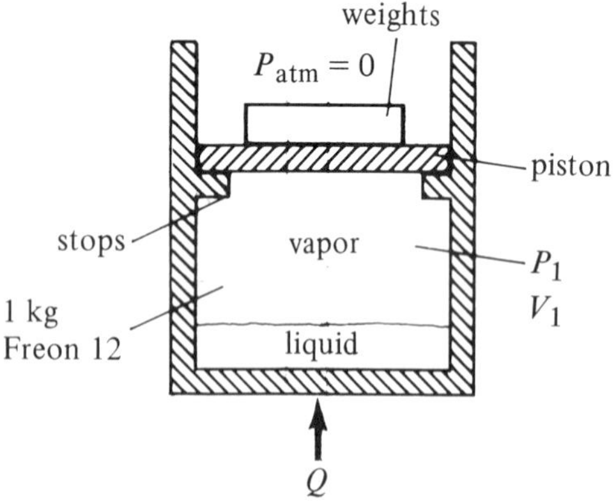

Figure 8P.9

- **(a)** Show the path of the process on *P-v* and *T-s* diagrams.
- **(b)** What is the temperature in the initial state?
- **(c)** What fraction of the total mass is liquid in the initial state?
- **(d)** What fraction of the initial volume is occupied by the liquid?
- **(e)** What is the temperature of the Freon as the piston just begins to lift off of the stops?
- **(f)** What is the heat transfer necessary to just lift the piston off the stops?
- **(g)** What is the work transfer experienced by the Freon during the process by which it is heated from its initial temperature to 170 C?

8.10 As shown in Figure 8P.10, an adiabatic cylinder is fitted with a frictionless, adiabatic piston with an area of 10^{-3} m^2 and a mass of 100.1 kg ($g = 9.80$ m/s^2). The volume trapped between the piston face and the end of the cylinder is partitioned off into two volumes (not equal) by a rigid, adiabatic membrane. The volume between the piston face and the membrane contains 1.25 kg of saturated liquid Freon-12 while the volume between the membrane and the end of the cylinder contains 0.01 kg of Freon at a pressure of 196.1×10^3 N/m^2 and a temperature of 10 C. At some

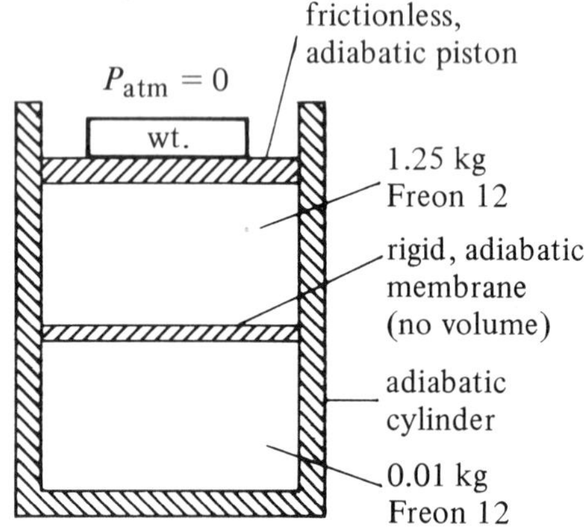

Figure 8P.10

instant of time, the membrane is destroyed and the system comes to a new equilibrium state.

(a) Determine the quality of the final equilibrium state.
(b) What is the final volume of the system?
(c) Is there any work transfer experienced by the 1.26 kg of Freon during the process by which the system comes to equilibrium? If so, how much?

8.11 Consider a H_2O system of 1 kg of vapor, 1 kg of liquid, and 1 kg of solid all in thermodynamic equilibrium (state 1). The mixture is stirred by a paddle wheel while the volume remains constant until all the ice has melted (state 2). In a second process the stirring is continued at constant volume until only vapor remains (state 3). In an alternative process the system at state 2 is stirred at constant pressure until only vapor remains (state 4).

(a) What are the mass proportions of liquid and vapor at state 2?
(b) Determine the temperature, pressure and volume for states 3 and 4.
(c) What are the paddle wheel work transfers for process 1-2, process 2-3, and process 2-4 if these processes are adiabatic?
(d) What is the entropy change for each of the three processes?

Triple Point Data for H_2O

$P_{TP} = 611.3\ \text{N/m}^2 \quad T_{TP} = 0.01\ \text{C}$

Phase	Enthalpy (kJ/kg)	Specific Volume (m^3/kg)	Entropy (kJ/kgK)
Solid	−333.40	1.0908−3	−1.221
Liquid	0.01	1.0002−3	0.0000
Vapor	2501.4	206.1	9.156

8.12 A cylinder containing 1 kg of steam is closed by a slowly moving piston traveling back and forth between two extreme positions. At one of these extreme positions the spécific volume available to the steam is 0.300 m^3/kg and at the other is 1.344 m^3/kg. Two heat reservoirs are available, one at 260 C and another at 140 C. During the progress of the piston from one extreme position to the other, the steam experiences heat transfer at such a rate as to keep the temperature of the steam constant and equal to the temperature of one of the heat reservoirs. With the piston at either end, the steam experiences heat transfer so that the temperature changes from the temperature of one heat reservoir to the temperature of the other heat reservoir.

(a) Show the cycle on P-v and T-s diagrams.
(b) Find the net work transfer per cycle for the engine.
(c) Find the energy conversion efficiency of the engine.

8.13 Consider the system shown in Figure 8P.13. Tank A initially contains 5 kg of steam at 600×10^3 N/m², 573.15 K, and is connected through a valve to cylinder B fitted with a frictionless piston. An internal pressure of 120×10^3 N/m² is required to balance the weight of the piston. Also, the piston rests on stops so that the initial volume of B is 0.5 m³. This volume is initially evacuated, $P = 0$. The connecting valve is opened until the pressure in A is reduced to 120×10^3 N/m². Assuming the entire process to be adiabatic and that the steam which remains in A has undergone a reversible adiabatic process, determine the work transfer to the piston and the final tem-

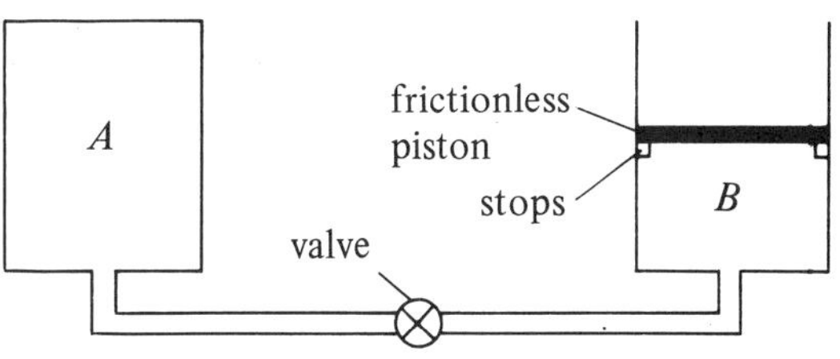

Figure 8P.13

perature of the steam in cylinder *B*. Note the final temperature in tank *A* and cylinder *B* are not equal.

8.14 A rigid, adiabatic container is divided into two compartments *A* and *B* which are separated by a rigid, adiabatic membrane as shown in Figure 8P.14. Compartment *A* contains 2 kg of saturated vapor CO_2 at 683.5×10^3 N/m^2 and compartment *B* contains 8 kg of saturated liquid CO_2 at 6.065×10^6 N/m^2. The rigid, adiabatic membrane is then destroyed, and all of the CO_2 comes into a new equilibrium state.

(a) Determine the temperature and pressure of the final equilibrium state.

(b) What is the quality of the CO_2 in the final equilibrium state?

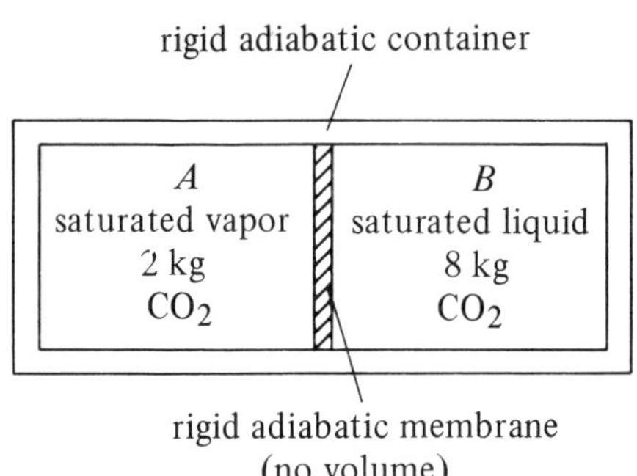

Figure 8P.14

HINT: Establish two independent equations for the final quality. Iterate with the saturation table to find the temperature which gives the same quality in both equations.

CHAPTER 9

Relations Between Properties of a Pure Substance

9.1 Introduction

In the previous chapter we treated the pure substance model from a phenomenological point of view. That is, we subjected a pure substance to a series of hypothetical experiments that enabled us to determine qualitatively the general behavior of the pure substance. We then showed how the various phases interact to maintain equilibrium and illustrated the use of the properties to determine the thermodynamic behavior of the system.

In this chapter, we shall show that the first and second laws of thermodynamics require that the properties of the pure substance model are interrelated by thermodynamic functions. Further, we will show the influence of this interrelationship on the equilibrium among phases. In concluding, we will demonstrate the use of the thermodynamic functions to determine all of the properties of the pure substance model from a minimum number of experimental measurements.

9.2 Thermodynamic Functions

Recall that for a closed system we have defined the canonical relation of thermodynamics, equation (8.1), to be

$$du = T\,ds - P\,dv \tag{9.1}$$

where we have rewritten equation (8.1) for a unit mass. The first law of thermodynamics guarantees that u is a property so that equation (9.1) can always be integrated to give the functional relation between the internal energy, the entropy, and the volume. Thus,

$$u = u(s,v) \tag{9.2}$$

where the specific entropy, s, and the specific volume, v, are the independent variables of the function. The state principle guarantees that equation (9.1) is complete so that the variables s and v are sufficient to specify completely every equilibrium state of the pure substance for which s and v are actually independent.

Equation (9.2) implies mathematically that we could write the total differential of the internal energy in the form

$$du = \left(\frac{\partial u}{\partial s}\right)_v ds + \left(\frac{\partial u}{\partial v}\right)_s dv \tag{9.3}$$

Upon comparing equations (9.1) and (9.3), we conclude that the following identities must exist:

$$T = \left(\frac{\partial u}{\partial s}\right)_v \quad \text{and} \quad -P = \left(\frac{\partial u}{\partial v}\right)_s \tag{9.4}$$

These identities are often taken as the *definitions of temperature and pressure* in an abstract mathematical sense. These definitions are abstract in that they presuppose that the actual function $u(s,v)$ is in hand. In reality, working definitions of temperature and pressure are necessary to actually obtain $u(s,v)$.

In the canonical relation, equation (9.1), the thermal variables (properties) T and s are associated in a single term (the reversible heat transfer interaction). Although so associated, the variables (properties) T and s are contrasted in that the temperature is intensive and dependent while the entropy is extensive and independent. Thus, the variables (properties) T and s are termed *conjugate*. For analogous reasons, the variables (properties) P and v are also termed conjugate.

If we examine equations (9.2), (9.4) and (9.1), we see that equation (9.2) contains all of the information necessary to completely specify all reversible interactions for a given path for the pure substance in terms of the independent properties s and v. The temperature and the pressure are obtained from equations (9.4) in terms of s and v. Thus, the reversible heat transfer, $\int T\,ds$, and the reversible work transfer, $\int P\,dv$, for a given path are completely specified in terms of s and v.

Since $u(s,v)$ interrelates both pairs of the conjugate variables (properties) for the two reversible energy transfer interactions, the function $u(s,v)$ may be thought of as a complete constitutive relation for the pure substance. In contrast, in order to have the same information as is implied by $u(s,v)$, we must have both T as a function of s and v, $T(s,v)$, and P as a function of s and v, $P(s,v)$. The function $T(s,v)$ is necessary to integrate $T\,ds$ (the reversible heat transfer) and the function $P(s,v)$ is necessary to integrate $P\,dv$ (the reversible work transfer).

A single thermodynamic function of one thermal variable (T or s) and one mechanical variable (P or v) which yields the two conjugate variables (s or T) and (v or P) by partial differentiation is called a *characteristic thermodynamic function*. Thus, $u(s,v)$ is one such function. A characteristic thermodynamic function contains complete information about the thermodynamic properties and is therefore a complete constitutive relation for the substance. The independent variables (properties) for the four possible characteristic functions are:

$$s,v \quad s,P \quad T,v \quad T,P$$

Each characteristic function is the integral of a total differential expression similar to the canonical relation, equation (9.1). The corresponding characteristic thermodynamic functions are the internal energy, u, the enthalpy, h, the Helmholtz free energy, f, and the Gibbs free energy, g.

We can think of the independent variables (properties) as a set of coordinates, so that the various characteristic thermodynamic functions should be obtainable from one another by means of a suitable transformation of coordinates. This transformation can be readily executed by employing the Legendre transformation.[1] The Legendre transformation leads us to the following rule for transforming the canonical relation:

> To replace one of the independent variables (properties) in an exact differential such as the canonical relation with its conjugate variable (property), subtract from the differential of the dependent variable the differential of the product of the two conjugate variables (properties).

Thus, starting with the differential of the internal energy and applying the rule to each variable (property) in turn, we obtain the three other characteristic functions. In particular, we obtain the second characteristic function by replacing v with P.

$$du - d[(-P)v] = T\,ds - P\,dv + P\,dv + v\,dP \tag{9.5}$$

This simplifies to

$$dh = T\,ds + v\,dP \tag{9.6}$$

since the specific enthalpy, h, was defined (see Chapter 8) as

$$h = u + Pv \tag{9.7}$$

Equation (9.6) can then be integrated to give the characteristic thermodynamic function

$$h = h(s,P) \tag{9.8}$$

By comparing the total differential of equation (9.8) with equation (9.6), we obtain the identities

$$T = \left(\frac{\partial h}{\partial s}\right)_P \text{ and } v = \left(\frac{\partial h}{\partial P}\right)_s \tag{9.9}$$

[1] Callen, H. B., *Thermodynamics* John Wiley, New York, 1960, pp. 90–101.

Observe that in applying the rule for the transformation of independent variables (properties) we have carefully associated the proper sign with the variable as required by the canonical form, equation (9.1). For example we used $(-P)$ instead of P.

The third characteristic function is obtained by replacing s with T.

$$du - d(Ts) = T\,ds - P\,dv - T\,ds - s\,dT \qquad \textbf{(9.10)}$$

which simplifies to

$$d(u - Ts) = df = -s\,dT - P\,dv \qquad \textbf{(9.11)}$$

Since the Helmholtz free energy, F, was defined by equation (7.85), we may define the specific Helmholtz free energy as

$$f = u - Ts \qquad \textbf{(9.12)}$$

Equation (9.11) then gives the characteristic function

$$f = f(T,v) \qquad \textbf{(9.13)}$$

and the identities

$$-s = \left(\frac{\partial f}{\partial T}\right)_v \text{ and } -P = \left(\frac{\partial f}{\partial v}\right)_T \qquad \textbf{(9.14)}$$

The fourth characteristic function and the associated relations are obtained by replacing v with P and s with T.

$$\begin{aligned} du - d[(-Pv] - d(Ts) &= T\,ds - P\,dv + P\,dv \\ &\quad + v\,dP - T\,ds - s\,dT \qquad \textbf{(9.15)} \\ dg &= -s\,dT + v\,dP \qquad \textbf{(9.16)} \\ g &= h - Ts \qquad \textbf{(9.17)} \\ g &= g(T,P) \qquad \textbf{(9.18)} \\ -s &= \left(\frac{\partial g}{\partial T}\right)_P \text{ and } v = \left(\frac{\partial g}{\partial P}\right)_T \qquad \textbf{(9.19)} \end{aligned}$$

Note that the specific Gibbs free energy, g, is defined from equation (7.88).

9.3 Maxwell Relations

The first and second laws of thermodynamics guarantee that the quantities u, h, f, and g are properties. Because these quantities are properties, a great deal of information can be derived from their total differentials. We have already seen how the conjugate variables (properties) can be obtained from the coefficients of the differentials of the independent variables (properties). We now make use of the fact that the total differential of a property is exact, and hence the mixed second-order partial derivatives of the dependent property are equal. This implies that the order of partial differentiation is immaterial. That is, given a function $z = z(x,y)$ with an exact differential dz,

$$\left\{\frac{\partial}{\partial x}\left[\left(\frac{\partial z}{\partial y}\right)_x\right]\right\}_y = \left\{\frac{\partial}{\partial y}\left[\left(\frac{\partial z}{\partial x}\right)_y\right]\right\}_x \tag{9.20}$$

If we apply this requirement to the canonical relation, equation (9.1), we obtain

$$\left\{\frac{\partial}{\partial v}\left[\left(\frac{\partial u}{\partial s}\right)_v\right]\right\}_s = \left\{\frac{\partial}{\partial s}\left[\left(\frac{\partial u}{\partial v}\right)_s\right]\right\}_v \tag{9.21}$$

Substituting equations (9.4) into equation (9.21), we have

$$\left(\frac{\partial T}{\partial v}\right)_s = -\left(\frac{\partial P}{\partial s}\right)_v \tag{9.22}$$

Similarly, for the enthalpy, equation (9.6), we have

$$\left\{\frac{\partial}{\partial P}\left[\left(\frac{\partial h}{\partial s}\right)_P\right]\right\}_s = \left\{\frac{\partial}{\partial s}\left[\left(\frac{\partial h}{\partial P}\right)_s\right]\right\}_P \tag{9.23}$$

Substituting equations (9.9) into equation (9.23), we obtain

$$\left(\frac{\partial T}{\partial P}\right)_s = \left(\frac{\partial v}{\partial s}\right)_P \tag{9.24}$$

For the Helmholtz free energy, equation (9.11), we have

$$\left\{\frac{\partial}{\partial T}\left[\left(\frac{\partial f}{\partial v}\right)_T\right]\right\}_v = \left\{\frac{\partial}{\partial v}\left[\left(\frac{\partial f}{\partial T}\right)_v\right]\right\}_T \tag{9.25}$$

Substitution of equations (9.14) into equation (9.25) yields

$$-\left(\frac{\partial P}{\partial T}\right)_v = -\left(\frac{\partial s}{\partial v}\right)_T \tag{9.26}$$

Finally, for the Gibbs free energy, equation (9.16), we have

$$\left\{\frac{\partial}{\partial P}\left[\left(\frac{\partial g}{\partial T}\right)_P\right]\right\}_T = \left\{\frac{\partial}{\partial T}\left[\left(\frac{\partial g}{\partial P}\right)_T\right]\right\}_P \tag{9.27}$$

Substituting equations (9.19) into equation (9.27), we obtain

$$-\left(\frac{\partial s}{\partial P}\right)_T = \left(\frac{\partial v}{\partial T}\right)_P \tag{9.28}$$

The four equations (9.22), (9.24), (9.26) and (9.28) are known as the Maxwell relations. These expressions show that the properties s, v, T, and P are interrelated while equations (9.4), (9.9), (9.14), and (9.19) relate these four properties to the characteristic thermodynamic functions (complete constitutive relation). These eight equations together enable us to evaluate a complete and consistent set of thermodynamic properties from experimental measurements of a minimal number of properties. They also make it possible for us to determine the values of certain thermodynamic properties which cannot be measured directly.

Since these results for the pure substance are so important in thermodynamics, it is useful to repeat these results in a systematic manner as shown in Table 9.1.

9.4 The Clapeyron Relation

As an example of the applicability of the characteristic thermodynamic function and Maxwell's relations, we once again consider the equilibrium between two phases. In Sections 8.3 and 8.5, we showed that the temperature and pressure of a two-phase equilibrium state are interrelated in a way established by equation (8.16). This equation requires equal specific Gibbs free energy for the two phases. Thus, g_{fg} is identically zero and the

TABLE 9.1
Summary of Thermodynamic Relations

I. The characteristic thermodynamic functions

Property	*Function*	*Total Derivative*	
$u = u$	$u = u(s,v)$	$du = T\,ds - P\,dv$	**(9.29)**
$h = u + Pv$	$h = h(s,P)$	$dh = T\,ds + v\,dP$	**(9.30)**
$f = u - Ts$	$f = f(T,v)$	$df = -s\,dT - P\,dv$	**(9.31)**
$g = h - Ts$	$g = g(T,P)$	$dg = -s\,dT + v\,dP$	**(9.32)**

II. The first partial derivatives of the characteristic functions

$$T \equiv \left(\frac{\partial u}{\partial s}\right)_v \equiv \left(\frac{\partial h}{\partial s}\right)_P \qquad \textbf{(9.33)}$$

$$-P \equiv \left(\frac{\partial u}{\partial v}\right)_s \equiv \left(\frac{\partial f}{\partial v}\right)_T \qquad \textbf{(9.34)}$$

$$-s \equiv \left(\frac{\partial f}{\partial T}\right)_v \equiv \left(\frac{\partial g}{\partial T}\right)_P \qquad \textbf{(9.35)}$$

$$v \equiv \left(\frac{\partial h}{\partial P}\right)_s \equiv \left(\frac{\partial g}{\partial P}\right)_T \qquad \textbf{(9.36)}$$

III. The mixed second partial derivatives of the characteristic functions, the Maxwell relations

$$\frac{\partial^2 u}{\partial s \partial v} = \left(\frac{\partial T}{\partial v}\right)_s = -\left(\frac{\partial P}{\partial s}\right)_v \qquad \textbf{(9.37)}$$

$$\frac{\partial^2 h}{\partial s \partial P} = \left(\frac{\partial T}{\partial P}\right)_s = \left(\frac{\partial v}{\partial s}\right)_P \qquad \textbf{(9.38)}$$

$$\frac{\partial^2 f}{\partial T \partial v} = -\left(\frac{\partial s}{\partial v}\right)_T = -\left(\frac{\partial P}{\partial T}\right)_v \qquad \textbf{(9.39)}$$

$$\frac{\partial^2 g}{\partial T \partial P} = -\left(\frac{\partial s}{\partial P}\right)_T = \left(\frac{\partial v}{\partial T}\right)_P \qquad \textbf{(9.40)}$$

average specific Gibbs free energy for the two phases taken together is equal to the value given in equation (8.15). This value is independent of the relative amounts of the two phases present and does not depend upon the quality, x (or the moisture content, y), of the two-phase system.

If the temperature of this two-phase system is now changed by an infinitesimal amount dT to a new value $T + dT$, the pressure will also change an infinitesimal amount dP in order to preserve the equilibrium between the two phases. These infinitesimal changes are related by the expression

$$dP = \left(\frac{dP}{dT}\right)_{sat} dT \tag{9.41}$$

Here $(dP/dT)_{sat}$ is the slope of the vapor pressure curve.

This infinitesimal change in the intensive state of the two-phase system causes the specific Gibbs free energy to change from g to $g + dg$. The infinitesimal change in the Gibbs free energy is given by equation (9.32). Substituting equation (9.41) into equation (9.32), we obtain

$$dg = \left[-s + v\left(\frac{dP}{dT}\right)_{sat}\right] dT \tag{9.42}$$

Since the states involved are two-phase states, we can write equation (9.42) in the form

$$dg = \left[-s_f - xs_{fg} + (v_f + xv_{fg})\left(\frac{dP}{dT}\right)_{sat}\right] dT \tag{9.43}$$

or

$$dg = -s_f + v_f\left(\frac{dP}{dT}\right)_{sat} + x\left[-s_{fg} + v_{fg}\left(\frac{dP}{dT}\right)_{sat}\right] dT \tag{9.44}$$

Since the Gibbs free energy for a two-phase equilibrium state must be independent of the quality x, the coefficient of the quality in equation (9.44) must vanish.

$$-s_{fg} + v_{fg}\left(\frac{dP}{dT}\right)_{sat} = 0 \tag{9.45}$$

Solving equation (9.45) for the slope of the vapor pressure curve, we obtain

$$\left(\frac{dP}{dT}\right)_{sat} = \frac{s_{fg}}{v_{fg}} \tag{9.46}$$

We can substitute equation (8.14) into equation (9.46) and obtain an alternate expression.

$$\left(\frac{dP}{dT}\right)_{\text{sat}} = \frac{h_{fg}}{Tv_{fg}} \tag{9.47}$$

Both equations (9.46) and (9.47) are known as the Clapeyron relation. Although we have derived this expression for liquid-vapor two-phase equilibrium states, this result is a general one and holds for any two phases in equilibrium.

One useful application of the Clapeyron relation is to determine the slope of the line representing the locus of two-phase equilibrium states in the pressure-temperature plane (cf. Figure 8.7). Notice that this slope is expressed in terms of the difference in the saturation properties for the two phases of the particular intensive state. Actually, a particular combination of temperature and pressure represents an infinite number of equilibrium two-phase states which differ in the relative amounts of the two phases present. As a consequence, each of these two-phase states will be represented by the same point with the same slope on the pressure-temperature plane. Thus, any thermodynamic surface that uses the pressure, temperature, and some specific extensive property as coordinates (for example the P-v-T surface) will be a surface of single curvature in the two-phase region as described in Section 8.10.

Equation (9.46) or (9.47) also illustrates the difference between Figures 8.7 (a) and 8.7 (b). For a pure substance that contracts upon freezing, the numerator and the denominator on the right side of equation (9.47) have the same sign. This indicates that the slope of the locus of solid-liquid states is positive as shown in Figure 8.7 (a). Comparatively, for a pure substance that expands upon freezing, the numerator and denominator on the right side are of opposite sign so that the slope of the locus of solid-liquid states is negative as in Figure 8.7 (b). In this sense the Clapeyron relation is a good example of the manner in which the equilibrium between phases restricts the behavior of the intensive state in order to preserve this equilibrium.

The equation of the vapor pressure curve can be obtained by integrating either equation (9.46) or (9.47). For example, the difference in saturation pressure between any two-phase equilibrium states at different temperatures is given by

$$P_2 - P_1 = \int_{T_1}^{T_2} \frac{h_{fg}}{v_{fg}} \frac{dT}{T} \tag{9.48}$$

Because the specific enthalpy and specific volume of saturated states depend upon the temperature, the integral in equation (9.48) is complicated and usually must be evaluated numerically.

There is one other useful application of the Clapeyron relation. Since pressure, temperature, and specific volume are readily measured, equation (9.48) can be used to obtain the latent heat from such measurements. In those cases for which the latent heat is also measured, the Clapeyron relation provides a useful check on the consistency of the data.

An alternate method of deriving the Clapeyron relation involves the use of one of Maxwell's relations, equation (9.39), which is

$$\left(\frac{\partial s}{\partial v}\right)_T = \left(\frac{\partial P}{\partial T}\right)_v \tag{9.49}$$

The left side of equation (9.49) represents the change in entropy per unit change in volume for an isothermal change of state. Thus for isothermal change of phase, the left side of equation (9.49) becomes

$$\left(\frac{\partial s}{\partial v}\right)_T = \frac{s_g - s_f}{v_g - v_f} \tag{9.50}$$

If we substitute equation (9.50) into equation (9.49) we obtain

$$\left(\frac{\partial P}{\partial T}\right)_v = \frac{s_g - s_f}{v_g - v_f} \tag{9.51}$$

For the case in which there are two phases in equilibrium, the left side of equation (9.51) becomes an ordinary derivative since the pressure is a function of the temperature only. Then

$$\left(\frac{dP}{dT}\right)_{\text{sat}} = \frac{s_{fg}}{v_{fg}} \tag{9.52}$$

Again, substitution of equation (8.14) into (9.52) yields equation (9.47).

9.5 Calculation of Thermodynamic Properties from Experimental Measurements

From a phenomenological point of view we have already discussed in considerable detail the thermodynamic behavior of the pure substance model. We have shown how this behavior can be demonstrated by experiment. We also have shown how we can use the properties of the pure substance model, in tabular or graphical form, to evaluate the response of the pure substance model to various processes. However, this approach presupposes that by some means we have already established the properties of the pure substance. The most obvious method to establish these properties is by direct measurement; however, there are certain properties that cannot be readily measured, and we must rely on some other indirect method for their evaluation. This indirect method takes advantage of the interrelationships that exist between properties that can be measured and those that cannot. In fact, the formation of this indirect method is one of the motives for developing the relations of Sections 9.2 and 9.3.

The properties most readily and accurately measured are the pressure, specific volume, and temperature. These three properties together form the P-v-T data and the P-v-T surface that we have already discussed in detail. In addition to measurements of P-v-T data, the various partial derivatives of the P-v-T surface are often measured directly. The derivatives are the coefficient of thermal expansion, $(1/v)(\partial v/\partial T)_P$, the isothermal compressibility, $-(1/v)(\partial v/\partial P)_T$, and the constant-volume pressure coefficient, $(1/P)(\partial P/\partial T)_v$.

Although P-v-T data are usually the most accurate and readily available thermodynamic properties, these data by themselves are not sufficient to formulate a complete constitutive relation (characteristic thermodynamic function) for the substance. A minimum amount of thermal data must be combined with the P-v-T data to complete the constitutive relations and specify the internal energy, the enthalpy, and the entropy. Normally the specific heats at zero pressure are the most reliable data for this purpose. Other thermal data which may be useful are the specific heat at constant pressure, the specific heat at constant volume, the Joule-Thomson coefficient, and the constant-temperature coefficient.

9.5.1 Definitions of Partial Differential Coefficients

A number of useful properties are defined in terms of various partial derivatives of the thermodynamic functions for the pure substance.

The *specific heat at constant volume* is a property defined by the partial derivative

$$c_v = \left(\frac{\partial u}{\partial T}\right)_v \tag{9.53}$$

For a quasi-static constant volume process in a pure substance, the first law of thermodynamics shows that the property c_v relates the differential heat transfer interaction to the differential change in the temperature. Thus,

$$\delta Q - mP\,dv = m\,du \tag{9.54}$$

but dv is zero and

$$du = \left(\frac{\partial u}{\partial T}\right)_v dT = c_v\,dT \tag{9.55}$$

$$[\delta Q]_{\substack{\text{quasi-static} \\ dv=0}} = mc_v\,dT \tag{9.56}$$

The *specific heat at constant pressure* is a property defined by the partial derivative

$$c_P = \left(\frac{\partial h}{\partial T}\right)_P \tag{9.57}$$

For a quasi-static process in a pure substance at constant pressure, the first law of thermodynamics shows that the property c_P

relates the differential heat transfer interaction to the differential temperature change.

$$\delta Q - mP\,dv = m\,du \tag{9.58}$$

Since the pressure is constant,

$$\delta Q = m[du + d(Pv)] = m\,dh \tag{9.59}$$

Since dP is zero,

$$dh = \left(\frac{\partial h}{\partial T}\right)_P dT = c_P\,dT \tag{9.60}$$

Therefore,

$$[\delta Q]_{\substack{\text{quasi-static}\\ dP\,=\,0}} = mc_P\,dT \tag{9.61}$$

The *Joule-Thomson coefficient,* μ, is the property that gives the increase in the temperature per unit increase in the pressure for a process in which the enthalpy remains constant. Thus,

$$\mu = \left(\frac{\partial T}{\partial P}\right)_h \tag{9.62}$$

The *constant-temperature coefficient,* c_T, is the property that gives the increase in the enthalpy per unit increase in the pressure for an isothermal process. Thus,

$$c_T = \left(\frac{\partial h}{\partial P}\right)_T \tag{9.63}$$

The *coefficient of thermal expansion,* β, is the property that gives the fractional increase in the specific volume per unit increase in the temperature while the substance remains at a constant pressure. Hence,

$$\beta = \frac{1}{v}\left(\frac{\partial v}{\partial T}\right)_P \tag{9.64}$$

This property, β, is a measure of the coupling between the thermal property temperature and the mechanical property volume.

The *isothermal compressibility, K*, is the property which gives the fractional decrease in the specific volume per unit increase in the pressure while the substance remains at a constant temperature. Thus,

$$K = -\frac{1}{v}\left(\frac{\partial v}{\partial P}\right)_T \tag{9.65}$$

This property, K, is a measure of the compressibility of the substance in the absence of any thermal effects due to temperature changes. In a sense, this property is the "isothermal spring constant".

9.5.2 Mathematical Relations Between Partial Derivatives

In addition to the thermodynamic relations which we have previously derived, several relations useful for the manipulation of thermodynamic properties can be taken from the differential calculus of a function of two variables. The most useful are:

the reciprocal relation

$$\left(\frac{\partial x}{\partial y}\right)_z = \frac{1}{(\partial y/\partial x)_z} \tag{9.66}$$

the chain rule

$$\left(\frac{\partial x}{\partial y}\right)_z = \left(\frac{\partial x}{\partial w}\right)_z \bigg/ \left(\frac{\partial y}{\partial w}\right)_z \tag{9.67}$$

the cyclical relation

$$\left(\frac{\partial x}{\partial y}\right)_z \left(\frac{\partial y}{\partial z}\right)_x \left(\frac{\partial z}{\partial x}\right)_y = -1 \tag{9.68}$$

and finally the substitution rule. For example, suppose there exists a function $z = z(x,y)$ with total differential

$$dz = \left(\frac{\partial z}{\partial x}\right)_y dx + \left(\frac{\partial z}{\partial y}\right)_x dy \tag{9.69}$$

If we wish to reformulate the function $z(x,y)$ in terms of properties w and y as independent variables, we first write the implicit relation for z in terms of the new variables

$$z = z(w,y) \tag{9.70}$$

and the total differential

$$dz = \left(\frac{\partial z}{\partial w}\right)_y dw + \left(\frac{\partial z}{\partial y}\right)_w dy \tag{9.71}$$

In order to eliminate the old independent variable x, we write x in terms of the new independent variables w and y

$$x = x(w,y) \tag{9.72}$$

with the total differential

$$dx = \left(\frac{\partial x}{\partial w}\right)_y dw + \left(\frac{\partial x}{\partial y}\right)_w dy \tag{9.73}$$

Substituting equation (9.73) into equation (9.69) we obtain

$$dz = \left(\frac{\partial z}{\partial x}\right)_y \left[\left(\frac{\partial x}{\partial w}\right)_y dw + \left(\frac{\partial x}{\partial y}\right)_w dy\right] + \left(\frac{\partial z}{\partial y}\right)_x dy \tag{9.74}$$

or

$$dz = \left[\left(\frac{\partial z}{\partial x}\right)_y \left(\frac{\partial x}{\partial w}\right)_y\right] dw + \left[\left(\frac{\partial z}{\partial x}\right)_y \left(\frac{\partial x}{\partial y}\right)_w + \left(\frac{\partial z}{\partial y}\right)_x\right] dy \tag{9.75}$$

But equation (9.75) must give the same value for dz as equation (9.71) for all values of dw and dy. Hence, the coefficients of dw and dy must be equal. Equating the coefficients, we obtain

$$\left(\frac{\partial z}{\partial w}\right)_y = \left(\frac{\partial z}{\partial x}\right)_y \left(\frac{\partial x}{\partial w}\right)_y \tag{9.76}$$

and

$$\left(\frac{\partial z}{\partial y}\right)_w = \left(\frac{\partial z}{\partial y}\right)_x + \left(\frac{\partial z}{\partial x}\right)_y \left(\frac{\partial x}{\partial y}\right)_w \tag{9.77}$$

9.5.3 Determination of the Characteristic Functions from Experimental Data

The problem of determining the complete and consistent constitutive relation for a pure substance will illustrate the use of the thermodynamic and mathematical relations that we have just enumerated. Normally, the starting point is the *P-v-T* data for the substance. These data are obtained by measuring the equilibrium pressure and the corresponding equilibrium temperature for a measured mass of the substance contained in a vessel of known volume. The *P-v-T* data can be measured with the best accuracy since the volume is the easiest extensive property to measure, and the intensive properties *P* and *T* are uniform at equilibrium. Next, curve fitting techniques are used (in three dimensions) to smooth the experimental *P-v-T* data. This must be done with accuracy in the derivatives of the *P-v-T* surface since these derivatives are necessary to determine the other properties.

In principle the *P-v-T* data are fitted with an explicit analytical function with sufficient accuracy for the determination of the entropy and the internal energy or the enthalpy. This explicit analytical function is the *P-v-T* constitutive relation and is usually called the equation of state for the substance. The ideal situation would be to have a single *P-v-T* function (equation of state) that can be valid for the solid, liquid, and vapor phases; however, the *P-v-T* behavior of the various phases is so significantly different that the development of a single *P-v-T* function (equation of state) valid for all phases is virtually impossible to achieve. The common approach is to develop *P-v-T* functions that are valid only for certain regions of the *P-v-T* surface. It is then necessary for us to insure that these different *P-v-T* functions match at the boundaries of the various regions—a difficult task that involves matching not only the properties themselves but their derivatives as well. More recently, improved computational capabilities have allowed the formulation of a characteristic function directly from experimental data.[2] With this technique it has been possible to describe both liquid and vapor phases with a single equation which yields all of the other thermodynamic properties by appropriate differentiation.

---

[2] Keenan, J. H., Keyes, F. G., Hill, P. G., and Moore, J. G., *Steam Tables* John Wiley, New York, 1969, p. 134.

9.6 Thermodynamic Properties from Simple *P-v-T* Surfaces

The problem of evaluation of properties can be simplified considerably if we use simple models that embody the essential features of certain limited regions of the *P-v-T* surface of the pure substance. These simple models are expressed as simplified equations of state (*P-v-T* functions) which actually form subclasses of the more general pure substance model.

9.6.1 Incompressible Fluid

For example, in the liquid phase region of the *P-v-T* surface we note that the specific volume is relatively insensitive to changes in pressure and temperature. That is, it is necessary to change the pressure or temperature by very large amounts in order to effect a significant change in volume. Since in many engineering situations the temperature and pressure changes are relatively small, we can safely model the behavior of liquids in these cases by a fluid of constant specific volume known as the *incompressible fluid.*

The equation of state which defines the incompressible fluid is simply

$$v = \text{constant} \tag{9.78}$$

The *P-v-T* surface is thus a plane perpendicular to the specific volume axis, and all equilibrium states of the fluid lie on this plane. Thus, the two derivatives $(\partial v/\partial T)_P$ and $(\partial v/\partial P)_T$ are both zero which means the coefficient of thermal expansion, β, and the isothermal compressibility, K, are identically zero. When we employ this model in actual engineering situations, we allow the intersection of this plane with the specific volume axis to shift from one location to another, depending upon the circumstances. In this sense, the specific volume at the point of intersection represents the average specific volume of the fluid suitable for the prevailing conditions of pressure and temperature.

In spite of the extreme simplicity of the model, it is very useful in many engineering situations; therefore, it is important to establish its properties. Specifically these are the enthalpy, the internal energy, and the entropy. Since we have an equation of state of the form $v(T,P)$, we seek expressions for the properties of the form $h(T,P)$, $u(T,P)$, and $s(T,P)$. To determine $h(P,T)$ we integrate

$$dh = \left(\frac{\partial h}{\partial P}\right)_T dP + \left(\frac{\partial h}{\partial T}\right)_P dT \tag{9.79}$$

The coefficient $(\partial h/\partial T)_P$ is by definition c_P, equation (9.57). The coefficient $(\partial h/\partial P)_T$ is expressed in terms of the P-v-T formulation by using the substitution technique, equations (9.69) through (9.77), on the characteristic function equation (9.30)

$$dh = T\,ds + v\,dP \tag{9.80}$$

Equation (9.30) implies $h(s,P)$ which we must reformulate as $h(T,P)$. To change from s independent to T independent, we consider $s(T,P)$ which implies the total differential of s is

$$ds = \left(\frac{\partial s}{\partial T}\right)_P dT + \left(\frac{\partial s}{\partial P}\right)_T dT \tag{9.81}$$

Substituting for ds in equation (9.80), we have

$$dh = \left[T\left(\frac{\partial s}{\partial P}\right)_T + v\right] dP + T\left(\frac{\partial s}{\partial T}\right)_P dP \tag{9.82}$$

Thus, comparing equations (9.79) and (9.82) we have

$$\left(\frac{\partial h}{\partial T}\right)_P = c_P = T\left(\frac{\partial s}{\partial T}\right)_P \tag{9.83}$$

and

$$\left(\frac{\partial h}{\partial P}\right)_T = T\left(\frac{\partial s}{\partial P}\right)_T + v \tag{9.84}$$

When we use the Maxwell relation, equation (9.40), we have

$$\left(\frac{\partial h}{\partial P}\right)_T = v - T\left(\frac{\partial v}{\partial T}\right)_P \tag{9.85}$$

Equation (9.79) then becomes

$$dh = \left[v - T\left(\frac{\partial v}{\partial T}\right)_P\right] dP + c_P\, dT \tag{9.86}$$

Since the coefficient of thermal expansion for the incompressible fluid is identically zero, equation (9.86) reduces further to

$$dh = v\, dP + c_P\, dT \tag{9.87}$$

Since v is constant and the second mixed partial derivatives are equal,

$$\frac{\partial}{\partial P}\left[\left(\frac{\partial h}{\partial T}\right)_P\right]_T = \frac{\partial}{\partial T}\left[\left(\frac{\partial h}{\partial P}\right)_T\right]_P \tag{9.88}$$

or

$$\left[\frac{\partial}{\partial P} c_P\right]_T = \left(\frac{\partial v}{\partial T}\right)_P = 0 \tag{9.89}$$

which implies that c_P is a function of temperature only. Equation (9.87) then integrates simply to give

$$h_2 - h_1 = v(P_2 - P_1) + \int_{T_1}^{T_2} c_P(T)\, dT \tag{9.90}$$

From the definition of h we get $u = h - Pv$. Thus,

$$u_2 - u_1 = \int_{T_1}^{T_2} c_P(T)\, dT \tag{9.91}$$

Equation (9.91) now can be used to evaluate c_v.

$$c_v = \left(\frac{\partial u}{\partial T}\right)_v = c_P \tag{9.92}$$

Hence, for the incompressible fluid the specific heat c_v and the specific heat c_P are the same. Therefore, the incompressible fluid has a specific heat, c, that must be a function of the temperature only. Equation (9.91) then becomes

$$u_2 - u_1 = \int_{T_1}^{T_2} c(T)\, dT \tag{9.93}$$

To determine the entropy of the incompressible fluid we integrate equation (9.81). When equation (9.83) and the Maxwell relation, equation (9.40), are substituted, equation (9.81) then becomes

$$ds = -\left(\frac{\partial v}{\partial T}\right)_P dP + \frac{c_P}{T}\, dT \tag{9.94}$$

Again since the coefficient of thermal expansion of the incompressible fluid is identically zero, equation (9.94) reduces to

$$ds = \frac{c(T)}{T}\, dT \tag{9.95}$$

Integrating equation (9.95), we have

$$s_2 - s_1 = \int_{T_1}^{T_2} \frac{c(T)}{T}\, dT \tag{9.96}$$

Equations (9.93) and (9.96) show that the equilibrium properties of the incompressible fluid are identical to those of a pure thermal system, equations (3.9) and (7.44).

Since the specific volume is constant, the reversible work transfer associated with the normal displacement of the boundary of an incompressible pure substance is zero. That is,

$$W_{1-2} = m \int_{v_1}^{v_2} P\, dv = 0 \tag{9.97}$$

Then the first law of thermodynamics for reversible processes in the incompressible fluid is identical to that of the pure thermal system

$$Q_{1-2} = U_2 - U_1 = m \int_{T_1}^{T_2} c\, dT \tag{9.98}$$

since the normal displacement of the boundary is the only relevant work transfer. In spite of its simplicity, the incompressible fluid model finds widespread use in predicting the thermodynamic behavior of liquid systems. It is particularly useful in the science of fluid mechanics since the compressibility effects are often negligible, and the uncoupled mechanical aspects associated with the bulk motion of the fluid are important.

9.6.2 Ideal Gas

The ideal gas is our simple model for the P-v-T behavior of the vapor phase of the pure substance. The ideal gas model is of approximation as the pressure approaches zero. In fact, it describes exactly the behavior of a vapor in the limit of zero pressure. As previously cited in Chapter 4, the P-v-T constitutive relation (equation of state) for a unit mass of an ideal gas is

$$Pv = RT \tag{9.99}$$

This P-v-T surface is a hyperbolic paraboloid.

The unusually simple form of the P-v-T constitutive relation, equation (9.99), is responsible for the simple thermodynamic behavior of the ideal gas model. One of the significant aspects of this behavior is the fact that both the internal energy and the enthalpy depend on the temperature only. We can illustrate this fact by the following arguments. Let us assume that the pressure and the internal energy are functions of the temperature and the volume. Then,

$$P = P(T,v) \quad u = u(T,v) \tag{9.100}$$

The total differential du is then

$$du = \left(\frac{\partial u}{\partial T}\right)_v dT + \left(\frac{\partial u}{\partial v}\right)_T dv \tag{9.101}$$

The coefficient $(\partial u/\partial T)_v$ is by definition c_v, equation (9.53). The coefficient $(\partial u/\partial v)_T$ is expressed in the P-v-T formulation by using the substitution rule on equation (9.29).

$$du = T\,ds - P\,dv \tag{9.102}$$

Assuming that there exists a function $s(T,v)$, we obtain

$$ds = \left(\frac{\partial s}{\partial T}\right)_v dT + \left(\frac{\partial s}{\partial v}\right)_T dv \tag{9.103}$$

Substituting equation (9.103) into equation (9.102), we obtain

$$du = T\left(\frac{\partial s}{\partial T}\right)_v dT + \left[T\left(\frac{\partial s}{\partial v}\right)_T - P\right] dv \tag{9.104}$$

Comparing equations (9.101) and (9.104), we obtain

$$\left(\frac{\partial u}{\partial T}\right)_v = c_v = T\left(\frac{\partial s}{\partial T}\right)_v \tag{9.105}$$

and

$$\left(\frac{\partial u}{\partial v}\right)_T = T\left(\frac{\partial s}{\partial v}\right)_T - P \tag{9.106}$$

Substituting Maxwell's relation, equation (9.39), into equation (9.106), we obtain

$$\left(\frac{\partial u}{\partial v}\right)_T = T\left(\frac{\partial P}{\partial T}\right)_v - P \tag{9.107}$$

From the ideal gas equation of state, equation (9.99), we get

$$\left(\frac{\partial P}{\partial T}\right)_v = \frac{R}{v} \tag{9.108}$$

Substituting equations (9.108) and (9.99) into equation (9.107), we obtain

$$\left(\frac{\partial u}{\partial v}\right)_T = \frac{RT}{v} - P = P - P = 0 \tag{9.109}$$

Thus, the internal energy is independent of the volume, and equation (9.101) reduces to

$$du = c_v\, dT \tag{9.110}$$

which permits us to write

$$c_v = \frac{du}{dT} \tag{9.111}$$

Hence, for the ideal gas model the partial derivative of equation (9.105) reduces to an ordinary derivative which can be integrated simply to give

$$u_2 - u_1 = \int_{T_1}^{T_2} c_v \, dT \tag{9.112}$$

From the definition of enthalpy $h = u + Pv$ we have from equation (9.112)

$$h_2 - h_1 = R(T_2 - T_1) + \int_{T_1}^{T_2} c_v \, dT \tag{9.113}$$

Differentiation of equation (9.113) with respect to T gives

$$\frac{dh}{dT} = R + c_v = c_P \tag{9.114}$$

which we have already stated without proof in equation (7.58).

In an alternate method of deriving the energy and enthalpy of an ideal gas, we assume that the volume and the enthalpy are functions of temperature and pressure.

$$v = v(T,P) \quad h = h(T,P) \tag{9.115}$$

The total differential dh is then

$$dh = \left(\frac{\partial h}{\partial T}\right)_P dT + \left(\frac{\partial h}{\partial P}\right)_T dP \tag{9.116}$$

This is the same formulation as we employed for the incompressible fluid. By using the substitution rule on the characteristic function, we obtained equations (9.83) through (9.86) which are generally applicable for this formulation in terms of T and P. We need only form the appropriate partial derivatives of the ideal gas $v(T,P)$ and substitute them in the equations. The ideal gas equation of state yields

$$\left(\frac{\partial v}{\partial T}\right)_P = \frac{R}{P} \tag{9.117}$$

Thus, equation (9.85) becomes

$$\left(\frac{\partial h}{\partial P}\right)_T = v - \frac{RT}{P} = v - v = 0 \tag{9.118}$$

Therefore, equation (9.86) reduces to

$$dh = c_P\,dT \tag{9.119}$$

which permits us to write

$$c_P = \frac{dh}{dT} \tag{9.120}$$

Thus, for the ideal gas model the partial derivative of equation (9.83) becomes an ordinary derivative which can be integrated to give

$$h_2 - h_1 = \int_{T_1}^{T_2} c_P\,dT \tag{9.121}$$

Generally, the specific heats c_P and c_v are functions of temperature. As a typical example, the specific heat c_v for a gas composed of diatomic molecules is shown in Figure 9.1 as a function of temperature.

The simple classical statistical mechanical model for the ideal gas shows that the specific heat c_v is related to the number of mechanical coordinates, n, required to specify the mechanical state of a single molecule by the relation

$$c_v = \frac{n}{2}R \tag{9.122}$$

In Figure 9.1 the three plateaus for which the specific heat is constant over a temperature range correspond to temperature

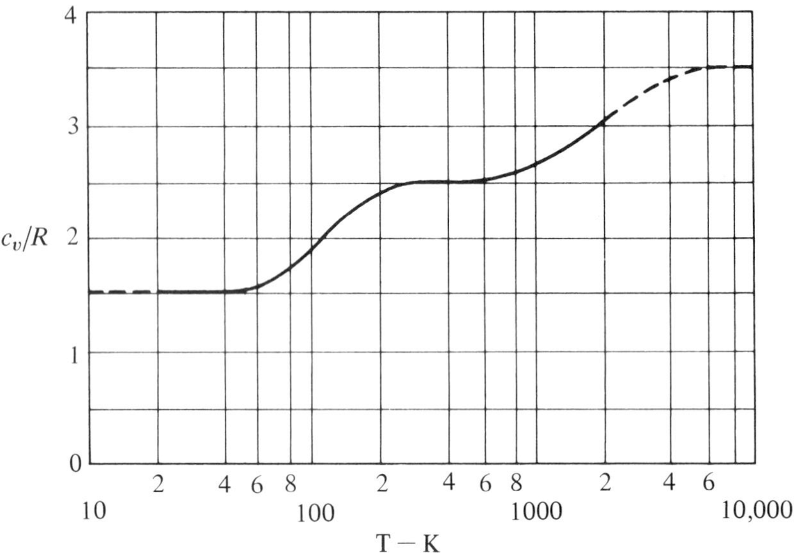

Figure 9.1 Specific heat for a diatomic ideal gas. Figure 9.1 reprinted from *Introduction to Modern Physics,* Richmyer and Kennard, 1947, by permission McGraw-Hill Book Co.

regions within which the bahavior of the gas is well represented by the classical statistical mechanics model. For the lowest temperature plateau, the required mechanical coordinates are simply the three translational coordinates. In the middle plateau, two rotational coordinates are necessary in addition to three translational coordinates. In the highest plateau, additional coordinates are necessary because of a vibration between the two atoms in the diatomic molecule. A quantum statistical mechanical model is necessary to represent adequately the behavior of the specific heat in the regions between the classical specific heat plateaus. Quantum statistics predicts the occupation of the energy levels associated with rotation and vibration of the molecule as the temperature of the gas is increased. In more complex molecules the regions of classical behavior are eliminated by overlapping regions of quantum mechanical behavior.

In spite of its limitations for more complex molecules, the classical microscopic model of the ideal gas gives useful results for ideal gases composed of simple molecules. For example the specific heats

$$c_v = \frac{3}{2} R \text{ and } c_P = \frac{5}{2} R \tag{9.123}$$

are quite accurate for monatomic ideal gases over a wide range of temperatures. In addition this model shows that the specific heat ratio should vary from a maximum of 5/3 for a monatomic gas to a limit of unity for a gas composed of complex molecules.

$$1.67 \geq \frac{c_P}{c_v} > 1 \tag{9.124}$$

For most of the common gases the quantum mechanical model has been worked out in detail from spectroscopic measurements. The evaluation of the ideal gas specific heat or internal energy for the substance from this microscopic model is a numerical procedure, and the results are presented in tabular form. These results for a number of common gases have been collected in the *Gas Tables*[3]. The ideal gas specific heats have also been empirically curve fitted with simple functions of temperature over a limited range of temperature. These specific functions are usually given in thermodynamic data books[4,5].

There are in common use two formulations of the entropy of the ideal gas that depend on the particular set of properties considered independent. In one case, we have $s(v,T)$ while in the other we have $s(P,T)$. Let us now consider the first possibility, $s(v,T)$.

$$ds = \left(\frac{\partial s}{\partial v}\right)_T dv + \left(\frac{\partial s}{\partial T}\right)_v dT \tag{9.125}$$

From equation (9.105) it follows that

$$\left(\frac{\partial s}{\partial T}\right)_v = \frac{c_v}{T} \tag{9.126}$$

---

[3] Keenan, J. H. and Kaye, J., *Gas Tables* John Wiley, New York, 1948.
[4] Hilsenrath, Joseph, et al., *Tables of Thermodynamic and Transport Properties* Pergamon Press, New York, 1960.
[5] *Data Book,* Thermophysical Properties Research Center Purdue University, Lafayette, Indiana, 1964, Vol. 2, Chapter 5.

From the Maxwell relation of equation (9.39) we have

$$\left(\frac{\partial s}{\partial v}\right)_T = \left(\frac{\partial P}{\partial T}\right)_v \tag{9.127}$$

Then combining equations (9.125), (9.126), and (9.127), we obtain

$$ds = \left(\frac{\partial P}{\partial T}\right)_v dv + \frac{c_v}{T}\,dT \tag{9.128}$$

For the ideal gas we can substitute equation (9.108) into equation (9.128).

$$ds = \frac{R}{v}\,dv + \frac{c_v}{T}\,dT \tag{9.129}$$

Integrating equation (9.129), we obtain

$$s_2 - s_1 = R \ln \frac{v_2}{v_1} + \int_{T_1}^{T_2} \frac{c_v}{T}\,dT \tag{9.130}$$

Now consider the second possibility $s(T,P)$. Then,

$$ds = \left(\frac{\partial s}{\partial T}\right)_P dT + \left(\frac{\partial s}{\partial P}\right)_T dP \tag{9.131}$$

From equation (9.83), it follows that

$$\left(\frac{\partial s}{\partial T}\right)_P = \frac{c_P}{T} \tag{9.132}$$

From the Maxwell relation of equation (9.40), we have

$$\left(\frac{\partial s}{\partial P}\right)_T = -\left(\frac{\partial v}{\partial T}\right)_P \tag{9.133}$$

Then combining equations (9.131), (9.132), and (9.133), we obtain

$$ds = -\left(\frac{\partial v}{\partial T}\right)_P dP + \frac{c_P}{T}\,dT \tag{9.134}$$

For the ideal gas we can substitute equation (9.117) into equation (9.134) to obtain

$$ds = -\frac{R}{P}\,dP + \frac{c_P}{T}\,dT \tag{9.135}$$

Integrating equation (9.135), we obtain

$$s_2 - s_1 = -R \ln \frac{P_2}{P_1} + \int_{T_1}^{T_2} \frac{c_P}{T}\,dT \tag{9.136}$$

For a given specific heat function, the integral in equation (9.130) or equation (9.136) depends only upon T_1 and T_2. If we define a function $\phi(T)$ by the relation

$$\phi(T) = \int_{T=T_0}^{T} \frac{c_P}{T}\,dT \tag{9.137}$$

then equation (9.136) becomes

$$s_2 - s_1 = -R \ln \frac{P_2}{P_1} + \phi(T_2) - \phi(T_1) \tag{9.138}$$

The function $\phi(T)$ has been evaluated for a number of common gases and is given in the *Gas Tables*[6]. The use of the function $\phi(T)$ and equation (9.138) makes the calculation of entropy changes for the ideal gas with variable specific heats as simple as those with constant specific heats. The *Gas Tables* also tabulate functions for rapid calculation of the pressure and volume ratios for isentropic (constant entropy) processes of an ideal gas with variable specific heats. A complete explanation and further reference in the use of these functions can be found in the *Gas Tables*.

If the specific heats of the ideal gas are constant then equations (9.112), (9.113), (9.130), and (9.136) become respectively

$$u_2 - u_1 = c_v(T_2 - T_1) \tag{9.139}$$

$$h_2 - h_1 = c_P(T_2 - T_1) \tag{9.140}$$

$$s_2 - s_1 = R \ln \frac{v_2}{v_1} + c_v \ln \frac{T_2}{T_1} \tag{9.141}$$

[6] Keenan, J. H. and Kaye, J., *Gas Tables,* John Wiley, New York, 1948.

and

$$s_2 - s_1 = -R \ln \frac{P_2}{P_1} + c_P \ln \frac{T_2}{T_1} \tag{9.142}$$

which we have previously stated.

9.7 Thermodynamic Properties from *P-v-T* Surfaces

In the previous section we developed techniques for calculating the thermodynamic properties of simple models of the *P-v-T* surface of the pure substance. In this section, we will show how these techniques can be applied to the more general case. The major difference between the simple and general case is the manner in which we integrate the differential expressions of the properties. For example, in Sections 9.6.1 and 9.6.2 we saw that certain partial differential coefficients, notably the specific heats, played an important role in the evaluation of other properties such as the internal energy, the enthalpy, and the entropy. For simple models like the incompressible fluid and the ideal gas, these partial differential coefficients assumed very simple forms (usually constants) which made the integration of relevant expressions trivial. In general, however, the formulations of the partial differential coefficients in terms of the properties pressure, volume, and temperature are exceedingly complex and render the integration of relevant expressions difficult. Fortunately, we can partially reduce these complexities by selection of a path of integration that passes through states for which the ideal gas model is valid. This will allow us to express the partial differential coefficients in very simple form. By employing such paths for integration, we can evaluate all of the thermodynamic properties from minimum experimental data. A complete set of *P-v-T* data and the specific heats for the ideal gas (low density) states are all that is required. The *P-v-T* data must provide accurate values for the partial derivatives of the *P-v-T* function.

The specific heats, $c_{v\infty}$ and c_{P0}, of the ideal gas states are themselves measured by indirect methods. As outlined earlier, there exist quantum mechanical models for the behavior of the ideal gas. When we use these models together, with the emission and absorption spectra for the substance at very low pressures, we obtain the specific heats $c_{v\infty}$ and c_{P0}.

There are two methods of proceeding with the calculations of the thermodynamic properties depending on the way that the P-v-T data have been formulated (curve fitted). One method is used for $P = P(v,T)$ and a second method is used for $v = v(P,T)$. When the formulation is $P = P(v,T)$, the convenient independent properties are v and T. The other thermodynamic properties are in the form $u = u(v,T)$, $h = h(v,T)$, and $s = s(v,T)$. To determine the internal energy, $u(v,T)$, we must integrate equation (9.101).

$$du = \left(\frac{\partial u}{\partial T}\right)_v dT + \left(\frac{\partial u}{\partial v}\right)_T dv \qquad \textbf{(9.143)}$$

When we perform the operations described in Section 9.6.2, equation (9.143) becomes

$$du = c_v\, dT + \left[T\left(\frac{\partial P}{\partial T}\right)_v - P\right] dv \qquad \textbf{(9.144)}$$

Here we have combined equations (9.105), (9.107), and (9.143).

If we select the path of integration for changes in u so that the temperature changes only when the substance is in ideal gas states ($P \rightarrow 0$, $v \rightarrow \infty$), the specific heat c_v becomes the ideal gas specific heat $c_{v\infty}$ and the volume contribution to the internal energy change vanishes. Then, the change in u is composed of an ideal gas term that depends on temperature only plus non-ideal gas correction terms. To perform the integration in this manner, we need explicit functions for $c_{v\infty}(T)$ and $P(v,T)$. Note, however, that the accuracy of the resulting internal energy will depend upon the accuracy of the derivative of $P(v,T)$.

For computing the difference in internal energy between two states 1 and 2 by this method, the most convenient path for integration of equation (9.144) is at constant temperature ($T = T_1$) from v_1 to $v \rightarrow \infty$ ($\rho = 0$). Next integrate at constant volume $v \rightarrow \infty$ ($\rho = 0$) from T_1 to T_2. For this part of the integration the substance is in ideal gas states. Finally, integrate at constant temperature ($T = T_2$) from $v \rightarrow \infty$ to v_2. This path is shown on the v-T plane in Figure 9.2 (a) and on the ρ-T plane in Fig-

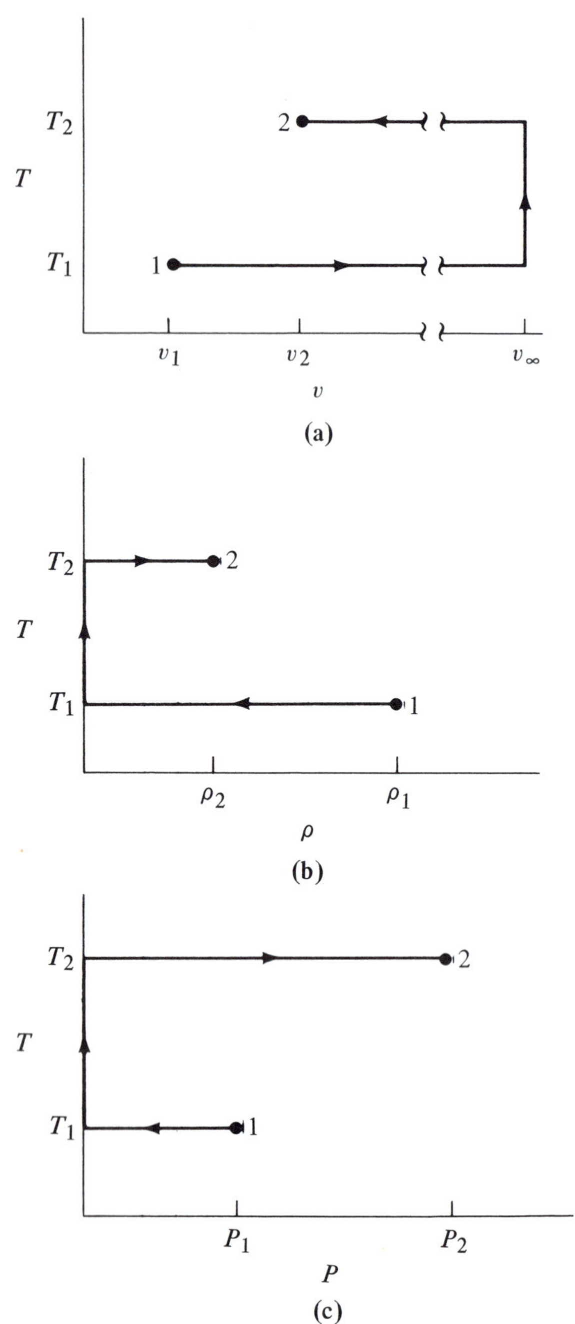

Figure 9.2 Path of integration for property calculations

ure 9.2 (b). The mathematical formulation for this path of integration is

$$u_2 - u_1 = \left[\int_{v_1}^{v\to\infty} \left[T\left(\frac{\partial P}{\partial T}\right)_v - P\right] dv\right]_{T=T_1} + \left[\int_{T_1}^{T_2} c_v\, dT\right]_{v\to\infty} + \left[\int_{v\to\infty}^{v_2} \left[T\left(\frac{\partial P}{\partial T}\right)_v - P\right] dv\right]_{T=T_2} \qquad \textbf{(9.145)}$$

In most practical cases the expressions for $P(T,v)$ and $c_{v\infty}$ are too complex to be integrated in closed form; therefore, the integrals of equation (9.145) are normally evaluated numerically with the aid of a digital computer.

The entropy, $s(v,T)$, can be evaluated in a manner similar to the internal energy, $u(v,T)$. In this case we must integrate equation (9.125) to get

$$ds = \left(\frac{\partial s}{\partial v}\right)_T dv + \left(\frac{\partial s}{\partial T}\right)_v dT \qquad \textbf{(9.146)}$$

Following the steps shown in Section 9.6.2, we obtain equation (9.128).

$$ds = \left(\frac{\partial P}{\partial T}\right)_v dv + \frac{c_v}{T}\, dT \qquad \textbf{(9.147)}$$

If we wish to employ the ideal gas specific heats for the integration, equation (9.147) must be modified to avoid the unbounded entropy at infinite volume, but the modification must permit equation (9.147) to be separated into the ideal gas entropy plus a correction term. From equation (9.129), it is apparent that adding and subtracting the term $(R/v)\, dv$ in equation (9.147) will achieve the desired result. Then

$$ds = \left[\left(\frac{\partial P}{\partial T}\right)_v - \frac{R}{v}\right] dv + \frac{R}{v}\, dv + \frac{c_v}{T}\, dT \qquad \textbf{(9.148)}$$

Here the last two terms are the ideal gas terms of equation (9.129), and the first term approaches zero as the volume becomes infinite.

For the difference in entropy between two states 1 and 2, the most convenient path for integration of equation (9.148) is the path shown in Figures 9.2 (a) and (b). That is, first integrate at constant temperature ($T = T_1$) from v_1 to $v \to \infty$ ($\rho = 0$), then integrate at constant volume $v \to \infty$ ($\rho = 0$) from T_1 to T_2. Finally, integrate at constant temperature ($T = T_2$) from $v \to \infty$ to v_2. Equation (9.148) then becomes

$$s_2 - s_1 = R \ln \frac{v_2}{v_1} + \left[\int_{T_1}^{T_2} \frac{c_v}{T}\, dT\right]_{v\to\infty} + \left[\int_{v_1}^{v\to\infty} \left[\left(\frac{\partial P}{\partial T}\right)_v - \frac{R}{v}\right] dv\right]_{T=T_1} + \left[\int_{v\to\infty}^{v_2} \left[\left(\frac{\partial P}{\partial T}\right)_v - \frac{R}{v}\right] dv\right]_{T=T_2} \tag{9.149}$$

Here again, the actual integrals are normally evaluated numerically.

The enthalpy can now be determined from the internal energy by means of the definition of h.

$$h(v,T) = u(v,T) + v[P(v,T)] \tag{9.150}$$

The free energies, f and g, can also be evaluated from the definitions

$$f(v,T) = u(v,T) - T[s(v,T)] \tag{9.151}$$

and

$$g(v,T) = h(v,T) - T[s(v,T)] \tag{9.152}$$

When the P-v-T data have been formulated as $v = v(P,T)$, a method very similar to the one previously described is normally employed to obtain the calculated thermodynamic properties. In this case the independent variables are P and T so the appropriate starting point is the evaluation of the enthalpy $h(P,T)$. Thus, we must integrate equation (9.79).

$$dh = \left(\frac{\partial h}{\partial P}\right)_T dP + \left(\frac{\partial h}{\partial T}\right)_P dT \tag{9.153}$$

When we follow the steps shown in Section 9.6.1, equation (9.153) becomes

$$dh = \left[v - T\left(\frac{\partial v}{\partial T}\right)_P\right] dP + c_P\, dT \tag{9.154}$$

Normally the path of integration used in integrating equation (9.154) is chosen so that the ideal gas specific heat c_{P0} is employed. Then, for the difference in enthalpy between two states 1 and 2, the path of integration is the one shown in Figure 9.2 (c). That is, $T = T_1$ from P_1 to $P = 0$, at $P = 0$ from T_1 to T_2, and at $T = T_2$ from $P = 0$ to P_2. For this path the integrated form of equation (9.154) is

$$h_2 - h_1 = \left[\int_{P_1}^{P=0} \left[v - T\left(\frac{\partial v}{\partial T}\right)_P\right] dP\right]_{T=T_1} + \left[\int_{T_1}^{T_2} c_{P0}\, dT\right]_{P=0} + \left[\int_{P=0}^{P_2} \left[v - T\left(\frac{\partial v}{\partial T}\right)_P\right] dP\right]_{T=T_2} \quad \textbf{(9.155)}$$

To determine the entropy $s(P,T)$, we must integrate equation (9.131).

$$ds = \left(\frac{\partial s}{\partial T}\right)_P dT + \left(\frac{\partial s}{\partial P}\right)_T dP \quad \textbf{(9.156)}$$

When we follow the steps shown in Section 9.6.2, equation (9.156) becomes

$$ds = -\left(\frac{\partial v}{\partial T}\right)_P dP + \frac{c_P}{T}\, dT \quad \textbf{(9.157)}$$

Again, the entropy becomes infinitely large at $P = 0$ unless we modify equation (9.157). Following our previous example, we make use of the ideal gas model, equation (9.135). Then equation (9.157) becomes

$$ds = \left[\frac{R}{P} - \left(\frac{\partial v}{\partial T}\right)_P\right] dP - \frac{R}{P}\, dP + \frac{c_P}{T}\, dT \quad \textbf{(9.158)}$$

When we integrate equation (9.158) along the path shown in Figure 9.2 (c), we obtain the difference in entropy between any two states 1 and 2.

$$s_2 - s_1 = -R \ln \frac{P_2}{P_1} + \left[\int_{T_1}^{T_2} \frac{c_{P0}}{T}\, dT\right]_{P=0} + \left[\int_{P_1}^{P=0} \left[\frac{R}{P} - \left(\frac{\partial v}{\partial T}\right)_P\right] dP\right]_{T=T_1} + \left[\int_{P=0}^{P_2} \left[\frac{R}{P} - \left(\frac{\partial v}{\partial T}\right)_P\right] dP\right]_{T=T_2} \quad \textbf{(9.159)}$$

The other properties can now be determined from the following definitions:

$$u(P,T) = h(P,T) - P[v(P,T)] \tag{9.160}$$

$$f(P,T) = u(P,T) - T[s(P,T)] \tag{9.161}$$

$$g(P,T) = h(P,T) - T[s(P,T)] \tag{9.162}$$

We have now developed methods for determining a characteristic thermodynamic function from the P-v-T formulation (and its partial derivatives) and a specific heat as a function of temperature at one particular pressure. Since this information provides us with a complete constitutive relation, it will always be possible to express any combination of thermodynamic properties and their derivatives in terms of the measurable thermodynamic properties. An adequate set of measurable properties is the P-v-T formulation (and its partial derivatives) plus a specific heat.

In order to express the derivatives of the properties in these terms, the following set of rules are useful[7]:

1. If any of the derivatives contain any of the properties f, g, h, and u, bring them one by one to the numerator and eliminate them by means of equations (9.33) through (9.36) or the substitution rule and equations (9.29) through (9.32).
2. If any of the derivatives contain the entropy, bring it to the numerator by means of one of the mathematical relations for partial derivatives. If one of the four Maxwell relations, equations (9.37) through (9.40), does not eliminate the entropy, use the chain rule for partial derivatives to obtain a partial derivative of the form $\partial s/\partial T$. This derivative is then one of the specific heats, equations (9.83) and (9.105).
3. Now eliminate the dependent variable (property) of the P-v-T formulation by bringing it to the numerator by means of the cyclic relation and the chain rule. If the P-v-T formulation is in the form

[7] Callen, H. B., *Thermodynamics* (John Wiley, New York, 1960), pp. 123–124.

$v(P,T)$, eliminate the volume in this manner. Conversely, if the P-v-T formulation is in the form $P(v,T)$, eliminate the pressure in this manner. The remaining derivatives are now expressed in terms of the derivatives of the P-v-T formulation.

4. The original derivatives have now been expressed in terms of c_v, c_P and the derivatives of the P-v-T formulation. If desired, one of the specific heats can be eliminated by the equation

$$c_P - c_v = \frac{T\beta^2 v}{K} \tag{9.163}$$

or

$$c_P - c_v = \frac{T[(\partial P/\partial T)_v]^2}{-(\partial P/\partial v)_T} \tag{9.164}$$

9.8 Special Formulations of the P-v-T Data for the Pure Substance Model

From the discussions of the preceding section, it is clear that the functional form used to represent the P-v-T data plays an important role in the calculation of thermodynamic properties. It should come as no surprise, then, that many (in excess of 85) formulations, or equations of state, have been reported in the literature. In general these formulations are restricted to the vapor phase, but some can be extrapolated to the liquid phase. They vary considerably in complexity, and with one exception, the van der Waals equation, have little or no physical significance. All have two or more adjustable constants; hence, they are quite useful for curve fitting and interpolating existing P-v-T data. In this section we present several of the more common equations of state in the order of their increasing complexity.

9.8.1 van der Waals Equation of State

The equation suggested by J. D. van der Waals in 1873 represents in a qualitative manner the P-v-T characteristics of both the liquid and vapor phases. In addition, the equation has a critical state and approaches ideal gas behavior as the pressure approaches zero. Although the van der Waals equation does not represent the properties of any given substance with precision, it is still very useful since it is the simplest model of a substance which explains the departures from ideal gas behavior.

Whereas in the ideal gas model the molecules exert no mutual influence except possibly during short time collisions, van der Waals suggested a molecular model in which the molecules each have a finite volume and exert long range attractive forces on one another. The net result of this model is that not all of the physical volume of the container is available to the molecules of the gas and that the force that the molecules exert on the container wall is reduced by the attractive force exerted on a molecule by its neighbors. The van der Waals equation of state then assumes the form

$$P = \frac{RT}{v - b} - \frac{a}{v^2} \tag{9.165}$$

Here the constant b is the volume excluded by the dimensions of the molecules themselves and the constant a accounts for the attractive forces between molecules. The numerical values of these two constants can be related to the critical state by recognizing that the critical isotherm must have a horizontal inflection point at the critical state. That is, in the critical state

$$\left(\frac{\partial P}{\partial v}\right)_T = \left(\frac{\partial^2 P}{\partial v^2}\right)_T = 0 \tag{9.166}$$

Applying this condition to the van der Waals equation of state, we obtain the results

$$a = \frac{27}{64}\frac{R^2 T_c^{\,2}}{P_c} \tag{9.167}$$

and

$$b = \frac{1}{8}\frac{RT_c}{P_c} \tag{9.168}$$

Here P_c is the critical pressure and T_c is the critical temperature.

At temperatures below the critical temperature, the van der Waals equation of state provides a qualitative description of liquid-vapor two-phase states. If we rewrite equation (9.165) in the form

$$Pv^3 - (bP + RT)v^2 + av - ab = 0 \qquad \textbf{(9.169)}$$

we have a polynomial which is third order in volume. At temperatures below T_c a given isotherm (line of constant temperature) has three volumes for each pressure over a range of pressure, as is shown in Figure 9.3. The multiple values of volume at a given T and P allow the van der Waals equation to represent two-phase states which have a range of volume for a given T and P. In fact, the van der Waals equation provides a simple model of the phase change. For the isotherm $T = T_1$ shown in Figure 9.3 the slope $(\partial P/\partial v)_T$ is positive in the region from m to n. This positive slope indicates that the gas is mechanically un-

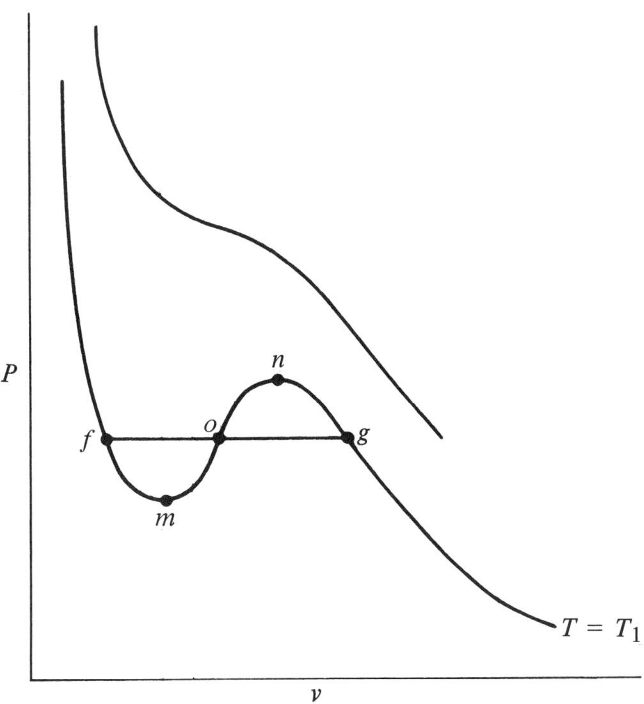

Figure 9.3 Isotherms of a van der Waals gas

stable in this region. If we have some gas at state o and a part of the gas happens to decrease in volume slightly, its pressure will decrease. The remainder of the gas expands and increases in pressure. This process accelerates, and finally a part of the gas has condensed to liquid at f and the remainder of the gas expands to vapor at g. The line f to g then represents the two-phase states composed of saturated liquid, state f, and saturated vapor, state g. For mechanical and thermal equilibrium the pressure and temperature must be the same for states f and g. In addition we showed in Section 8.9, equation (8.16), that the Gibbs free energy must be the same for states f and g to remain in equilibrium. This requirement for equilibrium with respect to mass transfer between phases determines the saturation pressure for each isotherm, that is, determines the vertical position of line f to g in Figure 9.3.

The saturation pressure for the isotherm may also be determined by considering a cycle proceeding along the path f-m-o-n-g and returning along g-o-f. If by some hypothetical means we could keep the unstable equilibrium states from collapsing, this would be a reversible cycle in communication with a single heat reservoir (all states in the cycle at the same T). The second law requires no net work, hence the isobar g-o-f is located so that the area between the curve f-m-o and the isobar is equal to the area between the curve o-n-g and the isobar.

The states represented by curve g-n are metastable vapor states which are stable except to the formation of a finite drop of the liquid phase. The states represented by curve f-m are metastable liquid states which are stable except to the formation of a finite bubble of the vapor phase.

9.8.2 Dieterici Equation of State

This equation is an empirical improvement upon the van der Waals equation of state. It has the form

$$P = \frac{RT}{v - b} \exp\left(-\frac{a}{vRT}\right) \tag{9.170}$$

Here the two constants may be expressed in terms of the critical temperature and pressure, or they may be adjusted to fit the

P-v-T data for two particular states. In this latter case the equation of state is used as an interpolation formula for the P-v-T data between the two fixed states.

9.8.3 Beattie-Bridgeman Equation of State

In order to obtain a fit to the P-v-T data of gases over a broader range of states, Beattie and Bridgeman proposed the following equation of state which contains five constants in addition to the gas constant R:

$$P = \frac{RT}{v^2}(1 - \epsilon)(V + B) - \frac{A_0}{v^2}\left(1 - \frac{a}{v}\right) \tag{9.171}$$

where

$$\epsilon = \frac{c}{vT^3} \text{ and } B = B_0\left(1 - \frac{b}{v}\right) \tag{9.172}$$

The values of the constants for some common gases are given in Table 9.2.

9.8.4 Benedict-Webb-Rubin Equation of State

Benedict, Webb, and Rubin have generalized the Beattie-Bridgeman equation of state to cover a broader range of states. The resulting equation of state contains eight constants in addition to the gas constant.

$$P = \frac{RT}{v} + \left(B_0RT - A_0 - \frac{c_0}{T^2}\right)\frac{1}{v^2} + (bRT - a)\frac{1}{v^3} + \frac{a}{v^6} + \frac{c(1 + \gamma/v^2)}{T^2}\frac{1}{v^3}\exp\left(-\frac{\gamma}{v^2}\right) \tag{9.173}$$

The values of the constants for a number of different gases are given in Table 9.3.

TABLE 9.2
Beattie-Bridgeman Constants

Gas	*Formula*	R J/kgK	A_0 Nm^4/kg^2	a m^3/kg	B_0 m^3/kg	b m^3/kg	C m^3K^3/kg
Ammonia	NH_3	488.15	836.16	1.00012 E−2	2.0052 E−3	1.1223 E−2	2.8003 E+5
Air		286.95	157.12	6.6674 E−4	1.5919 E−3	−3.801 E−5	1.498 E+3
Argon	A	208.14	81.99	5.8290 E−4	9.8451 E−4	0.0	1.499 E+3
n-Butane	C_4H_{10}	143.15	534.57	2.09388 E−3	4.2391 E−3	1.6225 E−3	6.0267 E+4
Carbon dioxide	CO_2	188.93	262.07	1.62129 E−3	2.3811 E−3	1.6444 E−3	1.4997 E+4
Carbon monoxide	CO	296.90	173.78	9.3457 E−4	1.8023 E−3	2.4660 E−4	1.498 E+3
Ethane	C_2H_6	276.62	659.89	1.95030 E−3	3.1283 E−3	6.3740 E−4	2.995 E+4
Ethylene	C_2H_4	296.58	793.50	1.77112 E−3	4.3370 E−3	1.2835 E−3	8.092 E+3
Ethyl ether	$C_4H_{10}O$	112.22	577.59	1.67748 E−3	6.1350 E−3	1.6137 E−3	4.499 E+3
Helium	He	2078.18	136.79	1.49581 E−2	3.5004 E−3	0.0	9.955
n-Heptane	C_7H_{16}	83.01	551.18	2.00398 E−3	7.0733 E−3	1.9159 E−3	3.992 E+4
Hydrogen	H_2	4115.47	4904.92	−2.50530 E−3	1.0376 E−2	−2.1582 E−2	2.4942 E+2
Methane	CH_4	518.60	897.96	1.15744 E−3	3.4854 E−3	−9.9013 E−4	8.003 E+3
Neon	Ne	411.98	52.89	1.08815 E−3	1.0207 E−3	0.0	50.10
Nitrogen	N_2	296.69	173.54	9.3394 E−4	1.8011 E−3	−2.4660 E−4	1.498 E+3
Nitrous oxide	N_2O	188.83	261.83	1.62005 E−3	2.3799 E−3	1.6431 E−3	1.498 E+4
Oxygen	O_2	259.79	147.56	8.0097 E−4	1.4452 E−3	1.3148 E−4	1.498 E+3
Propane	C_3H_8	188.67	622.24	1.66125 E−3	4.1082 E−3	9.7452 E−4	2.719 E+4

9.8.5 Virial Equation of State

Closer inspection of equations (9.165), (9.170), (9.171), and (9.173) reveals that all of these equations of state are of the form

$$\frac{Pv}{RT} = 1 + \frac{B(T)}{v} + \frac{C(T)}{v^2} + \frac{D(T)}{v^3} + \cdots \qquad \textbf{(9.174)}$$

Here $B(T)$, $C(T)$, $D(T)$, . . . are all functions of temperature only and are known as the second, third, fourth, . . . virial coefficients respectively. Equation (9.174) itself is known as the virial equation of state and was originally proposed by Clausius as an improvement over the ideal gas model. By arguments involving a microscopic model, it can be shown that the virial coefficients account for the interaction forces among molecules and that their magnitude depends upon the nature of the microscopic model used to describe the forces of interaction. Of course, equation (9.174) can also be used in an empirical fashion in which case the virial coefficients simply become parameters whose magnitudes can be adjusted to fit the P-v-T data of the particular gas.

9.9 The Principle of Corresponding States and the Generalized Equation of State

The principle of corresponding states is used in engineering practice when the thermodynamic behavior of the substance is not known and the ideal gas model is not adequate. J. D. van der Waals, who was the first individual to formulate this empirical principle, noted that all substances possess a critical state—the state on the critical isotherm for which the inflection point and the point of zero slope are coincident. van der Waals' idea was to express the P-v-T data of a pure substance in terms of the pressure, temperature, and the volume of the critical state. He postulated a universal function among these critical properties to establish the P-v-T data in any other state. The properties, which are normalized with respect to the critical properties, are called the reduced properties P_r, T_r, and v_r. They are defined by

$$P_r = \frac{P}{P_c}, \; T_r = \frac{T}{T_c}, \text{ and } v_r = \frac{v}{v_c} \tag{9.175}$$

where the critical properties are denoted by the subscript c. (Values of the critical properties for some common pure substances are listed in Table 9.4.) The empirical principle of corresponding states may then be expressed in the form:

> All substances obey the same equation of state expressed in terms of the reduced properties.

TABLE 9.3
Empirical Constants for Benedict-Webb-Rubin Equations

Gas	*Formula*	A_0 Nm^4/kg^2	B_0 m^3/kg	C_0 Nm^4K^2/kg^2
Methane	CH_4	7.31195 E+2	2.65735 E−3	8.89635 E+6
Ethylene	C_2H_4	4.30550 E+2	1.98649 E−3	1.69071 E+7
Ethane	C_2H_6	4.66269 E+2	2.08914 E−3	2.01509 E+7
Propylene	C_3H_6	3.50217 E+2	2.02308 E−3	2.51642 E+7
Propane	C_3H_8	3.58575 E+2	2.20855 E−3	2.65194 E+7
i-Butane	C_4H_{10}	3.07308 E+2	2.36826 E−3	2.55256 E+7
i-Butylene	C_4H_8	2.88571 E+2	2.06958 E−3	2.98871 E+7
n-Butane	C_4H_{10}	3.02865 E+2	2.14127 E−3	2.98168 E+7
i-Pentane	C_5H_{12}	2.49391 E+2	2.22006 E−3	3.40357 E+7
n-Pentane	C_5H_{12}	2.37376 E+2	2.17426 E−3	4.13424 E+7
n-Hexane	C_6H_{14}	1.97242 E+2	2.06498 E−3	4.53487 E+7
n-Heptane	C_7H_{16}	1.77041 E+2	1.98756 E−3	4.79543 E+7

According to the state principle, a stable, single-phase equilibrium state may be established by specifying any two of the three properties P, v, and T. Then, the principle of corresponding states implies that any dimensionless group of properties is a universal function of any two of the three reduced properties. In particular the compressibility factor, z, where

$$z = \frac{Pv}{RT} \tag{9.176}$$

can be written as a universal function of the reduced volume and temperature.

$$z = \frac{Pv}{RT} = f(T_r, v_r) \tag{9.177}$$

Although equation (9.177) is the form in which the principle of corresponding states was originally proposed, the critical volume has proven to be difficult to measure with any degree of

TABLE 9.3 *(cont'd.)*
Empirical Constants for Benedict-Webb-Rubin Equations

Gas	a Nm^7/kg^2	b m^6/kg^2	c Nm^7K^2/kg^3	α m^9/kg^3	γ m^6/kg^2
Methane	12.1466	1.31523 E−5	6.2577 E+4	3.01853 E−8	2.33469 E−5
Ethylene	1.19119	1.09451 E−5	9.7139 E+4	8.08173 E−9	1.17469 E−5
Ethane	1.28892	1.23191 E−5	1.22361 E+5	8.97220 E−9	1.30701 E−5
Propylene	1.05482	1.05806 E−5	1.39829 E+5	6.13014 E−9	1.03453 E−5
Propane	1.12224	1.15892 E−5	1.52759 E+5	7.09776 E−9	1.13317 E−5
i-Butane	1.00195	1.25806 E−5	1.47891 E+5	5.48279 E−9	1.00799 E−5
i-Butylene	0.97316	1.10774 E−5	1.58056 E+5	5.16963 E−9	9.41616 E−6
n-Butane	0.97334	1.18582 E−5	1.63610 E+5	5.62184 E−9	1.00799 E−5
i-Pentane	1.01546	1.28545 E−5	1.87887 E+5	4.53682 E−9	8.90805 E−6
n-Pentane	1.10159	1.28545 E−5	2.22807 E+5	4.83038 E−9	9.13893 E−6
n-Hexane	1.12913	1.47181 E−5	2.40013 E+5	4.40244 E−9	8.99353 E−6
n-Heptane	1.04602	1.51575 E−5	2.49275 E+5	4.33982 E−9	8.97754 E−6

accuracy. Therefore, it is more convenient to express the compressibility factor in terms of the reduced pressure and temperature. Thus,

$$z = \frac{Pv}{RT} = z(T_r, P_r) \tag{9.178}$$

In place of the actual reduced volume we use a pseudo-reduced volume, v_r', where

$$v'_r = \frac{v}{RT_c/P_c} = z\,\frac{T_r}{P_r} \tag{9.179}$$

Rather than attempting to formulate the analytical expression of equation (9.178), we find it more practical to establish the compressibility factor from experimental measurements and to present the data in the form of a graph of z as a function of P_r with T_r and v'_r as parameters. The original compressibility charts were prepared by averaging data for the following seven

TABLE 9.4
Critical Constants

Gas	*Formula*	*Molecular Weight*	*Critical Temperature, T_c* K	*Critical Temperature, T_c* R	*Critical Pressure, P_c* atm	*Critical Pressure, P_c* N/m^2	*Critical Volume, v_c* $m^3/kg\text{-}mole$	ω
Acetic acid	$C_2H_4O_2$	60.03	594.8	1070.6	57.2	5.79 E + 6	0.1711	0.454
Acetylene	C_2H_2	26.02	309	556	62	6.28 E + 6	0.1130	0.184
Ammonia	NH_3	17.03	405.5	729.8	111.3	1.127 E + 7	0.0724	0.250
Argon	A	39.944	151	272	48.0	4.86 E + 6	0.0749	−0.004
Benzene	C_6H_6	78.11	562	1012	48.6	4.92 E + 6	0.2604	0.212
Bromine	Br_2	159.832	584	1052	102	1.033 E + 7	0.1355	0.132
n-Butane	C_4H_{10}	58.120	425.2	765.2	37.5	3.80 E + 6	0.2547	0.193
Carbon dioxide	CO_2	44.01	304.2	547.2	72.9	7.38 E + 6	0.0943	0.225
Carbon monoxide	CO	28.01	133	240	34.5	3.49 E + 6	0.0930	0.049
Carbon tetrachloride	CCl_4	153.84	556.4	1001.5	45.0	4.56 E + 6	0.2760	0.194
Chlorine	Cl_2	70.914	417	751	76.1	7.71 E + 6	0.1243	0.073
Chloroform	$CHCl_3$	119.39	536.6	965.8	54.0	5.47 E + 6	0.2404	0.216
Decane	$C_{10}H_{22}$	142.17	619.4	1115	21.24	2.152 E + 6	0.6113	0.490
Deuterium (normal)	D_2	4.00	38.4	69.1	16.4	1.66 E + 6	—	−0.130
Dichlorodifluoro-methane Freon-12	CCl_2F_2	120.92	384.7	692.4	39.6	4.01 E + 6	0.2179	0.176
Dichlorofluoro-methane Freon-21	$CHCl_2F$	102.93	451.7	813.0	51.0	5.17 E + 6	0.1973	—
Ethane	C_2H_6	30.068	305.5	549.8	48.2	4.88 E + 6	0.1480	0.098
Ethyl alcohol	C_2H_5OH	46.07	516.0	929.0	63.0	6.38 E + 6	0.1673	0.635
Ethyl chloride	C_2H_5Cl	64.50	460.4	828.7	52	5.27 E + 6	0.1961	0.190
Ethyl ether	$C_4H_{10}O$	74.08	466.0	838.8	35.5	3.60 E + 6	0.2822	0.281
Ethylene	C_2H_4	28.052	282.4	508.3	50.5	5.12 E + 6	0.1243	0.085
Helium	He^4	4.003	5.3	9.5	2.26	2.29 E + 5	0.0578	−0.387
Helium^3	He^3	3.00	3.34	6.01	1.15	1.16 E + 5	—	—
Heptane	C_7H_{16}	100.12	540.17	972.31	27.00	2.735 E + 6	0.4108	0.351

TABLE 9.4 *(cont'd.)*
Critical Constants

Gas	*Formula*	*Molecular Weight*	*Critical Temperature, T_c* K	R	*Critical Pressure, P_c* atm	N/m^2	*Critical Volume, v_c* $m^3/kg\text{-}mole$	ω
n-Hexane	C_6H_{14}	86.172	507.9	914.2	29.9	3.03 E + 6	0.3678	0.296
Hydrogen (normal)	H_2	2.016	33.3	59.9	12.8	1.30 E + 6	0.0649	−0.220
Hydrogen cyanide	HCN	27.02	456.7	822.1	50	5.07 E + 6	0.1349	0.407
Krypton	Kr	83.7	209.4	376.9	54.3	5.50 E + 6	0.0924	−0.002
Methane	CH_4	16.042	191.1	343.9	45.8	4.64 E + 6	0.0993	0.008
Methyl alcohol	CH_3OH	32.04	513.2	923.7	78.5	7.95 E + 6	0.1180	0.559
Methyl chloride	CH_3Cl	50.49	416.3	749.3	65.9	6.68 E + 6	0.1430	0.156
Neon	Ne	20.183	44.5	80.1	26.9	2.72 E + 6	0.0417	0.0
Nitric oxide	NO	30.01	179	323	65	6.58 E + 6	0.0578	0.607
Nitrogen	N_2	28.016	126.2	227.1	33.5	3.39 E + 6	0.0899	0.040
Nitrous oxide	N_2O	44.02	309.7	557.4	71.7	7.26 E + 6	0.0962	0.160
Nonane	C_9H_{20}	128.16	596	1072	22.86	2.316 E + 6	0.5532	0.444
Octane	C_8H_{18}	114.14	569.4	1024.9	24.66	2.498 E + 6	0.4901	0.394
Oxygen	O_2	32.00	158.4	278.6	50.1	5.08 E + 6	0.0780	0.021
Pentane	C_5H_{12}	72.09	470.3	846.6	33.04	3.347 E + 6	0.3103	0.251
Propane	C_3H_8	44.094	370.0	665.9	42.0	4.25 E + 6	0.1998	0.152
Propene	C_3H_6	42.078	365.0	656.9	45.6	4.62 E + 6	0.1811	0.148
Propyne	C_3H_4	40.062	401	722	52.8	5.35 E + 6	—	0.218
Sulfur dioxide	SO_2	64.06	430.7	775.2	77.8	7.88 E + 6	0.1218	0.251
Sulfur trioxide	SO_3	80.06	491.5	884.7	83.6	8.47 E + 6	0.1268	0.410
Toluene	C_7H_8	92.06	593.8	1068.8	41.6	4.21 E + 6	0.3153	0.257
Trichlorofluoro-methane Freon-11	CCl_3F	137.38	471.2	848.1	43.2	4.38 E + 6	0.2479	0.188
Water	H_2O	18.016	647.4	1165.3	218.3	2.211 E + 7	0.0562	0.344
Xenon	Xe	131.3	289.75	521.55	58.0	5.88 E + 6	0.1186	0.002

gases: H_2, N_2, CO_2, NH_3, CH_4, C_3H_8, C_5H_{12}. The latest versions of the compressibility charts, for example Figure 9.4, have increased this number to thirty. Because the gases used in the averaging process include many types of molecules, the resulting chart cannot be expected to reproduce absolutely the properties of any one of the gases listed. However, more accurate charts can be prepared by averaging data of molecules which are more nearly the same type. In spite of these limitations, the use of Figure 9.4 to obtain a value of z at a given T_r and P_r should lead to errors less than 4 to 6 percent except near the critical state where z is a strong function of T_r and P_r.

In employing the compressibility chart for engineering calculations, any two of the three properties T_r, P_r, or v'_r are used to find the compressibility factor, z. This quantity is then used in the relation

$$Pv = zRT \tag{9.180}$$

to find the third property of state. Equation (9.180) is known as the generalized equation of state. As such, it is a two-parameter equation of state. In an attempt to improve the accuracy of this equation of state, it has been suggested that an additional third parameter be introduced into the expression for z given by equation (9.178). Lee and Kesler[8] have suggested that the Pitzer acentric factor, ω, defined by the expression

$$\omega = -\log\,(P_{\text{vap}})_{T_r=0.7} - 1.000 \tag{9.181}$$

be used for this purpose. In equation (9.181), $(P_{\text{vap}})_{T_r=0.7}$ is the vapor pressure of the fluid evaluated at $T_r = 0.7$. This factor which is a measure of the nonsphericity of the molecular force field ($\omega = 0$ for a rare gas with spherical symmetry) is then used in a linear expansion of the compressibility factor.

$$z = z^{(0)}\,(T_r,P_r) + \omega z^{(1)}(T_r,P_r) \tag{9.182}$$

where $z^{(0)}$ applies to spherical molecules and $z^{(1)}$ is a deviation function. Values of $z^{(0)}$ and $z^{(1)}$ are tabulated in Tables 9.5 and 9.6

[8] Lee, B. I. and Kesler, M. G., "A Generalized Thermodynamic Correlation Based on Three-Parameter Corresponding States," AIChE Journal, Vol. 21, No. 3, pp. 510–527, May, 1975.

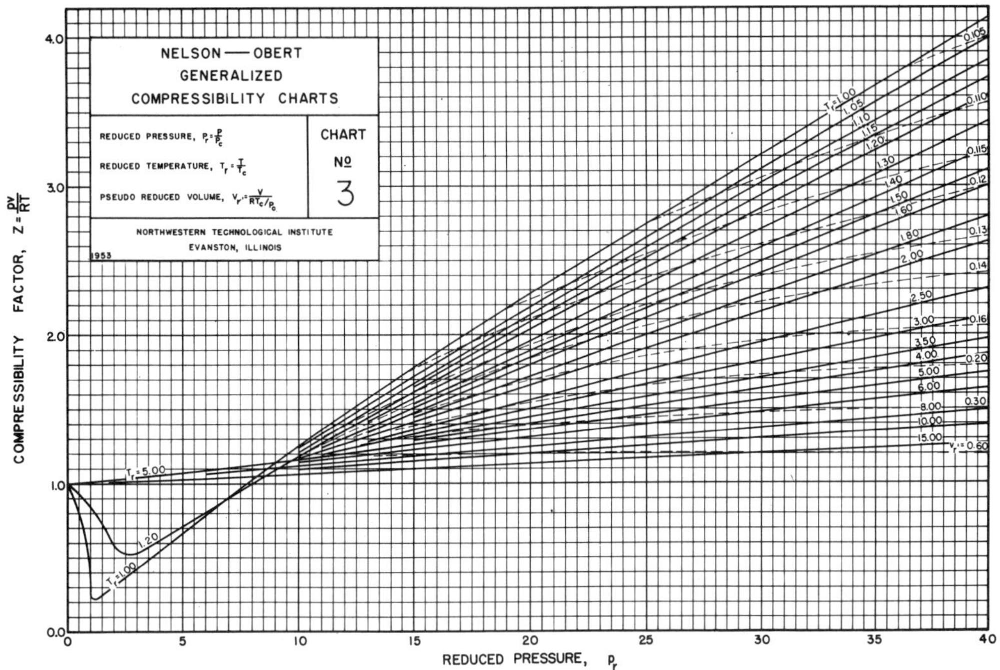

Figure 9.4(a) Generalized compressibility chart[9]

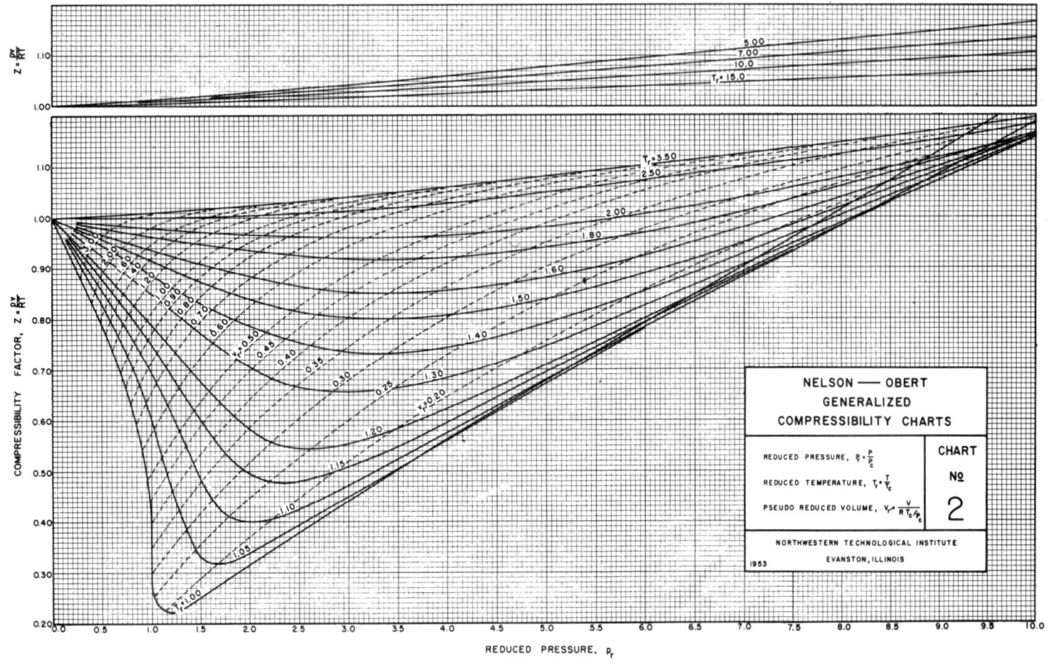

Figure 9.4(b) Generalized compressibility chart

[9] Figure 9.4 reprinted by permission of E. F. Obert, University of Wisconsin.

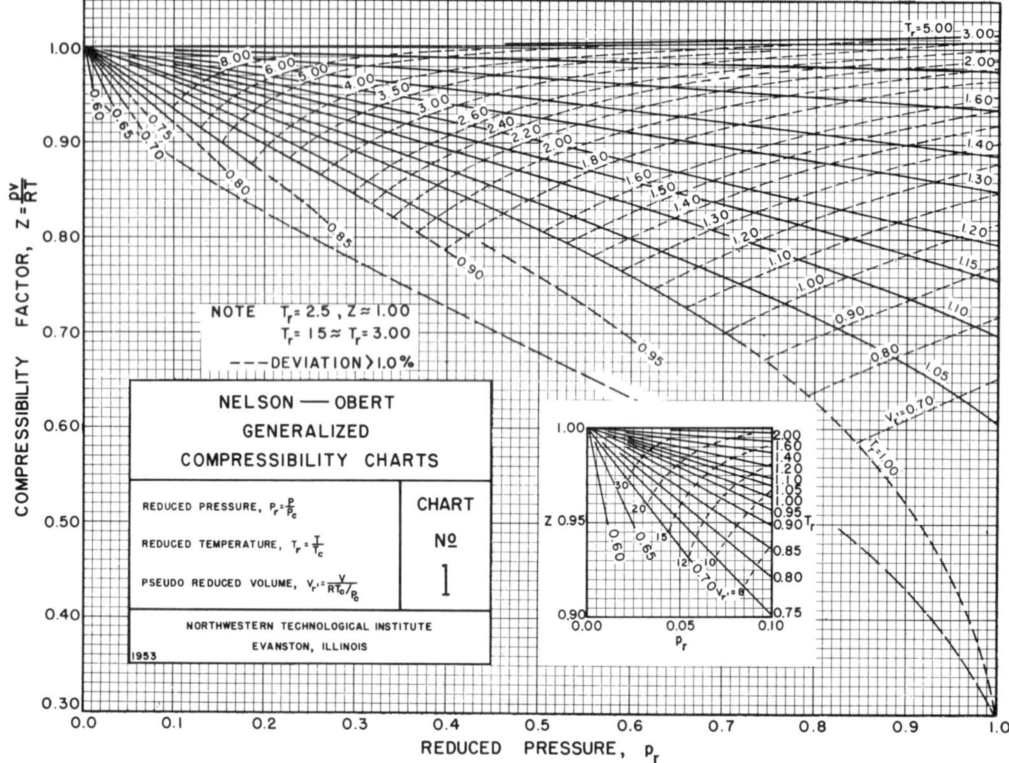

Figure 9.4(c) Generalized compressibility chart

for various values of T_r and P_r. The value of ω to be used in equation (9.182) is given for various fluids in Table 9.4.

Once we have generated the *P-v-T* data of a substance by means of the generalized compressibility factor, z, we can use these data and the methods of Section 9.7 to establish the other properties such as the enthalpy and entropy. The values of these two properties are most conveniently presented in terms of the deviation from the ideal gas model. Since these properties can be readily evaluated for the ideal gas model, the enthalpy and entropy of the generalized real substance can be evaluated by simply adding to the ideal gas result the contribution due to "real" effects as determined from the generalized compressibility factor. To evaluate the enthalpy we employ equation (9.155) since the independent variables in equation (9.182) are temperature and pressure. The "real" effect due to the deviation from the ideal gas model is defined as the enthalpy departure function. Specifically, the enthalpy departure function for a state of T_1 and P_1 is defined

$$(h_{\text{ideal}} - h_{\text{actual}})_{T_1} = h_{T_1,P=0} - h_{T_1,P_1} \tag{9.183}$$

Equation (9.155) then allows us to evaluate the enthalpy departure in terms of $v(T,P)$.

$$(h_{\text{ideal}} - h_{\text{actual}})_{T_1} = \int_{P_1}^{P=0} \left[v - T\left(\frac{\partial v}{\partial T}\right)_P \right]_{T=T_1} dP \tag{9.184}$$

We must now express equation (9.184) in dimensionless form in terms of P_r, T_r and the compressibility z. From the definitions of P_r, T_r and z we have

$$P = P_c P_r, \quad dP = P_c \, dP_r \tag{9.185}$$
$$T = T_c T_r, \quad dT = T_c \, dT_r \tag{9.186}$$

and

$$v = \frac{zRT_c}{P_c}\frac{T_r}{P_r} \tag{9.187}$$

With these relations the derivative $(\partial v/\partial T)_P$ becomes

$$\left(\frac{\partial v}{\partial T}\right)_P = \frac{zR}{P_c}\frac{1}{P_r} + \frac{R}{P_c}\frac{T_r}{P_r}\left(\frac{\partial z}{\partial T_r}\right)_{P_r} \tag{9.188}$$

The integrand of equation (9.184) then becomes

$$v - T\left(\frac{\partial v}{\partial T}\right)_P = \frac{zRT_c}{P_c}\frac{T_r}{P_r} - \frac{zRT_c}{P_c}\frac{T_r}{P_r} - \frac{RT_c}{P_c}\frac{T_r^2}{P_r}\left[\frac{\partial z}{\partial T_r}\right]_{P_r} \tag{9.189}$$

Equation (9.184) can then be written in terms of the reduced properties as

$$(h_{\text{ideal}} - h_{\text{actual}})_{T_1} = RT_c \left[\int_{P_r=0}^{P_r} T_r^2 \left[\frac{\partial z}{\partial T_r}\right]_{P_r} d(\ln P_r) \right]_{T=T_1} \tag{9.190}$$

where the limits of integration have been reversed. By using an analytical expression for the compressibility factor given in

TABLE 9.5
Values of $z^{(0)}$ from Lee and Kesler[10]

T_r	P_r 0.010	0.050	0.100	0.200	0.400	0.600	0.800
0.30	0.0029	0.0145	0.0290	0.0579	0.1158	0.1737	0.2315
0.35	0.0026	0.0130	0.0261	0.0522	0.1043	0.1564	0.2084
0.40	0.0024	0.0119	0.0239	0.0477	0.0953	0.1429	0.1904
0.45	0.0022	0.0110	0.0221	0.0442	0.0882	0.1322	0.1762
0.50	0.0021	0.0103	0.0207	0.0413	0.0825	0.1236	0.1647
0.55	0.9804	0.0098	0.0195	0.0390	0.0778	0.1166	0.1553
0.60	0.9849	0.0093	0.0186	0.0371	0.0741	0.1109	0.1476
0.65	0.9881	0.9377	0.0178	0.0356	0.0710	0.1063	0.1415
0.70	0.9904	0.9504	0.8958	0.0344	0.0687	0.1027	0.1366
0.75	0.9922	0.9598	0.9165	0.0336	0.0670	0.1001	0.1330
0.80	0.9935	0.9669	0.9319	0.8539	0.0661	0.0985	0.1307
0.85	0.9946	0.9725	0.9436	0.8810	0.0661	0.0983	0.1301
0.90	0.9954	0.9768	0.9528	0.9015	0.7800	0.1006	0.1321
0.93	0.9959	0.9790	0.9573	0.9115	0.8059	0.6635	0.1359
0.95	0.9961	0.9803	0.9600	0.9174	0.8206	0.6967	0.1410
0.97	0.9963	0.9815	0.9625	0.9227	0.8338	0.7240	0.5580
0.98	0.9965	0.9821	0.9637	0.9253	0.8398	0.7360	0.5887
0.99	0.9966	0.9826	0.9648	0.9277	0.8455	0.7471	0.6138
1.00	0.9967	0.9832	0.9659	0.9300	0.8509	0.7574	0.6353
1.01	0.9968	0.9837	0.9669	0.9322	0.8561	0.7671	0.6542
1.02	0.9969	0.9842	0.9679	0.9343	0.8610	0.7761	0.6710
1.05	0.9971	0.9855	0.9707	0.9401	0.8743	0.8002	0.7130
1.10	0.9975	0.9874	0.9747	0.9485	0.8930	0.8323	0.7649
1.15	0.9978	0.9891	0.9780	0.9554	0.9081	0.8576	0.8032
1.20	0.9981	0.9904	0.9808	0.9611	0.9205	0.8779	0.8330
1.30	0.9985	0.9926	0.9852	0.9702	0.9396	0.9083	0.8764
1.40	0.9988	0.9942	0.9884	0.9768	0.9534	0.9298	0.9062
1.50	0.9991	0.9954	0.9909	0.9818	0.9636	0.9456	0.9278
1.60	0.9993	0.9964	0.9928	0.9856	0.9714	0.9575	0.9439
1.70	0.9994	0.9971	0.9943	0.9886	0.9775	0.9667	0.9563
1.80	0.9995	0.9977	0.9955	0.9910	0.9823	0.9739	0.9659
1.90	0.9996	0.9982	0.9964	0.9929	0.9861	0.9796	0.9735
2.00	0.9997	0.9986	0.9972	0.9944	0.9892	0.9842	0.9796
2.20	0.9998	0.9992	0.9983	0.9967	0.9937	0.9910	0.9886
2.40	0.9999	0.9996	0.9991	0.9983	0.9969	0.9957	0.9948
2.60	1.0000	0.9998	0.9997	0.9994	0.9991	0.9990	0.9990
2.80	1.0000	1.0000	1.0001	1.0002	1.0007	1.0013	1.0021
3.00	1.0000	1.0002	1.0004	1.0008	1.0018	1.0030	1.0043
3.50	1.0001	1.0004	1.0008	1.0017	1.0035	1.0055	1.0075
4.00	1.0001	1.0005	1.0010	1.0021	1.0043	1.0066	1.0090

[10] Lee, B. I. and Kesler, M. G., "A Generalized Thermodynamic Correlation Based on Three-Parameter Corresponding States," AIChE Journal, Vol. 21, No. 3, pp. 510–527, May, 1975.

TABLE 9.5 (*cont'd.*)
Values of $z^{(0)}$

P_r							
1.000	*1.200*	*1.500*	*2.000*	*3.000*	*5.000*	*7.000*	*10.000*
0.2892	0.3470	0.4335	0.5775	0.8648	1.4366	2.0048	2.8507
0.2604	0.3123	0.3901	0.5195	0.7775	1.2902	1.7987	2.5539
0.2379	0.2853	0.3563	0.4744	0.7095	1.1758	1.6373	2.3211
0.2200	0.2638	0.3294	0.4384	0.6551	1.0841	1.5077	2.1338
0.2056	0.2465	0.3077	0.4092	0.6110	1.0094	1.4017	1.9801
0.1939	0.2323	0.2899	0.3853	0.5747	0.9475	1.3137	1.8520
0.1842	0.2207	0.2753	0.3657	0.5446	0.8959	1.2398	1.7440
0.1765	0.2113	0.2634	0.3495	0.5197	0.8526	1.1773	1.6519
0.1703	0.2038	0.2538	0.3364	0.4991	0.8161	1.1241	1.5729
0.1656	0.1981	0.2464	0.3260	0.4823	0.7854	1.0787	1.5047
0.1626	0.1942	0.2411	0.3182	0.4690	0.7598	1.0400	1.4456
0.1614	0.1924	0.2382	0.3132	0.4591	0.7388	1.0071	1.3943
0.1630	0.1935	0.2383	0.3114	0.4527	0.7220	0.9793	1.3496
0.1664	0.1963	0.2405	0.3122	0.4507	0.7138	0.9648	1.3257
0.1705	0.1998	0.2432	0.3138	0.4501	0.7092	0.9561	1.3108
0.1779	0.2055	0.2474	0.3164	0.4504	0.7052	0.9480	1.2968
0.1844	0.2097	0.2503	0.3182	0.4508	0.7035	0.9442	1.2901
0.1959	0.2154	0.2538	0.3204	0.4514	0.7018	0.9406	1.2835
0.2901	0.2237	0.2583	0.3229	0.4522	0.7004	0.9372	1.2772
0.4648	0.2370	0.2640	0.3260	0.4533	0.6991	0.9339	1.2710
0.5146	0.2629	0.2715	0.3297	0.4547	0.6980	0.9307	1.2650
0.6026	0.4437	0.3131	0.3452	0.4604	0.6956	0.9222	1.2481
0.6880	0.5984	0.4580	0.3953	0.4770	0.6950	0.9110	1.2232
0.7443	0.6803	0.5798	0.4760	0.5042	0.6987	0.9033	1.2021
0.7858	0.7363	0.6605	0.5605	0.5425	0.7069	0.8990	1.1844
0.8438	0.8111	0.7624	0.6908	0.6344	0.7358	0.8998	1.1580
0.8827	0.8595	0.8256	0.7753	0.7202	0.7761	0.9112	1.1419
0.9103	0.8933	0.8689	0.8328	0.7887	0.8200	0.9297	1.1339
0.9308	0.9180	0.9000	0.8738	0.8410	0.8617	0.9518	1.1320
0.9463	0.9367	0.9234	0.9043	0.8809	0.8984	0.9745	1.1343
0.9583	0.9511	0.9413	0.9275	0.9118	0.9297	0.9961	1.1391
0.9678	0.9624	0.9552	0.9456	0.9359	0.9557	1.0157	1.1452
0.9754	0.9715	0.9664	0.9599	0.9550	0.9772	1.0328	1.1516
0.9865	0.9847	0.9826	0.9806	0.9827	1.0094	1.0600	1.1635
0.9941	0.9936	0.9935	0.9945	1.0011	1.0313	1.0793	1.1728
0.9993	0.9998	1.0010	1.0040	1.0137	1.0463	1.0926	1.1792
1.0031	1.0042	1.0063	1.0106	1.0223	1.0565	1.1016	1.1830
1.0057	1.0074	1.0101	1.0153	1.0284	1.0635	1.1075	1.1848
1.0097	1.0120	0.0156	1.0221	1.0368	1.0723	1.1138	1.1834
1.0115	1.0140	1.0179	1.0249	1.0401	1.0747	1.1136	1.1773

TABLE 9.6
Value of $z^{(1)}$ from Lee and Kesler[11]

T_r	P_r 0.010	0.050	0.100	0.200	0.400	0.600	0.800
0.30	−0.0008	−0.0040	−0.0081	−0.0161	−0.0323	−0.0484	−0.0645
0.35	−0.0009	−0.0046	−0.0093	−0.0185	−0.0370	−0.0554	−0.0738
0.40	−0.0010	−0.0048	−0.0095	−0.0190	−0.0380	−0.0570	−0.0758
0.45	−0.0009	−0.0047	−0.0094	−0.0187	−0.0374	−0.0560	−0.0745
0.50	−0.0009	−0.0045	−0.0090	−0.0181	−0.0360	−0.0539	−0.0716
0.55	−0.0314	−0.0043	−0.0086	−0.0172	−0.0343	−0.0513	−0.0682
0.60	−0.0205	−0.0041	−0.0082	−0.0164	−0.0326	−0.0487	−0.0646
0.65	−0.0137	−0.0772	−0.0078	−0.0156	−0.0309	−0.0461	−0.0611
0.70	−0.0093	−0.0507	−0.1161	−0.0148	−0.0294	−0.0438	−0.0579
0.75	−0.0064	−0.0339	−0.0744	−0.0143	−0.0282	−0.0417	−0.0550
0.80	−0.0044	−0.0228	−0.0487	−0.1160	−0.0272	−0.0401	−0.0526
0.85	−0.0029	−0.0152	−0.0319	−0.0715	−0.0268	−0.0391	−0.0509
0.90	−0.0019	−0.0099	−0.0205	−0.0442	−0.1118	−0.0396	−0.0503
0.93	−0.0015	−0.0075	−0.0154	−0.0326	−0.0763	−0.1662	−0.0514
0.95	−0.0012	−0.0062	−0.0126	−0.0262	−0.0589	−0.1110	−0.0540
0.97	−0.0010	−0.0050	−0.0101	−0.0208	−0.0450	−0.0770	−0.1647
0.98	−0.0009	−0.0044	−0.0090	−0.0184	−0.0390	−0.0641	−0.1100
0.99	−0.0008	−0.0039	−0.0079	−0.0161	−0.0335	−0.0531	−0.0796
1.00	−0.0007	−0.0034	−0.0069	−0.0140	−0.0285	−0.0435	−0.0588
1.01	−0.0006	−0.0030	−0.0060	−0.0120	−0.0240	−0.0351	−0.0429
1.02	−0.0005	−0.0026	−0.0051	−0.0102	−0.0198	−0.0277	−0.0303
1.05	−0.0003	−0.0015	−0.0029	−0.0054	−0.0092	−0.0097	−0.0032
1.10	−0.0000	0.0000	0.0001	0.0007	0.0038	0.0106	0.0236
1.15	0.0002	0.0011	0.0023	0.0052	0.0127	0.0237	0.0396
1.20	0.0004	0.0019	0.0039	0.0084	0.0190	0.0326	0.0499
1.30	0.0006	0.0030	0.0061	0.0125	0.0267	0.0429	0.0612
1.40	0.0007	0.0036	0.0072	0.0147	0.0306	0.0477	0.0661
1.50	0.0008	0.0039	0.0078	0.0158	0.0323	0.0497	0.0677
1.60	0.0008	0.0040	0.0080	0.0162	0.0330	0.0501	0.0677
1.70	0.0008	0.0040	0.0081	0.0163	0.0329	0.0497	0.0667
1.80	0.0008	0.0040	0.0081	0.0162	0.0325	0.0488	0.0652
1.90	0.0008	0.0040	0.0079	0.0159	0.0318	0.0477	0.0635
2.00	0.0008	0.0039	0.0078	0.0155	0.0310	0.0464	0.0617
2.20	0.0007	0.0037	0.0074	0.0147	0.0293	0.0437	0.0579
2.40	0.0007	0.0035	0.0070	0.0139	0.0276	0.0411	0.0544
2.60	0.0007	0.0033	0.0066	0.0131	0.0260	0.0387	0.0512
2.80	0.0006	0.0031	0.0062	0.0124	0.0245	0.0365	0.0483
3.00	0.0006	0.0029	0.0059	0.0117	0.0232	0.0345	0.0456
3.50	0.0005	0.0026	0.0052	0.0103	0.0204	0.0303	0.0401
4.00	0.0005	0.0023	0.0046	0.0091	0.0182	0.0270	0.0357

[11] Lee and Kesler, ibid.

TABLE 9.6 (*cont'd.*)
Values of $z^{(1)}$

P_r							
1.000	*1.200*	*1.500*	*2.000*	*3.000*	*5.000*	*7.000*	*10.000*
−0.0806	−0.0966	−0.1207	−0.1608	−0.2407	−0.3996	−0.5572	−0.7915
−0.0921	−0.1105	−0.1379	−0.1834	−0.2738	−0.4523	−0.6279	−0.8863
−0.0946	−0.1134	−0.1414	−0.1879	−0.2799	−0.4603	−0.6365	−0.8936
−0.0929	−0.1113	−0.1387	−0.1840	−0.2734	−0.4475	−0.6162	−0.8606
−0.0893	−0.1069	−0.1330	−0.1762	−0.2611	−0.4253	−0.5831	−0.8099
−0.0849	−0.1015	−0.1263	−0.1669	−0.2465	−0.3991	−0.5446	−0.7521
−0.0803	−0.0960	−0.1192	−0.1572	−0.2312	−0.3718	−0.5047	−0.6928
−0.0759	−0.0906	−0.1122	−0.1476	−0.2160	−0.3447	−0.4653	−0.6346
−0.0718	−0.0855	−0.1057	−0.1385	−0.2013	−0.3184	−0.4270	−0.5785
−0.0681	−0.0808	−0.0996	−0.1298	−0.1872	−0.2929	−0.3901	−0.5250
−0.0648	−0.0767	−0.0940	−0.1217	−0.1736	−0.2682	−0.3545	−0.4740
−0.0622	−0.0731	−0.0888	−0.1138	−0.1602	−0.2439	−0.3201	−0.4254
−0.0604	−0.0701	−0.0840	−0.1059	−0.1463	−0.2195	−0.2862	−0.3788
−0.0602	−0.0687	−0.0810	−0.1007	−0.1374	−0.2045	−0.2661	−0.3516
−0.0607	−0.0678	−0.0788	−0.0967	−0.1310	−0.1943	−0.2526	−0.3339
−0.0623	−0.0669	−0.0759	−0.0921	−0.1240	−0.1837	−0.2391	−0.3163
−0.0641	−0.0661	−0.0740	−0.0893	−0.1202	−0.1783	−0.2322	−0.3075
−0.0680	−0.0646	−0.0715	−0.0861	−0.1162	−0.1728	−0.2254	−0.2989
−0.0879	−0.0609	−0.0678	−0.0824	−0.1118	−0.1672	−0.2185	−0.2902
−0.0223	−0.0473	−0.0621	−0.0778	−0.1072	−0.1615	−0.2116	−0.2816
−0.0062	0.0227	−0.0524	−0.0722	−0.1021	−0.1556	−0.2047	−0.2731
0.0220	0.1059	0.0451	−0.0432	−0.0838	−0.1370	−0.1835	−0.2476
0.0476	0.0897	0.1630	0.0698	−0.0373	−0.1021	−0.1469	−0.2056
0.0625	0.0943	0.1548	0.1667	0.0332	−0.0611	−0.1084	−0.1642
0.0719	0.0991	0.1477	0.1990	0.1095	−0.0141	−0.0678	−0.1231
0.0819	0.1048	0.1420	0.1991	0.2079	0.0875	0.0176	−0.0423
0.0857	0.1063	0.1383	0.1894	0.2397	0.1737	0.1008	0.0350
0.0864	0.1055	0.1345	0.1806	0.2433	0.2309	0.1717	0.1058
0.0855	0.1035	0.1303	0.1729	0.2381	0.2631	0.2255	0.1673
0.0838	0.1008	0.1259	0.1658	0.2305	0.2788	0.2628	0.2179
0.0816	0.0978	0.1216	0.1593	0.2224	0.2846	0.2871	0.2576
0.0792	0.0947	0.1173	0.1532	0.2144	0.2848	0.3017	0.2876
0.0767	0.0916	0.1133	0.1476	0.2069	0.2819	0.3097	0.3096
0.0719	0.0857	0.1057	0.1374	0.1932	0.2720	0.3135	0.3355
0.0675	0.0803	0.0989	0.1285	0.1812	0.2602	0.3089	0.3459
0.0634	0.0754	0.0929	0.1207	0.1706	0.2484	0.3009	0.3475
0.0598	0.0711	0.0876	0.1138	0.1613	0.2372	0.2915	0.3443
0.0565	0.0672	0.0828	0.1076	0.1529	0.2268	0.2817	0.3385
0.0497	0.0591	0.0728	0.0949	0.1356	0.2042	0.2584	0.3194
0.0443	0.0527	0.0651	0.0849	0.1219	0.1857	0.2378	0.2994

TABLE 9.7

Lee-Kesler Residual Enthalpy,[12] Values of $\left(\frac{h_{ideal} - h_{actual}}{RT_c}\right)^{(0)}$

T_r	P_r 0.010	0.050	0.100	0.200	0.400	0.600	0.800
0.30	6.045	6.043	6.040	6.034	6.022	6.011	5.999
0.35	5.906	5.904	5.901	5.895	5.882	5.870	5.858
0.40	5.763	5.761	5.757	5.751	5.738	5.726	5.713
0.45	5.615	5.612	5.609	5.603	5.590	5.577	5.564
0.50	5.465	5.463	4.459	5.453	5.440	5.427	5.414
0.55	0.032	5.312	5.309	5.303	5.290	5.278	5.265
0.60	0.027	5.162	5.159	5.153	5.141	5.129	5.116
0.65	0.023	0.118	5.008	5.002	4.991	4.980	4.968
0.70	0.020	0.101	0.213	4.848	4.838	4.828	4.818
0.75	0.017	0.088	0.183	4.687	4.679	4.672	4.664
0.80	0.015	0.078	0.160	0.345	4.507	4.504	4.499
0.85	0.014	0.069	0.141	0.300	4.309	4.313	4.316
0.90	0.012	0.062	0.126	0.264	0.596	4.074	4.094
0.93	0.011	0.058	0.118	0.246	0.545	0.960	3.920
0.95	0.011	0.056	0.113	0.235	0.516	0.885	3.763
0.97	0.011	0.054	0.109	0.225	0.490	0.824	1.356
0.98	0.010	0.053	0.107	0.221	0.478	0.797	1.273
0.99	0.010	0.052	0.105	0.216	0.466	0.773	1.206
1.00	0.010	0.051	0.103	0.212	0.455	0.750	1.151
1.01	0.010	0.050	0.101	0.208	0.445	0.728	0.102
1.02	0.010	0.049	0.099	0.203	0.434	0.708	1.060
1.05	0.009	0.046	0.094	0.192	0.407	0.654	0.955
1.10	0.008	0.042	0.086	0.175	0.367	0.581	0.827
1.15	0.008	0.039	0.079	0.160	0.334	0.523	0.732
1.20	0.007	0.036	0.073	0.148	0.305	0.474	0.657
1.30	0.006	0.031	0.063	0.127	0.259	0.399	0.545
1.40	0.005	0.027	0.055	0.110	0.224	0.341	0.463
1.50	0.005	0.024	0.048	0.097	0.196	0.297	0.400
1.60	0.004	0.021	0.043	0.086	0.173	0.261	0.350
1.70	0.004	0.019	0.038	0.076	0.153	0.231	0.309
1.80	0.003	0.017	0.034	0.068	0.137	0.206	0.275
1.90	0.003	0.015	0.031	0.062	0.123	0.185	0.246
2.00	0.003	0.014	0.028	0.056	0.111	0.167	0.222
2.20	0.002	0.012	0.023	0.046	0.092	0.137	0.182
2.40	0.002	0.010	0.019	0.038	0.076	0.114	0.150
2.60	0.002	0.008	0.016	0.032	0.064	0.095	0.125
2.80	0.001	0.007	0.014	0.027	0.054	0.080	0.105
3.00	0.001	0.006	0.011	0.023	0.045	0.067	0.088
3.50	0.001	0.004	0.007	0.015	0.029	0.043	0.056
4.00	0.000	0.002	0.005	0.009	0.017	0.026	0.033

[12] Lee and Kesler, ibid.

TABLE 9.7 (*cont'd.*)

Lee-Kesler Residual Enthalpy, Values of $\left(\frac{h_{ideal} - h_{actual}}{RT_c}\right)^{(0)}$

P_r							
1.000	*1.200*	*1.500*	*2.000*	*3.000*	*5.000*	*7.000*	*10.000*
5.987	5.975	5.957	5.927	5.868	5.748	5.628	5.446
5.845	5.833	5.814	5.783	5.721	5.595	5.469	5.278
5.700	5.687	5.668	5.636	5.572	5.442	5.311	5.113
5.551	5.538	5.519	5.486	5.421	5.288	5.154	4.950
5.401	5.388	5.369	5.336	5.270	5.135	4.999	4.791
5.252	5.239	5.220	5.187	5.121	4.986	4.849	4.638
5.104	5.091	5.073	5.041	4.976	4.842	4.704	4.492
4.956	4.945	4.927	4.896	4.833	4.702	4.565	4.353
4.808	4.797	4.781	4.752	4.693	4.566	4.432	4.221
4.655	4.646	4.632	4.607	4.554	4.434	4.303	4.095
4.494	4.488	4.478	4.459	4.413	4.303	4.178	3.974
4.316	4.316	4.312	4.302	4.269	4.173	4.056	3.857
4.108	4.118	4.127	4.132	4.119	4.043	3.935	3.744
3.953	3.976	4.000	4.020	4.024	3.963	3.863	3.678
3.825	3.865	3.904	3.940	3.958	3.910	3.815	3.634
3.658	3.732	3.796	3.853	3.890	3.856	3.767	3.591
3.544	3.652	3.736	3.806	3.854	3.829	3.743	3.569
3.376	3.558	3.670	3.758	3.818	3.801	3.719	3.548
2.584	3.441	3.598	3.706	3.782	3.774	3.695	3.526
1.796	3.283	3.516	3.652	3.744	3.746	3.671	3.505
1.627	3.039	3.442	3.595	3.705	3.718	3.647	3.484
1.359	2.034	3.030	3.398	3.583	3.632	3.575	3.420
1.120	1.487	2.203	2.965	3.353	3.484	3.453	3.315
0.968	1.239	1.719	2.479	3.091	3.329	3.329	3.211
0.857	1.076	1.443	2.079	2.807	3.166	3.202	3.107
0.698	0.860	1.116	1.560	2.274	2.825	2.942	2.899
0.588	0.716	0.915	1.253	1.857	2.486	2.679	2.692
0.505	0.611	0.774	1.046	1.549	2.175	2.421	2.486
0.440	0.531	0.667	0.894	1.318	1.904	2.177	2.285
0.387	0.466	0.583	0.777	1.139	1.672	1.953	2.091
0.344	0.413	0.515	0.683	0.996	1.476	1.751	1.908
0.307	0.368	0.458	0.606	0.880	1.309	1.571	1.736
0.276	0.330	0.411	0.541	0.782	1.167	1.411	1.577
0.226	0.269	0.334	0.437	0.629	0.937	1.143	1.295
0.187	0.222	0.275	0.359	0.513	0.761	0.929	1.058
0.155	0.185	0.228	0.297	0.422	0.621	0.756	0.858
0.130	0.154	0.190	0.246	0.348	0.508	0.614	0.689
0.109	0.129	0.159	0.205	0.288	0.415	0.495	0.545
0.069	0.081	0.099	0.127	0.174	0.239	0.270	0.264
0.041	0.048	0.058	0.072	0.095	0.116	0.110	0.061

TABLE 9.8

Lee-Kesler Residual Enthalpy,[13] Values of $\left(\dfrac{h_{\text{ideal}} - h_{\text{actual}}}{RT_c}\right)^{(1)}$

T_r	P_r 0.010	0.050	0.100	0.200	0.400	0.600	0.800
0.30	11.098	11.096	11.095	11.091	11.083	11.076	11.069
0.35	10.656	10.655	10.654	10.653	10.650	10.646	10.643
0.40	10.121	10.121	10.121	10.120	10.121	10.121	10.121
0.45	9.515	9.515	9.516	9.517	9.519	9.521	9.523
0.50	8.868	8.869	8.870	8.872	8.876	8.880	8.884
0.55	0.080	8.211	8.212	8.215	8.221	8.226	8.232
0.60	0.059	7.568	7.570	7.573	7.579	7.585	7.591
0.65	0.045	0.247	6.949	6.952	6.959	6.966	6.973
0.70	0.034	0.185	0.415	6.360	6.367	6.373	6.381
0.75	0.027	0.142	0.306	5.796	5.802	5.809	5.816
0.80	0.021	0.110	0.234	0.542	5.266	5.271	5.278
0.85	0.017	0.087	0.182	0.401	4.753	4.754	4.758
0.90	0.014	0.070	0.144	0.308	0.751	4.254	4.248
0.93	0.012	0.061	0.126	0.265	0.612	1.236	3.942
0.95	0.011	0.056	0.115	0.241	0.542	0.994	3.737
0.97	0.010	0.052	0.105	0.219	0.483	0.837	1.616
0.98	0.010	0.050	0.101	0.209	0.457	0.776	1.324
0.99	0.009	0.048	0.097	0.200	0.433	0.722	1.154
1.00	0.009	0.046	0.093	0.191	0.410	0.675	1.034
1.01	0.009	0.044	0.089	0.183	0.389	0.632	0.940
1.02	0.008	0.042	0.085	0.175	0.370	0.594	0.863
1.05	0.007	0.037	0.075	0.153	0.318	0.498	0.691
1.10	0.006	0.030	0.061	0.123	0.251	0.381	0.507
1.15	0.005	0.025	0.050	0.099	0.199	0.296	0.385
1.20	0.004	0.020	0.040	0.080	0.158	0.232	0.297
1.30	0.003	0.013	0.026	0.052	0.100	0.142	0.177
1.40	0.002	0.008	0.016	0.032	0.060	0.083	0.100
1.50	0.001	0.005	0.009	0.018	0.032	0.042	0.048
1.60	0.000	0.002	0.004	0.007	0.012	0.013	0.011
1.70	0.000	0.000	0.000	−0.000	−0.003	−0.009	−0.017
1.80	−0.000	−0.001	−0.003	−0.006	−0.015	−0.025	−0.037
1.90	−0.001	−0.003	−0.005	−0.011	−0.023	−0.037	−0.053
2.00	−0.001	−0.003	−0.007	−0.015	−0.030	−0.047	−0.065
2.20	−0.001	−0.005	−0.010	−0.020	−0.040	−0.062	−0.083
2.40	−0.001	−0.006	−0.012	−0.023	−0.047	−0.071	−0.095
2.60	−0.001	−0.006	−0.013	−0.026	−0.052	−0.078	−0.104
2.80	−0.001	−0.007	−0.014	−0.028	−0.055	−0.082	−0.110
3.00	−0.001	−0.007	−0.014	−0.029	−0.058	−0.086	−0.114
3.50	−0.002	−0.008	−0.016	−0.031	−0.062	−0.092	−0.122
4.00	−0.002	−0.008	−0.016	−0.032	−0.064	−0.096	−0.127

[13] Lee and Kesler, ibid.

TABLE 9.8 (*cont'd.*)

Lee-Kesler Residual Enthalpy, Values of $\left(\frac{h_{\text{ideal}} - h_{\text{actual}}}{RT_c}\right)^{(1)}$

P_r							
1.000	*1.200*	*1.500*	*2.000*	*3.000*	*5.000*	*7.000*	*10.000*
11.062	11.055	11.044	11.027	10.992	10.935	10.872	10.781
10.640	10.637	10.632	10.624	10.609	10.581	10.554	10.529
10.121	10.121	10.121	10.122	10.123	10.128	10.135	10.150
9.525	9.527	9.531	9.537	9.549	9.576	9.611	9.663
8.888	8.892	8.899	8.909	8.932	8.978	9.030	9.111
8.238	8.243	8.252	8.267	8.298	8.360	8.425	8.531
7.596	7.603	7.614	7.632	7.669	7.745	7.824	7.950
6.980	6.987	6.997	7.017	7.059	7.147	7.239	7.381
6.388	6.395	6.407	6.429	6.475	6.574	6.677	6.837
5.824	5.832	5.845	5.868	5.918	6.027	6.142	6.318
5.285	5.293	5.306	5.330	5.385	5.506	5.632	5.824
4.763	4.771	4.784	4.810	4.872	5.008	5.149	5.358
4.249	4.255	4.268	4.298	4.371	4.530	4.688	4.916
3.934	3.937	3.951	3.987	4.073	4.251	4.422	4.662
3.712	3.713	3.730	3.773	3.873	4.068	4.248	4.497
3.470	3.467	3.492	3.551	3.670	3.885	4.077	4.336
3.332	3.327	3.363	3.434	3.568	3.795	3.992	4.257
3.164	3.164	3.223	3.313	3.464	3.705	3.909	4.178
2.471	2.952	3.065	3.186	3.358	3.615	3.825	4.100
1.375	2.595	2.880	3.051	3.251	3.525	3.742	4.023
1.180	1.723	2.650	2.906	3.142	3.435	3.661	3.947
0.877	0.878	1.496	2.381	2.800	3.167	3.418	3.722
0.617	0.673	0.617	1.261	2.167	2.720	3.023	3.362
0.459	0.503	0.487	0.604	1.497	2.275	2.641	3.019
0.349	0.381	0.381	0.361	0.934	1.840	2.273	2.692
0.203	0.218	0.218	0.178	0.300	1.066	1.592	2.086
0.111	0.115	0.108	0.070	0.044	0.504	1.012	1.547
0.049	0.046	0.032	−0.008	−0.078	0.142	0.556	1.080
0.005	−0.004	−0.023	−0.065	−0.151	−0.082	0.217	0.689
−0.027	−0.040	−0.063	−0.109	−0.202	−0.223	−0.028	0.369
−0.051	−0.067	−0.094	−0.143	−0.241	−0.317	−0.203	0.112
−0.070	−0.088	−0.117	−0.169	−0.271	−0.381	−0.330	−0.092
−0.085	−0.105	−0.136	−0.190	−0.295	−0.428	−0.424	−0.255
−0.106	−0.128	−0.163	−0.221	−0.331	−0.493	−0.551	−0.489
−0.120	−0.144	−0.181	−0.242	−0.356	−0.535	−0.631	−0.645
−0.130	−0.156	−0.194	−0.257	−0.376	−0.567	−0.687	−0.754
−0.137	−0.164	−0.204	−0.269	−0.391	−0.591	−0.729	−0.836
−0.142	−0.170	−0.211	−0.278	−0.403	−0.611	−0.763	−0.899
−0.152	−0.181	−0.224	−0.294	−0.425	−0.650	−0.827	−1.015
−0.158	−0.188	−0.233	−0.306	−0.442	−0.680	−0.874	−1.097

TABLE 9.9

Lee-Kesler Residual Enthalpy,[14] Values of $\left(\frac{S_{\text{ideal}} - S_{\text{actual}}}{R}\right)^{(0)}$

T_r	P_r 0.010	0.050	0.100	0.200	0.400	0.600	0.800
0.30	11.614	10.008	9.319	8.635	7.961	7.574	7.304
0.35	11.185	9.579	8.890	8.205	7.529	7.140	6.869
0.40	10.802	9.196	8.506	7.821	7.144	6.755	6.483
0.45	10.453	8.847	8.157	7.472	6.794	6.404	6.132
0.50	10.137	8.531	7.841	7.156	6.479	6.089	5.816
0.55	0.038	8.245	7.555	6.870	6.193	5.803	5.531
0.60	0.029	7.983	7.294	6.610	5.933	5.544	5.273
0.65	0.023	0.122	7.052	6.368	5.694	5.306	5.036
0.70	0.018	0.096	0.206	6.140	5.467	5.082	4.814
0.75	0.015	0.078	0.164	5.917	5.248	4.866	4.600
0.80	0.013	0.064	0.134	0.294	5.026	4.649	4.388
0.85	0.011	0.054	0.111	0.239	4.785	4.418	4.166
0.90	0.009	0.046	0.094	0.199	0.463	4.145	3.912
0.93	0.008	0.042	0.085	0.179	0.408	0.750	3.723
0.95	0.008	0.039	0.080	0.168	0.377	0.671	3.556
0.97	0.007	0.037	0.075	0.157	0.350	0.607	1.056
0.98	0.007	0.036	0.073	0.153	0.337	0.580	0.971
0.99	0.007	0.035	0.071	0.148	0.326	0.555	0.903
1.00	0.007	0.034	0.069	0.144	0.315	0.532	0.847
1.01	0.007	0.033	0.067	0.139	0.304	0.510	0.799
1.02	0.006	0.032	0.065	0.135	0.294	0.491	0.757
1.05	0.006	0.030	0.060	0.124	0.267	0.439	0.656
1.10	0.005	0.026	0.053	0.108	0.230	0.371	0.537
1.15	0.005	0.023	0.047	0.096	0.201	0.319	0.452
1.20	0.004	0.021	0.042	0.085	0.177	0.277	0.389
1.30	0.003	0.017	0.033	0.068	0.140	0.217	0.298
1.40	0.003	0.014	0.027	0.056	0.114	0.174	0.237
1.50	0.002	0.011	0.023	0.046	0.094	0.143	0.194
1.60	0.002	0.010	0.019	0.039	0.079	0.120	0.162
1.70	0.002	0.008	0.017	0.033	0.067	0.102	0.137
1.80	0.001	0.007	0.014	0.029	0.058	0.088	0.117
1.90	0.001	0.006	0.013	0.025	0.051	0.076	0.102
2.00	0.001	0.006	0.011	0.022	0.044	0.067	0.089
2.20	0.001	0.004	0.009	0.018	0.035	0.053	0.070
2.40	0.001	0.004	0.007	0.014	0.028	0.042	0.056
2.60	0.001	0.003	0.006	0.012	0.023	0.035	0.046
2.80	0.000	0.002	0.005	0.010	0.020	0.029	0.039
3.00	0.000	0.002	0.004	0.008	0.017	0.025	0.033
3.50	0.000	0.001	0.003	0.006	0.012	0.017	0.023
4.00	0.000	0.001	0.002	0.004	0.009	0.013	0.017

[14] Lee and Kesler, ibid.

TABLE 9.9 (*cont'd.*)

Lee-Kesler Residual Entropy, Values of $\left(\dfrac{S_{\text{ideal}} - S_{\text{actual}}}{R}\right)^{(0)}$

P_r							
1.000	*1.200*	*1.500*	*2.000*	*3.000*	*5.000*	*7.000*	*10.000*
7.099	6.935	6.740	6.497	6.182	5.847	5.683	5.578
6.663	6.497	6.299	6.052	5.728	5.376	5.194	5.060
6.275	6.109	5.909	5.660	5.330	4.967	4.772	4.619
5.924	5.757	5.557	5.306	4.974	4.603	4.401	4.234
5.608	5.441	5.240	4.989	4.656	4.282	4.074	3.899
5.324	5.157	4.956	4.706	4.373	3.998	3.788	3.607
5.066	4.900	4.700	4.451	4.120	3.747	3.537	3.353
4.830	4.665	4.467	4.220	3.892	3.523	3.315	3.131
4.610	4.446	4.250	4.007	3.684	3.322	3.117	2.935
4.399	4.238	4.045	3.807	3.491	3.138	2.939	2.761
4.191	4.034	3.846	3.615	3.310	2.970	2.777	2.605
3.976	3.825	3.646	3.425	3.135	2.812	2.629	2.463
3.738	3.599	3.434	3.231	2.964	2.663	2.491	2.334
3.569	3.444	3.295	3.108	2.860	2.577	2.412	2.262
3.433	3.326	3.193	3.023	2.790	2.520	2.362	2.215
3.259	3.188	3.081	2.932	2.719	2.463	2.312	2.170
3.142	3.106	3.019	2.884	2.682	2.436	2.287	2.148
2.972	3.010	2.953	2.835	2.646	2.408	2.263	2.126
2.178	2.893	2.879	2.784	2.609	2.380	2.239	2.105
1.391	2.736	2.798	2.730	2.571	2.352	2.215	2.083
1.225	2.495	2.706	2.673	2.533	2.325	2.191	2.062
0.965	1.523	2.328	2.483	2.415	2.242	2.121	2.001
0.742	1.012	1.557	2.081	2.202	2.104	2.007	1.903
0.607	0.790	1.126	1.649	1.968	1.966	1.897	1.810
0.512	0.651	0.890	1.308	1.727	1.827	1.789	1.722
0.385	0.478	0.628	0.891	1.299	1.554	1.581	1.556
0.303	0.372	0.478	0.663	0.990	1.303	1.386	1.402
0.246	0.299	0.381	0.520	0.777	1.088	1.208	1.260
0.204	0.247	0.312	0.421	0.628	0.913	1.050	1.130
0.172	0.208	0.261	0.350	0.519	0.773	0.915	1.013
0.147	0.177	0.222	0.296	0.438	0.661	0.799	0.908
0.127	0.153	0.191	0.255	0.375	0.570	0.702	0.815
0.111	0.134	0.167	0.221	0.325	0.497	0.620	0.733
0.087	0.105	0.130	0.172	0.251	0.388	0.492	0.599
0.070	0.084	0.104	0.138	0.201	0.311	0.399	0.496
0.058	0.069	0.086	0.113	0.164	0.255	0.329	0.416
0.048	0.058	0.072	0.094	0.137	0.213	0.277	0.353
0.041	0.049	0.061	0.080	0.116	0.181	0.236	0.303
0.029	0.034	0.042	0.056	0.081	0.126	0.166	0.216
0.021	0.025	0.031	0.041	0.059	0.093	0.123	0.162

TABLE 9.10

Lee-Kesler Residual Entropy,[15] Values of $\left(\frac{S_{\text{ideal}} - S_{\text{actual}}}{R}\right)^{(1)}$

T_r	P_r 0.010	0.050	0.100	0.200	0.400	0.600	0.800
0.30	16.782	16.774	16.764	16.744	16.705	16.665	16.626
0.35	15.413	15.408	15.401	15.387	15.359	15.333	15.305
0.40	13.990	13.986	13.981	13.972	13.953	13.934	13.915
0.45	12.564	12.561	12.558	12.551	12.537	12.523	12.509
0.50	11.202	11.200	11.197	11.192	11.182	11.172	11.162
0.55	0.115	9.948	9.946	9.942	9.935	9.928	9.921
0.60	0.078	8.828	8.826	8.823	8.817	8.811	8.806
0.65	0.055	0.309	7.832	7.829	7.824	7.819	7.815
0.70	0.040	0.216	0.491	6.951	6.945	6.941	6.937
0.75	0.029	0.156	0.340	6.173	6.167	6.162	6.158
0.80	0.022	0.116	0.246	0.578	5.475	5.468	5.462
0.85	0.017	0.088	0.183	0.408	4.853	4.841	4.832
0.90	0.013	0.068	0.140	0.301	0.744	4.269	4.249
0.93	0.011	0.058	0.120	0.254	0.593	1.219	3.914
0.95	0.010	0.053	0.109	0.228	0.517	0.961	3.697
0.97	0.010	0.048	0.099	0.206	0.456	0.797	1.570
0.98	0.009	0.046	0.094	0.196	0.429	0.734	1.270
0.99	0.009	0.044	0.090	0.186	0.405	0.680	1.098
1.00	0.008	0.042	0.086	0.177	0.382	0.632	0.977
1.01	0.008	0.040	0.082	0.169	0.361	0.590	0.883
1.02	0.008	0.039	0.078	0.161	0.342	0.552	0.807
1.05	0.007	0.034	0.069	0.140	0.292	0.460	0.642
1.10	0.005	0.028	0.055	0.112	0.229	0.350	0.470
1.15	0.005	0.023	0.045	0.091	0.183	0.275	0.361
1.20	0.004	0.019	0.037	0.075	0.149	0.220	0.286
1.30	0.003	0.013	0.026	0.052	0.102	0.148	0.190
1.40	0.002	0.010	0.019	0.037	0.072	0.104	0.133
1.50	0.001	0.007	0.014	0.027	0.053	0.076	0.097
1.60	0.001	0.005	0.011	0.021	0.040	0.057	0.073
1.70	0.001	0.004	0.008	0.016	0.031	0.044	0.056
1.80	0.001	0.003	0.006	0.013	0.024	0.035	0.044
1.90	0.001	0.003	0.005	0.010	0.019	0.028	0.036
2.00	0.000	0.002	0.004	0.008	0.016	0.023	0.029
2.20	0.000	0.001	0.003	0.006	0.011	0.016	0.021
2.40	0.000	0.001	0.002	0.004	0.008	0.012	0.015
2.60	0.000	0.001	0.002	0.003	0.006	0.009	0.012
2.80	0.000	0.001	0.001	0.003	0.005	0.008	0.010
3.00	0.000	0.001	0.001	0.002	0.004	0.006	0.008
3.50	0.000	0.000	0.001	0.001	0.003	0.004	0.006
4.00	0.000	0.000	0.001	0.001	0.002	0.003	0.005

[15] Lee and Kesler, ibid.

TABLE 9.10 *(cont'd.)*

Lee-Kesler Residual Entropy, Values of $\left(\frac{s_{\text{ideal}} - s_{\text{actual}}}{R}\right)^{(1)}$

P_r							
1.000	*1.200*	*1.500*	*2.000*	*3.000*	*5.000*	*7.000*	*10.000*
16.586	16.547	16.488	16.390	16.195	15.837	15.468	14.925
15.278	15.251	15.211	15.144	15.011	14.751	14.496	14.153
13.896	13.877	13.849	13.803	13.714	13.541	13.376	13.144
12.496	12.482	12.462	12.430	12.367	12.248	12.145	11.999
11.153	11.143	11.129	11.107	11.063	10.985	10.920	10.836
9.914	9.907	9.897	9.882	9.853	9.806	9.769	9.732
8.799	8.794	8.787	8.777	8.760	8.736	8.723	8.720
7.810	7.807	7.801	7.794	7.784	7.779	7.785	7.811
6.933	6.930	6.926	6.922	6.919	6.929	6.952	7.002
6.155	6.152	6.149	6.147	6.149	6.174	6.213	6.285
5.458	5.455	5.453	5.452	5.461	5.501	5.555	5.648
4.826	4.822	4.820	4.822	4.839	4.898	4.969	5.082
4.238	4.232	4.230	4.236	4.267	4.351	4.442	4.578
3.894	3.885	3.884	3.896	3.941	4.046	4.151	4.300
3.658	3.647	3.648	3.669	3.728	3.851	3.966	4.125
3.406	3.391	3.401	3.437	3.517	3.661	3.788	3.957
3.264	3.247	3.268	3.318	3.412	3.569	3.701	3.875
3.093	3.082	3.126	3.195	3.306	3.477	3.616	3.796
2.399	2.868	2.967	3.067	3.200	3.387	3.532	3.717
1.306	2.513	2.784	2.933	3.094	3.297	3.450	3.640
1.113	1.655	2.557	2.790	2.986	3.209	3.369	3.565
0.820	0.831	1.443	2.283	2.655	2.949	3.134	3.348
0.577	0.640	0.618	1.241	2.067	2.534	2.767	3.013
0.437	0.489	0.502	0.654	1.471	2.138	2.428	2.708
0.343	0.385	0.412	0.447	0.991	1.767	2.115	2.430
0.226	0.254	0.282	0.300	0.481	1.147	1.569	1.944
0.158	0.178	0.200	0.220	0.290	0.730	1.138	1.544
0.115	0.130	0.147	0.166	0.206	0.479	0.823	1.222
0.086	0.098	0.112	0.129	0.159	0.334	0.604	0.969
0.067	0.076	0.087	0.102	0.127	0.248	0.456	0.775
0.053	0.060	0.070	0.083	0.105	0.195	0.355	0.628
0.043	0.049	0.057	0.069	0.089	0.160	0.286	0.518
0.035	0.040	0.048	0.058	0.077	0.136	0.238	0.434
0.025	0.029	0.035	0.043	0.060	0.105	0.178	0.322
0.019	0.022	0.027	0.034	0.048	0.086	0.143	0.254
0.015	0.018	0.021	0.028	0.041	0.074	0.120	0.210
0.012	0.014	0.018	0.023	0.035	0.065	0.104	0.188
0.010	0.012	0.015	0.020	0.031	0.058	0.093	0.158
0.007	0.009	0.011	0.015	0.024	0.046	0.073	0.122
0.006	0.007	0.009	0.012	0.020	0.038	0.060	0.100

equation (9.182), Lee and Kesler[16] have been able to express the enthalpy departure function in the form

$$(h_{\text{ideal}} - h_{\text{actual}}) = \frac{RT_c}{M}\left[\left(\frac{h_{\text{ideal}} - h_{\text{actual}}}{RT_c}\right)^{(0)} + \omega\left(\frac{h_{\text{ideal}} - h_{\text{actual}}}{RT_c}\right)^{(1)}\right] \quad \textbf{(9.191)}$$

where the terms $[(h_{\text{ideal}} - h_{\text{actual}})/RT_c]^{(0)}$ and $[(h_{\text{ideal}} - h_{\text{actual}})/RT_c]^{(1)}$ are given in Tables 9.7 and 9.8 and have the same meaning as $z^{(0)}$ and $z^{(1)}$ in equation (9.182). A pseudo-analytical technique suitable for programming this approach on a computer is described in detail by Lee and Kesler.[16]

An actual change in enthalpy between two states is expressed as the sum of an ideal gas term and two enthalpy departure terms.

$$(h_2 - h_1)_{\text{actual}} = (h_2 - h_1)_{\text{ideal}} - [(h_2)_{\text{ideal}} - (h_2)_{\text{actual}}]_{T=T_2} + [(h_1)_{\text{ideal}} - (h_1)_{\text{actual}}]_{T=T_1} \quad \textbf{(9.192)}$$

where the ideal gas term is given by

$$(h_2 - h_1)_{\text{ideal}} = \int_{T_1}^{T_2} c_{P0}\, dT \quad \textbf{(9.193)}$$

We can execute a similar set of calculations for the generalized entropy departure. When the ideal gas terms of equation (9.159) are transposed to the left side of the equation, we can write the entropy departure.

$$(s_{\text{ideal}} - s_{\text{actual}})_T = \left[\int_P^{P=0}\left[\frac{R}{P} - \left(\frac{\partial v}{\partial T}\right)_P\right] dP\right]_T \quad \textbf{(9.194)}$$

When equations (9.185) through (9.188) are used to express equation (9.194) in dimensionless form, we have

$$(s_{\text{ideal}} - s_{\text{actual}})_T = \frac{R}{M}\left[\int_{P_r=0}^{P_r}\left\{(z-1) + T_r\left[\frac{\partial z}{\partial T_r}\right]_{P_r}\right\} d(\ln P_r)\right]_T \quad \textbf{(9.195)}$$

[16] Lee, B. I. and Kesler, M. G., "A Generalized Thermodynamic Correlation Based on Three-Parameter Corresponding States," AIChE Journal, Vol. 21, No. 3, pp. 510–527, May, 1975.

In a manner similar to the enthalpy departure, the entropy departure can be written

$$(s_{\text{ideal}} - s_{\text{actual}})_T = \frac{R}{M}\left[\left(\frac{s_{\text{ideal}} - s_{\text{actual}}}{R}\right)^{(0)} + \omega\left(\frac{s_{\text{ideal}} - s_{\text{actual}}}{R}\right)^{(1)}\right] \tag{9.196}$$

where the terms $[(s_{\text{ideal}} - s_{\text{actual}})/R]^{(0)}$ and $[(s_{\text{ideal}} - s_{\text{actual}})/R]^{(1)}$ represent the spherical and non-spherical contribution to the entropy departure. Values of these terms are tabulated in Tables 9.9 and 9.10 taken from the data of Lee and Kesler.[16] An actual change in entropy between two states is calculated in terms of the entropy departures in the same way that the change in enthalpy is computed in terms of the enthalpy departures, equation (9.192). The ideal gas contribution to the entropy change is

$$(s_2 - s_1)_{\text{ideal}} = \int_{T_1}^{T_2} \frac{c_{P0}}{T}\, dT - R \ln \frac{P_2}{P_1} \tag{9.197}$$

As an alternate approach, tabulated values of ideal gas entropies can be used in place of equation (9.197). In such cases, care must be exercised to account for differences in pressure between the ideal gas states of interest and the tabulated states since tabulated values are often for pressures of 1 atm or 1 bar.

Problems

9.1 One kg of liquid is compressed in a reversible, adiabatic process from P_1 to P_2.

(a) If the liquid is modeled as a fluid with constant specific heats and a constant coefficient of thermal expansion, β, show that the final temperature after compression is approximately

$$T_2 = T_1 \exp\left[\frac{\beta v}{c_P}(P_2 - P_1)\right]$$

where $\beta = (1/v)(\partial v/\partial T)_P$.

(b) Experimental measurements of β and c_P show that for saturated liquid water at 7×10^3 N/m²,

$$\beta = 3.708 \times 10^{-4}(\text{K})^{-1}$$
$$c_P = 4.124 \text{ kJ/kgK}$$

What is the change in temperature for a compression from this saturated state to a final pressure of 5×10^5 N/m²?

(c) Suppose that saturated vapor at the same initial pressure of 7×10^3 N/m² is compressed in a reversible adiabatic manner to 5×10^5 N/m^2. What would the change in temperature be for this compression process? (Note from the temperature-entropy diagram that the initial state lies in a region where the vapor can be well represented by the ideal gas model.) How do you account for the difference between your answers to parts (a) and (b)?

9.2 From the following data, determine the heat of vaporization of liquid helium at 4.00 K. Compare your result with the experimentally determined value of 21.9 J/g.

HINT: Use a graphical-numerical method to evaluate dP/dT and assume that the vapor can be modeled as an ideal gas ($R = 8.314$ J/g mole K).

Liquid Helium

Vapor Pressure P(mm Hg)	Saturation Temperature T(K)
1000	4.516
900	4.396
800	4.266
750	4.197
700	4.125
650	4.049
600	3.970
550	3.885
590	3.795
450	3.699
400	3.596

ρ of liquid at 4.0 K = 0.129 g/cm^3

9.3 Prove that the specific heat at constant pressure, c_P, for an ideal gas is a function of temperature only.

HINT: Use proof by contradiction. That is, assume $c_P = c_P(T,P)$ and show that $(\partial c_P/\partial P)_T = 0$.

9.4 Consider the classical example of a free expansion. A rigid, insulated container is divided into two compartments by a diaphragm. One compartment of volume V_1 contains a gas at a temperature T_1 and a pressure P_1. The other compartment of volume V_2 is evacuated. If the diaphragm separating the compartments is suddently fractured, the gas will spontaneously expand to the volume of the whole container.

(a) Show that the temperature change for such a process is

$$T_2 - T_1 = \int_{V_1}^{V_1+V_2} \frac{1}{m}\left[\frac{P}{c_v} - \frac{T\beta}{c_v \mathrm{K}}\right] dV$$

where β is the coefficient of thermal expansion and K is the coefficient of isothermal compressibility,

$$\beta = \frac{1}{V}\left(\frac{\partial V}{\partial T}\right)_P$$

$$\mathrm{K} = -\frac{1}{V}\left(\frac{\partial V}{\partial P}\right)_T$$

(b) Show that for an ideal gas there is no temperature change.

9.5 One method of liquefying gases is to compress the gas to a very high pressure and then expand the gas to a much lower pressure by means of a throttle. Since in a throttling process the enthalpy at inlet is identical to the enthalpy at outlet, the relevant performance parameter is the Joule-Thomson coefficient defined as $\mu = (\partial T/\partial P)_h$.

(a) Show that in general

$$\mu = \frac{1}{c_P}\left[T\left(\frac{\partial v}{\partial T}\right)_P - v\right]$$

(b) Can an ideal gas be liquefied in this manner?

(c) Determine the Joule-Thomson coefficient for a fluid of constant density.

9.6 The vapor pressure, P, in N/m^2, for solid ammonia (molecular weight = 17) is given by

$$\ln P(\mathrm{N/m^2}) = 35.82 - \frac{3750(\mathrm{K})}{T}$$

where T is the temperature in K. The vapor pressure for liquid ammonia is given by

$$\ln P(\mathrm{N/m^2}) = 32.28 - \frac{3061(\mathrm{K})}{T}$$

(a) What is the temperature of the triple point?
(b) What are the latent heats of sublimation and vaporization at the triple point? (The answer to this question requires a good deal of modeling.)
(c) What is the latent heat of fusion at the triple point?

9.7 Show that the difference in the specific heats $c_P - c_v$ may be expressed as

$$c_P - c_v = \frac{Tv\beta^2}{\mathrm{K}}$$

where β = coefficient of thermal expansion
v = specific volume
T = temperature
K = isothermal compressibility.

9.8 The pressure of a 1 kg block of copper at a temperature of 0 C is increased isothermally and reversibly from 10^5 N/m² to 10^8 N/m². Assume that the coefficient of thermal expansion β, the isothermal compressibility K and the density ρ are constant.

(a) Calculate the work transfer and heat transfer for the copper.
(b) Does your answer violate the first law of thermodynamics?
(c) What would have been the rise in temperature if the process had been adiabatic instead of isothermal?

$$\beta = 5.04 \times 10^{-5}\ \mathrm{K^{-1}}$$
$$\mathrm{K} = 8.56 \times 10^{-12}\ \mathrm{m^2/N}$$
$$\rho = 8009\ \mathrm{kg/m^3}$$

9.9 Assume that the changes in enthalpy and specific volume upon melting of ice are independent of pressure.

(a) Calculate the melting temperature of ice at a pressure of 10^8 N/m² by using the triple point data given below.

Triple Point Data for H_2O
$P_{TP} = 611.3$ N/m² $T_{TP} = 0.01$ C

Phase	Enthalpy (kJ/kg)	Specific Volume (m³/kg)	Entropy (kJ/kgK)
Solid	−333.40	1.0908 − 3	1.221
Liquid	0.01	1.0002 − 3	0.0000
Vapor	2501.4	206.1	9.156

(b) A man who weighs 90 kg is ice skating on "hollow" ground blades which have a total area in contact with the ice of 21.3 mm². If the temperature of the ice is −2 C, will the ice melt under the blades?

9.10 An elastic rod has a length X and a temperature T when the tensile force on its ends is F. For this system, the energy U and the entropy S are given by the relations

$$U = C_X T$$

and

$$S = C_X \ln T - KX^2$$

where C_X and K are constants.

(a) Find the equation of state $f(F,X,T) = 0$ for this system.
(b) The elastic rod with $|K| = 17.2$ N/m²K and $C_X = 9.76$ J/K is taken through a Carnot cycle between heat reservoirs at 335 K and 315 C. The minimum length at the higher temperature is 3 m and the work per cycle is 70 J. Show the Carnot cycle on the T-S and the F-X diagrams.

9.11 In a particular combustion test rig it is necessary to provide a supply of ethane at a constant pressure of 7.3×10^6 N/m² and a constant temperature of 340 C. To meet these requirements the research team built a piston-cylinder apparatus, shown in Figure 9P.11, which uses a weighted piston to maintain con-

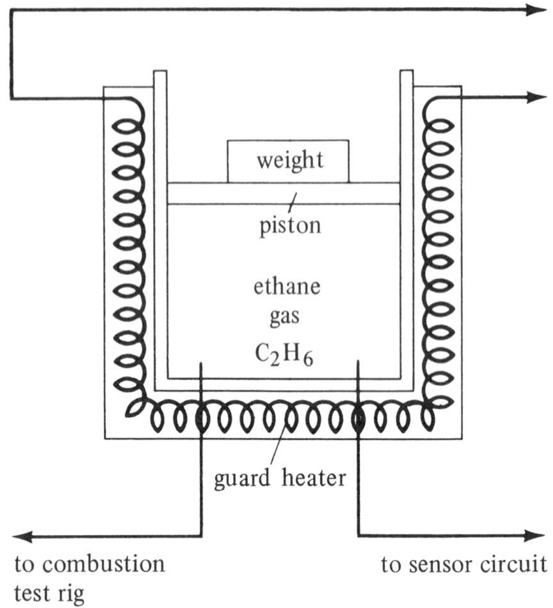

Figure 9P.11

stant pressure and a sensor and guard heater to maintain constant temperature. During one night while the supply valve is closed (no mass flow), the janitor unknowingly turns off the power supply to the guard heaters, and the temperature of the ethane falls due to heat losses to the environment. By the time the error is discovered the following morning, the temperature of the ethane has dropped to 63 C. At this time, the research assistant turns on the guard heaters; however, unknown to him, the piston is now seized in the cylinder resulting in a gas volume corresponding to the temperature and pressure at the time the guard heaters were turned on. With the piston now locked in this position, the guard heaters return the temperature to 340 C. The zero pressure specific heat of ethane is $c_{P0} =$ 1.767 kJ/kgK.

(a) Show the sequence of events on a P-v diagram assuming all processes are quasi-static.

(b) What is the pressure of the gas at the operating temperature of 340 C if the piston remains seized?

(c) What is the heat transfer (per unit mass of ethane) for the surroundings during the cool down time while the power is off?

(d) What is the heat transfer for the guard heaters while the piston is seized?

9.12 The speed of sound "a" in a homogeneous medium is defined in the following manner:

$$a^2 = \left(\frac{\partial P}{\partial \rho}\right)_s$$

where ρ is the density and $\rho = 1/v$.

(a) Show that the above expression can be written in the following form:

$$a^2 = \frac{v\gamma}{K}$$

where $\gamma = c_P/c_v$ and K is the coefficient of isothermal compressibility and $K = -(1/v)(\partial v/\partial P)_T$.

(b) Show that for an ideal gas the speed of sound is given by

$$a = (\gamma RT)^{1/2}$$

9.13 A pure substance is in a two-phase equilibrium state for which the saturated liquid is in equilibrium with the saturated vapor. Determine the equation of the vapor pressure curve. That is, find P as a function of T for all such liquid-vapor two-phase states. Assume that the latent heat of vaporization h_{fg} is a constant, that the vapor can be modeled as an ideal gas with gas constant R, and that the liquid can be modeled as an incompressible fluid with a specific volume v_f much less than the specific volume of the vapor v_g.

9.14 The equilibrium radiation within an evacuated uniform temperature enclosure is called blackbody radiation and Planck's law for blackbody radiation can be applied to this radiation. With the aid of Planck's law it can be shown that the energy density U/V of the radiation is

$$u = \frac{U}{V} = \frac{4}{c}\sigma T^4$$

where c is the speed of light in the medium and σ is the Stefan-Boltzman constant. The formulation of macroscopic thermodynamics can be applied to this equilibrium radiation field.

(a) Starting with the Maxwell relation, $(\partial s/\partial u)_v = 1/T$, derive an expression for the entropy of the enclosure in terms of T and V. Take $s = 0$ at $T = 0$.

(b) What are the Helmholtz free energy and the pressure of the equilibrium radiation field?

(c) What are the Gibbs free energy and the specific heats c_P and c_v for the equilibrium radiation field? Explain the significance of each result.

CHAPTER 10

Open Thermodynamic Systems and Control Volumes

10.1 Introduction

The principles of thermodynamics were developed in the preceding chapters by considering a system of fixed identity which always contains the same mass and is called a closed system. For example, the gas in a piston-cylinder apparatus is a closed system as long as the apparatus is leak free. From everyday experience we are familiar with equipment, especially fluid machinery, with mass flowing in, changing state, and then flowing out. For example, in an automobile engine, a fuel-air mixture flows in through valves, experiences a series of changes of state including combustion, and exits as combustion products through another set of valves. The mass within the apparatus is changing as air flows in and gases flow out the exhaust. Thus, a closed system model is not appropriate for the analysis of the engine.

In principle we could select a mass element of fixed identity and determine the time history of the states as it flows through the apparatus by applying the thermodynamic principles in a closed system form. However, we are not usually interested in the behavior of a single mass element, but rather in the behavior of the physical apparatus itself which is determined by the average behavior of a large number of mass elements. In principle, we could obtain the information necessary for the average, but the effort required would be generally prohibitive.

An apparatus with mass flow across the boundary is more conveniently modeled as an open system with the coordinate frame for analysis fixed in space. This description is a distinct contrast to the closed system model with the coordinate frame attached to a particular quantity of matter. With the open system model we station ourselves in space and observe the situation as matter passes through a given volume defined by a boundary (real or imagined) fixed in space. Then we describe the performance of the apparatus in terms of the states of the mass entering and leaving the fixed volume and in terms of the heat transfers and work transfers across the boundary of this volume.

10.2 The Bulk Flow Model

The analysis of the open system is simplified when the velocity and other thermodynamic properties can be defined at every point in space. When the equilibrium thermodynamic properties, which were defined in terms of the equilibrium states of a closed system, are a satisfactory description of the state of the fluid at any given point in space, the resulting system model is known as the *bulk flow model*. Although an idealization of the actual situation, the bulk flow model provides a suitable description of the majority of fluid flow apparatuses which are used in engineering systems. The bulk flow model is not without its limitations. In particular, the bulk flow model may be inadequate for flow on a molecular scale as might occur in the molecular diffusion of one gas into a second gas or the flow of a gas at very low pressure (free molecular flow). The bulk flow model may also be inadequate where the property gradients are extremely large as in the interior of a strong shock wave.[1] In the present treatment we shall limit our discussions to bulk flow.

The open system model starts with a fixed region of space called a *control volume* that is delineated by a well defined boundary called a *control surface*. The control surface is permeable to the flow of mass so that the identity of the mass contained within the control volume varies with time. The basic technique of the bulk flow, open system model is to relate the interactions between the control volume (the mass instantaneously contained within it) and the environment to the equilibrium states of the fluid at those points on the control surface where mass crosses (the inlet and outlet ports).[2]

The desired information is obtained from the states of the fluid as it crosses the control surface without reference to the detailed behavior within the control volume. Thus, the control volume becomes a "black box" with various inlet and outlet ports through which mass may cross the control surface. The interactions between this "black box" and its environment occur through heat transfers and work transfers across the control surface.

In the following sections we will systematically develop conservation of mass and the laws of thermodynamics for a control volume from the corresponding closed system equations. With the bulk flow of mass across the control surface, an equation for the conservation of mass is necessary to "keep track" of the mass within the control volume. In a closed system analysis, conservation of mass is satisfied in the definition of the system. The first law of thermodynamics must account for energy con-

---

[1] For detailed examples of physical situations for which the bulk flow model is not adequate see (1) Bond, J. W., Watson, K. M., and Welch, J. A., *Atomic Theory of Gas Dynamics,* Addison-Wesley Pub. Co., Reading, Mass., 1965. (2) Vincenti, W. G. and Kruger, C. H., *Introduction to Physical Gas Dynamics,* John Wiley, New York, 1965. (3) DeGroot, S. R. and Mazur, P., *Non-Equilibrium Thermodynamics,* North-Holland Pub. Co., Amsterdam, 1962.

[2] The equations to be developed in the present treatment apply only to a control volume fixed relative to a Newtonian frame of reference, i.e., a frame of reference fixed in space. For control volumes that accelerate or change shape the reader should consult a more advanced treatment such as Housner, G. W. and Hudson, D. E., *Applied Mechanics: Dynamics,* D. Van Nostrand Co., New York, 1959, Chapter 8.

vected across the control surface by the bulk flow in addition to the energy transferred by work transfer and heat transfer. The second law must include entropy convected across the control surface in addition to the entropy transfer associated with heat transfer.

To further simplify our model for the open system with bulk flow, we shall restrict our attention to the special case of *one-dimensional bulk flow*.[3] That is, we shall consider only control volumes with bulk flow at a finite number of discrete ports with an insignificant variation of the properties across each port. At any given instant each property of the fluid has a single value at a given port. In many engineering situations, this single value assigned to a property at a port is the appropriate average value of the property which actually varies across the port. In these cases the use of the average value is adequate because only the gross or average performance of the apparatus is necessary. The restriction to one-dimensional ports may be removed by extending the analysis for a finite number of one-dimensional ports, as developed in the following sections, to an infinite number of infinitesimal ports.

10.3 Conservation of Mass in the Bulk Flow Model

Consider an open system, such as Figure 10.1, which can be adequately described by the one-dimensional bulk flow model. Since our concern here is with mass flow, we disregard any other interactions involving the system and its environment. The control volume of interest is defined by means of the control surface shown as the heavy dashed line in Figure 10.1. In most cases, the location of this surface is obvious from the character of the apparatus; however, as we shall see later, there are some cases in which a great deal of care must be exercised in locating the control surface.

[3] For the more general derivation for a fluid flow field with continuous spatial variation of properties see Shapiro, A. H., *The Dynamics and Thermodynamics of Compressible Fluid Flow,* Vol. I., Ronald Press, New York, New York, 1954, Chapter 1.

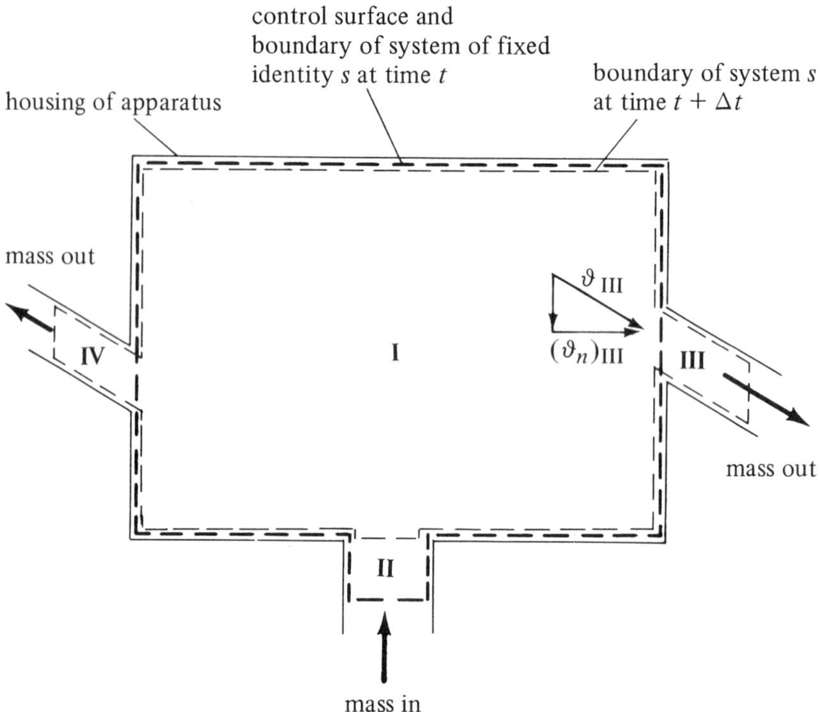

Figure 10.1 Open system with mass flow across system boundary

We will develop conservation of mass for a control volume from conservation of mass for a system of fixed identity. At some instant of time t, let us define a system of fixed identity s as the mass within the control volume at that time. The boundary of this system of fixed identity coincides with the control surface at the time t.

At a later instant of time, $t + \Delta t$, where Δt is some small interval of time, some of this initial mass in the control volume has moved outside. Simultaneously, some mass initially outside has moved inside. Under these circumstances, the boundary of the system of fixed identity assumes the new form shown by the light dashed line of Figure 10.1 while the control surface remains fixed in space.

Since the system s is of fixed identity, the conservation of mass for the system requires that the mass is constant in time. Thus,

$$m_{s)t+\Delta t} - m_{s)t} = 0 \tag{10.1}$$

Conservation of mass for the control volume is obtained by expressing $m_{s)t}$ and $m_{s)t+\Delta t}$ in terms of the mass within the control volume m_{cv} and the mass Δm which crosses the control surface at each port. Since the system coincides with the control volume at time t, we have

$$m_{s)t} = m_{cv)t} \tag{10.2}$$

Using the symbol I to denote the region occupied by the mass within the innermost combination of dashed lines (Figure 10.1) and the symbols II, III, and IV to denote the regions occupied by the mass between the dashed lines, we can then express the mass of the system at time $t + \Delta t$ as

$$m_{s)t+\Delta t} = m_{cv)t+\Delta t} + \Delta m_{\text{III}} + \Delta m_{\text{IV}} - \Delta m_{\text{II}} \tag{10.3}$$

If we now substitute equations (10.2) and (10.3) into equation (10.1), we obtain

$$m_{cv)t+\Delta t} + \Delta m_{\text{III}} + \Delta m_{\text{IV}} - \Delta m_{\text{II}} - m_{cv)t} = 0 \tag{10.4}$$

Thus, the change of mass of the control volume during the time interval Δt becomes

$$m_{cv)t+\Delta t} - m_{cv)t} = \Delta m_{\text{II}} - \Delta m_{\text{III}} - \Delta m_{\text{IV}} \tag{10.5}$$

If we divide equation (10.5) by Δt and take the limit as Δt approaches zero, we obtain

$$\frac{\partial m_{cv}}{\partial t} = \dot{m}_{\text{II}} - \dot{m}_{\text{III}} - \dot{m}_{\text{IV}} \tag{10.6}$$

Here the rate of change of the mass within the control volume is

$$\frac{\partial m_{cv}}{\partial t} = \lim_{\Delta t \to 0} \frac{m_{cv)t+\Delta t} - m_{cv)t}}{\Delta t} \tag{10.7}$$

and the mass flow rate at a port is

$$\dot{m} = \lim_{\Delta t \to 0} \frac{\Delta m}{\Delta t} \tag{10.8}$$

Closer inspection of the flow across the control surface reveals an inflow of mass in region II and an outflow of mass in regions III and IV. Thus, for the most general case equation (10.6) may be written

$$\frac{\partial m_{cv}}{\partial t} = \sum \dot{m}_{\text{in}} - \sum \dot{m}_{\text{out}} \qquad \textbf{(10.9)}$$

Equation (10.9) states that the rate of accumulation of mass within the control volume is equal to the excess of the incoming rate of mass flow over the outgoing rate.

The rate of mass flow at a port may be expressed in terms of the density and the fluid motion as shown in Figures 10.1 and 10.2. The mass crossing the control surface at port i is contained in region i and is given by

$$\Delta m_i = (\rho \Delta V)_i \qquad \textbf{(10.10)}$$

Here ρ_i is the density of the fluid at port i and ΔV_i is the volume of region i. From Figure 10.2 this volume is in general a non-right cylinder with a volume given by the product of the area of its base, A_i, and its altitude, $(\vartheta_n \Delta t)$.

$$\Delta V_i = (A\vartheta_n)_i \Delta t \qquad \textbf{(10.11)}$$

Here $(\vartheta_n)_i$ is the velocity normal to the area A_i of port i on the control surface and is given by $\vartheta_i \cos \theta$ where θ is the angle between the outward normal to the control surface and the velocity vector. Thus,

$$\dot{m}_i = \lim_{\Delta t \to 0} \left(\frac{\Delta m}{\Delta t}\right)_i = (\rho A \vartheta_n)_i \qquad \textbf{(10.12)}$$

10.4 First Law of Thermodynamics for the Bulk Flow Model

We now consider energy interactions between the control volume and its environment. The heat transfer and the work transfer are the same as for a closed system. The flow of mass across the control surface carries or convects energy into the

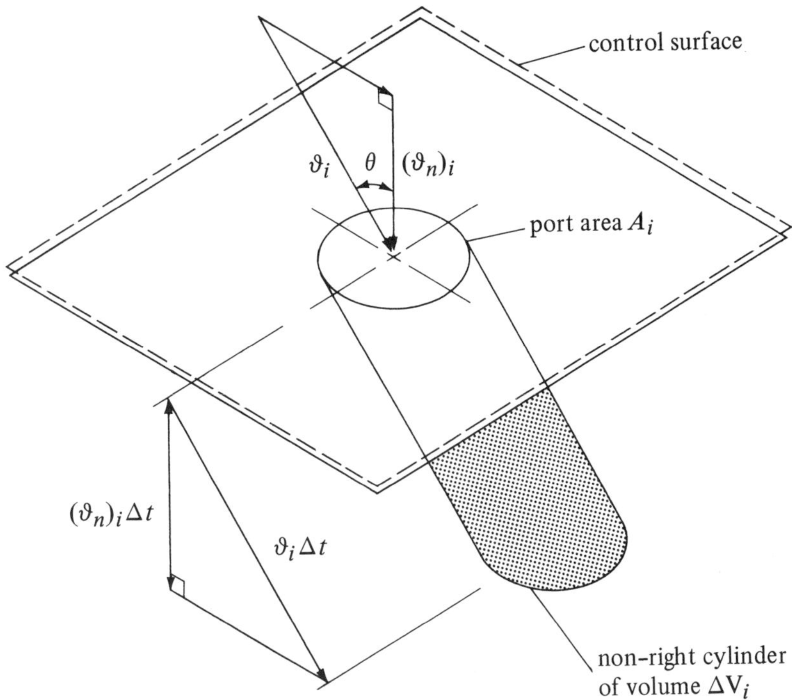

Figure 10.2 Flow at a port of a control surface

control volume. Thus, the mass flow creates another energy transfer interaction. The relation between the mass flow and the energy interactions is determined by modifying the first law of thermodynamics for application to a control volume. Consider the physical situation represented by Figure 10.3. The control volume is established by means of the control surface represented by the heavy dashed line in Figure 10.3. The situation is modeled as bulk flow with the following interactions across the control surface: (1) mass flow at ports II, III and IV, (2) work transfer by means of a rotating shaft cut by the control surface, (3) a support strut cut by the control surface, and (4) heat transfer across the control surface. A fan blade is shown on the end of the rotating shaft to indicate an internal work transfer between the shaft and the fluid within the control volume.

As before we start with a system, *s*, of fixed identity which consists of the mass within the control volume of Figure 10.3 at

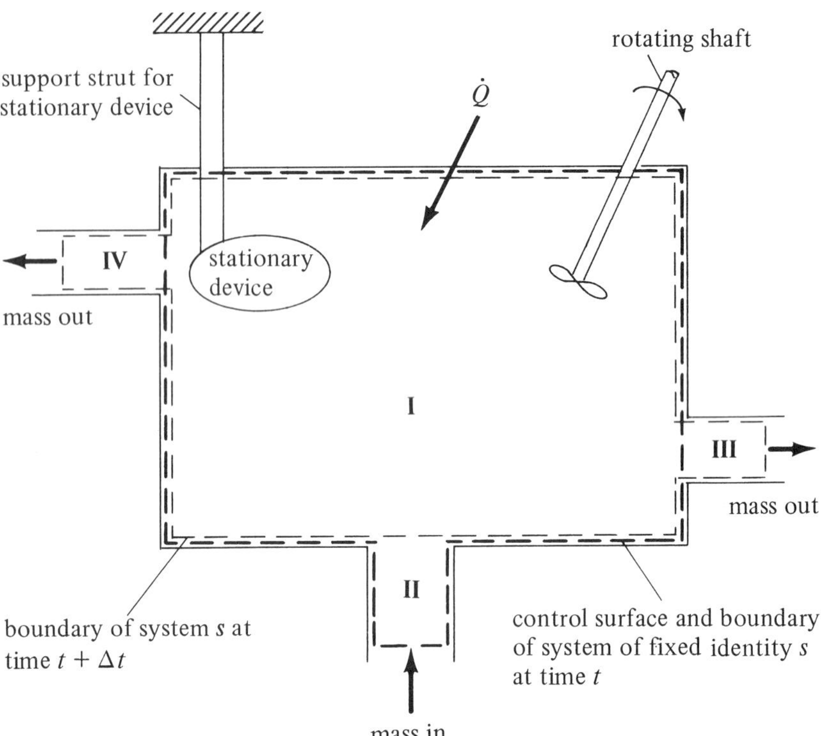

Figure 10.3 Open system with mass, force, heat, and work transfer across system boundary

some instant of time, t. We write the first law for changes in state of this system during a short time interval Δt. The first law for the control volume is then obtained by relating changes in state and interactions for the system to changes in state and interactions for the control volume.

The first law for the changes in state of the system, s, during the short time interval Δt is

$$E_{s)t+\Delta t} - E_{s)t} = \delta Q_s - \delta W_s \tag{10.13}$$

At time t, the system s coincides with the control volume. Thus,

$$E_{s)t} = E_{cv)t} \tag{10.14}$$

After the short time interval Δt, the system has moved relative to the control volume because of the mass flow at the ports II, III, and IV. We can express the energy of the system at time $t + \Delta t$ in terms of the energy stored within the control volume and the energy stored within the mass in the regions II, III, and IV, Figure 10.3. Thus,

$$E_{s)t+\Delta t} = E_{cv)t+\Delta t} + (e\Delta_m)_{\text{III}} + (e\Delta m)_{\text{IV}} - (e\Delta m)_{\text{II}} \qquad \textbf{(10.15)}$$

where e_i is the specific energy of mass Δm_i within region i.

The next step in deriving the first law for the control volume is to substitute equations (10.14) and (10.15) into the first law, equation (10.13).

$$E_{cv)t+\Delta t} - E_{cv)t} = \delta Q_s - \delta W_s + (e\Delta m)_{\text{II}} - (e\Delta m)_{\text{III}} - (e\Delta m)_{\text{IV}} \qquad \textbf{(10.16)}$$

Thus, during the time interval Δt the change of the energy contained within the control volume is the result of the heat transfer and the work transfer interactions for system s together with the energies of masses Δm which are convected across the control surface.

In order to complete the derivation, we must express the heat transfer and the work transfer for the system s in terms of the interactions for the control volume. Heat transfer and work transfer across the control surface are the same as for the system s, except when the interactions are at ports with mass flow. Since significant work transfer for system s is always associated with the motion of the mass through the ports, we must consider the work transfer in detail.

In defining the work transfer for a control volume, we first subdivide the work transfer for the system into two parts. They are $\delta W_{s)\text{normal}}$ which is associated with motion normal to the system boundary, and $\delta W_{s)\text{shear}}$, the other work transfer. Thus,

$$\delta W_s = \delta W_{s)\text{normal}} + \delta W_{s)\text{shear}} \qquad \textbf{(10.17)}$$

Work transfer $\delta W_{s)\text{shear}}$ usually arises from shear stresses and motion in the plane of the system boundary, although the term is general and may include other effects, for example, electrical work transfer. This work transfer will be discussed after work transfer due to pressure.

The work transfer $\delta W_{s)\text{normal}}$ is due to the pressure acting on the various ports of the open system and the motion of the boundary of system s. Since the flow is one dimensional the work transfer for system s at each port i is

$$\delta W_i = \pm P_i \Delta V_i \tag{10.18}$$

where P_i is the uniform pressure at the port and ΔV_i is the volume of region i. The work transfer is positive for outflow ports (system s expanding) and negative for inflow ports (system s contracting). The volume ΔV_i is given by equation (10.11). To express the work transfer, equation (10.18), in terms of the mass of region i we use equation (10.10) to obtain

$$\delta W_i = \pm (Pv\Delta m)_i \tag{10.19}$$

For the particular open system shown in Figure 10.3, we find the work transfer associated with normal motion of the boundary of system s is

$$\delta W_{s)\text{normal}} = (Pv\Delta m)_{\text{III}} + (Pv\Delta m)_{\text{IV}} - (Pv\Delta m)_{\text{II}} \tag{10.20}$$

The heat transfer δQ_s and the work transfer $\delta W_{s)\text{shear}}$ for the system s during the time interval Δt may be simply related to the interactions for the control volume if we neglect δQ_s and $\delta W_{s)\text{shear}}$ at the ports with mass flow. In this case the system boundary and the control surface remain coincident during time Δt at the locations of δQ and δW_{shear}. Thus,

$$\delta Q_s = \delta Q_{cv} \tag{10.21}$$

and

$$\delta W_{s)\text{shear}} = \delta W_{cv)\text{shear}} \tag{10.22}$$

The simple definitions, equations (10.21) and (10.22), may lead to difficulties when there are heat transfer or shear work transfer interactions at a port with mass flow. These complex cases are considered in detail in the study of coupled fluxes in irreversible thermodynamics, but are beyond the scope of this text.

If we substitute equations (10.22) and (10.20) into (10.17), we obtain the total work transfer for system s during time interval Δt.

$$\delta W_s = \delta W_{cv)\text{shear}} + (Pv\Delta m)_{\text{III}} + (Pv\Delta m)_{\text{IV}} - (Pv\Delta m)_{\text{II}} \qquad \textbf{(10.23)}$$

When we combine equations (10.21) and (10.23) with the first law, equation (10.16), we have

$$E_{cv)t+\Delta t} - E_{cv)t} = \delta Q_{cv} - \delta W_{cv)\text{shear}} + (Pv\Delta m)_{\text{II}} - (Pv\Delta m)_{\text{III}} - (Pv\Delta m)_{\text{IV}} + (e\Delta m)_{\text{II}} - (e\Delta m)_{\text{III}} - (e\Delta m)_{\text{IV}} \qquad \textbf{(10.24)}$$

If we now divide by Δt and take the limit as Δt approaches zero, we obtain

$$\frac{\partial E_{cv}}{\partial t} = \dot{Q}_{cv} - \dot{W}_{cv)\text{shear}} + \sum \dot{m}(e + Pv)_{\text{in}} - \sum \dot{m}(e + Pv)_{\text{out}} \qquad \textbf{(10.25)}$$

Here we have employed the following definitions for heat transfer and shear work transfer rates for the control volume:

$$\dot{Q}_{cv} = \lim_{\Delta t \to 0} \frac{\delta Q_{cv}}{\Delta t} \qquad \textbf{(10.26)}$$

$$\dot{W}_{cv)\text{shear}} = \lim_{\Delta t \to 0} \frac{\delta W_{cv)\text{shear}}}{\Delta t} \qquad \textbf{(10.27)}$$

It is common, but by no means universal practice to refer to the $Pv\dot{m}$ terms as the rate of "flow work" since they represent the rates of work transfer for the system of fixed identity, s, due to the mass flows across the control surface. Thus, the "flow work" is an energy flux across the control surface as the result of the pressure in the fluid and the motion associated with the mass flow.

The terms $e\dot{m}$ in equation (10.25) represent the rate at which energy is transported, or convected, into and out of the control volume by virtue of the energy stored in the mass so transported. For a thermodynamic system, the stored energy is the sum of the internal energy and any stored energy forms that are uncoupled in the thermodynamic sense. The two uncoupled stored energy forms most commonly encountered in engineering practice are the translational kinetic energy, E_t, due to the bulk motion of mass in a Newtonian frame of reference and the gravitational potential energy, E_g, due to the displacement of mass in a conservative gravitational field. Thus, for a unit mass

$$e = u + \frac{\vartheta^2}{2} + gz \tag{10.28}$$

If we substitute equation (10.28) into equation (10.24), the first law for the control volume becomes

$$\frac{\partial E_{cv}}{\partial t} = \dot{Q}_{cv} - \dot{W}_{cv)\text{shear}} + \sum \dot{m}\left(h + \frac{\vartheta^2}{2} + gz\right)_{\text{in}} - \sum \dot{m}\left(h + \frac{\vartheta^2}{2} + gz\right)_{\text{out}} \tag{10.29}$$

Here we have used the property enthalpy, h, for the combination of properties $(u + Pv)$, equation (8.10). Equation (10.29) is the first law of thermodynamics applicable to a control volume or open system, and as such illustrates one of the important uses of the property enthalpy. Equation (10.29) relates the instantaneous rate of change of the total energy stored within the control volume (the term on the left side) to the rate of energy transfer across the control surface in all forms (heat transfer, work transfer and convection). For this reason, equation (10.29) is sometimes referred to as the ''energy equation'', but the specific limitations of this equation must be remembered.

It is important that we understand the relation between the term $\dot{W}_{\text{shear}}$ in equation (10.29) and the physical situation. The shear work, $\dot{W}_{\text{shear}}$, includes all work transfer interactions which have been defined for a closed system, except work transfers resulting from pressure and normal displacement of the boundary ($P\,dV$ work transfer). Since the control volume is fixed in space, all motion normal to the control surface results in a mass flow across the control surface. The energy transfer resulting from this motion is proportional to the mass flow and is accounted for by the Pv part of the enthalpy.

Although body forces due to conservative force fields must be included with surface forces in the determination of the accelerations of systems, the body forces are not included in the determination of work transfers. These forces do not contribute to the work transfer associated with relative motion at a system boundary in the sense of Section 4.4. As a result, the energy effects of body forces are accounted for as uncoupled stored energies, for example, gravitational potential energy and translational kinetic energy.

The other work transfers which are included in the general term shear work, W_{shear}, include electrical work transfer due to a charge flow and an electric potential difference, as well as work transfer by shear stresses and mass motions in the plane of the control surface. Two types of shear work transfer that arise from shear stresses are of importance. The first is due to the rotating shaft passing through the system boundary. When a shaft is cut by a control surface normal to the axis of the shaft, the torque on the shaft and the angular motion of the shaft produce a shear work transfer. The torque in the shaft results from the moment of the shear stresses in the shaft caused by the torsional strain. Each differential area of the shaft contributes a work transfer proportional to its area, shear stress, and velocity in the plane of the control surface. The integral over the cross section of the shaft gives the total rate of shear work transfer along the shaft which is equal to the product of the torque and the angular velocity of the shaft. This rate of shaft work transfer is called *shaft power* and is included as part of the shear work $\dot{W}_{shear}$.

The second form of shear stress work transfer is due to the shear stresses that develop through the action of viscosity when two adjacent fluid layers slide over one another. Since there is a relative motion of one fluid layer with respect to another, there is a viscous shear stress. If the mass points at the control surface have a component of displacement in the direction of the shear stress, the action of the shear stress results in a work transfer. The rate at which this work transfer occurs is called *fluid shear power*[4] and is included as part of the shear work.

Note that both of these contributions to the work transfer require boundary motion. If there is no boundary motion, there is no work transfer. For example, a stationary shaft can have

[4] For a discussion of shear work transfer in fluids see Shapiro, A. H., *The Dynamics and Thermodynamics of Compressible Fluid Flow*, Vol. II, Ronald Press, New York, 1954, Chapter 26.

shear stresses due to static loads, but these stresses do not result in a work transfer unless the shaft moves. Similarly, if the system boundary coincides with a stationary wall of some apparatus, the fluid shear stresses acting on the wall do not result in a work transfer since the surface is not in motion. Of course, if the system boundary were situated in the viscous boundary layer parallel to the velocity of the fluid, a definite work transfer would result because of the fluid shear stress and motion of the fluid at the boundary.

10.5 Second Law of Thermodynamics for the Bulk Flow Model

We now proceed to develop the second law of thermodynamics for a control volume with bulk flow. Figure 10.3 is useful for this purpose. Following our usual procedure, we define a system, s, which consists of the mass within the control volume of Figure 10.3 at some instant of time t. We write the second law for the changes in state during a short time interval Δt. The second law for the control volume is then obtained by relating changes and interactions for the system to changes and interactions for the control volume.

The second law for the system s during Δt is

$$S_{s)t+\Delta t} - S_{s)t} \geq \sum_j \left(\frac{\delta Q_s}{T}\right)_j \tag{10.30}$$

Here $(\delta Q_s/T)_j$ is the entropy transfer (Section 7.2) across the system boundary at location j where the heat transfer is δQ_j and the boundary temperature is T_j.

At time t the system coincides with the control volume. Thus,

$$S_{s)t} = S_{cv)t} \tag{10.31}$$

After a short time interval Δt the system has moved relative to the control volume because of the mass flow at ports II, III, and IV. We can express the entropy of the system at time $t + \Delta t$ in terms of the entropy of the control volume and the entropy of the mass in regions II, III, and IV. Thus,

$$S_{s)t+\Delta t} = S_{cv)t+\Delta t} + (s\Delta m)_{\text{III}} + (s\Delta m)_{\text{IV}} - (s\Delta m)_{\text{II}} \quad \textbf{(10.32)}$$

If we use the definition of δQ_{cv}, equation (10.21), the entropy transfer for the system during Δt becomes

$$\sum_j \left(\frac{\delta Q_s}{T}\right)_j = \sum_j \left(\frac{\delta Q_{cv}}{T}\right)_j \quad \textbf{(10.33)}$$

As in the case of the first law, simultaneous heat transfer and mass flow at a port may lead to complexities not considered in equation (10.33).

The second law for the control volume is obtained by substituting equations (10.31), (10.32), and (10.33) into equation (10.30).

$$S_{cv)t+\Delta t} - S_{cv)t} \geq \sum_j \left(\frac{\delta Q_{cv}}{T}\right)_j + \sum (s\Delta m)_{\text{in}} - \sum (s\Delta m)_{\text{out}} \quad \textbf{(10.34)}$$

If we divide by Δt and take the limit as Δt approaches zero, we obtain

$$\left(\frac{\partial S_{cv}}{\partial t}\right) \geq \sum_j \left(\frac{\dot{Q}_{cv}}{T}\right)_j + \sum (\dot{m}s)_{\text{in}} - \sum (\dot{m}s)_{\text{out}} \quad \textbf{(10.35)}$$

In equation (10.35) we have defined the rate of entropy flux to the control volume as

$$\sum_j \left(\frac{\dot{Q}_{cv}}{T}\right)_j = \lim_{\Delta t \to 0} \sum_j \left(\frac{\delta Q_{cv}}{T}\right)_j \frac{1}{\Delta t} \quad \textbf{(10.36)}$$

The summation appearing for the entropy flux in equation (10.35) is over all areas of the control surface participating in the heat transfer interaction at one particular instant of time. Each term in the summation represents a location on the control surface where the heat flux is $\dot{Q}_j$ and the temperature of the control surface is T_j. In the event that these locations become infinitesimally small, the summation becomes a surface integral over the control surface. The surface integral would be required for complex cases with the heat flux and the surface temperature continuously variable over the control surface.

Physically, equation (10.35) states that the rate at which entropy appears within the control volume exceeds (equals in the limit of reversible processes) the sum of the rates at which entropy flows across the control surface. These entropy flows are the result of heat transfer and mass flow across the control surface. For a closed adiabatic system, equation (10.35) requires not only a positive entropy change (or in the reversible limit, no change) as required by equation (7.35), but also a positive rate of change. Thus, it is not possible for a closed adiabatic system to first decrease in entropy and then increase in entropy so that the net change in entropy is positive (or zero in the reversible limit). Equation (10.35) requires that the entropy increase (or remain constant) throughout the adiabatic process.

Equation (10.35) may be expressed as an equality if a rate of entropy generation is included as a source of entropy to account for all irreversibility inside the control volume.

$$\frac{\partial S_{cv}}{\partial t} = \sum_j \left(\frac{\dot{Q}_{cv}}{T}\right)_j + \sum (\dot{m}s)_{\text{in}} - \sum (\dot{m}s)_{\text{out}} + \dot{S}_{\text{gen}} \qquad \textbf{(10.37)}$$

Thus, the entropy generated is the amount by which the increase in entropy stored inside the control volume exceeds the net entropy transferred in across the control surface.

$$\dot{S}_{\text{gen}} = \frac{\partial (S_{cv})}{\partial t} - \left[\sum_j \left(\frac{\dot{Q}_{cv}}{T}\right)_j + \sum (\dot{m}s)_{\text{in}} - \sum (\dot{m}s)_{\text{out}}\right] \qquad \textbf{(10.38)}$$

The second law requires that $\dot{S}_{\text{gen}}$ is positive or zero in the limit of a completely reversible process.

10.6 Limitations Imposed by the Second Law of Thermodynamics on the Heat Transfer and Shear Work Transfer for a Control Volume in Thermal Communication with Only One Heat Reservoir

As we have seen in Chapter 7, the second law of thermodynamics establishes limits on the heat transfer and work transfer for a closed system. Such limits for an open system also depend

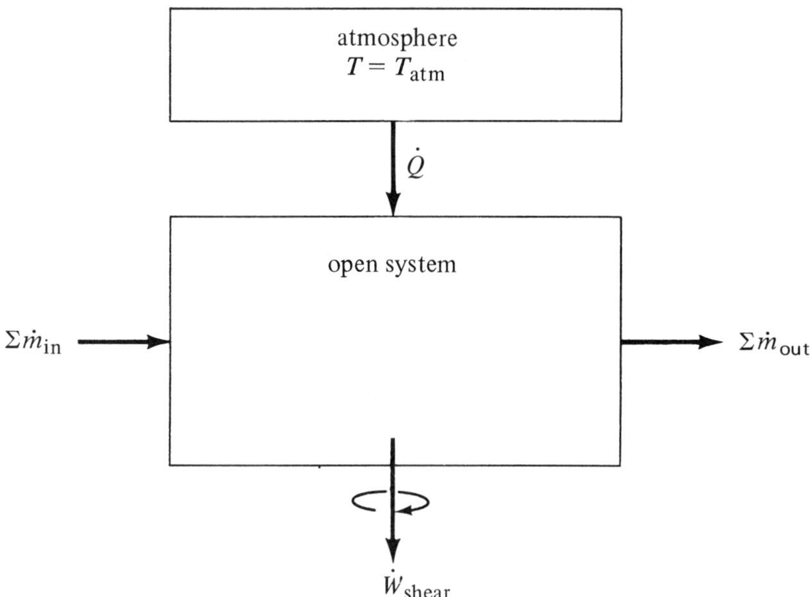

Figure 10.4 Open system in thermal communication with an atmosphere

upon the conditions. To illustrate these limits we restrict our attention to an open system in thermal communication only with an atmosphere as shown schematically in Figure 10.4. The atmosphere is assumed to be so large that its temperature is not affected by the interactions with the open system and thus may be modeled as a heat reservoir with a temperature T_{atm}.

Equation (10.35) can be written in the form

$$\sum_j \left(\frac{\dot{Q}_{cv}}{T}\right)_j \leq \frac{\partial}{\partial t}(S_{cv}) + \sum (\dot{m}s)_{out} - \sum (\dot{m}s)_{in} \qquad \mathbf{(10.39)}$$

If the open system can experience heat transfer only with a heat reservoir at T_{atm}, then $T \leq T_{atm}$ when $\dot{Q}_{cv}$ is positive and $T \geq T_{atm}$ when $\dot{Q}_{cv}$ is negative. Thus,

$$\frac{\dot{Q}_{cv}}{T} \geq \frac{\dot{Q}_{cv}}{T_{atm}} \qquad \mathbf{(10.40)}$$

and we may write equation (10.37) in the form

$$\frac{\dot{Q}_{cv}}{T_{\text{atm}}} \leq \frac{\partial}{\partial t}(S_{cv}) + \sum (\dot{m}s)_{\text{out}} - \sum (\dot{m}s)_{\text{in}} \tag{10.41}$$

where $\sum (\dot{Q}_{cv})_j = \dot{Q}_{cv}$. Multiplying equation (10.41) through by the positive quantity T_{atm}, we obtain the limit imposed by the second law of thermodynamics on the rate of heat transfer between a control volume and a heat reservoir.

$$\dot{Q}_{cv} \leq \frac{\partial}{\partial t}(T_{\text{atm}}S_{cv}) + \sum (\dot{m}T_{\text{atm}}s)_{\text{out}} - \sum (\dot{m}T_{\text{atm}}s)_{\text{in}} \tag{10.42}$$

By combining the first and second laws of thermodynamics, we find a similar limit exists for the shaft work transfer. In particular, if we substitute equation (10.42) into equation (10.29), we obtain

$$\begin{aligned}\frac{\partial}{\partial t}(E_{cv}) \leq &- \dot{W}_{\text{shear}} + \frac{\partial}{\partial t}(T_{\text{atm}}S_{cv}) + \sum \left[\dot{m}\left(h + \frac{\vartheta^2}{2} + gz\right)\right]_{\text{in}} \\ &- \sum (\dot{m}T_{\text{atm}}s)_{\text{in}} - \sum \left[\dot{m}\left(h + \frac{\vartheta^2}{2} + gz\right)\right]_{\text{out}} \\ &+ \sum (\dot{m}T_{\text{atm}}s)_{\text{out}}\end{aligned} \tag{10.43}$$

Rearranging equation (10.43), we obtain the limit imposed by the second law of thermodynamics on the shear work transfer rate for a control volume in thermal communication only with one heat reservoir.

$$\begin{aligned}\dot{W}_{\text{shear}} \leq &- \frac{\partial}{\partial t}(E_{cv} - T_{\text{atm}}S_{cv}) + \sum \left[\dot{m}\left(h - T_{\text{atm}}s + \frac{\vartheta^2}{2} + gz\right)\right]_{\text{in}} \\ &- \sum \left[\dot{m}\left(h - T_{\text{atm}}s + \frac{\vartheta^2}{2} + gz\right)\right]_{\text{out}}\end{aligned} \tag{10.44}$$

For the sake of simplicity we introduce the property b of the open system-atmosphere combination where[5]

$$b = h - T_{\text{atm}}s \tag{10.45}$$

[5] Keenan, J. H., *Thermodynamics,* John Wiley, 1941, Chapter 17.

Then equation (10.44) becomes

$$W_{\text{shear}} \leq -\frac{\partial}{\partial t}(E_{cv} - T_{\text{atm}} S_{cv}) + \sum\left[\dot{m}\left(b + \frac{\vartheta^2}{2} + gz\right)\right]_{\text{in}} - \sum\left[\dot{m}\left(b + \frac{\vartheta^2}{2} + gz\right)\right]_{\text{out}} \qquad \textbf{(10.46)}$$

Note that although there is a similarity between the property b and the property Φ (Section 7.9), the two are in general different because the pressure appearing implicitly in the enthalpy term of equation (10.46) is not necessarily the pressure of the atmosphere. That is,

$$b = u + Pv - T_{\text{atm}}s \qquad \textbf{(10.47)}$$

but

$$\Phi = u + P_{\text{atm}}v - T_{\text{atm}}s \qquad \textbf{(10.48)}$$

10.7 Steady Flow in the Bulk Flow Model

In our analysis up to this point, we have developed the conservation of mass, the first law, and the second law for an unsteady bulk flow, that is, a flow with all quantities varying with time. These expressions involve the instantaneous rates so the evaluation of a net change for the control volume requires time integration of the various rates.

In many practical engineering situations, the flow through the system is steady or at worst, periodic. Under these circumstances, each of the expressions developed for mass, energy, and entropy can be integrated directly over a finite time interval. For the flow in an open system to be steady or periodic, it must satisfy the following requirements:

1. The state of the material at every point inside the control volume must be independent of time or at worst, must be periodic with time.
2. The mass flow rate, velocity, and thermodynamic properties of the fluid crossing the control surface

in either direction must be constant or periodic. Of course, these quantities may be different at various locations of entry and exit, but they must be independent of time.

3. The forces, work transfer rates, and heat transfer rates at each point on the control surface must be constant or periodic.

For a flow which satisfies these restrictions, the conservation of mass, equation (10.49), becomes

$$\sum \dot{m}_{\text{in}} = \sum \dot{m}_{\text{out}} \tag{10.49}$$

In many engineering systems, there is only one port of entry and one port of exit. Then equation (10.49) becomes

$$\dot{m}_{\text{in}} = \dot{m}_{\text{out}} = \dot{m} = \rho A \vartheta_n \tag{10.50}$$

In the case of periodic flow, equations (10.49) and (10.50) must be interpreted in terms of the average mass flow rate for one period.

Under the circumstances of steady or periodic flow, the first law of thermodynamics, equation (10.29), becomes

$$\dot{Q}_{cv} - \dot{W}_{\text{shear}} = \sum \left[\dot{m}\left(h + \frac{\vartheta^2}{2} + gz\right)\right]_{\text{out}} - \sum \left[\dot{m}\left(h + \frac{\vartheta^2}{2} + gz\right)\right]_{\text{in}} \tag{10.51}$$

For the special case of one entry port and one exit port, equation (10.51) can be divided by the mass flow rate $\dot{m}$ from equation (10.50).

$$\frac{\dot{Q}}{\dot{m}} - \frac{\dot{W}_{\text{shear}}}{\dot{m}} = \left(h + \frac{\vartheta^2}{2} + gz\right)_{\text{out}} - \left(h + \frac{\vartheta^2}{2} + gz\right)_{\text{in}} \tag{10.52}$$

As we shall see in Chapter 11, this form of the first law of thermodynamics is applicable to the majority of open systems encountered in engineering practice. Finally, under the conditions of steady or periodic flow, the second law of thermodynamics becomes

$$\sum_j \left(\frac{\dot{Q}_{cv}}{T}\right)_j + \sum (\dot{m}s)_{\text{in}} - \sum (\dot{m}s)_{\text{out}} \leq 0 \tag{10.53}$$

or

$$\sum (\dot{m}s)_{\text{out}} \geq \sum_j \left(\frac{\dot{Q}_{cv}}{T}\right)_j + \sum (\dot{m}s)_{\text{in}} \tag{10.54}$$

For a steady irreversible process the second law of thermodynamics requires the net outflow of entropy to be larger than the net inflow of entropy (note that the term $\sum (\dot{Q}_{cv}/T)_j$ is the entropy flow due to heat transfer which can be positive or negative). This inequality is due to the entropy generated within the control volume which must be transferred out of the control volume to meet the steady flow requirement that entropy not accumulate inside the system. Such a system is said to be internally irreversible. Clearly in the reversible case there is no internal entropy generation and the entropy flow out of the system is equal to the entropy flow into the system.

Equation (10.54) can be expressed as an equality if a rate of entropy generation is included to account for irreversibility.

$$\sum (\dot{m}s)_{\text{out}} = \sum_j \left(\frac{\dot{Q}_{cv}}{T}\right)_j + \sum (\dot{m}s)_{\text{in}} + \dot{S}_{\text{gen}} \tag{10.55}$$

where $\dot{S}_{\text{gen}} \geq 0$. Then,

$$\dot{S}_{\text{gen}} = \sum (\dot{m}s)_{\text{out}} - \sum_j \left(\frac{\dot{Q}_{cv}}{T}\right)_j - \sum (\dot{m}s)_{\text{in}} \tag{10.56}$$

For the case of steady flow with a single entry port and a single exit port, equation (10.56) reduces to

$$s_{\text{out}} - s_{\text{in}} \geq \sum_j \left(\frac{\dot{Q}/\dot{m}}{T}\right)_j \tag{10.57}$$

For steady flow in an open system in thermal communication

only with an atmosphere at T_{atm}, the rate of shear work transfer, equation (10.46), would be limited by

$$\dot{W}_{shear} \leq \sum\left[\dot{m}\left(b + \frac{\vartheta^2}{2} + gz\right)\right]_{in} - \sum\left[\dot{m}\left(b + \frac{\vartheta^2}{2} + gz\right)\right]_{out} \quad \textbf{(10.58)}$$

where b is $h + T_{atm}s$. Note the similarity between equation (10.58) and equation (7.92). The combination of terms $(b + \vartheta^2/2 + gz)$ can be referenced to a datum state. At the datum state the temperature and pressure of the fluid are equal to the temperature and pressure of the atmosphere, the velocity of the fluid is zero, and the elevation is some minimum elevation. This new combination of properties $[(b + \vartheta^2/2 + gz) - (b_{atm} + gz_{min})]$ then represents the availability of the open system-atmosphere combination when the steady stream leaving the open system is to be exhausted into the atmosphere. Then equation (10.58) shows that the rate of shear work transfer must be less than (in the reversible limit, equal to) the decrease in this availability. Note that this availability neglects any shear work transfer which could be produced by mixing the exhaust stream with the atmosphere at constant temperature and pressure.

10E.1 Sample Problem: Superheated steam at $P = 5 \times 10^5$ N/m^2 and $T = 300$ C is desuperheated to saturated vapor by spraying liquid water into the steady stream of steam as shown in Figure 10E.1. The liquid enters at $P = 10^6$ N/m^2 and $T = 40$ C. All of the liquid evaporates so that the leaving state is saturated vapor at $P = 5 \times 10^5$ N/m^2 and $x = 1.0$. The mass flow rate of the steam is 1 kg/s, and the heat transfer rate from the outside of the desuperheater to the ambient air is 3 kJ/s. What is the required mass flow rate of liquid if changes in kinetic and gravitational energy are negligible?

Solution: For the control volume shown in Figure 10E.1, conservation of mass, equation (10.49), requires

$$m_1 + \dot{m}_2 = \dot{m}_3 \qquad \text{or} \qquad \frac{\dot{m}_2}{\dot{m}_1} + 1 = \frac{\dot{m}_3}{\dot{m}_1}$$

The first law, equation (10.51), for the control volume requires

$$\dot{Q} = \dot{m}_1 h_1 + \dot{m}_2 h_2 - \dot{m}_3 h_3$$

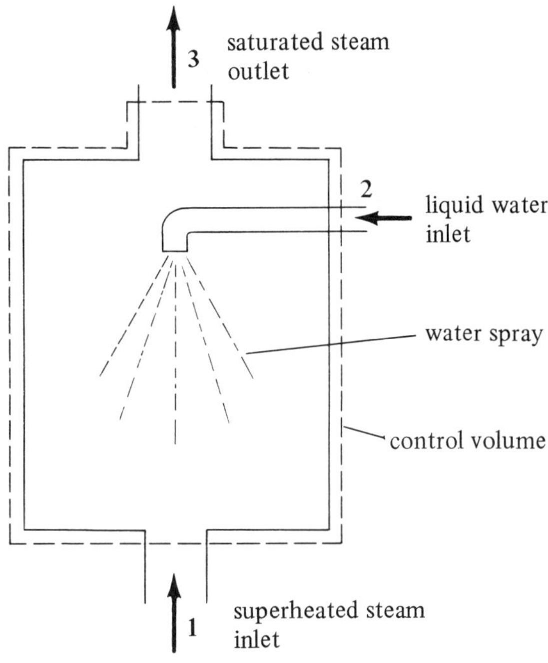

Figure 10E.1 Steady flow steam desuperheater

where $\Delta\vartheta^2$ and Δgz have been neglected and $\dot{W}_{shear} = 0$ since the walls of the desuperheater are rigid.

$$\frac{\dot{Q}}{\dot{m}_1} = h_1 + \left(\frac{\dot{m}_2}{\dot{m}_1}\right) h_2 - \left(\frac{\dot{m}_2}{\dot{m}_1} + 1\right) h_3$$

$$\frac{\dot{m}_2}{\dot{m}_1} = \frac{h_1 - h_3 - \dot{Q}/\dot{m}_1}{h_3 - h_2}$$

$$\frac{\dot{m}_2}{\dot{m}_1} = \frac{3064.2 - 2748.7 - 3/1}{2748.7 - 168.44} = 0.121$$

$$\dot{m}_2 = 0.121 \text{ kg/s}$$

10E.2 Sample Problem: A Hilsch tube is a rigid-walled device which employs a vortex flow to separate a high-pressure stream of gas into two low-pressure streams at different temperatures as shown in Figure 10E.2. The internal interaction which causes the separation is a fluid shear work transfer between the

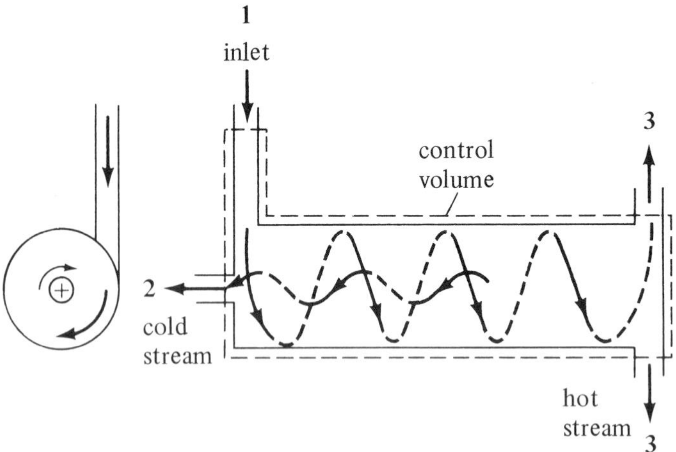

Figure 10E.2 Hilsch tube

flow in the inner vortex and the outer vortex. The steady operating conditions are: The gas is air, $R = 0.287$ kJ/kgK, $c_P = 1.003$ kJ/kgK. Inlet state, $P_1 = 10^6$ N/m², $T_1 = 300$ K, and $\dot{m}_1 = 0.1$ kg/s. Cold stream outlet state, $P_2 = 10^5$ N/m², $T_2 = 230$ K, and $\dot{m}_2 = 0.01$ kg/s. Hot stream outlet state, $P_3 = 10^5$ N/m². The walls of the Hilsch tube are adiabatic. The kinetic energies of the gas streams are negligible at the inlet and at the outlets. Changes in height are small.

(a) What is the hot stream outlet temperature?

(b) Is the operation reversible?

(c) If not reversible, what is the value for P_1 for reversible operation, with the same flows, temperatures and outlet pressures as given?

Solution:

(a) For the control volume shown in Figure 10E.2, conservation of mass, equation (10.49), requires

$$\dot{m}_1 = \dot{m}_2 + \dot{m}_3 \qquad \text{or} \qquad \dot{m}_3 = \dot{m}_1 - \dot{m}_2$$
$$\dot{m}_3 = 0.1 - 0.01 = 0.09 \text{ kg/s}$$

The first law, equation (10.51), for the control volume requires

$$0 = \dot{m}_1 h_1 - \dot{m}_2 h_2 - \dot{m}_3 h_3$$
$$\dot{m}_3 c_P T_3 = \dot{m}_1 c_P T_1 - \dot{m}_2 c_P T_2$$

$$T_3 = \left(\frac{\dot{m}_1}{\dot{m}_3}\right) T_1 - \left(\frac{\dot{m}_2}{\dot{m}_3}\right) T_2$$

$$T_3 = \left(\frac{0.1}{0.09}\right) (300) - \left(\frac{0.01}{0.09}\right) (230) = 307.78 \text{ K}$$

(b) The second law, equation (10.53), for the control volume requires

$$\dot{m}_1 s_1 - \dot{m}_2 s_2 - \dot{m}_3 s_3 \leq 0$$

since the heat transfer and thus the entropy transfer Q/T are zero.

$$\dot{m}_2(s_1 - s_2) + \dot{m}_3(s_1 - s_3) \leq 0$$

The entropy changes are given by equation (7.25).

$$\dot{m}_2 \left[c_P \ln \frac{T_1}{T_2} - R \ln \frac{P_1}{P_2}\right] + \dot{m}_3 \left[c_P \ln \frac{T_1}{T_3} - R \ln \frac{P_1}{P_3}\right] \leq 0$$

$$0.01 \left[1.003 \ln \frac{300}{230} - 0.287 \ln 10\right] + 0.09 \left[1.003 \ln \frac{300}{307.78} - 0.287 \ln 10\right]$$

$$-0.00394 + (-0.0618) = -0.657 \text{ kJ/Ks} < 0$$

The negative sign indicates a net entropy transfer out of the control volume as required for an irreversible process.

(c) For reversible operation, entropy is conserved so the second law requirement becomes an equality. The entropy convected in equals the entropy convected out, and P_1 becomes the unknown.

$$\dot{m}_2 \left[c_P \ln \frac{T_1}{T_2} - R \ln \frac{P_1}{P_2}\right] + \dot{m}_3 \left[c_P \ln \frac{T_1}{T_3} - R \ln \frac{P_1}{P_2}\right] = 0$$

since $P_2 = P_3$.

$$\dot{m}_1 R \ln \frac{P_1}{P_2} = \dot{m}_2 c_P \ln \frac{T_1}{T_2} + \dot{m}_3 c_P \ln \frac{T_1}{T_3}$$

$$\ln \frac{P_1}{P_2} = \frac{0.01}{0.1} \frac{1.003}{0.287} \ln \frac{300}{230} + \frac{0.09}{0.1} \frac{1.003}{0.287} \ln \frac{300}{307.78}$$

$$\ln \frac{P_1}{P_2} = 0.0929 - 0.0805 = 0.0124$$

$$P_1 = 1.0124 P_2 = 1.0124 \times 10^5 \text{ N/m}^2$$

10.8 Unsteady Flow in the Bulk Flow Model

Many engineering systems operate in the steady state, that is, to satisfy the requirements of steady flow; however, there are situations of interest which are not steady flow. Among these is a special class called unsteady flow with a uniform state. For example, filling a vessel with a fluid cannot be regarded as steady since the mass, and the state of the mass contained with the vessel, changes with time. The complete form of the first law of thermodynamics for an open system, equation (10.29), must be used to analyze the unsteady system behavior.

When the fluid within the control volume is spatially uniform in state, the solution of equation (10.29) is simplified since a single set of independent properties describes the state of all the fluid within the control volume. The energy of the control volume is simply the mass times the uniform specific energy of the fluid. The specific energy is the sum of the internal energy and the uncoupled energies. Thus,

$$E_{cv} = m_{cv}e_{cv} = m_{cv}\left(u + \frac{\vartheta^2}{2} + gz\right)_{cv} \tag{10.59}$$

Substituting equation (10.59) into equation (10.29), we obtain the first law of thermodynamics for an open system in unsteady flow with a uniform state.

$$\frac{\partial}{\partial t}\left[m\left(u + \frac{\vartheta^2}{2} + gz\right)\right]_{cv} = \dot{Q}_{cv} - \dot{W}_{\text{shear}} + \sum\left[\dot{m}\left(h + \frac{\vartheta^2}{2} + gz\right)\right]_{\text{in}} - \sum\left[\dot{m}\left(h + \frac{\vartheta^2}{2} + gz\right)\right]_{\text{out}} \tag{10.60}$$

Since the flow work terms are only associated with the fluid crossing the control surface, the left side of equation (10.60) includes the internal energy but the right side, which accounts for convection of energy across the control surface, includes the enthalpy.

For a given fluid model, the internal energy is uniquely related to the state of the fluid within the control volume. Therefore, equation (10.60) can be used to determine the temporal behavior of the state of the fluid within the control volume in terms of the state of the fluid crossing the control surface. In the analysis of systems of this class, the continuity equation is used to relate the mass flow across the control surface to the rate of accumulation of mass within the control volume.

For the case of unsteady flow with a uniform state, the second law of thermodynamics, equation (10.35), is unchanged with the exception that the entropy can now be expressed in terms of the specific entropy which is uniform throughout the control volume. For the shear power, we can combine equations (10.46) and (10.59) to obtain

$$\begin{aligned}\dot{W}_{\text{shear}} \leq & -\frac{\partial}{\partial t}\left[m\left(u - T_{\text{atm}}s + \frac{\vartheta^2}{2} + gz\right)\right]_{cv} \\ & + \sum\left[\dot{m}\left(b + \frac{\vartheta^2}{2} + gz\right)\right]_{\text{in}} \\ & - \sum\left[\dot{m}\left(b + \frac{\vartheta^2}{2} + gz\right)\right]_{\text{out}}\end{aligned} \quad \textbf{(10.61)}$$

Here again the interpretation of equation (10.61) in terms of the availability is obvious.

Unsteady flow with a uniform state is used to model a flow into a well mixed tank. Mixing may be by mechanical stirring or by the jet action of the incoming stream. A further simplification in the solution is possible if the state of the entering fluid is constant, as in filling an evacuated tank with ambient air. With this state constant, the inflow term in equation (10.60) can be integrated in time since it is just a constant times the mass flow rate into the tank. The time integral of the continuity equation gives the change in mass in the volume in terms of the time integral of the mass flow rate.

Unsteady flow with a uniform state is used to model the blow down of a well-mixed tank. Since the state inside the tank is uniform at any given time, the fluid flowing out of the tank has the same intensive and specific properties as the fluid inside. Thus, u in the control volume term on the left of equation (10.60) is evaluated at the same T and P as the h in the outflow term. This again simplifies the integration of equation (10.60). The uniform state model may also be used for the blow down of a tank that is initially in a uniform state, provided that the heat transfer with the walls is not significant (not too slow a blow down) and the spatial variation of the pressure in the tank is small (not too fast a blow down). Since the pressure and the entropy of the fluid remain spatially uniform the state remains uniform without fluid mixing.

10E.3 Sample Problem: Air inside an adiabatic tank with a volume of 1 m^3 is initially at $P_0 = 10^6$ N/m², and $T_0 = 500$ K. The air in the tank is to be reduced in temperature while at constant pressure by admitting air at $P_{in} = 10^6$ N/m² and $T_{in} = 300$ K. The pressure in the tank is held constant at $P = 10^6$ N/m² by discharging air from the tank at the uniformly mixed temperature of the air inside the tank. The mass flow rate of entering air is constant at 0.01 kg/s and the mass flow rate out varies to keep constant pressure inside the tank. For air $R = 0.287$ kJ/kgK and $c_v = 0.716$ kJ/kgK.

Determine the temperature of the air inside the tank (well mixed) as a function of time for this constant pressure, unsteady flow process.

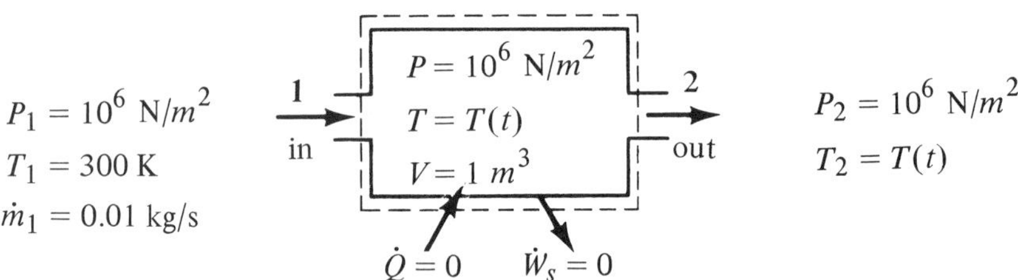

Figure 10E.3 Convective cooling of a tank of air

Solution: The control volume selected for analysis is shown in Figure 10E.3. Conservation of mass, equation (10.9), requires

$$\frac{\partial m_{cv}}{\partial t} = \dot{m}_1 - \dot{m}_2$$

The first law, equation (10.29), requires

$$\frac{\partial E_{cv}}{\partial t} = \dot{m}_1 h_1 - \dot{m}_2 h_2$$

where kinetic energy and changes in height are neglected. The mass in the control volume at any time is

$$m_{cv} = \frac{P_0 V}{RT}$$

Here P_0 and V are the constant pressure and volume of the tank, and T is the variable temperature of the air in the tank. The energy of the control volume is

$$E_{cv} = m_{cv} c_v T = \frac{P_0 V}{RT} c_v T = \frac{P_0 V c_v}{R}$$

From conservation of mass the flow out is

$$\dot{m}_2 = \dot{m}_1 - \frac{P_0 V}{R} \frac{d}{dt}\left(\frac{1}{T}\right)$$

From the first law

$$\frac{\partial}{\partial t}\left[\frac{P_0 V c_v}{R}\right] = 0 = \dot{m}_1 c_P T_1 - c_P T \left[\dot{m}_1 - \frac{P_0 V}{R}\frac{d}{dt}\left(\frac{1}{T}\right)\right]$$

$$-\dot{m}_1(T_1 - T) = \frac{P_0 V}{R} T \frac{d}{dt}\left(\frac{1}{T}\right)$$

$$-(T_1 - T) = \frac{P_0 V}{\dot{m}_1 R} T \left(-\frac{1}{T^2}\right)\frac{dT}{dt}$$

$$\frac{\dot{m}_1 R}{P_0 V} dt = \frac{dT}{T(T_1 - T)}$$

$$\frac{\dot{m}_1 R}{P_0 V} t = -\frac{1}{T_1} \ln \frac{T_1 - T}{T} + C$$

At time $t = 0$, $T = T_0 = 500$ K, the initial temperature,

$$C = \frac{1}{T_1} \ln \frac{T_1 - T_0}{T_0}$$

$$-\frac{\dot{m}_1 R T_1}{P_0 V} t = \ln \frac{(T_1 - T)}{(T_1 - T_0)} \frac{T_0}{T} = \ln \left(\frac{T_1}{T} - 1\right)\left(\frac{T_0}{T_1 - T_0}\right)$$

$$\frac{T_1}{T} - 1 = \frac{T_1 - T_0}{T_0} e^{-t/\tau}$$

where

$$\tau = \frac{P_0 V}{\dot{m} R T_1} = \frac{10^6(1)}{(0.01)(287)(300)} = 1161 \text{ s} = 19.3 \text{ min}$$

$$\frac{T_1}{T} = 1 - \frac{T_0 - T_1}{T_0} e^{-t/\tau}$$

$$T = \frac{T_1}{1 - [(T_0 - T_1)/T_0]e^{-t/\tau}}$$

$$T = \frac{300 \text{ K}}{1 - 0.6667 e^{-t/1161 \text{ s}}}$$

Problems

10.1 As shown in Figure 10P.1, air flows steadily at a rate of 0.5 kg/s through a porous plug with negligible inlet and outlet velocities. The inlet condition is $P = 7 \times 10^4$ N/m² and $T = 540$ C; the outlet condition is $P = 3.5 \times 10^4$ N/m² and $T = 450$ C.

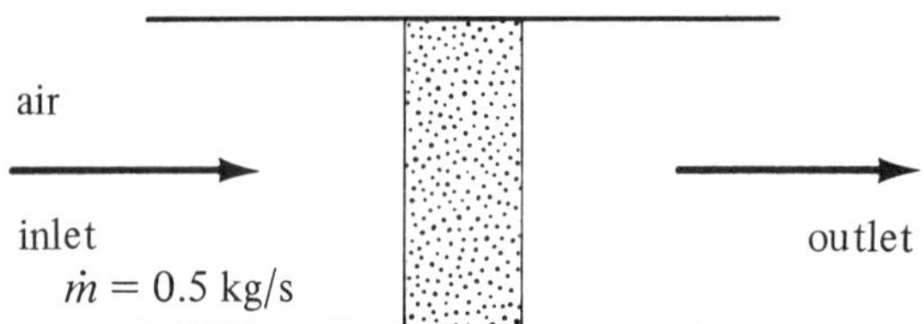

Figure 10P.1

(a) Determine the rate of heat transfer and its direction.
(b) Determine the change in entropy per unit mass flow rate through the plug.

10.2 In an air conditioning system, two streams of air are mixed in a steady flow process as shown in Figure 10P.2 to form a third stream. Assume that the air can be modeled as an ideal gas with $c_v = 715.9$ J/kgK and $R = 287$ J/kgK.
(a) Determine the temperature at discharge from the mixing section.
(b) Determine the volume flow rate (m³/s) at discharge from the mixing section.

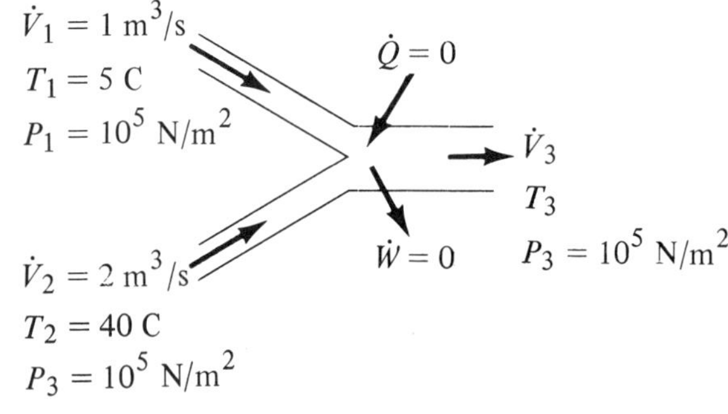

Figure 10P.2

10.3 The scheme shown in Figure 10P.3 has been proposed for recovering the energy presently being wasted in the exhaust gas from a combustion apparatus. Assume that the exhaust gas can be modeled as an ideal gas with $c_P = 1400$ J/kgK and $c_v = 1000$ J/kgK. Assume that all heat transfer processes are reversible and that all work transfer processes are reversible and adiabatic.
(a) Determine the heat transfer rate Q_H from the exhaust gas to the engine.
(b) Determine the heat transfer rate Q_L necessary to insure that the engine operates in a reversible manner.
(c) Evaluate the power produced by the heat engine.
(d) Determine the energy conversion efficiency of the engine and compare it with

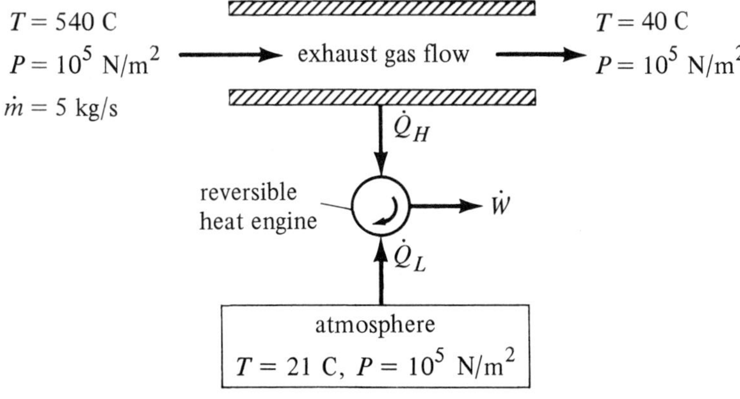

Figure 10P.3

the efficiency of a Carnot engine operating between heat reservoirs at 540 C and 21 C.

10.4 The steady-flow, adiabatic steam ejector shown in Figure 10P.4 uses a jet of high pressure steam to pump cold water from a low pressure to a higher pressure. The streams have small velocity at each port.

(a) For the inlet states 1 and 2 shown, determine the outlet state 3 for the given outlet pressure.

(b) Show numerically whether the ejector is operating reversibly or irreversibly.

10.5 In the manufacture of ethyl alcohol, a rectification column is commonly used and may be arranged as shown in Figure 10P.5. The alcohol vapor, which is formed at the top of the column, must be condensed for subsequent processing and shipment. In order to minimize the size of the condenser, the ethyl alcohol will be condensed by boiling water. The heat transfer from the condensing alcohol boils the water. Unfortunately, the boiling points of these two fluids are *not* the same at one atmosphere pressure, so the two sides of the heat exchanger (condenser) must operate at different pressures. The alcohol side operates at 10^5 N/m^2, the water side operates at a lower pressure achieved by means of an adiabatic, zero work pumping device attached to the water side of the heat exchanger.

(a) Assuming no temperature differential is necessary for operation of the condenser, determine the temperature and pressure of the water side of the condenser.

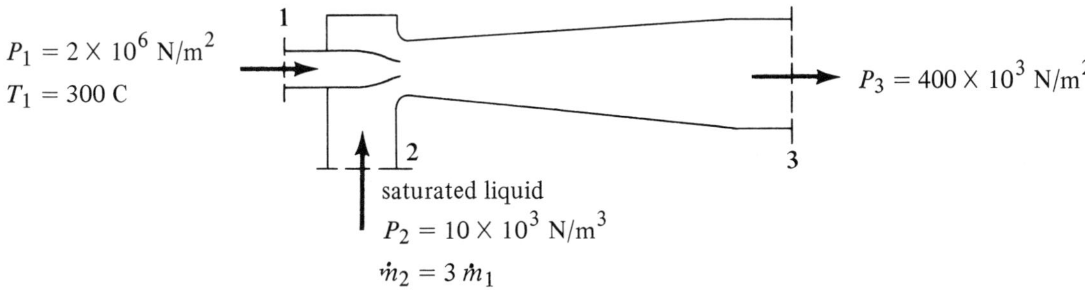

Figure 10P.4

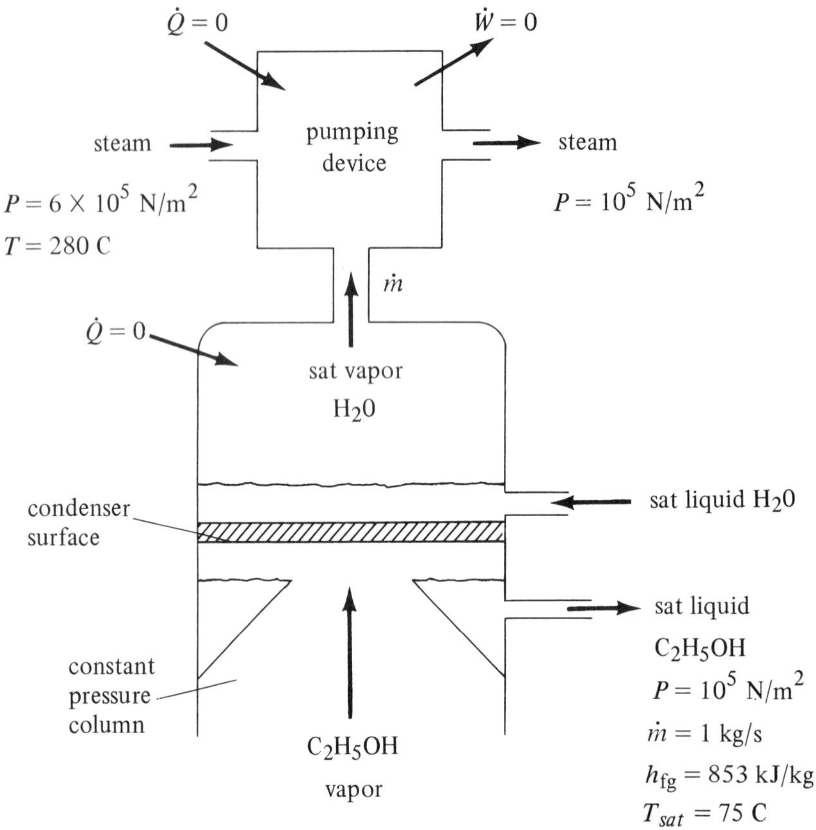

Figure 10P.5

(b) What is the heat transfer across the condenser surface?

(c) Assuming steady state operation, what are the vapor flow rate and the liquid flow rate on the water side of the condenser?

(d) What is the minimum 6×10^5 N/m^2 steam flow rate in the pumping device?

(e) What is the temperature of the discharge stream from the pumping device?

10.6 An adiabatic steady-flow pumping device shown in Figure 10P.6 employs a high pressure stream of gas, port 1, to pump a low pressure stream of gas, port 2. All of the gas leaves the device at port 3. Determine the maximum ratio $\dot{m}_2/\dot{m}_1$. The gas may be modeled as an ideal gas with $R = 2.077$ kJ/kgK, $c_P = 5.234$ kJ/kgK.

10.7 The adiabatic steady-flow device shown in Figure 10P.7 has no shaft or shear work transfer with the environment and the states of the gas at the two ports are as shown.

(a) Determine the mass flow rate per unit of cross section $\dot{m}/A_1$, and the velocity of the fluid at sections 1 and 2.

(b) Determine whether the flow is from section 1 to section 2 or from section 2 to section 1.

The gas can be modeled as an ideal gas, $R = 287$ J/kgK, $c_v = 715.9$ J/kgK.

10.8 As shown in Figure 10P.8, a steady-flow machine of unknown internal construc-

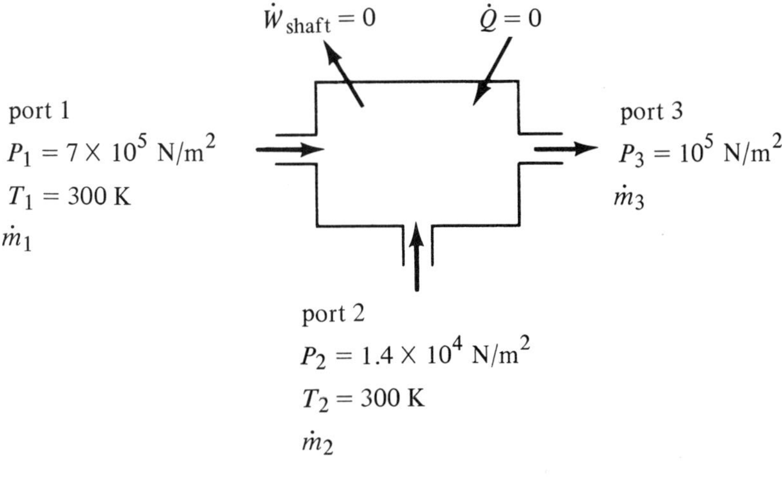

Figure 10P.6

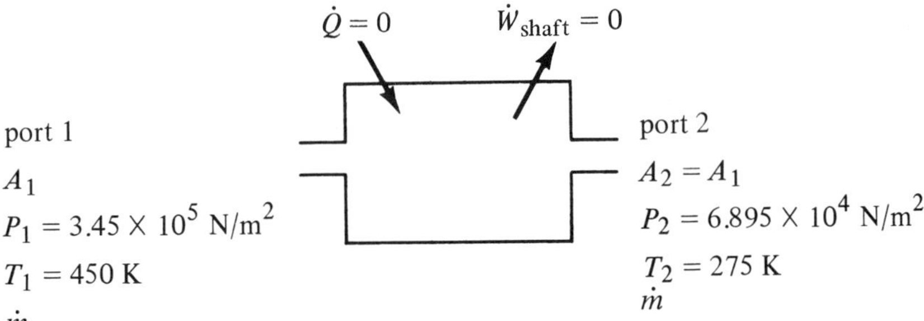

Figure 10P.7

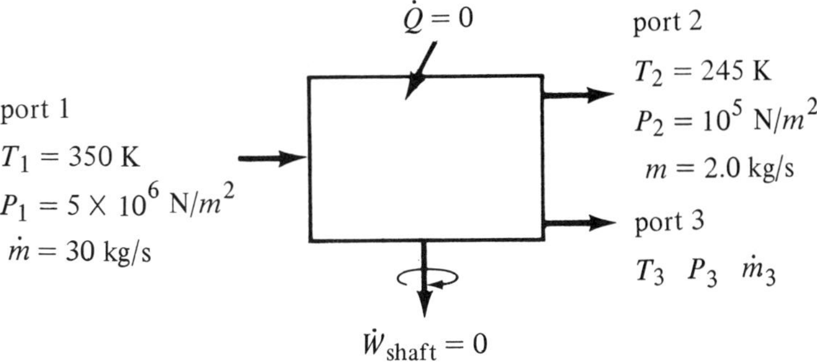

Figure 10P.8

tion has one inlet port, port 1, and two exit ports, port 2 and port 3. An incompressible fluid, with specific heat $c = 4.187$ kJ/kgK and specific volume $v = 0.001$ m^3/kg, flows through the device. There is no net heat transfer or shaft work transfer between the machine and the environment, and changes in kinetic and potential energies of the fluid are negligible.

(a) For the states shown in Figure 10P.8, determine the mass flow rate and the temperature and pressure of the fluid at port 3 if the device operates in a reversible manner.

(b) Describe the nature of the changes produced in the unknown quantities at port 3 if the device operates in an irreversible manner.

10.9 An empty greenhouse with a volume of 60 m^3 has cracks around doors and windows so that the pressure inside is always equal to that of the environment. On a particular day when the atmospheric pressure is 10^5 N/m^2 the air in the greenhouse changes temperature from 10 C to 40 C by means of radiation heat transfer from the sun. Assume that the air can be modeled as an ideal gas with $c_v =$ 715.9 J/kgK and $R = 286$ J/kgK.

(a) Treating the greenhouse as an open system, determine the heat transfer necessary to produce this temperature change.

(b) What is the change in the stored energy of this system?

10.10 A storage bottle of constant volume initially contains 11.3 grams of air at a temperature of 21 C. A valve on the bottle is opened and air flows into the bottle from the atmosphere where the air temperature is 21 C and the absolute pressure is 1.013×10^5 N/m^2. After 10 minutes have passed, the mass of the air in the bottle is 34 grams and its temperature is 40 C. Determine the *average* rate (over the 10 minute period) of heat transfer across the inside surface of the bottle. The air can be modeled as an ideal gas with $c_v =$ 715.9 J/kgK and $R = 287$ J/kgK.

10.11 A large, well insulated tank is being charged from an air line in which the temperature of the air is 33 C, the absolute pressure is 7×10^5 N/m^2, and the specific volume is 0.127 m^3/kg. (All line conditions are constant.) At the instant when the charging rate is 1 kg/min., the weight of air in the tank is 15 kg and the temperature is 21 C. The air line enters the tank at the same elevation as the center of mass of the air in the tank. The behavior of the air can be described by the ideal gas model with $c_v = 715.9$ J/kgK and $R = 287$ J/kgK. Assuming kinetic energies to be negligible, determine the instantaneous rate of temperature rise in the tank.

10.12 An insulated, rigid vessel of volume V is initially evacuated. At one end of the tank is an opening which is covered by a diaphragm. Outside the tank is the environment air at a pressure P and a temperature T. At some instant, the diaphragm ruptures and the vessel fills with air until the pressure inside the vessel reaches P. Show that if the air can be modeled as an ideal gas with constant specific heats, the ratio of the temperature of the air inside the tank immediately after mechanical equilibrium is established to that of the environment is simply the ratio of the specific heats c_P/c_v.

10.13 A rigid, insulated vessel which is initially evacuated is to be filled with steam from a steam line at 6×10^5 N/m^2 and 260 C. Steam flows into the vessel until the pressure in the vessel is 6×10^5 N/m^2. At this instant the valve on the charging line is closed.

(a) What is the temperature of the steam in the vessel at the instant the valve is closed if the filling process is adiabatic?

(b) What is the final pressure of the steam in the vessel after it has reached thermal equilibrium with the environment at 40 C?

(c) What is the heat transfer per kg of steam

for the process by which the steam and the environment come into thermal equilibrium?

10.14 A mass of air confined to a rigid, insulated tank of volume 2 m^3 is heated by dissipating electrical energy in an electrical resistor at the constant rate of 100 J/s. In order to maintain the air inside the tank at a constant temperature of 100 C, air is bled from the system at a controlled rate. Assume the air can be modeled as an ideal gas with c_v = 715.9 J/kgK and R = 287 J/kgK. Determine the instantaneous rate of mass flow from the system when the pressure of the air inside the tank is 7 × 10^5 N/m^2. State all the assumptions that must be made in order to effect a solution.

10.15 A meteorological balloon filled with helium receives energy by thermal radiation from the sun as it rises through the atmosphere. Since the pressure of the atmosphere decreases with altitude, the balloon will ultimately rupture. To prevent rupture, gas is released from the balloon at such a rate that the volume of the balloon is constant. For a balloon with a volume of 0.1 m^3, the calculated heat transfer rate to the helium gas in the balloon is 0.25 kJ/s. At some particular altitude the gas pressure is decreasing at the instantaneous rate of 2 × 10^3 N/m^{2}s. If the pressure of the helium gas at that instant is 9 × 10^4 N/m^2 while its temperature is 150 C, determine the instantaneous rate at which mass must be released to maintain the volume constant. Assume that the helium can be modeled as an ideal gas with c_v = 3.14 kJ/kgK and R = 2.077 kJ/kgK.

10.16 The adiabatic container shown in Figure 10P.16 is connected by a tube to a pump. Initially 1 kg of saturated triple point liquid nitrogen and a negligible amount of saturated triple point vapor are in the container. No triple point solid is present initially. The pump is started and vapor is removed from the container until all of the liquid has either boiled away or frozen to saturated triple point solid. Unknown amounts of saturated triple point solid and saturated triple point vapor remain in the container. (No liquid is present.)

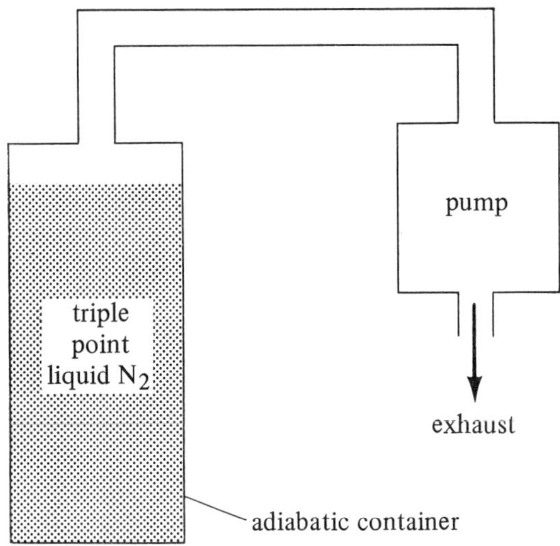

Figure 10P.16

(a) Determine the mass of the triple point solid which remains in the container. Assume that the mass of triple point vapor remaining in the container is negligible. Further assume that equilibrium is maintained as the vapor is pumped away so that no metastable states or violent boiling occurs.

(b) Estimate the mass of triple point vapor remaining in the container at the final state to show that the assumption of negligible vapor mass used in part (a) is justified.

Triple Point Data for Nitrogen
Temperature = 63 K
Pressure = 0.1249 × 10^5 N/m^2

Phase	Enthalpy (kJ/kg)	Specific Volume (m^3/kg)
Solid	−25.75	1.129 × 10^{-3}
Liquid	0	1.152 × 10^{-3}
Vapor	216.0	1.487

CHAPTER 11

Application of the Principles of Thermodynamics to Steady-Flow Components of Engineering Systems

11.1 Introduction

Many complex engineering systems operate with steady or periodic flow which simplifies the thermodynamic analysis as was shown in the preceding chapter. The systems are constructed by interconnecting steady-flow components which are grouped into four classes according to function: (1) shaft work machines, (2) nozzles and diffusers, (3) throttles, and (4) heat exchangers. An understanding of the behavior of these components is the key to understanding the thermodynamic plants discussed in the next chapter.

Our objective in this chapter is to determine the thermodynamic behavior of the components of each class by applying the principles of thermodynamics to a control volume containing the component. This approach yields the "black box" characteristics of the component that must be known in order to evaluate the performance of the complete system. This does not mean that the principles of thermodynamics do not apply or are not useful in determining the detailed internal processes of the component. On the contrary, in designing such a component, thorough consideration must be given to the complex internal processes; however, the analysis of these internal processes is outside the scope of our present objective.

The analysis of this chapter is based on the equations for steady flow for the control volume with one inlet port and one exit port, equations (10.50), (10.52) and (10.57). In addition it will be useful to consider a control volume of infinitesimal extent in the direction of flow. The fluid experiences an infinitesimal change of state between inlet and outlet in response to infinitesimal rates of heat transfer and shear work transfer. Equations (10.52) and (10.57) then become differential relations:

$$\delta\dot{Q}/\dot{m} - \delta\dot{W}_{\text{shear}}/\dot{m} = dh + \vartheta\, d\vartheta - g\, dz \tag{11.1}$$

$$ds \geq \frac{\delta\dot{Q}/\dot{m}}{T} \tag{11.2}$$

The analysis will be illustrated by application to the flow of an incompressible fluid, the flow of an ideal gas, and to the flow of a pure substance in two phase states. These cases illustrate the basic behavior of each of four classes of steady-flow components.

11.2 Shaft Work Machines[1]

The first class of components is comprised of machines which change the state of a stream by positive or negative shaft work

[1] For more detailed information than is presented here, the following references should be consulted: (1) Stodola, A., *Steam and Gas Turbines*, Vols. I and II, translated by L. C. Loewenstein, McGraw-Hill, N.Y., 1927. (2) Hill, P. G., and Peterson, C. R., *Mechanics and Thermodynamics of Propulsion*, Addison-Wesley, Reading Mass., 1965. (3) Horlock, J. H., *Axial Flow Turbines; Fluid Mechanics and Thermodynamics*, Butterworths, London, 1966.

transfer. Machines with positive shaft work transfer are commonly called turbines, reciprocating engines, expanders, or fluid motors, depending upon the application and the method of developing the pressure forces that produce the shaft work transfer. Machines with negative shaft work transfer are commonly called compressors, pumps, or fans depending upon the application.

The operation of a shaft work machine does not depend upon the attainment of thermal equilibrium between the flowing fluid and the walls of the apparatus; consequently, the rate of work transfer is not limited by the relatively slow thermal conduction process. Rather, the work transfer depends upon the pressure forces on the moving surfaces internal to the apparatus (turbine or compressor blades or piston faces). Thus, the limit on the rate of work transfer is related to the velocity of propagation of pressure waves (sound) in the fluid. This rate is often fast enough that the rate of work transfer for the machine is actually limited by the forces which result from the accelerations of the solid parts of the machine (inertia stress limit).

Since the work transfer rate is rapid, the shaft work machine is small enough that the fluid remains within the machine for a time period that is small compared to the time period required to attain thermal equilibrium. Thus, the apparatus is essentially adiabatic. Note that some heat transfer does occur in virtually every case; however, the magnitude is negligible compared to the shaft work transfer. This situation is especially true for turbo machines in which the high work transfer rate is the result of internal pressure differences (across the blades) produced by accelerating (deflecting) the moving stream of fluid. In reciprocating or positive displacement machines, the internal forces are the result of equilibrium pressure (spatially uniform) acting on a moving piston face or the equivalent moving surface.

A second result of the relatively rapid work transfer rate and the resulting small size of the machine is that the change in gravitational potential energy is usually negligible. This is usually true even for the large water turbines used in hydroelectric power stations provided the control volume does not include the penstock.

The preceding discussion indicates that shaft work machines can be reasonably modeled as adiabatic devices with negligible changes in gravitational potential energy. Further, in many practical cases the change in kinetic energy of the bulk flow is also negligible. Thus the first law of thermodynamics, equation (10.51), applied to a control volume representing a machine of this type reduces to

$$-\dot{W}_{\text{shaft}} = \dot{m}(h_{\text{out}} - h_{\text{in}}) \tag{11.3}$$

For a machine with a unit mass flow rate of a fluid experiencing an infinitesimal change of state, equation (11.3) reduces to

$$-\delta\left(\frac{\dot{W}}{\dot{m}}\right)_{\text{shaft}} = dh \tag{11.4}$$

Equations (11.3) and (11.4) indicate that positive shaft work transfer occurs at the expense of the enthalpy of the flowing fluid and that negative shaft work transfer increases the enthalpy of the flowing fluid. In addition to equations (11.3) and (11.4), mechanical equilibrium requires that in a machine with a positive shaft work transfer, e.g., a turbine, the pressure of the flowing fluid must decrease as it passes through the device. On the other hand, in order to increase the pressure of a fluid stream in a shaft work machine (a pump or compressor), mechanical equilibrium requires a negative work transfer.

In applying equation (11.3) or (11.4) to a physical situation, one must always be careful to insure that the actual physical circumstances are consistent with the assumptions associated with these equations. For example, in gas turbines it is not uncommon for the kinetic energy of the flowing fluid to be significant. Also in a steam turbine exhausting steam at low pressure, the kinetic energy of the exhaust steam is often significant. In these more complex cases, equation (10.52) must be used to include changes in kinetic energy.

The evaluation of the performance of a shaft work machine is especially simple in the limiting case of the adiabatic machine that operates in a reversible manner with a negligible net change in the kinetic energy of the fluid as it passes through the machine. From equation (10.57) or (11.2) it is clear that the entropy of the fluid leaving the control volume is equal to the entropy of the fluid entering the control volume. According to the state principle (cf. Section 8.2), we need only specify two independent properties to establish the state of the fluid. Thus the inlet entropy is fixed by the inlet pressure and the inlet temperature. The outlet pressure and the outlet entropy (equal to the inlet entropy) fix the outlet state of the reversible adiabatic shaft work machine.

To be more specific, consider the reversible adiabatic machine involving an infinitesimal change of state of the working fluid. By means of equation (9.30), the change in enthalpy can be related to the changes in entropy and pressure.

$$dh = T\,ds + v\,dP \tag{11.5}$$

Since in the present case the entropy is constant, equation (11.5) reduces to

$$(dh)_{\substack{\text{constant}\\ \text{entropy}}} = v\,dP \tag{11.6}$$

Then equation (11.4) reduces to

$$-[\delta(\dot{W}/\dot{m})_{\text{shaft}})]_{\substack{\text{reversible}\\ \text{adiabatic}}} = v\,dP \tag{11.7}$$

Thus, the shaft work transfer rate is

$$-(\dot{W}_{\text{shaft}})_{\substack{\text{reversible}\\ \text{adiabatic}}} = \dot{m}\left[\int_{P_{\text{in}}}^{P_{\text{out}}} v\,dP\right]_{s=s_{\text{in}}} \tag{11.8}$$

Because the path of the process is specified (constant entropy), the volume is uniquely related to the pressure and the integral in equation (11.8) can be evaluated directly for a specific model for the fluid properties.

If the machine is adiabatic but irreversible, the second law of thermodynamics, equation (11.2), requires that the outlet entropy, s_{out}, be greater than the inlet entropy, s_{in}. As a result, the irreversible adiabatic machine has a smaller work transfer rate and a larger outlet enthalpy than the reversible machine (with the same mass flow rate, inlet state, and outlet pressure).

We may show this result of the irreversibility from the second law of thermodynamics and the relations for the properties of a pure substance. Let us compare an irreversible adiabatic machine with a reversible adiabatic machine, both with the same inlet state and the same exit pressure. In this case, the exit state of the irreversible machine will have the same pressure but a higher entropy than the exit state of the reversible machine. Equation (9.33) requires that

$$\left(\frac{\partial h}{\partial s}\right)_P = T \tag{11.9}$$

which is always positive. As a result, the lines of constant pressure have a positive slope on the h-s plane (see Figure 11.1). Thus the exit enthalpy of the irreversible machine will be greater than the exit enthalpy of the reversible machine. From equation (11.3), the shaft work is proportional to $(h_{in} - h_{out})$ for both machines. Thus for a given inlet state and exit pressure

$$(\dot{W}_{shaft})_{rev} - (\dot{W}_{shaft})_{irrev} = \dot{m}[(h_{out})_{irrev} - (h_{out})_{rev}] \tag{11.10}$$

$$(\dot{W}_{shaft})_{rev} - (\dot{W}_{shaft})_{irrev} = \dot{m}\left[\int_{s_{in}}^{s_{out}} T\,ds\right]_{P=P_{out}} \tag{11.11}$$

The irreversible work transfer rate is smaller in an algebraic sense. Consequently, the irreversible turbine will have a smaller positive work transfer than the reversible turbine; however, the irreversible compressor will require a larger work transfer into the machine than the reversible compressor.

In practice, it has been found convenient to measure the work transfer rate of an actual adiabatic shaft work transfer machine in terms of the performance of a reversible adiabatic machine with the same inlet state and the same exit pressure as the actual machine. It is conventional to express the ratio of these two work transfer rates as an efficiency; therefore, it is necessary to have one efficiency for machines with positive work transfer rates and a second efficiency for machines with negative work transfer rates. For the machine involving a positive work transfer, we use the conventional definition

$$\eta_+ = \frac{\dot{W}_{actual}}{\dot{W}_{reversible}} \tag{11.12}$$

Here η_+ is frequently known as the adiabatic turbine efficiency or the adiabatic expander efficiency, $\dot{W}_{actual}$ is the power output

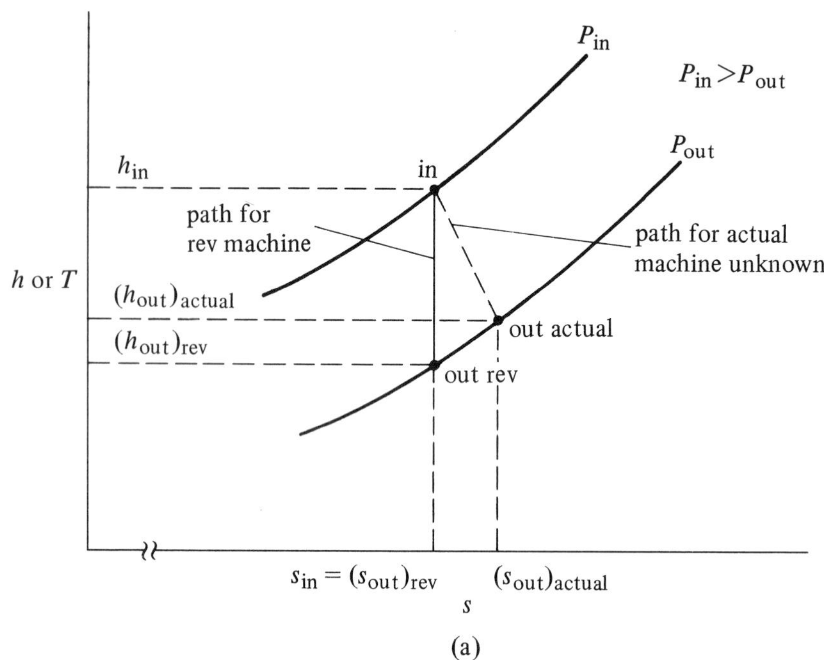

Figure 11.1(a) States of ideal gas in adiabatic positive shaft work machine

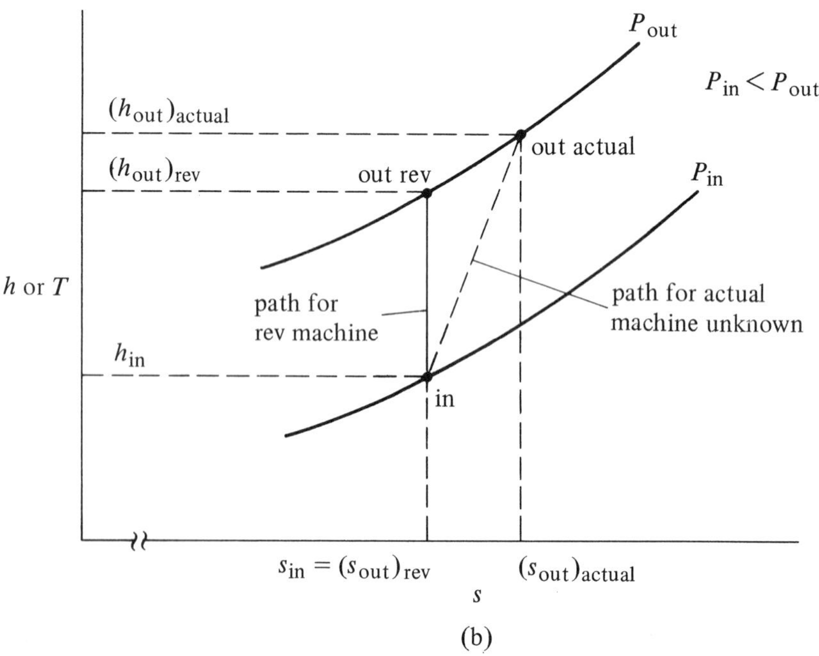

Figure 11.1(b) States of ideal gas in adiabatic negative shaft work machine

of the actual machine, and $\dot{W}_{\text{reversible}}$ is the power output of the reversible machine with the same inlet state and outlet pressure. Substituting equation (11.3) into equation (11.12), we obtain

$$\eta_+ = \frac{[\dot{m}(h_{\text{in}} - h_{\text{out}})]_{\text{actual}}}{[\dot{m}(h_{\text{in}} - h_{\text{out}})]_{\text{rev}}} = \frac{(h_{\text{in}} - h_{\text{out}})_{\text{actual}}}{(h_{\text{in}} - h_{\text{out}})_{\text{rev}}} \tag{11.13}$$

For the machine involving a negative work transfer, we use the conventional definition

$$\eta_- = \frac{\dot{W}_{\text{rev}}}{\dot{W}_{\text{actual}}} \tag{11.14}$$

where η_- is frequently known as the adiabatic compressor efficiency or pump efficiency. Again use of equation (11.3) yields

$$\eta_- = \frac{(h_{\text{in}} - h_{\text{out}})_{\text{rev}}}{(h_{\text{in}} - h_{\text{out}})_{\text{actual}}} \tag{11.15}$$

By formulating the definitions of efficiency as in equations (11.12) and (11.15), we always obtain a value of the efficiency which lies between zero and unity. Further, as the actual machine becomes more nearly reversible, its efficiency improves and approaches unity. Both of these features are consistent with our intuitive interpretations of efficiency.

Although most authors use the term efficiency to mean the adiabatic efficiencies as previously defined, other definitions for efficiency are sometimes used. It is prudent to verify the exact definition for the efficiency being employed. For example, the term isothermal efficiency is often used to compare the actual machine to a reversible isothermal machine.

11.2.1 Shaft Work Machines Processing an Ideal Gas

These conclusions can now be illustrated in detail for machines processing fluids with properties which can be adequately described by a simple model of the constitutive relations. If the fluid can be modeled as an ideal gas with constant specific heats, equation (11.3) becomes

$$-(\dot{W}_{\text{shaft}})_{\substack{\text{ideal}\\\text{gas}}} = \dot{m}c_P(T_{\text{out}} - T_{\text{in}}) \tag{11.16}$$

which shows that in this case the rate of shaft work transfer (for a given mass flow rate) is determined solely by the temperature change of the gas. For the limiting case of the reversible adiabatic machine, the outlet temperature is related to the inlet temperature and the pressure ratio through the second law of thermodynamics. Thus, from equation (9.142)

$$s_{out} - s_{in} = c_P \ln \frac{T_{out}}{T_{in}} - R \ln \frac{P_{out}}{P_{in}} = 0 \qquad \textbf{(11.17)}$$

Equation (11.17) then establishes the outlet temperature T_{out} from the pressure ratio P_{out}/P_{in} and the inlet temperature T_{in}. The rate of shaft work transfer is then given by equation (11.16) in terms of the inlet and outlet temperatures, T_{in} and T_{out}.

Alternatively, we could have used equation (11.8) and the fact that the pressure and volume of an ideal gas in a reversible adiabatic process are related by the expression

$$Pv^{\gamma} = \text{constant} \qquad \textbf{(11.18)}$$

Here γ is the ratio of specific heats, c_P/c_v. Then equation (11.8) can be integrated to give

$$-(\dot{W}_{shaft})_{\substack{\text{reversible}\\ \text{adiabatic}\\ \text{ideal gas}}} = \frac{\dot{m}\gamma}{\gamma - 1}[P_{out}v_{out} - P_{in}v_{in}] \qquad \textbf{(11.19)}$$

where v_{out} is uniquely related to v_{in} through the pressure ratio and equation (11.18). Note that equation (11.19) could have been obtained directly from equation (11.16) simply by substituting the ideal gas constitutive relation and by making use of the definition of γ. Since equation (11.16) holds for all adiabatic machines whether reversible or irreversible, it must be the case that for the ideal gas in general

$$-(\dot{W}_{shaft})_{\substack{\text{adiabatic}\\ \text{ideal gas}}} = \frac{\dot{m}\gamma}{\gamma - 1}[P_{out}v_{out} - P_{in}v_{in}] \qquad \textbf{(11.20)}$$

The significance of the reversibility is that only in the reversible case can we reduce equation (11.20) to the form

$$-(\dot{W}_{shaft})_{\substack{\text{reversible}\\ \text{adiabatic}\\ \text{ideal gas}}} = \frac{\dot{m}\gamma}{\gamma - 1} P_{in}v_{in}\left[\left(\frac{P_{out}}{P_{in}}\right)^{(\gamma-1)/\gamma} - 1\right] \qquad \textbf{(11.21)}$$

That is, for a given mass flow rate the inlet state and pressure ratio uniquely determine the shaft work transfer rate for the reversible adiabatic machine. In an irreversible machine, both the inlet state and the outlet state must be known before equation (11.20) can be evaluated.

This point is perhaps more clearly revealed in Figure 11.1 which shows the locus of states for the ideal gas as it passes through the shaft work machine. The information of Figure 11.1 is presented in terms of the coordinates h and s. The locus of states would have exactly the same appearance in terms of the coordinates T and s since the enthalpy and the temperature are directly related for the ideal gas. Note that in the case of the irreversible (actual) machine the path of the process cannot be specified since some of the states involved may be nonequilibrium states. It is clear from Figure 11.1 that knowing the inlet state, the path, and the outlet pressure is the equivalent of knowing both the inlet state and the outlet state.

For the ideal gas, the adiabatic shaft work efficiencies reduce to

$$(\eta_+)_{\substack{\text{ideal}\\\text{gas}}} = \frac{(T_{out} - T_{in})_{actual}}{(T_{out} - T_{in})_{rev}} \qquad \textbf{(11.22)}$$

and

$$(\eta_-)_{\substack{\text{ideal}\\\text{gas}}} = \frac{(T_{out} - T_{in})_{rev}}{(T_{out} - T_{in})_{actual}} \qquad \textbf{(11.23)}$$

These efficiencies can be geometrically interpreted as the ratios of the vertical distances between the inlet and outlet states on Figure 11.1.

11E.1 Sample Problem: Modern airliners are powered by turbo fan engines consisting of a gas turbine engine driving a single-stage axial flow fan. At take off power, the gas turbine engine produces 25 MW. The conditions of the air at the fan inlet are $P_1 = 10^5$ N/m^2, $T_1 = 300$ K, and $\vartheta_1 = 250$ m/s. The exit conditions from the fan are $P_2 = 1.36 \times 10^5$ N/m^2, and $\vartheta_2 = 250$ m/s. The efficiency of the fan is 0.85 based on the same kinetic energy at the inlet and exit states.

(a) What is the air flow through the fan?
(b) What flow area is required at the fan inlet?
(c) What is the exit air temperature?
(d) What is the entropy generation rate in the fan?

Solution: (a) The control volume selected for the analysis is shown in Figure 11E.1. For the reversible adiabatic fan the entropy is constant. Thus from equation (5.22) and the ideal gas constitutive relation

$$T_{2r} = T_1 \left(\frac{P_2}{P_1}\right)^{(\gamma-1)/\gamma}$$

$$T_{2r} = 300(1.36)^{0.4/1.4} = 327.55 \text{ K}$$

where $P_{2r} = P_2$ and γ for air is 1.4.

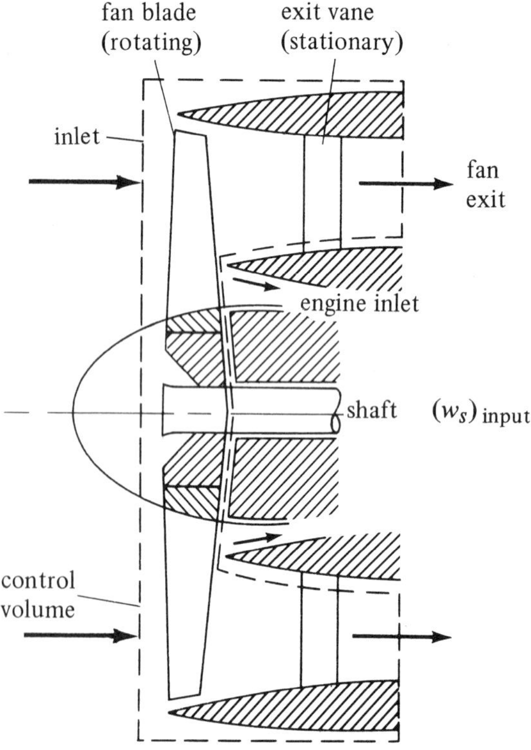

Figure 11E.1 Turbo fan

The first law for the control volume, equation (10.52), around the reversible fan gives

$$-\left(\frac{\dot{W}_{\text{shaft}}}{\dot{m}}\right)_{\text{rev}} = \left(h_{2r} + \frac{\vartheta_{2r}^2}{2}\right) - \left(h_1 + \frac{\vartheta_1^2}{2}\right)$$

where $\dot{Q} = 0$. For this case with $\vartheta_{2r} = \vartheta_2 = \vartheta_1$,

$$-\left(\frac{\dot{W}_{\text{shaft}}}{\dot{m}}\right)_{\text{rev}} = c_P(T_{2r} - T_1)$$

From the definition of efficiency, equation (11.14),

$$-\left(\frac{\dot{W}_{\text{shaft}}}{\dot{m}}\right)_{\text{actual}} = \frac{1}{\eta} c_P(T_{2r} - T_1)$$

Then

$$\dot{m} = \frac{-(\dot{W}_s)_{\text{actual}}\eta-}{c_P(T_2 - T_1)}$$

$$\dot{m} = \frac{(25 \times 10^6 \text{ J/s})(0.85)}{(1003 \text{ J/kgK})(327.55 - 300 \text{ K})} = 769.0 \text{ kg/s}$$

(b) The volume flow is given by $v\dot{m}$ and the flow area is the volume flow divided by the velocity.

$$A = \frac{v\dot{m}}{\vartheta} = \frac{\dot{m}RT}{P\vartheta}$$

$$A = \frac{(769.0)(287)(300)}{(10^5)(250)} = 2.65 \text{ m}^2$$

(c) The exit temperature is given by the first law for the control volume around the actual fan. Thus,

$$-\left(\frac{\dot{W}_{\text{shaft}}}{\dot{m}}\right) = c_P(T_2 - T_1)$$

since $\dot{Q} = 0$ and $\vartheta_1 = \vartheta_2$

$$T_2 = T_1 - \frac{\dot{W}_{\text{shaft}}}{\dot{m}c_P}$$

$$T_2 = 300 - \left[\frac{-25 \times 10^6 \text{ J/s}}{(769.0 \text{ kg/s})(1003 \text{ J/kgK})}\right] = 332.4 \text{ K}$$

(d) Since the flow is steady and adiabatic the entropy generated is the net entropy convected out of the control volume.

$$\dot{S}_{gen} = \dot{m}(s_2 - s_1) = \dot{m}\left(c_P \ln \frac{T_2}{T_1} - R \ln \frac{P_2}{P_1}\right)$$

$$\dot{S}_{gen} = 769.0\left(1003 \ln \frac{332.4}{300} - 287 \ln 1.36\right) = 11{,}240 \text{ J/Ks}$$

11.2.2 Shaft Work Machines Processing an Incompressible Fluid

If the fluid can be modeled as incompressible with a constant specific heat c, the enthalpy can [according to equation (9.90)] be separated into a thermal energy cT and an uncoupled mechanical property Pv. Then the enthalpy can be written

$$h = cT + Pv \tag{11.24}$$

and equation (11.3) becomes

$$-(\dot{W}_{shaft})_{incompressible} = \dot{m}c(T_{out} - T_{in}) + \dot{m}v(P_{out} - P_{in}) \tag{11.25}$$

According to equation (9.95), the entropy of the incompressible fluid is related only to its thermal behavior. Application of equation (9.95) to the limiting case of the reversible adiabatic machine yields

$$s_{out} - s_{in} = c \ln \frac{T_{out}}{T_{in}} = 0 \tag{11.26}$$

Thus the outlet temperature is equal to the inlet temperature. Then the shaft work transfer rate is directly related to the pressure difference and the volume flow rate mv. Thus

$$-(\dot{W}_{shaft})_{\substack{reversible\\ adiabatic\\ incompressible}} = \dot{m}v(P_{out} - P_{in}) \tag{11.27}$$

Equation (11.27) is often written as the product of the volume flow rate of the fluid ($\dot{m}v$) and the pressure difference. In incompressible fluid mechanics the same term is often expressed as the product $\dot{m}(P_{out} - P_{in})/\rho$ where ρ is the density of the incompressible fluid.

If the adiabatic machine is irreversible, the entropy of the fluid stream increases as it passes through the machine. From equation (11.26), it follows that the outlet temperature is greater than the inlet temperature. Thus, the first term in equation (11.25) is always positive for this adiabatic class of machines. This term represents the decrease in the shaft work transfer rate because of internal irreversibility which in this case is simply mechanical energy dissipation within the working fluid.

The locus of states for the incompressible fluid as it passes through the shaft work machine is shown in Figure 11.2. Again since some of the states in the actual machine are nonequilibrium states, the path cannot be specified for the irreversible (actual) machine. Also note that since the entropy and temperature of the incompressible fluid are directly related (equation (11.26)), the locus of states cannot be presented in terms of the coordinates T and s.

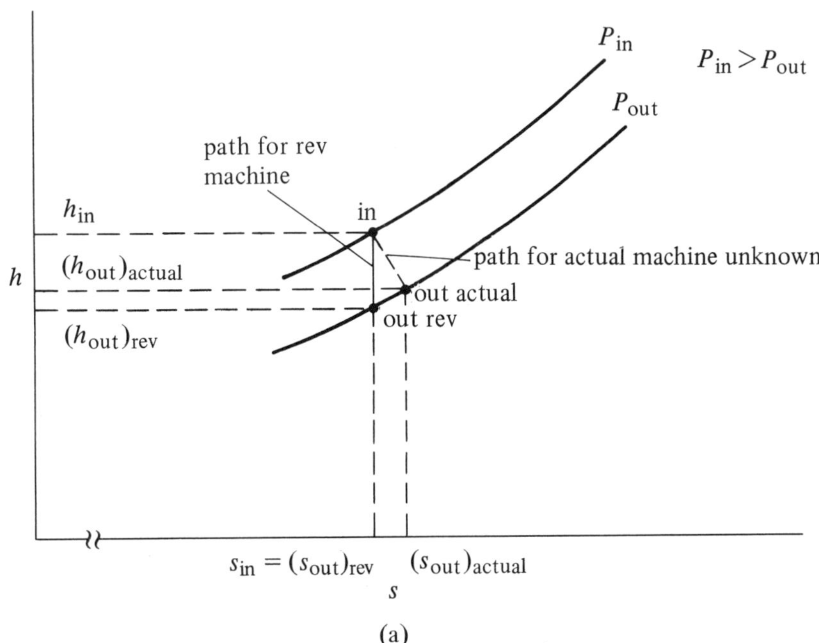

Figure 11.2(a) States of an incompressible fluid in a positive shaft work machine

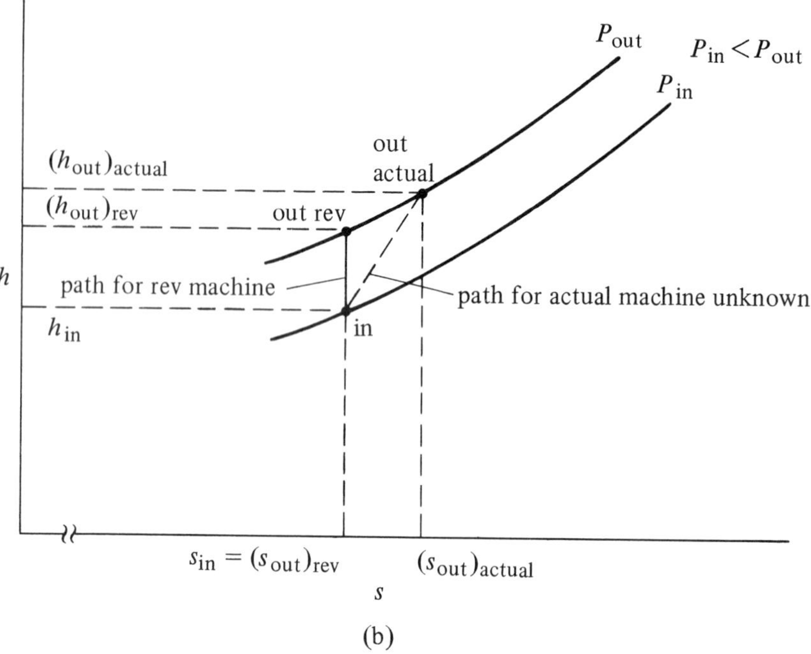

Figure 11.2(b) States of an incompressible fluid in a negative shaft work machine

For the incompressible fluid, the shaft work efficiencies reduce to a particularly simple form obtained by substituting equations (11.25) and (11.27) into equations (11.13) and (11.15).

$$(\eta_{+})_{\text{incompressible}} = 1 - \frac{c(T_{out} - T_{in})}{v(P_{in} - P_{out})} \tag{11.28}$$

and

$$(\eta_{-})_{\text{incompressible}} = 1 - \frac{c(T_{out} - T_{in})}{c(T_{out} - T_{in}) + v(P_{out} - P_{in})} \tag{11.29}$$

In both cases, the outlet temperature is greater than the inlet temperature because of the irreversibility. For the positive shaft work machine the inlet pressure is greater than outlet pressure. However, for the negative shaft work machine the inlet pressure is less than the outlet pressure. Thus, in both equations (11.28) and (11.29) the second term on the right side is positive and less than unity and represents approximately the fraction of reversible shaft work transfer lost to irreversibility.

11.2.3 Shaft Work Machines Processing a Pure Substance

The locus of states of the pure substance as it passes through various shaft work machines is shown in Figures 11.3 and 11.4. Note that one effect of the irreversibility present in the actual machine is to increase the amount of vapor resulting from the expansion process to a two-phase state in a positive shaft work machine. If the outlet state is a single-phase superheated vapor state, the irreversibilities increase the degree of superheat. Similarly, the vapor at the outlet of a negative shaft work machine is also superheated as a result of the irreversibilities present during

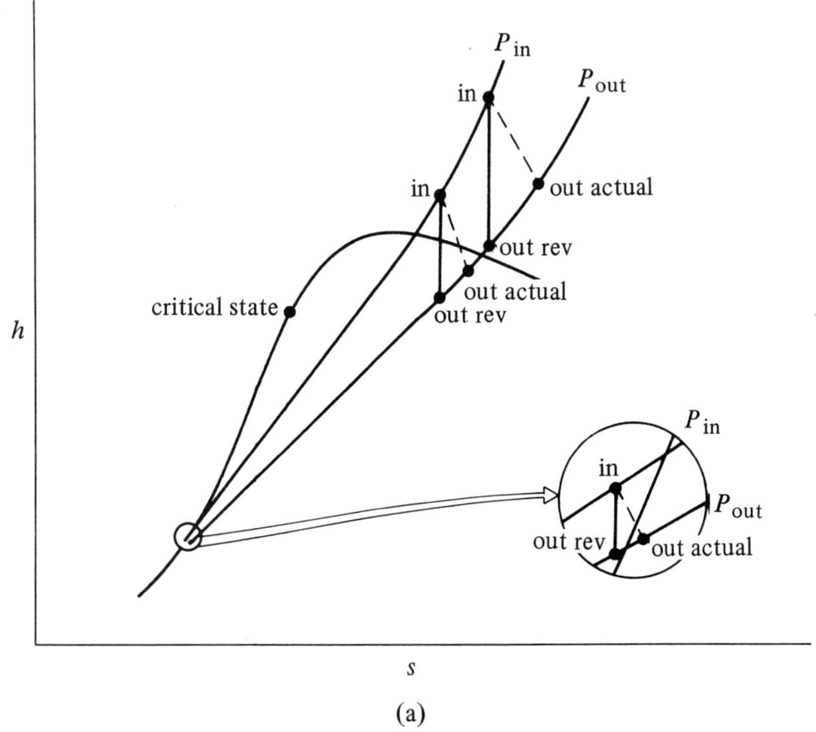

(a)

Figure 11.3(a) States of a pure substance in a positive shaft work machine

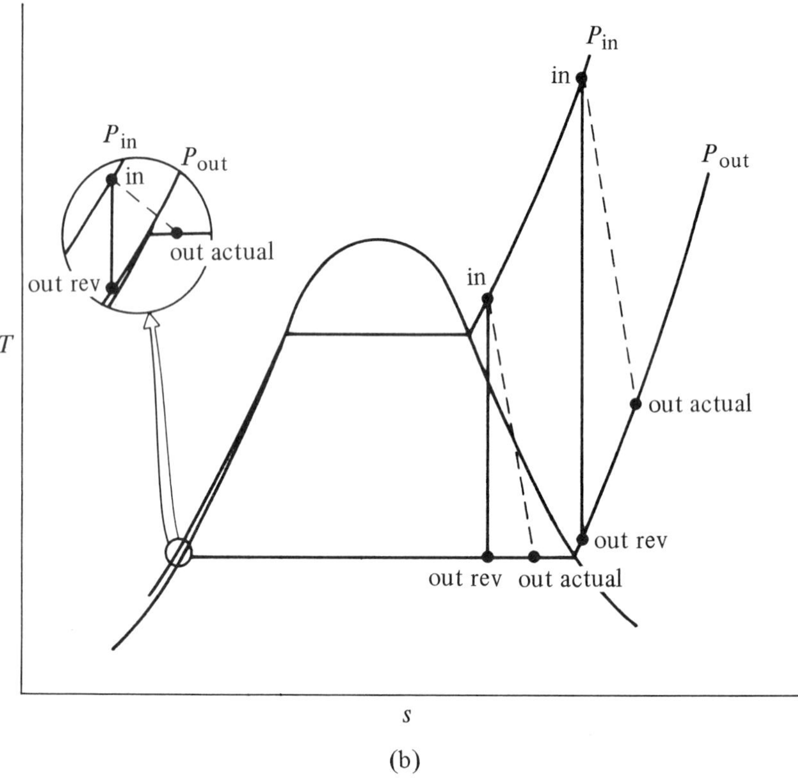

Figure 11.3(b) States of a pure substance in a positive shaft work machine

the compression process. For both negative and positive shaft work machines operating in the subcooled liquid region, the irreversibility of the process reduces the degree of subcooling.

11E.2 Sample Problem: A steam turbine is to provide 1 MW to drive a boiler feed pump. The steam conditions at the turbine inlet are $P_1 = 3.0 \times 10^6$ N/m² and $T_1 = 260$ C. The turbine exhaust pressure is $P_2 = 2 \times 10^5$ N/m². The turbine efficiency is 0.85.

(a) What steam flow is required?
(b) What is the quality of the exit steam?
(c) What is the entropy generation rate in the turbine?

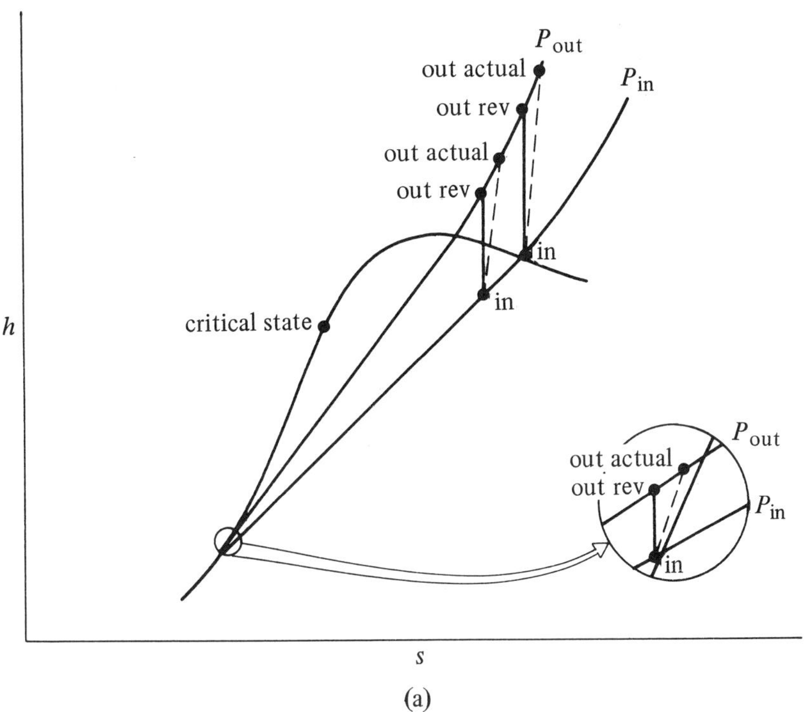

Figure 11.4(a) States of a pure substance in a negative shaft work machine

Solution: (a) The control volume selected for analysis of the turbine is shown in Figure 11E.2. The steam properties at the inlet, state 1, are

$$P_1 = 3 \times 10^6 \text{ N/m}^2,\ T_1 = 260 \text{ C}$$
$$h_1 = 2885.5 \text{ kJ/kg},\ s_1 = 6.3435 \text{ kJ/kgK}$$

The state $2r$ at the exit of reversible adiabatic turbine with the same state 1 and exhaust pressure is given by

$$s_{2r} = s_1 = 6.3435 \text{ kJ/kgK}$$
$$P_{2r} = P_2 = 2 \times 10^5 \text{ N/m}^2$$

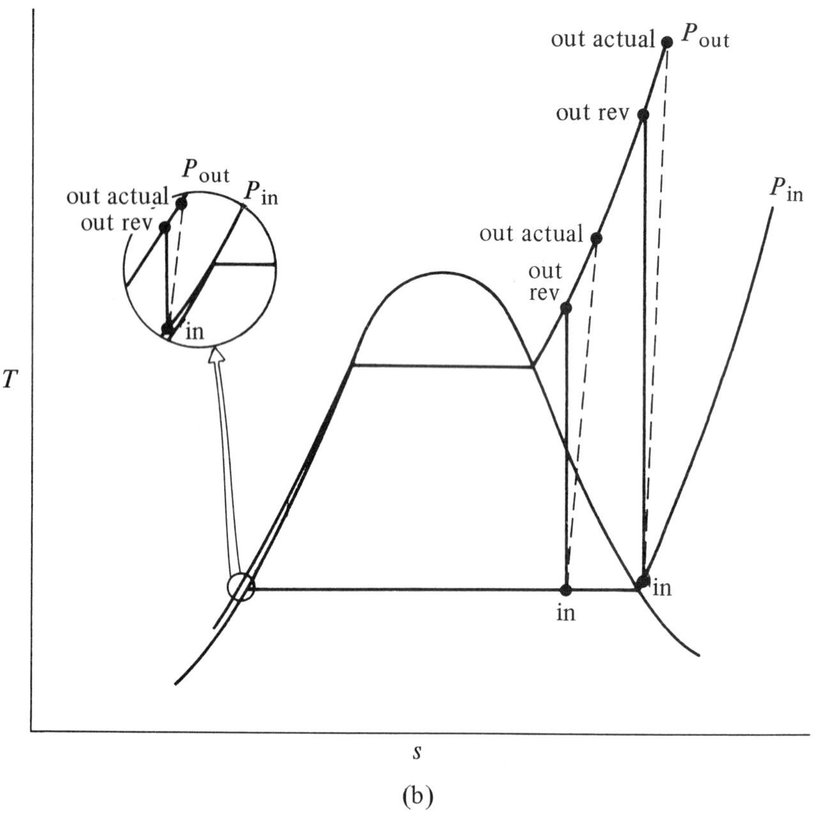

Figure 11.4(b) States of a pure substance in a negative shaft work machine

The state $2r$ is two phase since $s_f < s_{2r} < s_g$ at $P = 2 \times 10^5$. The exit quality is

$$x_{2r} = \frac{s_{2r} - s_f}{s_{fg}} = \frac{6.3435 - 1.5301}{5.5970}$$

$$x_{2r} = 0.8600$$

The exhaust enthalpy is

$$h_{2r} = h_f + x_{2r}h_{fg} = 504.70 + (0.8600)(2202.0)$$

$$h_{2r} = 2398.4 \text{ kJ/kg}$$

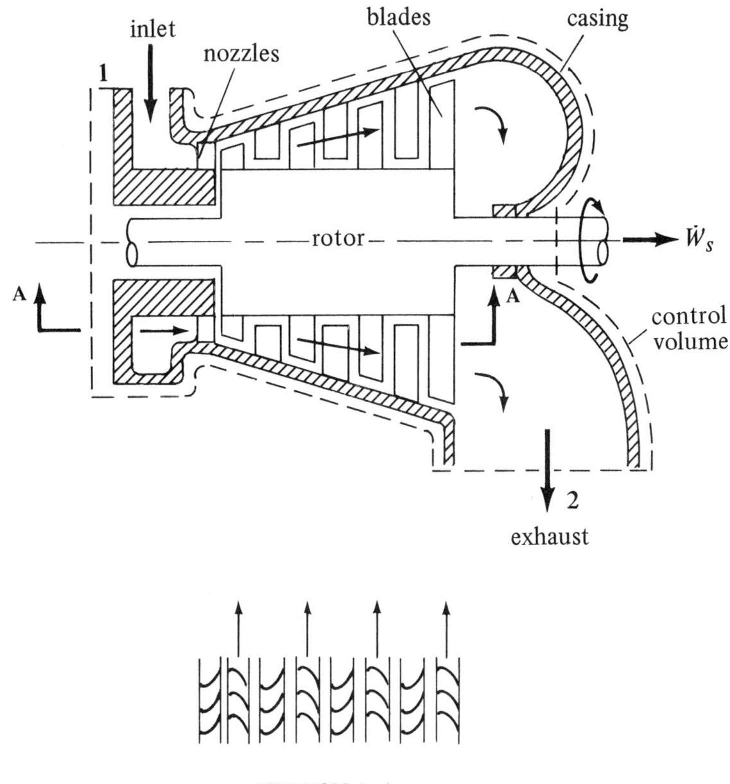

Figure 11E.2 Four-stage steam turbine

From equation (11.3), where kinetic energy and changes in height are neglected

$$\left(\frac{\dot{W}_{\text{shaft}}}{\dot{m}}\right)_{\text{rev}} = h_1 - h_{2r}$$

$$\left(\frac{\dot{W}_{\text{shaft}}}{\dot{m}}\right)_{\text{rev}} = 2885.5 - 2398.4 = 487.1 \text{ kJ/kg}$$

From the definition of turbine efficiency, equation (11.12),

$$\left(\frac{\dot{W}_{\text{shaft}}}{\dot{m}}\right)_{\text{actual}} = \eta_+ \left(\frac{\dot{W}_{\text{shaft}}}{\dot{m}}\right)_{\text{rev}} = 0.85(487.1) = 414.0 \text{ kJ/kg}$$

The mass flow rate for 1 MW is

$$\dot{m} = \frac{\dot{W}_{\text{shaft}}}{(\dot{W}_{\text{shaft}}/\dot{m})} = \frac{1000 \text{ kJ/s}}{414.0 \text{ kJ/kg}} = 4.83 \text{ kg/s}$$

(b) The quality of the steam at the turbine exit, state 2, is calculated from the enthalpy, h_2. From equation (11.3)

$$h_2 = h_1 - \left(\frac{\dot{W}_{\text{shaft}}}{\dot{m}}\right)_{\text{actual}}$$

$$h_2 = 2885.5 - 414.0 = 2471.5 \text{ kJ/kg}$$

$$x_2 = \frac{h_2 - h_f}{h_{fg}} = \frac{2471.5 - 504.70}{2202.0} = 0.8932$$

(c) To calculate the entropy generated in the turbine we need the entropy at state 2.

$$s_2 = s_f + x_2 s_{fg} = 1.5301 + 0.8932(5.5970) = 6.5292 \text{ kJ/kgK}$$

The rate of entropy generation is

$$\dot{S}_{\text{gen}} = \dot{m}(s_2 - s_1) = 4.83(6.5292 - 6.3435)$$

$$\dot{S}_{\text{gen}} = 0.897 \text{ kJ/Ks}$$

11.3 Nozzles and Diffusers[2]

The operation of many engineering systems requires a fluid stream of high kinetic energy. For example, the power output of a turbine is due to internal forces produced by the deflection of high velocity fluid streams. These high velocities can be achieved by accelerating the fluid stream in a device called a nozzle.

[2] For a more comprehensive treatment of the thermodynamics and fluid mechanics of nozzles and diffusers, the reader should consult a specialized treatment such as Shapiro, A. H., *The Dynamics and Thermodynamics of Compressible Fluid Flow*. Vols. I and II, Ronald Press, N. Y., 1954.

11.3.1 Nozzles

A typical nozzle with a single inlet and a single outlet is shown in Figure 11.5. The fluid enters the nozzle with a low velocity at a specified temperature and pressure, and leaves the nozzle with a high velocity at a lower pressure. The pressure difference imposed across the nozzle accelerates the fluid. The mass flow rate is related to the nozzle cross sectional area by the continuity equation. The distribution of fluid velocity and state along the length of the nozzle is determined by its shape.

The fluid velocity in the nozzle is high so the fluid does not attain thermal equilibrium with the nozzle walls. In addition, the surface area of contact between the fluid and the nozzle walls is relatively small. As a result, the heat transfer experienced by the fluid is negligibly small relative to the change in kinetic energy of the fluid stream, and the flow through the nozzle can be modeled as adiabatic. Also, the size of a typical nozzle is such that regardless of its orientation, the changes in the gravitational potential energy are negligible.

Since the flow through the nozzle is steady and adiabatic, and there is no shaft work transfer for a control surface coincident with the nozzle walls, the first law of thermodynamics, equation (10.51), applied to a control volume bounded by the inlet port and the outlet port becomes

$$\dot{m}\left(\frac{\vartheta_{\text{out}}^2 - \vartheta_{\text{in}}^2}{2}\right) = \dot{m}(h_{\text{in}} - h_{\text{out}}) \qquad \textbf{(11.30)}$$

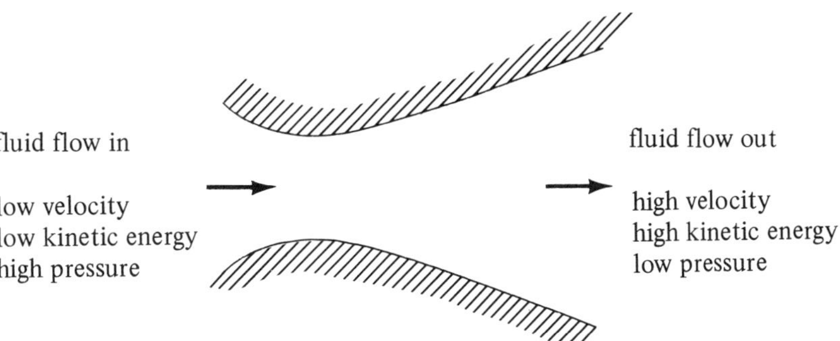

Figure 11.5 Converging-diverging nozzle

For the nozzle in which unit mass flow rate experiences an infinitesimal change of state, the first law of thermodynamics, equation (11.1), becomes

$$\vartheta \, d\vartheta = -dh \tag{11.31}$$

Although equation (11.30) describes the performance of most nozzles in steady flow operation, one must always be certain that in a given situation all of the underlying assumptions of its derivation are satisfied.

Physically, equation (11.30) requires that the increase in the kinetic energy of the fluid stream is produced at the expense of the enthalpy of the fluid. From the definition of enthalpy, equation (11.30) can also be written

$$\frac{\vartheta_{out}^2 - \vartheta_{in}^2}{2} = (u_{in} - u_{out}) + (P_{in}v_{in} - P_{out}v_{out}) \tag{11.32}$$

or for a control volume of infinitesimal extent in the direction of flow

$$\vartheta \, d\vartheta = -du - d(Pv) \tag{11.33}$$

That is, any increase in the kinetic energy of the fluid may be the result of a decrease in the internal energy of the fluid as well as the net flow work.

As in the case of the shaft work machine, the evaluation of the performance of a nozzle is especially simple in the limiting case of reversible adiabatic operation. Since in this case the entropy of the fluid remains constant as it passes through the machine, it is necessary to specify only the inlet state and the outlet pressure in order to fix the states throughout the nozzle. The inlet state is normally specified by the pressure, temperature, and velocity. The pressure and the entropy (which is the same as the inlet entropy) then completely determine the exit state.

Consider for example the nozzle described by equation (11.33). For the reversible adiabatic case, the enthalpy is related to the pressure and volume by equation (11.5). Thus, equation (11.31) becomes

$$(\vartheta \, d\vartheta)_{\substack{\text{reversible}\\ \text{adiabatic}}} = -v \, dP \tag{11.34}$$

Integrating equation (11.34), we obtain

$$\dot{m}\left(\frac{\vartheta_{out}^2 - \vartheta_{in}^2}{2}\right)_{\substack{\text{reversible}\\ \text{adiabatic}}} = -m\left[\int_{P_{in}}^{P_{out}} v\, dP\right]_{s=s_{in}} \tag{11.35}$$

where the integral on the right side is evaluated along the path of constant entropy. For a nozzle, the value of the integral is clearly less than zero. The physical interpretation of equations (11.34) and (11.35) is that the change in the kinetic energy of each fluid particle at each point as it passes through the nozzle is produced only by the net normal pressure forces on the particle at that point. In addition, the local change in kinetic energy is proportional to the local specific volume of the fluid. This implies that for reversible fluid acceleration, the fluid shear stress must be zero everywhere throughout the fluid. The requirement of zero shear stress for reversible flow can be further verified by deriving equation (11.35) by integrating the momentum equation (with no shear stress) along the stream from inlet to outlet.[3] As a result, the irreversible nozzle produces a smaller increase in kinetic energy of the fluid than the reversible nozzle (with the same inlet state and outlet pressure).

This result may be shown from Figure 11.6 and equations (9.33) and (11.30). In Figure 11.6 $(s_{out})_{actual}$ is greater than $(s_{out})_{rev}$ and $(\partial h/\partial s)_P > 0$ by equation (9.33). Since the decrease in h is smaller in the actual nozzle, the increase in kinetic energy is smaller than in the reversible nozzle.

$$(h_{in} - h_{out})_{actual} = (h_{in} - h_{out})_{rev} - \left[\int_{s_{in}}^{s_{out}} \left(\frac{\partial h}{\partial s}\right)_P ds\right]_{P=P_{out}} \tag{11.36}$$

$$\frac{1}{2}(\vartheta_{out}^2 - \vartheta_{in}^2)_{actual} = \frac{1}{2}(\vartheta_{out}^2 - \vartheta_{in}^2)_{rev} - \left[\int_{s_{in}}^{s_{out}} T\, ds\right]_{P=P_{out}} \tag{11.37}$$

The origin of the irreversibility is the shear stress in the fluid which results from the fluid viscosity and the relative velocity between the fluid and the wall. In a nozzle with smooth gradual contours, the effects of the fluid shear stresses are usually confined to these boundary layers near the walls. (The flow of a very

[3] Hunsaker, J. C., and Rightmire, B. G., *Engineering Applications of Fluid Mechanics*, McGraw-Hill, N. Y., 1947, Chapters 4, 5, and 6.

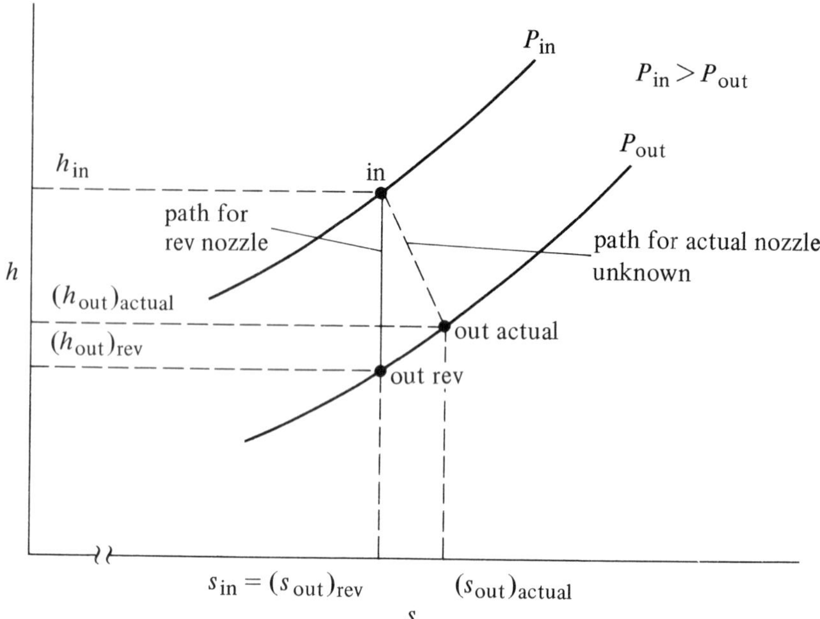

Figure 11.6 States of an ideal gas flowing through a nozzle

viscous fluid is of course an exception.) As a result, the major part of the stream outside these boundary layers is accelerated essentially reversibly.

We can define a nozzle efficiency that will provide us with an indication of the magnitude of the irreversibility of an irreversible (actual) nozzle. This efficiency is simply the ratio of the actual increase in kinetic energy of the fluid to the increase in kinetic energy that would have resulted from reversible adiabatic operation between the same inlet state and the same outlet pressure. Thus,

$$\eta_N = \frac{(\vartheta_{out}^2 - \vartheta_{in}^2)_{actual}}{(\vartheta_{out}^2 - \vartheta_{in}^2)_{rev}} \tag{11.38}$$

11.3.2 Nozzles Accelerating an Ideal Gas

If the fluid being accelerated can be modeled as an ideal gas with constant specific heats, equation (11.30) can be written

$$\dot{m}\left(\frac{\vartheta_{out}^2 - \vartheta_{in}^2}{2}\right)_{\substack{ideal\\gas}} = -\dot{m}c_P(T_{out} - T_{in}) \tag{11.39}$$

By making use of the relation between the specific heats and the gas constant, equation (9.114), and the ideal gas constitutive relation, we can write equation (11.39) in the form

$$\dot{m}\left(\frac{\vartheta_{out}^2 - \vartheta_{in}^2}{2}\right)_{\substack{\text{ideal}\\ \text{gas}}} = -\frac{\dot{m}\gamma}{\gamma - 1}(P_{out}v_{out} - P_{in}v_{in}) \quad \textbf{(11.40)}$$

Equation (11.40) shows that for an ideal gas with constant specific heats, the increase in kinetic energy of the fluid stream depends upon the nature of the expansion process (how v_{out} depends upon P_{out}) and upon how strongly the fluid is coupled in the thermodynamic sense (the value of γ).

For example, if the acceleration of the fluid through the nozzle is reversible as well as adiabatic, we can make use of equation (11.18) for the ideal gas. Then equation (11.40) becomes

$$\dot{m}\left(\frac{\vartheta_{out}^2 - \vartheta_{in}^2}{2}\right)_{\substack{\text{reversible}\\ \text{adiabatic}\\ \text{ideal gas}}} = -\frac{\dot{m}\gamma}{\gamma - 1}P_{in}v_{in}\left[\left(\frac{P_{out}}{P_{in}}\right)^{(\gamma-1)/\gamma} - 1\right] \quad \textbf{(11.41)}$$

Thus for a particular ideal gas, the increase in kinetic energy in a reversible adiabatic nozzle is uniquely determined by the inlet state and the outlet pressure. Note that by means of the ideal gas constitutive relation, we can write equation (11.41) in the form

$$\dot{m}\left(\frac{\vartheta_{out}^2 - \vartheta_{in}^2}{2}\right)_{\substack{\text{reversible}\\ \text{adiabatic}\\ \text{ideal gas}}} = -\frac{\dot{m}\gamma}{\gamma - 1}RT_{in}\left[\left(\frac{P_{out}}{P_{in}}\right)^{(\gamma-1)/\gamma} - 1\right] \quad \textbf{(11.42)}$$

Thus for a fixed inlet temperature and pressure and a fixed outlet pressure, the ideal gas with the largest value of γ and R will provide the largest increase in kinetic energy. From Chapter 8 it is clear that a monatomic gas with a low molecular weight provides the desired combination. It is for this reason that helium is often used as the working fluid in a high velocity wind tunnel. (Actually hydrogen would be a better choice because of its extremely low molecular weight; however, helium is inert and therefore much safer to work with than the potentially explosive hydrogen.)

If the nozzle operates in a adiabatic but irreversible manner, the increase in kinetic energy of the ideal gas will be less than in the reversible case as shown in equation (11.37). For the ideal gas, the nozzle efficiency, equation (11.38) becomes

$$(\eta_N)_{\substack{\text{ideal}\\ \text{gas}}} = \frac{(T_{out} - T_{in})_{actual}}{(T_{out} - T_{in})_{rev}} \quad \textbf{(11.43)}$$

when combined with equation (11.39). Clearly the outlet temperature for reversible operation is lower than the outlet temperature for irreversible operation. Since $T_{out} < T_{in}$ regardless of whether the process is reversible or irreversible, $\eta_N < 1$.

The locus of states of the ideal gas as it passes through the nozzle is shown in Figure 11.6 in terms of the enthalpy and entropy. Because of the relationship between temperature and enthalpy in the ideal gas, the locus of states shown in Figure 11.6 would have exactly the same appearance in terms of temperature and entropy. Geometrically the nozzle efficiency of equation (11.43) represents the ratio of the vertical distances between the inlet and outlet states for reversible and irreversible operation as shown in Figure 11.6.

11.3.3 Nozzles Accelerating an Incompressible Fluid

If the fluid flowing through the nozzle can be modeled as an incompressible fluid, the situation is somewhat different than previously described. The difference arises because the thermal and mechanical aspects of an incompressible fluid are not coupled as they are in a gas. In fact, the temperature and pressure are completely uncoupled so the enthalpy can be separated into a thermal energy and an uncoupled mechanical property, Pv, in the manner of equation (11.24). Then for the incompressible fluid, equation (11.30) can be written

$$\dot{m}\left(\frac{\vartheta_{out}^2 - \vartheta_{in}^2}{2}\right)_{incompressible} = \dot{m}[(u_{in} - u_{out}) + v(P_{in} - P_{out})] \quad \textbf{(11.44)}$$

where the specific volume v is constant. According to equation (9.93), the internal energy depends on temperature only.

$$\dot{m}\left(\frac{\vartheta_{out}^2 - \vartheta_{in}^2}{2}\right)_{incompressible} = \dot{m}c(T_{in} - T_{out}) + \dot{m}v(P_{in} - P_{out}) \quad \textbf{(11.45)}$$

The change in entropy experienced by the fluid is given by equation (11.26). Since in a reversible adiabatic expansion the entropy is constant, equation (11.26) requires that the temperature is also constant. For this situation, equation (11.45) reduces to

$$\dot{m}\left(\frac{\vartheta_{out}^2 - \vartheta_{in}^2}{2}\right)_{\substack{\text{reversible}\\\text{adiabatic}\\\text{incompressible}}} = \dot{m}v(P_{in} - P_{out}) \tag{11.46}$$

Note that since the specific volume is constant, equation (11.46) could have been obtained directly from equation (11.34) by means of a simple integration. In any event, equation (11.46) again indicates that in the reversible adiabatic nozzle the fluid is accelerated and increases in kinetic energy solely as a result of the net normal pressure forces on the fluid at the control surface (no shear stresses).

If in equation (11.46) we were to replace the specific volume with the density and were to consider unit flow rate, equation (11.46) becomes the familiar Bernoulli equation of incompressible fluid mechanics.

$$\frac{\vartheta_{out}^2 - \vartheta_{in}^2}{2} = \frac{P_{in}}{\rho} - \frac{P_{out}}{\rho} \tag{11.47}$$

For a flow in which the acceleration process is irreversible and adiabatic, the second law of thermodynamics requires that the entropy should increase. Since the inlet state and outlet pressure are externally imposed, equation (11.45) shows that the increase in temperature and the associated increase in internal energy can occur only by dissipating uncoupled mechanical energy to uncoupled thermal energy. Thus, as shown in equation (11.37), the irreversible adiabatic acceleration of an incompressible fluid in a nozzle does not, for a given inlet state and outlet pressure, produce as large an increase in kinetic energy as a reversible adiabatic expansion. This result becomes more obvious in the calculation of the nozzle efficiency. For the incompressible fluid, equation (11.38) reduces to

$$(\eta_N)_{\text{incompressible}} = 1 - \frac{c(T_{out} - T_{in})}{v(P_{in} - P_{out})} \tag{11.48}$$

Physically the second term on the right side of equation (11.48) represents the fraction of potentially available kinetic energy dissipated into internal energy through the action of irreversible processes.

The locus of states for the incompressible fluid flowing through the nozzle is shown in Figure 11.7. The nozzle efficiency has the usual geometric interpretation on this diagram.

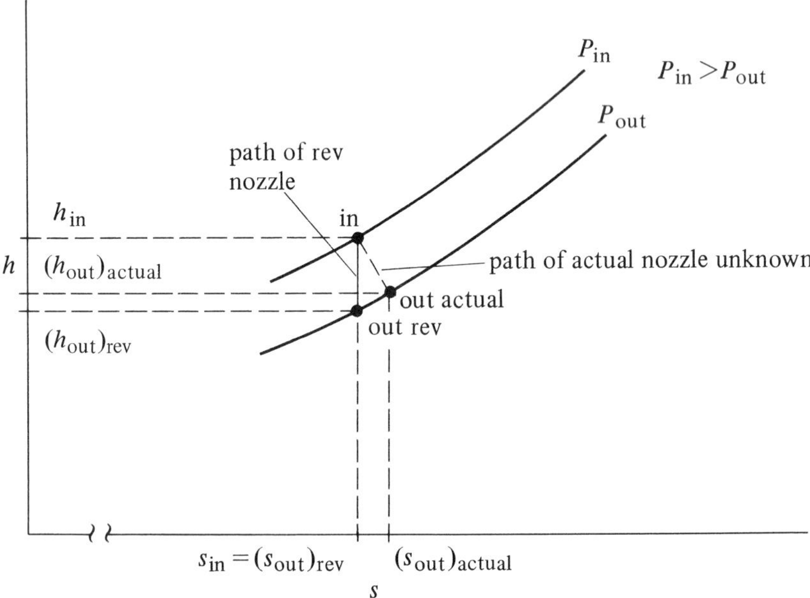

Figure 11.7 States of an incompressible fluid flowing through a nozzle

11E.3 Sample Problem: The water in a fire hose flows steadily at $P = 10^6$ N/m^2 and $T = 25$ C, and approaches the nozzle with negligible kinetic energy. The stream leaving the nozzle is at atmospheric pressure, $P_2 = 10^5$ N/m^2. The nozzle efficiency is 0.95.

(a) What is the velocity of the water leaving the nozzle?

(b) What is the temperature change of the water as it passes through the nozzle?

(c) What is the maximum height of fire that could be reached by the stream from the fire hose?

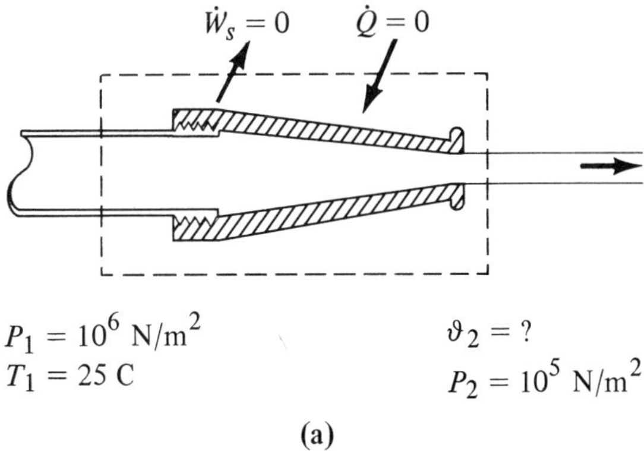

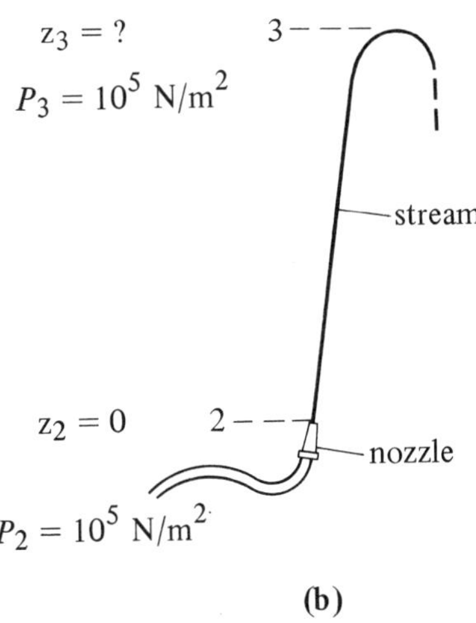

Figure 11E.3 Fire nozzle

Solution: (a) The first law applied to the control volume shown in Figure 11E.3 (a) gives the exit velocity from a reversible nozzle with the same conditions as the actual nozzle.

$$0 = \dot{m}(h_1) - \dot{m}\left(h_{2r} + \frac{\vartheta_{2r}^2}{2}\right)$$

where $\vartheta_1^2/2$ has been neglected, the nozzle has been modeled as adiabatic, and the shear work transfer is zero. Changes in height are also neglected for this part of the problem.

$$\frac{\vartheta_{2r}^2}{2} = h_1 - h_{2r}$$

For the reversible nozzle, $s_{2r} = s_1$ since the nozzle is adiabatic. If we model the water as an incompressible fluid, then $T_{2r} = T_1$ since the entropy of an incompressible fluid depends only on temperature (equation (11.26) or (9.96)).For an incompressible fluid $h = cT + Pv$. Thus for water with $v = 10^{-3}$ m^3/kg,

$$\frac{\vartheta_{2r}^2}{2} = c(T_1 - T_{2r}) + v(P_1 - P_2)$$

$$\vartheta_{2r}^2 = 2v(P_1 - P_{2r}) = 2(0.001)(10^6 - 10^5)$$

$$\vartheta_{2r}^2 = 1800 \text{ Nm/kg} = 1800 \text{ m}^2/\text{s}^2$$

The actual leaving kinetic energy is determined from the definition of nozzle efficiency, equation (11.38), with $\vartheta_1^2 = 0$.

$$\vartheta_2^2 = \eta_N \vartheta_{2r}^2 = 0.95(1800) = 1710 \text{ m}^2/\text{s}^2$$

$$\vartheta_2 = 41.35 \text{ m/s}$$

(b) The temperature change is determined by applying the first law to the actual nozzle.

$$0 = \dot{m}h_1 - \dot{m}\left(h_2 + \frac{\vartheta_2^2}{2}\right)$$

$$h_1 - h_2 = \frac{\vartheta_2^2}{2}$$

$$c(T_1 - T_2) + v(P_1 - P_2) = \frac{\vartheta_2^2}{2}$$

$$T_2 - T_1 = \frac{v}{c}(P_1 - P_2) - \frac{\vartheta_2^2}{2c}$$

$$T_2 - T_1 = \frac{0.001}{4180}(10^6 - 10^5) - \frac{1710}{2(4180)}$$

where $c = 4180$ J/kgK for water.

$$T_2 - T_1 = \frac{45}{4180} = 0.011 \text{ C}$$

(c) The maximum height is determined from the first law applied to the reversible constant pressure deceleration of the free stream from the nozzle exit (state 2) to the maximum height where the velocity is zero, state 3 in Figure 11E.3 (b). For a reversible adiabatic deceleration at constant pressure, the first law becomes

$$0 = \dot{m}\left(h_2 + \frac{\vartheta_2^2}{2} + gz_2\right) - \dot{m}\left(h_3 + \frac{\vartheta_3^2}{2} + gz_3\right)$$

Since $s_2 = s_3$, $T_2 = T_3$. With $P_2 = P_3$ we have $h_3 = h_2$ since h is fixed by T and P.

$$0 = \frac{\vartheta_2^2}{2} - gz_3$$

$$z_3 = \frac{\vartheta_2^2}{2g} = \frac{1710}{2(9.80)} = 87.2 \text{ m}$$

11.3.4 Nozzles Accelerating a Pure Substance

The behavior of a pure substance when accelerated in a nozzle is between the two extremes represented by the incompressible fluid and the ideal gas models. The loci of states for a pure substance flowing through nozzles with various inlet states are shown in Figure 11.8.

11.3.5 Diffusers

In some engineering systems, it is desirable to decelerate a fluid stream from a high velocity to a low velocity. This situation occurs at the inlet to a jet aircraft engine. Here it is necessary to decelerate the air entering the engine so that it can be efficiently compressed in the engine compressor. The device in which this deceleration is effected is called a diffuser.

The operation of the diffuser is essentially the inverse of the operation of the nozzle. The fluid is decelerated in the diffuser by increasing the pressure acting on the fluid in the direction of motion. As a result, some of the kinetic energy of bulk motion is converted into enthalpy of the flowing fluid. Thus equation

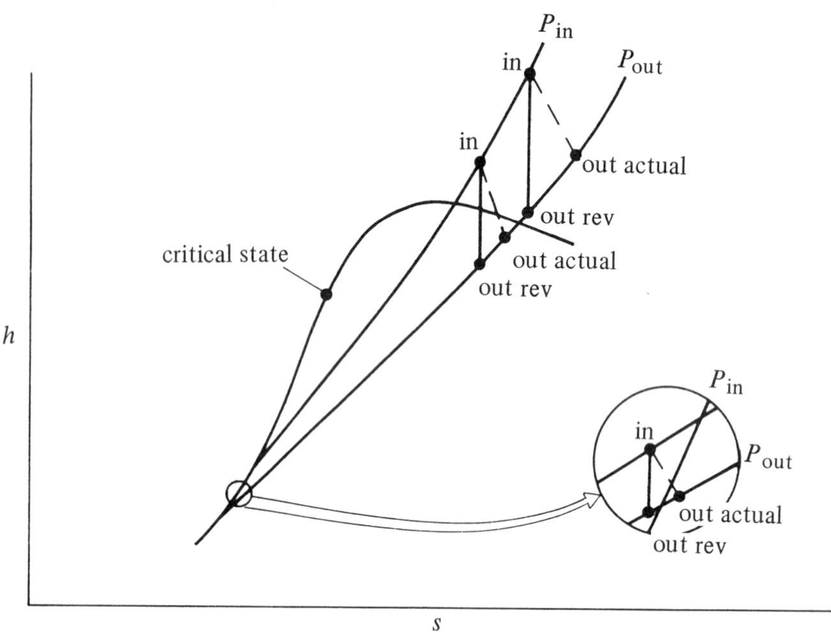

Figure 11.8(a) States of a pure substance flowing through a nozzle

(11.30) or (11.31) describes the operation of a diffuser as well as that of the nozzle. In fact, the reversible adiabatic diffuser is exactly the inverse process of the reversible adiabatic nozzle and the same relations apply to either. In practice, the difference between the nozzle and the diffuser results from the irreversibility of the processes, especially in the magnitude of the irreversibility. Ultimately, the difference can be traced to the dynamical behavior of the fluid itself.

The acceleration of a fluid produced by decreasing the pressure in the direction of motion as in a nozzle is inherently stable. The fluid will accelerate smoothly through any reasonably smooth convergent flow passage in an almost reversible manner. The design of the supersonic (divergent) portion of a nozzle does, however, require the proper passage contour to obtain reasonably reversible flow.

We might then infer that by simply inverting the nozzle design, a diffuser should result in which it would be possible to increase the pressure and thereby decelerate the fluid almost re-

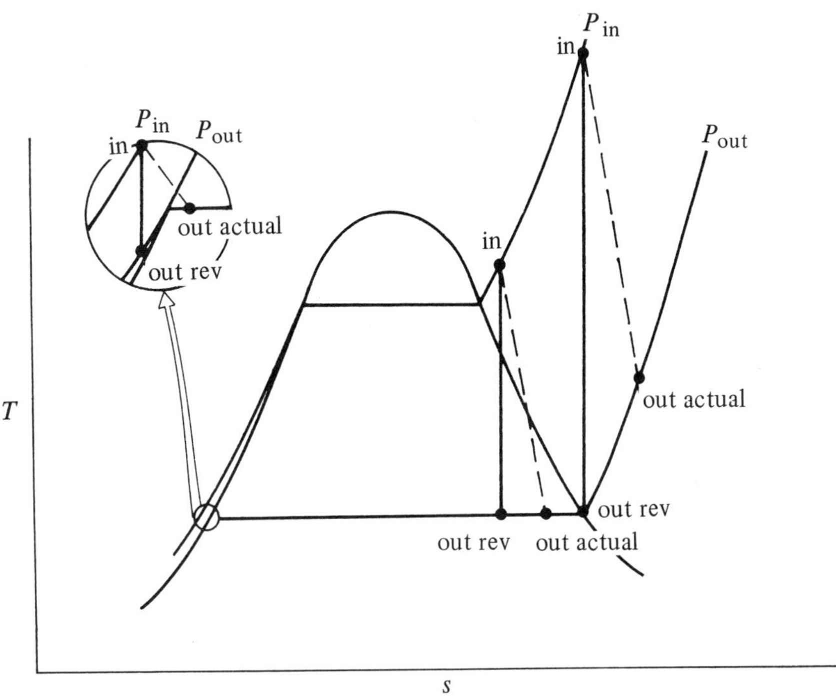

Figure 11.8(b) States of a pure substance flowing through a nozzle

versibly. Although certainly true in principle, the realization of this situation is in fact another matter entirely.

The difficulty is that the deceleration of the fluid is inherently unstable. The fluid is decelerated by an increase in the pressure forces which oppose its motion. Further, in the boundary layer near the wall of the diffuser, the fluid is being further decelerated by the action of viscous shear stresses. Unless careful consideration is given to the boundary layer of the flow, it is decelerated at a higher rate than the rest of the flow. The pressure forces and viscous shear stresses that oppose the fluid motion eventually overcome the forward motion in the boundary layer. The pressure gradient established by continued deceleration of the main flow then produces a region of backflow near the wall causing the main flow to separate from the wall. In the backflow, eddies are formed in which kinetic energy of the fluid motion is dissipated irreversibly resulting in an increase in entropy. Thus, the separated mode of diffuser operation is quite irreversible and does not realize as large a pressure increase for a given decrease in velocity as the unseparated mode.

Specifically, equation (11.30) shows that if the deceleration process is without work or heat transfer, the change in kinetic energy is the same for a given change in enthalpy regardless of the irreversibility of the process. However, note that equation (11.30) does not relate the change in kinetic energy directly to the pressure change. This relationship depends upon the irreversibility and the constitutive relations for the fluid. The effect of the irreversibility is shown in Figure 11.9. Since the slope of the constant pressure lines must be positive, an increase in entropy during the deceleration results in a lower final pressure for the same increase in enthalpy (or the same decrease in kinetic energy), Figure 11.9 (a). Stated in another way, the increase in entropy due to irreversibility requires a larger increase in en-

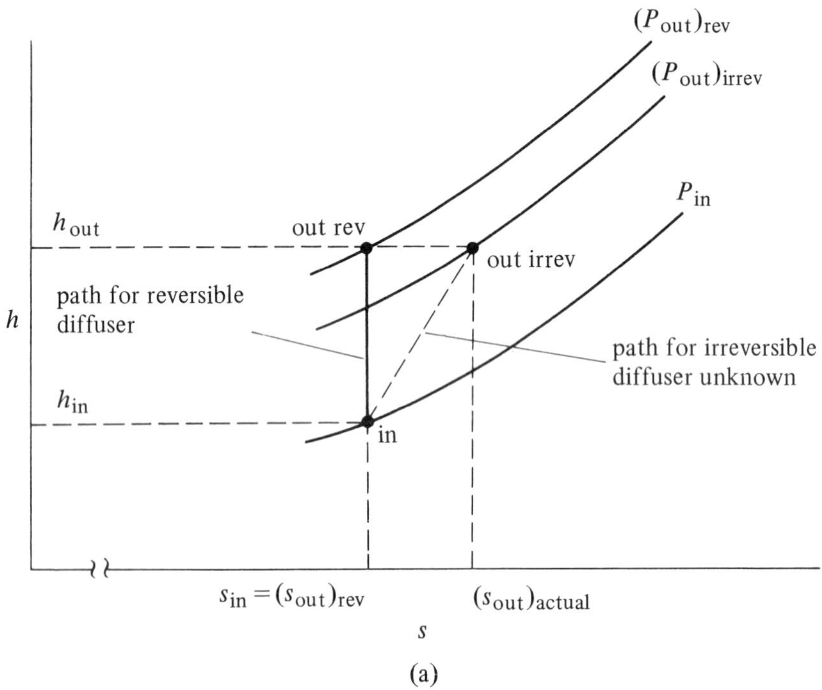

Figure 11.9(a) Pressure increase for an ideal gas flowing through a diffuser

thalpy (decrease in kinetic energy) to reach the same final pressure. The reason is the irreversible dissipation of some of the kinetic energy increased the entropy of the fluid as well as the enthalpy. If the objective is simply to eliminate the kinetic energy, the irreversible diffuser is as satisfactory as the reversible diffuser.

However, in many applications the objective is to utilize the kinetic energy to produce a low velocity, high pressure fluid stream from a high velocity, low pressure fluid stream. In effect, the diffuser acts as a ram compressor. The reversible diffuser utilizes more effectively the kinetic energy to create a pressure rise than the irreversible diffuser. The diffuser efficiency expresses the effectiveness of the diffuser in converting the kinetic energy into a pressure increase. The diffuser efficiency is the ratio of the kinetic energy decrease in an ideal reversible adiabatic diffuser to the kinetic energy decrease in an actual irrevers-

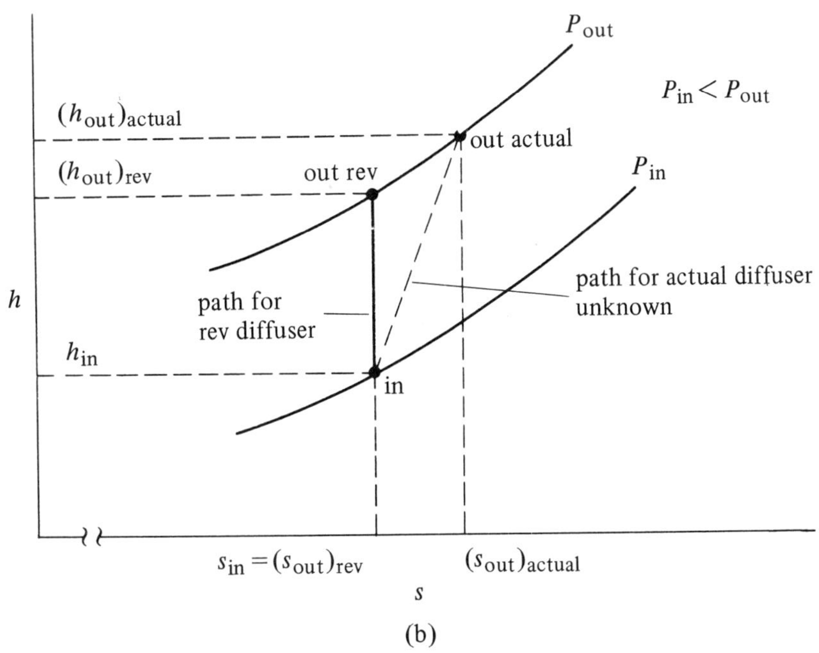

Figure 11.9(b) Kinetic energy decrease for an ideal gas flowing through a diffuser

ible diffuser where the two diffusers have the same inlet state and outlet pressure. Thus,

$$\eta_D = \frac{(\vartheta_{in}^2 - \vartheta_{out}^2)_{rev}}{(\vartheta_{in}^2 - \vartheta_{out}^2)_{actual}} \tag{11.49}$$

Substituting equation (11.30) into equation (11.49), we obtain

$$\eta_D = \frac{(h_{out} - h_{in})_{rev}}{(h_{out} - h_{in})_{actual}} \tag{11.50}$$

Note: Several definitions of diffuser efficiency other than equation (11.49) are currently in use. This must be kept in mind when reviewing the literature.

11.3.6 Diffusers Decelerating an Ideal Gas

The effect of the constitutive relations for the fluid on the performance of the diffuser can be illustrated by considering the two simple models—the ideal gas and the incompressible fluid. If the fluid can be modeled as an ideal gas, the change in kinetic energy can be determined from equation (11.40) in the general adiabatic case or from equation (11.41) in the reversible adiabatic case. For the irreversible case the diffuser efficiency, equation (11.50), reduces to

$$(\eta_D)_{\substack{ideal\\gas}} = \frac{(T_{out} - T_{in})_{ideal}}{(T_{out} - T_{in})_{actual}} \tag{11.51}$$

Figure 11.9 (a) illustrates the pressure increase that results from a given change in kinetic energy in a diffuser. Clearly $(P_{out})_{rev} > (P_{out})_{irrev}$. Figure 11.9 (b) illustrates the change in kinetic energy that results for a given pressure increase in the fluid for both reversible and irreversible deceleration. The diffuser efficiency can be interpreted geometrically as the ratio of the vertical distances between the inlet and outlet states on Figure 11.9 (b).

11.3.7 Diffusers Decelerating an Incompressible Fluid

For those cases in which the fluid can be modeled as an incompressible fluid, reversible diffuser operation can be described by equation (11.46). For the irreversible case, equation (11.45) must be used. From equation (11.45), it is apparent that in the irreversible case some of the mechanical kinetic energy is dissipated into uncoupled thermal energy. For this case the diffuser efficiency reduces to

$$(\eta_D)_{\text{incompressible}} = 1 - \frac{c(T_{\text{out}} - T_{\text{in}})}{c(T_{\text{out}} - T_{\text{in}}) + v(P_{\text{out}} - P_{\text{in}})} \tag{11.52}$$

Equation (11.52) has much the same interpretation as equation (11.48). The locus of states for the incompressible fluid flowing through the diffuser is shown in Figure 11.10.

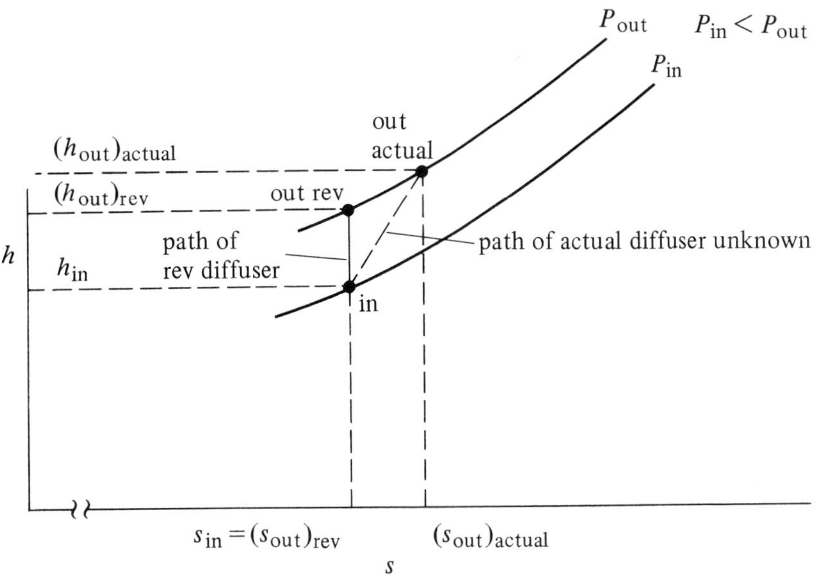

Figure 11.10 States of an incompressible fluid flowing through a diffuser

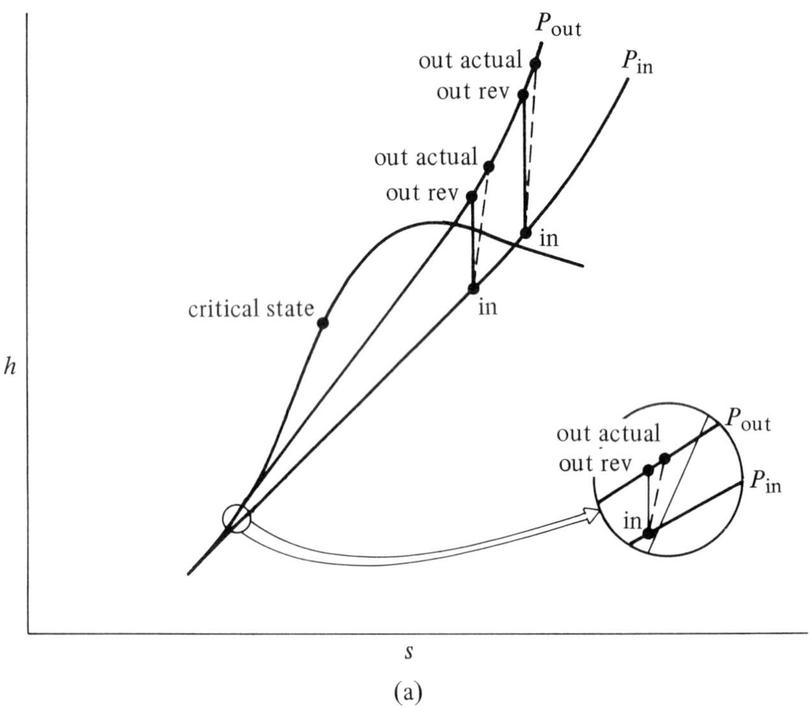

Figure 11.11(a) States of a pure substance flowing through a diffuser

11.3.8 Diffusers Decelerating a Pure Substance

The behavior of a pure substance when decelerated in a diffuser is between the two extremes represented by the incompressible fluid and the ideal gas models. The loci of states for a pure substance flowing through diffusers with various inlet states are shown in Figure 11.11.

11.4 Throttle

In the two previous sections, it was stated that the pressure difference (or equivalently the pressure ratio) across the device was externally imposed. The throttle is the most common system component for controlling the pressures of a flowing

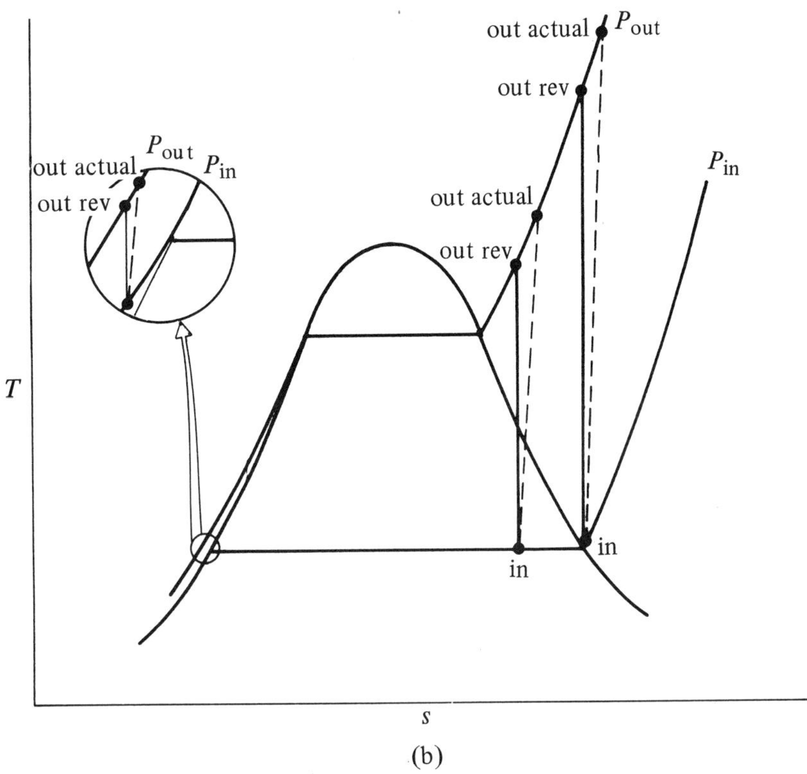

Figure 11.11(b) States of a pure substance flowing through a diffuser

stream. For specific applications the throttle is also known by various other names such as control valve, expansion valve, Joule-Thomson valve, porous plug, orifice, and flow resistance. Although the names are different, the principle of operation is the same.

In its simplest form, the throttle is nothing more than a port with an adjustable flow area which provides an adjustable resistance to the flow of fluid through the system. An increase in flow resistance requires a greater pressure difference across the throttle for a given flow rate. By adjusting the flow resistance of a throttle on the inlet of a device, we may adjust the inlet pressure while the supply pressure is fixed. Typically, the increase in flow resistance is produced by decreasing the cross sectional area of the flow at one section inside the throttle.

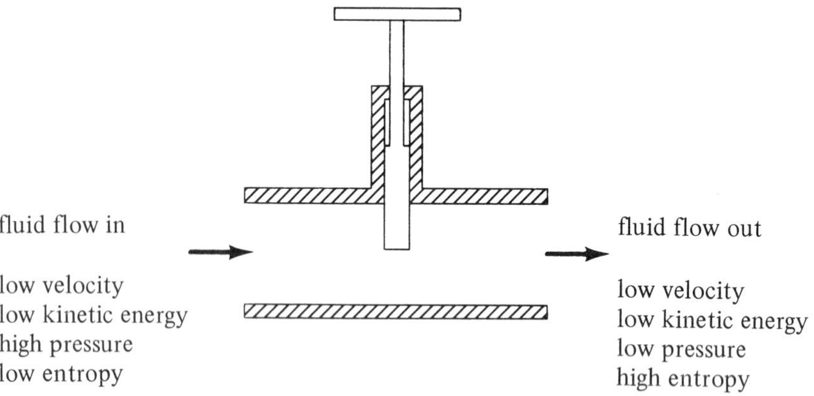

Figure 11.12 Throttle

A typical throttle installation is shown schematically in Figure 11.12. The residence time of the fluid within the throttle is insufficient to permit significant heat transfer. The throttle can then be adequately modeled as an adiabatic device. In principle, the throttle is a nozzle in series with a very poor diffuser. In the nozzle section, the flow is accelerated to some velocity while the pressure is decreased to a value P. In the diffuser section, the flow is decelerated to the original velocity ϑ_{in}, but because the diffuser section operates in the separated mode, the stream returns to the inlet enthalpy with only a small pressure rise. Essentially all of the flow irreversibility associated with the throttle is located in the diffuser section. By combining the governing equations for the nozzle and the diffuser, we obtain

$$\dot{m}\,[(h_{in} - h) + (h - h_{out})] = \dot{m}\left[\frac{\vartheta^2 - \vartheta_{in}^2}{2} + \frac{\vartheta_{out}^2 - \vartheta^2}{2}\right] \quad \textbf{(11.53)}$$

or

$$h_{in} - h_{out} = \frac{\vartheta_{out}^2 - \vartheta_{in}^2}{2} \quad \textbf{(11.54)}$$

In typical applications of the throttle, we locate the outlet port of the control volume far enough downstream so that the outlet velocity has essentially reached its final value. Typical values of the inlet and outlet velocities are such that the kinetic energies at the inlet and outlet ports are negligible (compared to the enthalpy). Thus,

$$\vartheta_{\text{in}}^2 \simeq \vartheta_{\text{out}}^2 \simeq 0 \qquad \textbf{(11.55)}$$

Then equation (11.54) becomes

$$h_{\text{out}} = h_{\text{in}} \qquad \textbf{(11.56)}$$

For the case of unit mass flow rate with an infinitesimal change of state, equation (11.1) reduces to

$$dh = 0 \qquad \textbf{(11.57)}$$

Since

$$dh = T\,ds + v\,dP \qquad \textbf{(11.58)}$$

it follows from equation (11.57) that

$$T\,ds = -v\,dP \qquad \textbf{(11.59)}$$

Thus in order for the pressure to decrease in the direction of flow, the entropy of the fluid must increase. Since the throttle is adiabatic, it follows from the second law of thermodynamics that the operation of the throttle must be irreversible if the throttle is to operate in the desired manner.

Note that because the throttle operates in an irreversible manner, the locus of states of the fluid as it passes through the throttle is unknown. That is, even though the inlet and outlet enthalpies of the fluid are identical, the path is not necessarily one of constant enthalpy. The irreversibility internal to the throttle makes it impossible to identify the intermediate states.

11.4.1 Throttling an Ideal Gas

If the fluid can be modeled as an ideal gas, equation (11.56) reduces to

$$T_{\text{out}} = T_{\text{in}} \qquad \textbf{(11.60)}$$

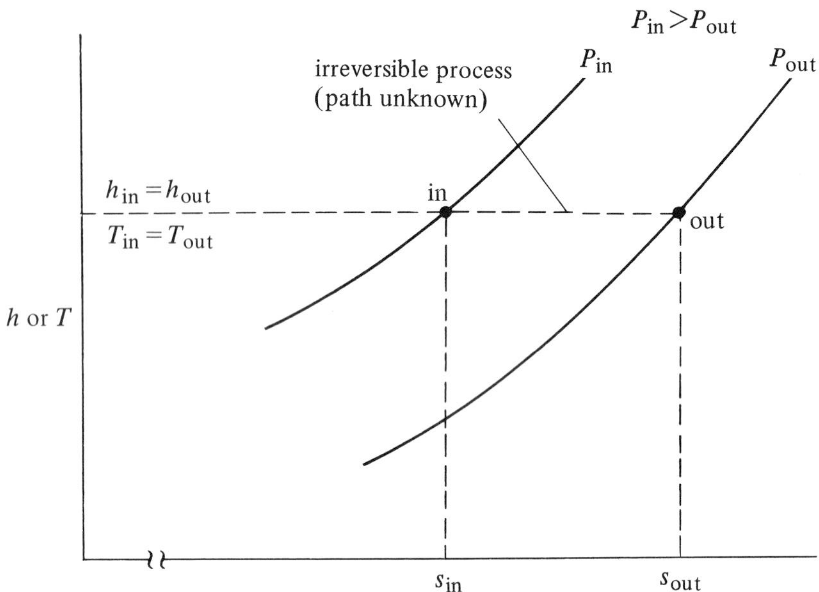

Figure 11.13 Inlet and outlet states for an ideal gas flowing through a throttle

The inlet and outlet states for the ideal gas flowing through the throttle are shown in Figure 11.13.

11.4.2 Throttling an Incompressible Fluid

If the fluid can be modeled as an incompressible fluid, equation (11.56) reduces to

$$u_{out} - u_{in} = v(P_{in} - P_{out}) \qquad \textbf{(11.61)}$$

or

$$c(T_{out} - T_{in}) = v(P_{in} - P_{out}) \qquad \textbf{(11.62)}$$

Since $P_{in} > P_{out}$, equation (11.62) shows that the temperature of the incompressible fluid increases as the fluid passes through the throttle. The source of this temperature increase is the irreversibility located in the diffuser section of the throttle. The inlet and outlet states for this case are shown in Figure 11.14.

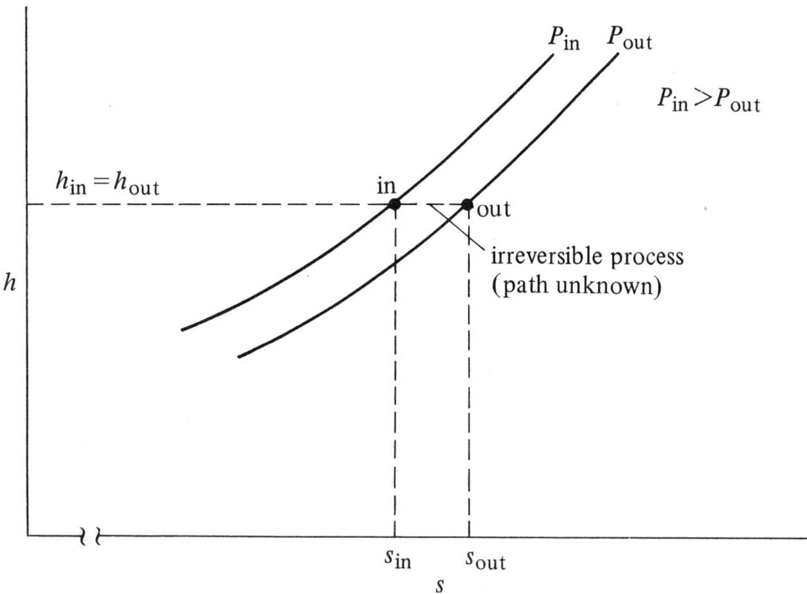

Figure 11.14 Inlet and outlet states for an incompressible fluid flowing through a throttle

11.4.3 Throttling a Pure Substance

In the case of a pure substance, the changes of state that may occur as the fluid passes through the throttle may be significantly different from either of the cases just discussed. The pure substance may either increase or decrease in temperature. If both the liquid and vapor phases are present, either at the inlet or at the outlet, the temperature of the pure substance will decrease. However, if only the vapor phase is present, at both inlet and outlet, the temperature may increase or decrease depending upon the inlet state. In Figure 11.15 we have plotted the locus of states of constant enthalpy as a function of temperature and pressure for the vapor phase of a pure substance. Also shown in Figure 11.15 is the locus of the maxima in the curves of constant enthalpy. This locus is called the inversion curve for the Joule-Thomson coefficient, which was defined by equation (9.62) as $\mu = (\partial T/\partial P)_h$. For states to the left of the inversion curve, μ is positive, while for states to the right of the inversion

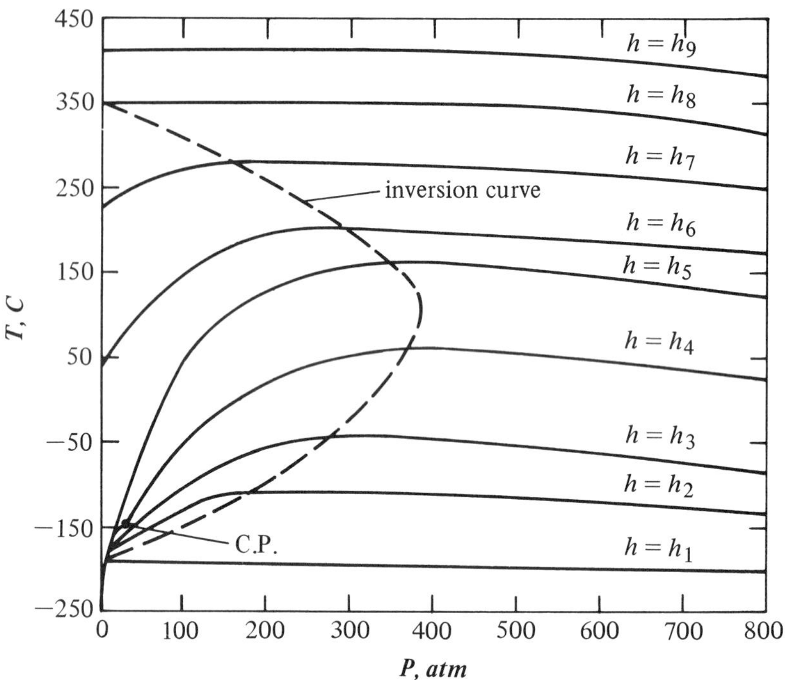

Figure 11.15 Inversion curve for the vapor phase of a pure substance (nitrogen)

curve, μ is negative. Thus, the inversion curve is the locus of states for which $\mu = 0$.

If both of the end states for a throttle are to the left of the inversion curve on Figure 11.15, the temperature must decrease as the fluid passes through the throttle. If both of the end states are on the right of the inversion curve in Figure 11.15, the temperature must increase. If the two end states are on opposite sides, the temperature may increase or decrease depending upon the exact states. Some typical throttling processes in the pure substance are shown in Figure 11.16.

In the region of positive μ, the internal energy of the pure substance depends strongly on the volume as well as the temperature. (The internal energy increases with volume at constant temperature.) This accounts for the behavior which is different from that of the ideal gas and the incompressible fluid which both have an internal energy which depends only upon T.

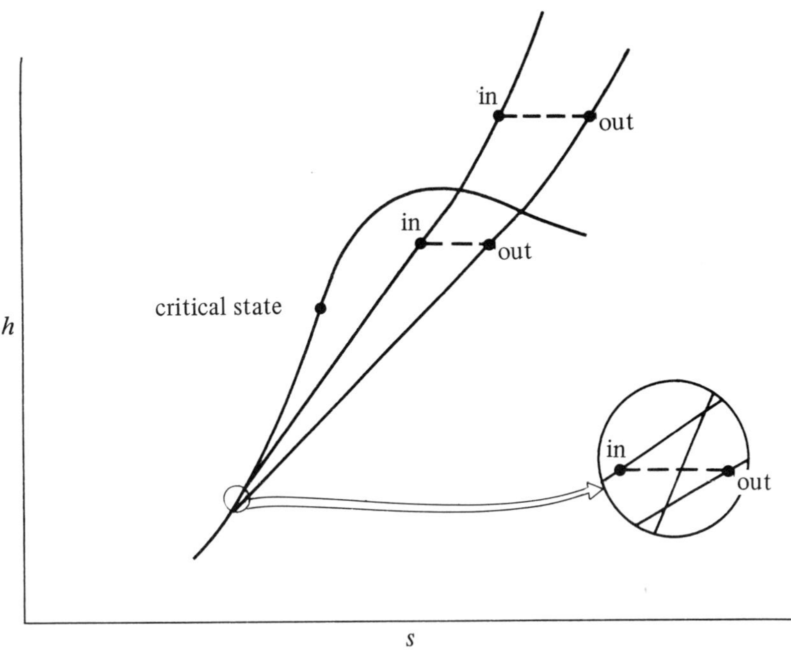

Figure 11.16(a) Throttling processes for the pure substance

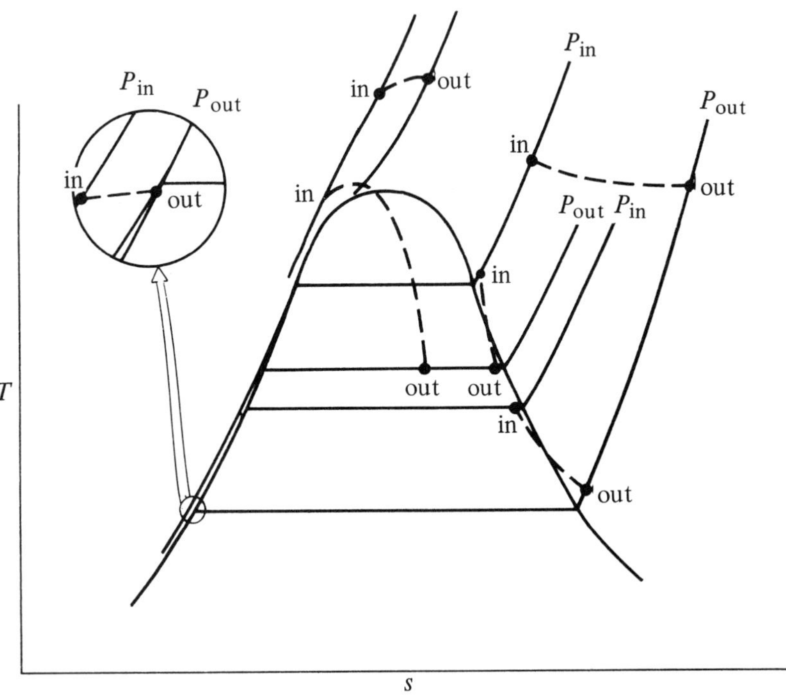

Figure 11.16(b) Throttling processes for the pure substance

11.5 Heat Exchangers[4]

All of the system components considered previously operate effectively in an adiabatic manner. Many engineering systems require heat transfer between a solid wall and a flowing stream for proper operation. Such a situation is shown schematically in Figure 11.17.

The apparatus designed for such heat transfer is called a heat exchanger. Heat exchangers usually serve one of two primary functions. They are to change the thermodynamic state of the fluid or to remove or add energy to the solid wall.

Examples of the first function are the heat exchangers (boiler and superheater) used in a steam power cycle to increase the temperature (and enthalpy) of the steam before it enters the turbine. The change of state of the steam produced by the heat transfer is very important in producing a large work transfer as the steam expands through the turbine. In a similar manner, the heat transfer to the propellant gas in the heat exchanger of a nuclear rocket is of utmost importance in obtaining a maximum acceleration of the gas in the nozzle and thereby achieving maximum thrust. Conversely, the cooling of a gas stream in a heat exchanger prior to compression can result in a significant reduc-

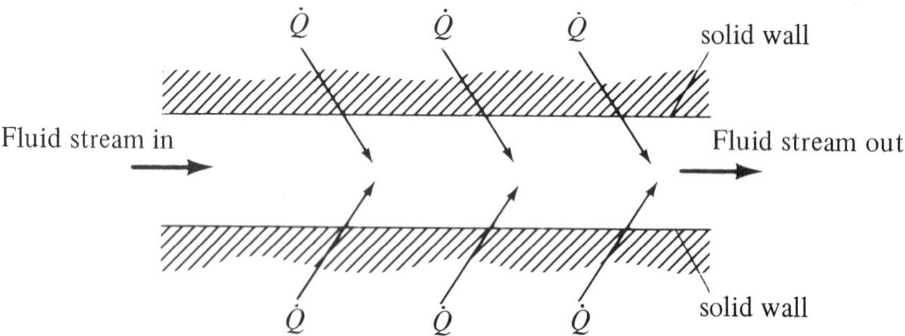

Figure 11.17 Heat transfer from solid wall to fluid stream

[4] For more detailed information than is presented here, the following references are suggested: (1) Rohsenow, W. M. and Choi, H. Y., *Heat, Mass, and Momentum Transfer*, Prentice-Hall, N. J., 1961. (2) Kays, W. M. and London, A. L., *Compact Heat Exchangers*, McGraw-Hill, N. Y., 1964.

tion in the work transfer required. Pre-cooling heat exchangers are also important in refrigeration cycles.

The second function of heat exchangers is to promote heat transfer to (or at times from) the solid wall of the heat exchanger to maintain a desired operating wall temperature. The cooling jacket or cooling fins of an internal combustion engine are a common example of this type of heat exchanger. Another example is the heat sink used for mounting power transistors. In reality the heat sink is a heat exchanger which promotes heat transfer from the transistor to the ambient air so that the dissipation (irreversibilities) in the transistor will not cause an excessive operating temperature. In large rotating electrical machines, the electrical windings are carefully designed to be heat exchangers for heat transfer to ambient air which is circulated by fans on the rotor. In fact, the power rating of the machine is determined by the effectiveness of the heat transfer from the windings and the maximum operating temperature for the insulation.

As an example of a situation in which both heat exchanger functions are equally important, consider the nuclear power reactor. The temperature of the nuclear fuel elements must be kept at a safe level while the thermodynamic state of the coolant (power cycle fluid) must be changed significantly by the heat transfer.

As is indicated by this brief discussion, the range of applications of heat exchangers is very wide, and heat exchangers are often known by other names which denote their application. These are terms such as boiler, evaporator, condenser, regenerator, recuperator, heater, cooler, water jacket, and heat sink.

In those heat exchangers in which the change of state of the fluid is of prime importance, it is usually desirable to effect the heat transfer in a reversible manner. Thus, the fluid must achieve thermal equilibrium with the heating or cooling element. In general, since the rate of heat transfer decreases with a decrease in temperature difference between the two media, the time required to achieve thermal equilibrium becomes inconveniently large. Therefore, thermal equilibrium can only be approached as a limit. The closeness of the approach to thermal equilibrium is a consequence of a suitable compromise between the fluid velocity, the temperature difference for heat transfer, and the surface area for heat transfer.

The compromise varies with the application depending upon the relative importance of weight, size, irreversibility, cost, and other factors. The fluid velocity is usually low so that the pressure difference due to fluid flow is negligible to the first order. Thus, the heat exchanger operates at essentially constant pressure.

As the fluid flows through the heat exchanger the change in kinetic energy is negligible to the first order. Also, the change in the gravitational potential energy is usually negligibly small compared to the heat transfer. The shear work transfer is zero for a control surface coincident with the walls. Thus, the first law is

$$\dot{Q} = \dot{m}(h_{out} - h_{in}) \tag{11.63}$$

For the case of unit mass flow rate and infinitesimal change of state, equation (11.1) reduces to

$$\frac{\delta\dot{Q}}{\dot{m}} = dh \tag{11.64}$$

It is interesting to note that equation (11.63) confirms that for a given heat transfer rate $\dot{Q}$, the fluid velocity should be as low as possible if a maximum increase in enthalpy is to be achieved. That is, if the fluid mass flow rate is increased while the heat transfer rate remains constant, the change in the enthalpy of the fluid as it passes through the heat exchanger decreases, and the outlet enthalpy approaches the inlet enthalpy. Thus, to use the heat exchanger to increase the enthalpy of the stream as effectively as possible, it is necessary to maintain a relatively low mass flow rate for a given heat transfer rate and surface area. These low mass flow rates can be achieved with low fluid velocities which insure that the pressure drop due to fluid friction is negligible.

When the temperature of the apparatus is of prime importance, the situation is different since the main objective is to establish heat transfer at a maximum rate through a given surface area. Under these circumstances, a large mass flow is used with a small change in the enthalpy of the fluid. In such cases, the pressure drop due to fluid friction is usually significantly larger than in the previous case.

11.5.1 Heat Transfer to an Ideal Gas

If the fluid can be modeled as an ideal gas with constant specific heats, equation (11.63) becomes

$$\dot{Q} = \dot{m}c_P(T_{out} - T_{in}) \tag{11.65}$$

Thus, if $T_{out} > T_{in}$, the heat transfer is positive (into the fluid from the walls). On the other hand, if $T_{out} < T_{in}$, the heat transfer is negative (out of the fluid and into the walls). Figure 11.18 shows the changes of state for constant pressure heat transfer to an ideal gas.

For the case of the ideal gas we can now examine the validity of the assumption of negligible change in kinetic energy associated with the change in density of the gas flowing at constant pressure in a tube of constant cross sectional area.

From the continuity equation, we have

$$(\rho A \vartheta_n)_{in} = (\rho A \vartheta_n)_{out} \tag{11.66}$$

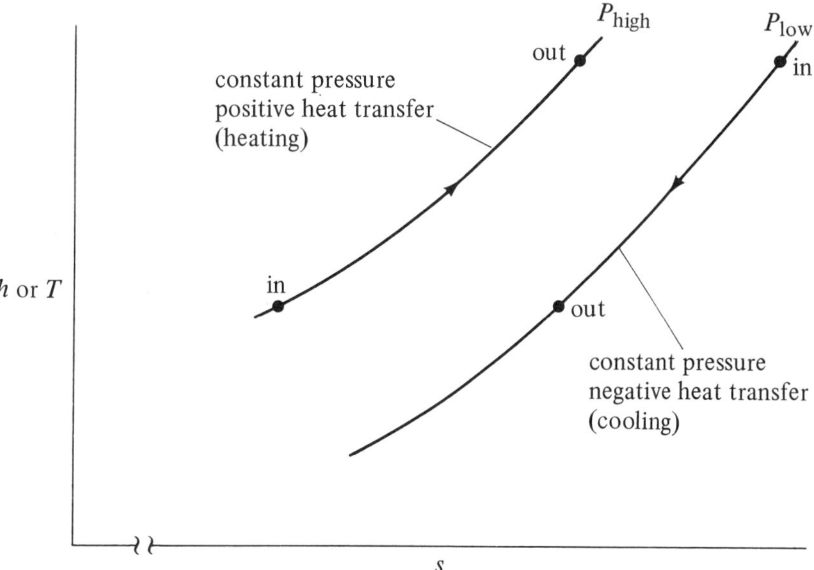

Figure 11.18 Heat transfer processes in constant pressure heat exchangers with an ideal gas flowing

But the area is constant and the flow is normal to this area so that

$$\rho_{\text{in}}\vartheta_{\text{in}} = \rho_{\text{out}}\vartheta_{\text{out}} \tag{11.67}$$

Substituting the ideal gas constitutive relation into equation (11.67), we obtain

$$\frac{P_{\text{in}}}{RT_{\text{in}}}\vartheta_{\text{in}} = \frac{P_{\text{out}}}{RT_{\text{out}}}\vartheta_{\text{out}} \tag{11.68}$$

Since the pressure is constant, equation (11.68) reduces to

$$\frac{\vartheta_{\text{in}}}{T_{\text{in}}} = \frac{\vartheta_{\text{out}}}{T_{\text{out}}} \tag{11.69}$$

while the change in kinetic energy becomes

$$\frac{\vartheta_{\text{in}}^2 - \vartheta_{\text{out}}^2}{2} = \frac{\vartheta_{\text{in}}^2}{2}\left(1 - \frac{\vartheta_{\text{out}}^2}{\vartheta_{\text{in}}^2}\right) \tag{11.70}$$

Combining equations (11.69) and (11.70), we obtain

$$\frac{\vartheta_{\text{in}}^2 - \vartheta_{\text{out}}^2}{2} = \frac{\vartheta_{\text{in}}^2}{2}\left[1 - \left(\frac{T_{\text{out}}}{T_{\text{in}}}\right)^2\right] \tag{11.71}$$

If we define a temperature difference along the stream to be

$$\Delta T_s = T_{\text{out}} - T_{\text{in}} \tag{11.72}$$

then equation (11.71) becomes

$$\frac{\vartheta_{\text{in}}^2 - \vartheta_{\text{out}}^2}{2} = \frac{\vartheta_{\text{in}}^2}{2}\left[1 - \left(1 + \frac{\Delta T_s}{T_{\text{in}}}\right)^2\right] \tag{11.73}$$

Although ΔT_s can be large, the ratio of this temperature difference to the themodynamic temperature is usually small. Further, the inlet velocity is also small so that the combination of terms appearing in equation (11.73) is negligibly small compared to the heat transfer. Thus, equation (11.65) is the simplest form of the first law of thermodynamics for a heat exchanger with an ideal gas as the fluid. However, in some situations the change in kinetic energy may be significant and equation (11.65) must be suitably modified.[5]

11.5.2 Heat Transfer to an Incompressible Fluid

If the fluid can be modeled as an incompressible fluid, equation (11.24) can be substituted for the enthalpy in equation (11.63). Then

$$\dot{Q} = \dot{m}c(T_{\text{out}} - T_{\text{in}}) + \dot{m}v(P_{\text{out}} - P_{\text{in}}) \qquad \textbf{(11.74)}$$

Since the pressure is usually constant, equation (11.74) reduces to

$$\dot{Q} = \dot{m}c(T_{\text{out}} - T_{\text{in}}) \qquad \textbf{(11.75)}$$

Note that in the case of the incompressible fluid, the temperature and pressure are uncoupled so that there are no changes in velocity due to the density change of the fluid (cf. equation (11.66)). Thus, equation (11.75) is the correct form of the first law of thermodynamics for this case.

The canonical relation requires for any fluid

$$dh = T\,ds + v\,dP \qquad \textbf{(11.76)}$$

Equations (11.64) and (11.76) show that for constant pressure heat transfer the entropy of the fluid will increase when the heat transfer is into the fluid stream and will decrease when the heat transfer is out of the fluid stream. Figures 11.18 and 11.19 show the entropy changes for constant pressure heat transfer to the ideal gas and to the incompressible fluid.

11.5.3 Two-Stream Heat Exchangers

Many practical heat exchangers are constructed for heat transfer between two fluid streams. The state of one stream is changed at

---

[5] See Kays, W. M. and London, A. L., *op. cit.*

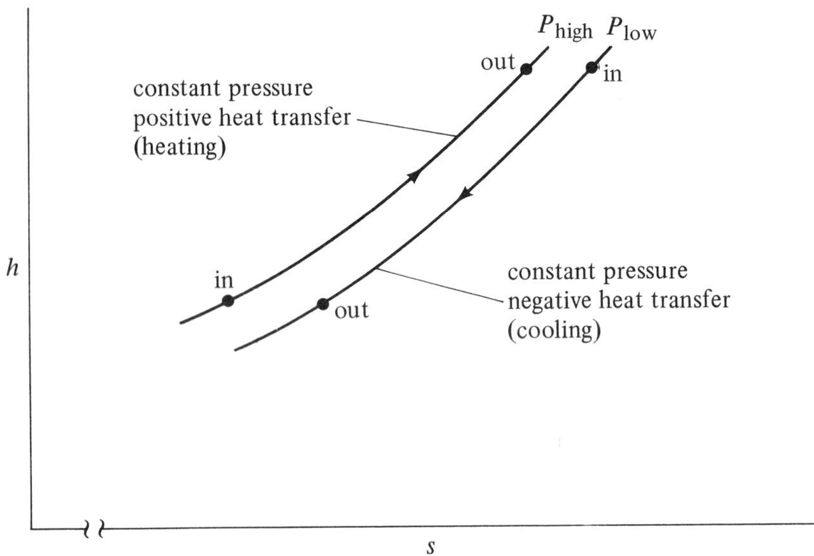

Figure 11.19 Heat transfer processes in constant pressure heat exchangers with an incompressible fluid flowing

the expense of changes in the second. Thus, the heat transfer which increases the enthalpy of the stream in Figure 11.17 would be provided by another fluid stream which decreases in enthalpy. The conduit wall across which the heat transfer takes place usually has a more complicated geometry than that shown in Figure 11.17. We analyze the heat exchanger as a "black box" with two entering fluid streams and two exiting fluid streams as in Figure 11.20. Since the heat transfer takes place between the two fluid streams and not with the environment, such a heat exchanger is commonly referred to as an *adiabatic heat exchanger*. Admittedly the nomenclature is somewhat confusing, but the term "adiabatic" refers to the interaction between the heat exchanger and the environment and not the internal interactions.

Then for an adiabatic heat exchanger with two fluid streams a and b, the first law of thermodynamics to the same order of approximation as equation (11.63) becomes

$$\dot{m}_a(h_a)_{\text{in}} + \dot{m}_b(h_b)_{\text{in}} = \dot{m}_a(h_a)_{\text{out}} + \dot{m}_b(h_b)_{\text{out}} \qquad \textbf{(11.77)}$$

or

$$\dot{m}_a[(h_a)_{\text{in}} - (h_a)_{\text{out}}] = \dot{m}_b[(h_b)_{\text{out}} - (h_b)_{\text{in}}] \qquad \textbf{(11.78)}$$

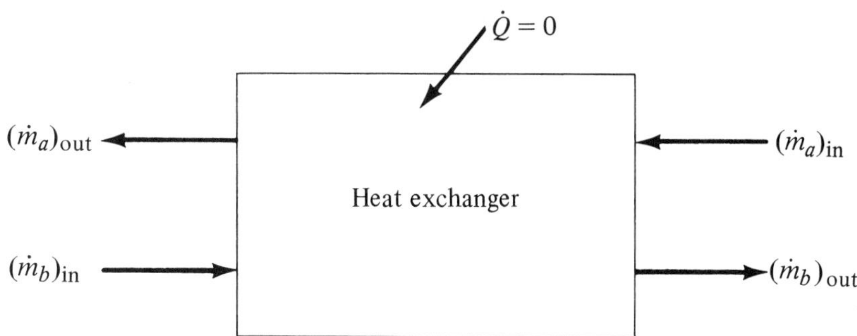

Figure 11.20 Adiabatic heat exchanger

To simplify analysis we consider simple models for the fluid properties, a simple model for the heat transfer rate, and co-linear flow geometries as shown in Figure 11.21.

In many applications the fluids can be modeled as having constant specific heats so that the enthalpy varies linearly with the temperature. Heat transfer to ideal gases and incompressible fluids are usually analyzed in this way. In this case equation (11.78) becomes

$$\begin{aligned}\dot{Q}_a &= \dot{m}_a c_{Pa}[(T_a)_{out} - (T_a)_{in}] \\ &= -\dot{m}_b c_{Pb}[(T_b)_{out} - (T_b)_{in}] = -\dot{Q}_b \end{aligned} \qquad \textbf{(11.79)}$$

Here $\dot{Q}_a$ and $\dot{Q}_b$ are the heat transfers for the two streams considered individually. The quantities $\dot{m}_a c_{Pa}$ and $\dot{m}_b c_{Pb}$ are often called the *capacity rates* for the two streams.

In a two-stream heat exchanger the extent of approach to thermal equilibrium depends upon the resistance to heat transfer between the streams. The simplest model for the kinetics of the heat transfer process is that the heat transfer rate is proportional to the temperature difference between the two streams and the surface area available for heat transfer.

$$\dot{Q}_b = U(T_a - T_b)A \qquad \textbf{(11.80)}$$

Here A is the heat exchanger surface area normal to the direction of heat transfer, $(T_a - T_b)$ is some suitable difference in temperature between streams a and b, and U is the constant of

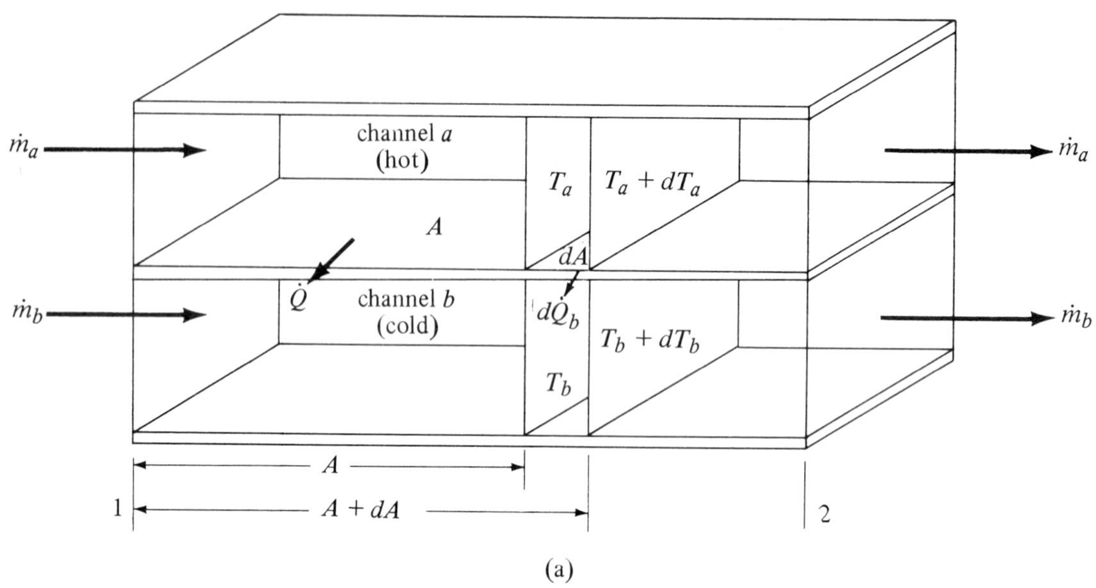

(a)

Figure 11.21(a) Parallel flow configuration for a two-channel heat exchanger

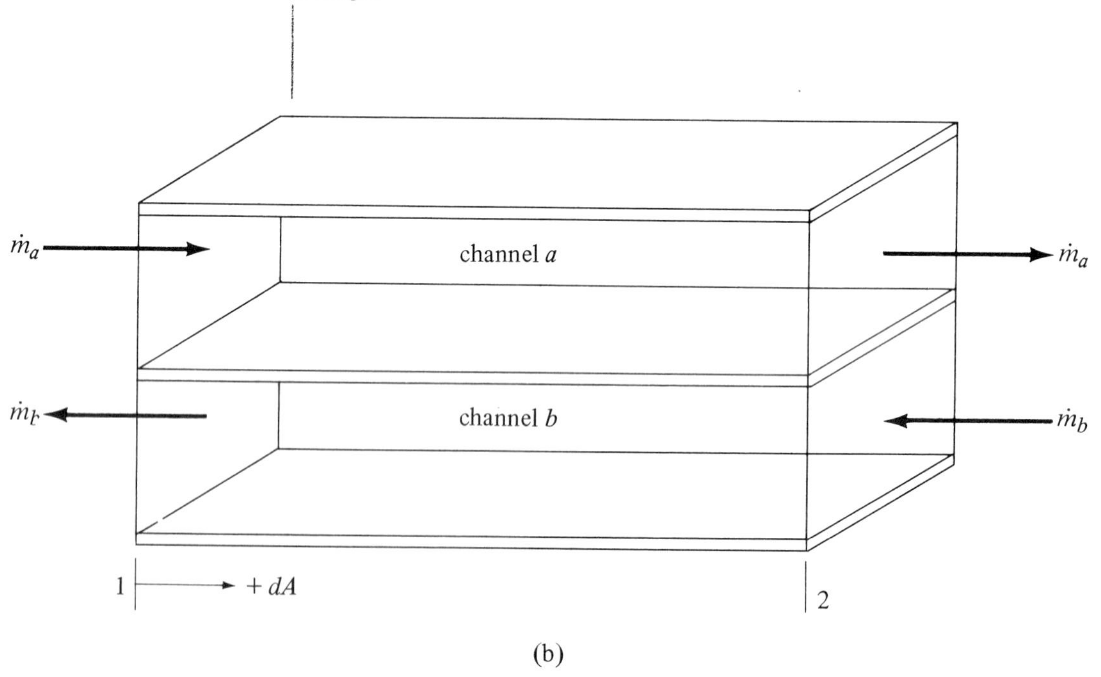

(b)

Figure 11.21(b) Counterflow configuration for a two-channel heat exchanger

proportionality known as the *overall heat transfer coefficient*. The coefficient U is the thermal conductance (reciprocal of thermal resistance) per unit of surface area normal to the direction of heat transfer. The overall heat transfer coefficient is a complicated function of the heat exchanger geometry, the fluid flow velocity, the fluid properties, and the heat exchanger material properties. (For a detailed consideration consult a text on heat transfer.[6]) For further simplification of the model, we assume that the overall heat transfer coefficient U has the same value over the entire area of the heat exchanger.

In the especially simple case where the temperature difference, $(T_a - T_b)$, is uniform along the length of the heat exchanger, equations (11.79) and (11.80) completely specify the states of the heat exchanger. For example, if the two inlet temperatures, mass flow rates, and specific heats are given together with the surface area and the overall heat transfer coefficient, then equations (11.79) and (11.80) determine the total heat transfer rate and the two exit temperatures.

As we shall illustrate, the temperature difference is uniform along the heat exchanger only when $\dot{m}c_{Pa} = \dot{m}c_{Pb}$ and the two streams are flowing in opposite directions, Figure 11.21 (b), as opposed to flow in the same direction, Figure 11.21 (a). The situation in which the two streams flow in the same direction is termed *parallel flow* while flow in opposite directions is termed *counter flow*.

We must now consider the axial temperature distribution in the heat exchanger so that we may determine the proper average temperature difference to use in equation (11.80). The temperature distribution is determined by considering the control volumes of extent dA as shown in Figure 11.21 (a). The first law of thermodynamics is

$$\delta\dot{Q}_a = \dot{m}_a c_{Pa}\, dT_a \qquad \textbf{(11.81)}$$

where $\delta\dot{Q}_a$ denotes the heat transfer rate for stream a in the con-

[6] Rohsenow, W. M. and Choi, H. Y., *Heat, Mass, and Momentum Transfer* Prentice-Hall, N.J., 1961; Kays, W. M. and London, A. L., *Compact Heat Exchangers* McGraw-Hill, N.Y., 1964; McAdams, W. H., *Heat Transmission* McTraw-Hill, N.Y., 1954.

trol volume of infinitesimal extent dA. For stream b, the first law of thermodynamics becomes

$$\delta \dot{Q}_b = \dot{m}_b c_{Pb}\, dT_b \tag{11.82}$$

Since the two control volumes combined are adiabatic,

$$\delta \dot{Q}_a = -\delta \dot{Q}_b \tag{11.83}$$

Stream a is the hot stream; therefore $\delta \dot{Q}_b$ is positive, dT_a negative, and dT_b positive. The local rate of heat transfer between the two streams is also obtained by applying equation (11.80) to the surface area dA.

$$\delta \dot{Q}_b = U(T_a - T_b)\, dA \tag{11.84}$$

At this point it is convenient to define the temperature difference ΔT as

$$\Delta T = T_a - T_b \tag{11.85}$$

where T_a and T_b are at the same axial position in the heat exchanger. Thus

$$d(\Delta T) = dT_a - dT_b \tag{11.86}$$

Substituting equations (11.81), (11.82), and (11.83) into (11.86), we obtain

$$d(\Delta T) = \frac{\delta \dot{Q}_a}{\dot{m} c_{Pa}} - \frac{\delta \dot{Q}_b}{\dot{m} c_{Pb}} = -\delta \dot{Q}_b \left[\frac{1}{\dot{m} c_{Pa}} + \frac{1}{\dot{m} c_{Pb}}\right] \tag{11.87}$$

Substituting for δQ_b from equation (11.84), we obtain

$$\frac{d(\Delta T)}{\Delta T} = -\left[\frac{1}{\dot{m} c_{Pa}} + \frac{1}{\dot{m} c_{Pb}}\right] U\, dA \tag{11.88}$$

The integrated form of equation (11.88) is

$$\left[\ln \frac{\Delta T}{\Delta T_1}\right] = -\left[\frac{1}{\dot{m} c_{Pa}} + \frac{1}{\dot{m} c_{Pb}}\right] UA \tag{11.89}$$

where A is the surface area to the left of the location of ΔT in Figure 11.21 (a). For the complete heat exchanger, equation (11.89) becomes

$$\left[\ln \frac{\Delta T_2}{\Delta T_1}\right] = -\left[\frac{1}{\dot{m}_a c_{Pa}} + \frac{1}{\dot{m}_b c_{Pb}}\right] UA_T \tag{11.90}$$

Using equation (11.79) to eliminate $\dot{m}_a c_{Pa}$ and $\dot{m}_b c_{Pb}$, we obtain

$$\ln \frac{\Delta T_2}{\Delta T_1} = [(T_{b2} - T_{b1}) - (T_{a2} - T_{a1})] \frac{UA_T}{Q_b} \tag{11.91}$$

On rearranging equation (11.91) and applying equation (11.85) at each end of the heat exchanger, we obtain

$$\dot{Q}_b = \left[\frac{\Delta T_1 - \Delta T_2}{\ln (\Delta T_1/\Delta T_2)}\right] UA_T \tag{11.92}$$

If we define the log mean temperature difference, ΔT_{LM}, as

$$\Delta T_{LM} = \left[\frac{\Delta T_1 - \Delta T_2}{\ln (\Delta T_1/\Delta T_2)}\right] \tag{11.93}$$

equation (11.92) takes the same form as equation (11.80).

$$\dot{Q}_b = \Delta T_{LM} UA_T \tag{11.94}$$

If we consider the counter flow arrangement, Figure 11.21 (b), all of the equations are the same except that a negative sign appears in front of each $\dot{m}_b$. The principal equations which change are (11.82), (11.89), and (11.90), which becomes

$$\ln \frac{\Delta T_2}{\Delta T_1} = -\left[\frac{1}{\dot{m}_a c_{Pa}} - \frac{1}{\dot{m}_b c_{Pb}}\right] UA_2 \tag{11.95}$$

It is important to note that equations (11.93) and (11.94) remain unchanged. Equation (11.95) shows that a counter flow exchanger with $\dot{m}_a c_{Pa} = \dot{m}_b c_{Pb}$ has the same ΔT at each end of the exchanger. Equation (11.89) with the sign change for counter flow shows that ΔT is constant along the length of the heat exchanger for equal capacity rates.

The actual temperature distributions for the two streams can be obtained from equations (11.89), (11.85), and (11.79). After some manipulation, the temperature distributions are:

Parallel flow

$$T_b - T_{b1} =$$

$$\frac{(T_{a1} - T_{b1})}{(1 + \dot{m}_b c_{Pb}/\dot{m}_a c_{Pa})}\left[1 - \exp - \left(\frac{1}{\dot{m}_a c_{Pa}} + \frac{1}{\dot{m}_b c_{Pb}}\right) UA\right] \quad \textbf{(11.96)}$$

$$T_{a1} - T_a =$$

$$\frac{\dot{m}_b c_{Pb}}{\dot{m}_a c_{Pa}} \frac{(T_{a1} - T_{b1})}{(1 + \dot{m}_b c_{Pb}/\dot{m}_a c_{Pa})}\left[1 - \exp - \left(\frac{1}{\dot{m}_a c_{Pa}} + \frac{1}{\dot{m}_b c_{Pb}}\right) UA\right] \quad \textbf{(11.97)}$$

Counter flow (with sign change for $\dot{m}_b$)

$$T_{b1} - T_b =$$

$$\frac{(T_{a1} - T_{b1})}{(\dot{m}_b c_{Pb}/\dot{m}_a c_{Pa} - 1)}\left[1 - \exp - \left(\frac{1}{\dot{m}_a c_{Pa}} - \frac{1}{\dot{m}_b c_{Pb}}\right) UA\right] \quad \textbf{(11.98)}$$

$$T_{a1} - T_a =$$

$$\frac{\dot{m}_b c_{Pb}}{\dot{m}_a c_{Pa}} \frac{(T_{a1} - T_{b1})}{(\dot{m}_b c_{Pb}/\dot{m}_a c_{Pa} - 1)}\left[1 - \exp - \left(\frac{1}{\dot{m}_a c_{Pa}} - \frac{1}{\dot{m}_b c_{Pb}}\right) UA\right] \quad \textbf{(11.99)}$$

Figure 11.22 shows these temperature distributions plotted for a given case. Figure 11.22 illustrates two generalizations of the behavior of the temperature distribution. First the ΔT always decreases in the direction of flow of the stream with the smaller capacity rate, $\dot{m}c_P$, and second, the temperature change from inlet to outlet is always largest for the stream with the smallest capacity rate, $\dot{m}c_P$.

In calculating the states for a heat exchanger, the method depends upon the given information. If the four temperatures

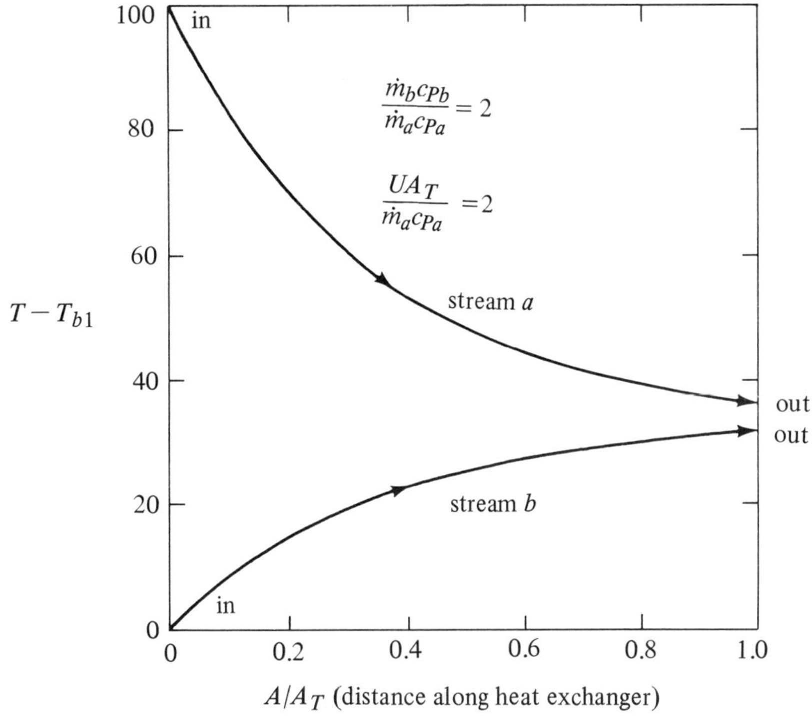

Figure 11.22(a) Temperature distributions for two-stream, parallel flow heat exchanger

and the capacity rates are given, equation (11.93) gives the ΔT_{LM}, equation (11.79) gives the heat transfer rate, and equation (11.94) gives the required UA_T product. On the other hand, if the two inlet temperatures, the capacity rates and the UA_T product are given, equation (11.79) must be solved simultaneously with equation (11.90) or (11.95) for the two unknown temperatures. Equation (11.79) then gives the heat transfer rate.

The performance of a heat exchanger is normally expressed in terms of the heat exchanger effectiveness which is arbitrarily based on the stream with the minimum capacity rate. The *heat exchanger effectiveness*, ϵ, is defined by the ratio of the actual heat transfer (absolute value) to the heat transfer (absolute value) that would occur if the stream with the minimum capacity

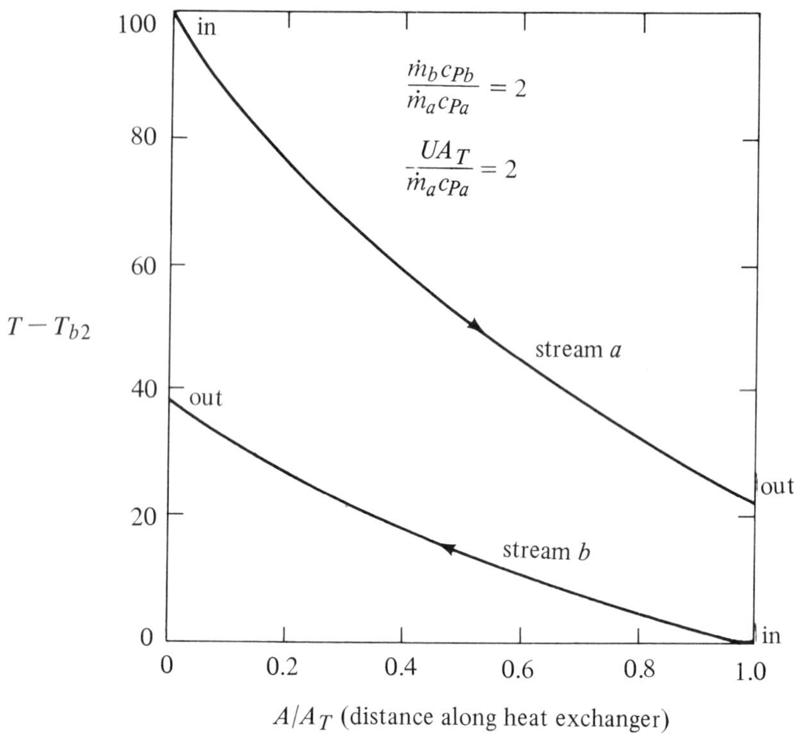

Figure 11.22(b) Temperature distributions for two-stream, counter flow heat exchanger

rate were heated (or cooled) from its inlet temperature to the inlet temperature of the other stream.

$$\epsilon = \frac{\dot{Q}_{\text{actual}}}{(\dot{m}c_P)_{\min}\,[(T_a)_{\text{in}} - (T_b)_{\text{in}}]}$$

$$= \frac{\dot{m}_a c_{Pa}\,[(T_a)_{\text{out}} - (T_a)_{\text{in}}]}{(\dot{m}c_P)_{\min}\,[(T_a)_{\text{in}} - (T_b)_{\text{in}}]}$$

$$\epsilon = \frac{\dot{m}_b c_{Pb}\,[(T_b)_{\text{out}} - (T_b)_{\text{in}}]}{(\dot{m}c_P)_{\min}\,[(T_a)_{\text{in}} - (T_b)_{\text{in}}]} \qquad \textbf{(11.100)}$$

In more complex cases than the simple models which we have considered, the heat exchanger effectiveness is expressed in terms of enthalpy changes. Equation (11.100) is therefore sometimes termed the temperature effectiveness equation.

The heat exchanger effectiveness approaches unity only for a counter flow exchanger with equal capacity rates for the two streams and with an infinite UA_T product. Thus, an effectiveness of unity implies a reversible exchanger which must have a zero ΔT throughout. The effectiveness may be thought of as the ratio of the actual heat transfer to the heat transfer in a reversible exchanger with the same two inlet states, the same minimum capacity rate, but a different maximum capacity rate.

The effectiveness is less than unity for two reasons. First, if UA_T is finite, some of the temperature difference between the streams is the result of insufficient time for thermal equilibrium to be established. Second, if the exchanger is parallel flow or if the capacity rates are not the same, some of the ΔT is the result of first law requirements on the temperature distributions. A heat exchanger with infinite surface area is not necessarily reversible.

11E.4 Sample Problem: Saturated steam at $P = 10^5$ N/m^2, $T =$ 99.63 C is used in a steady counter flow heat exchanger to heat 1.0 m^3/s of air at $P = 10^5$ N/m^2, $T = 20$ C. The H_2O leaves the heat exchanger as saturated liquid at $P = 10^5$ N/m^2. The air leaves the heat exchanger at $P = 10^5$ N/m^2.

(a) What is the leaving temperature for the air corresponding to the maximum heat transfer for the given states for the air and steam? What is the maximum rate of heat transfer? What flow rate of steam is required?

(b) If the effectiveness of the heat exchanger is 0.80 what is the leaving air temperature? What flow rate of steam is required?

(c) If the overall heat exchanger coefficient for the heat exchanger surface is 1.0 kW/m^{2}K, what surface area is required for the heat exchanger with $\epsilon = 0.80$?

Solution: (a) The maximum heat transfer brings the air up to the constant temperature of the two-phase H_2O stream. Any additional heat transfer from the H_2O would have to be to the air at a higher temperature than the H_2O. Thus $T_4 = 99.63$ C. The heat transfer rate is given by the first law for a control volume around the air stream as shown in Figure 11E.4.

$$0 = \dot{Q}_{\text{air}} + \dot{m}h_3 - \dot{m}h_4$$

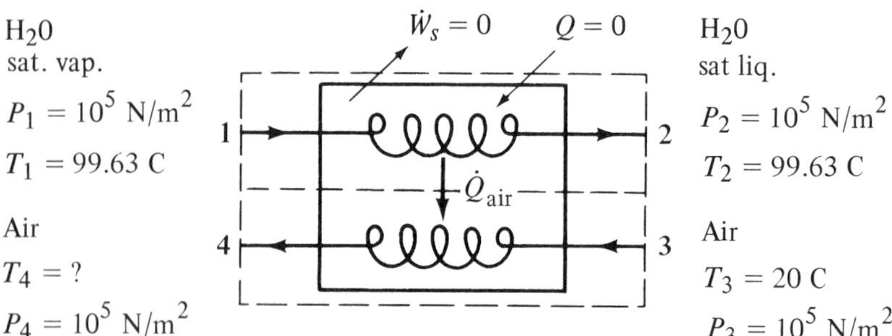

Figure 11E.4 Heat exchanger

where $\dot{W}_{\text{shaft}} = 0$ and kinetic and gravitational energies are neglected. The mass flow rate of air is

$$\dot{m} = \frac{\vartheta_1}{v_1} = \frac{\vartheta_1 P_1}{RT_1}$$

$$\dot{m} = \frac{(1.0 \text{ m}^3/\text{s})(10^5 \text{ N/m}^2)}{(287 \text{ Nm/kgK})(293.16 \text{ K})} = 1.188 \text{ kg/s}$$

$$\dot{Q}_{\text{air}} = \dot{m}(h_4 - h_3) = \dot{m}c_P(T_4 - T_3)$$

$$\dot{Q}_{\text{air}} = (1.188 \text{ kg/s})(1.003 \text{ kJ/kgK})(79.63 \text{ K}) = 94.93 \text{ kJ/s}$$

The required flow rate of stream is given by the first law for a control volume around the H_2O stream, Figure 11E.4.

$$0 = \dot{Q}_w + \dot{m}_w(h_1 - h_2)$$

$$\dot{m}_w = \frac{-\dot{Q}_w}{h_1 - h_2} = \frac{-(-94.93 \text{ kJ/s})}{2258.0 \text{ kJ/kg}} = 0.0420 \text{ kg/s}$$

(b) The heat transfer rate for the heat exchanger with an effectiveness of 0.80 is given by the definition of effectiveness, equation (11.100).

$$\dot{Q}_{\text{actual}} = \epsilon(\dot{Q}_{\text{air}})_{\text{max}}$$

Note that for this heat exchanger the specific heat for the H_2O stream is infinite.

$$\dot{Q}_{\text{actual}} = 0.80(94.93) = 75.94 \text{ kJ/s}$$

The first law for the air stream then gives T_4.

$$T_4 = T_3 + \frac{\dot{Q}_{\text{actual}}}{\dot{m}c_P} = 20 \text{ C} + \frac{75.94}{(1.188)(1.003)} = 83.73 \text{ C}$$

The first law for the H_2O stream gives $\dot{m}_w$.

$$\dot{m}_w = \frac{-\dot{Q}_w}{h_1 - h_2} = \frac{-(-75.94)}{2258.0} = 0.0336 \text{ kg/s}$$

(c) Since the H_2O stream has a constant temperature of 99.63 C, the ΔT at the air inlet end of the heat exchanger is

$$\Delta T = 99.63 - 20 = 79.63 \text{ C}$$

and the ΔT at the air outlet end of the heat exchanger is

$$\Delta T = 99.63 - 88.73 = 15.90 \text{ C}$$

The log mean temperature difference is given by equation (11.93).

$$\Delta T_{LM} = \frac{79.63 - 15.90}{\ln (79.63/15.90)} = 39.56 \text{ C}$$

The required surface area is given by equation (11.94).

$$A_T = \frac{\dot{Q}}{\Delta T_{LM} U} = \frac{75.94 \text{ kJ/s}}{(39.56 \text{ C})(1.0 \text{ kW/m}^2 \text{ K})}$$

$$A_T = 1.92 \text{ m}^2$$

Problems

11.1 The steady operating conditions of a water-cooled air compressor are shown in Figure 11P.1. The heat transfer between the compressor and the environment is negligible. If kinetic and potential energy effects are neglected, determine the power required to drive the compressor. The air can be modeled as an ideal gas with $c_v = 715.9$ J/kgK, $R =$

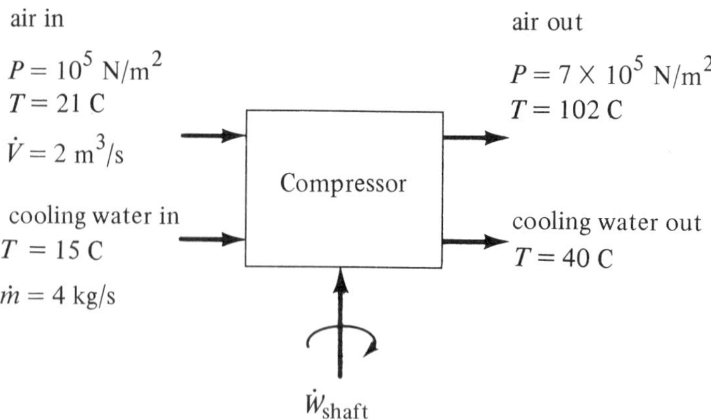

Figure 11P.1

287 J/kgK. A unit mass of the cooling water may be modeled as a pure thermal system, with c = 4.187 kJ/kgK.

11.2 A steady flow pump receives saturated liquid H_2O at 27 C. The pump delivers the liquid at 7×10^6 N/m^2.

(a) Determine the shaft work transfer rate per unit mass flow rate if the pump is reversible adiabatic.

(b) Estimate the outlet temperature from the pump if the efficiency is 0.5. (State clearly the model used for the constitutive relation for the fluid.)

(c) Repeat parts (a) and (b) for a pump discharge state with the exit 400 m above the pump inlet and the pressure at the elevated exit equal 7×10^6 N/m^2.

11.3 The centrifugal air compressor of a gas turbine receives air from the ambient atmosphere where the pressure is 9×10^4 N/m^2 and the temperature is 27 C. At the discharge of the compressor, the pressure is 3.7×10^5 N/m^2, the temperature is 205 C, and the velocity is 100 m/s. The mass rate of flow into the compressor is a steady 20 kg/s.

(a) If the air can be modeled as an ideal gas with constant specific heats, determine the power required to drive the compressor.

(b) Determine the compressor efficiency.

11.4 In the gas processing industry, nitrogen is separated from atmospheric air as a liquid at 10^5 N/m^2. The nitrogen is then stored as a liquid until needed. The liquid is then vaporized and pumped into tanks at 2×10^7 N/m^2 and 27 C for use in industrial facilities, welding operations, etc. Three schemes have been suggested for the vaporization and tank charging process as sketched in Figure 11P.4. The first involves compressing the liquid from 10^5 to 2×10^7 N/m^2 with a reversible, adiabatic pump and then vaporizing the high pressure liquid by heat transfer with the atmosphere. The second involves first vaporizing the liquid and then compressing the gas in a reversible, adiabatic compressor so that the gas leaves the compressor at 2×10^7 N/m^2, 27 C. Finally, the third scheme involves vaporizing the liquid to produce a gas at 10^5 N/m^2 and 27 C and subsequently compressing the gas to 2×10^7 N/m^2 in a reversible isothermal compressor.

(a) Determine the heat transfer and work transfer per unit mass of nitrogen in each of the schemes suggested.

(b) Which scheme requires the minimum work transfer and why?

11.5 The compressor of a turbo-charger shown in Figure 11P.5 has an inlet flow rate of 7.5 m^3/minute of air as measured at the inlet condition. This condition is $P_1 = 1.013 \times$

Scheme #1
sat liq N_2, $P = 10^5$ N/m^2
$\dot{Q}_P = 0$
pump
$\dot{W}$
$\dot{Q}_v$
gaseous N_2
vaporizer
$P = 2 \times 10^7$ N/m^2
$P = 2 \times 10^7$ N/m^2
$T = 27$ C
Scheme #2
sat liq N_2, $P = 10^5$ N/m^2
vaporizer
$\dot{Q}_c = 0$
$\dot{Q}_v$
compressor
$\dot{W}$
gaseous N_2
$P = 2 \times 10^7$ N/m^2
$T = 27$ C
Scheme #3
sat liq N_2, $P = 10^5$ N/m^2
vaporizer
$P = 10^5$ N/m^2
$\dot{Q}_c$
$T = 27$ C
$\dot{Q}_v$
compressor
$\dot{W}$
gaseous N_2
$P = 2 \times 10^7$ N/m^2
$T = 27$ C

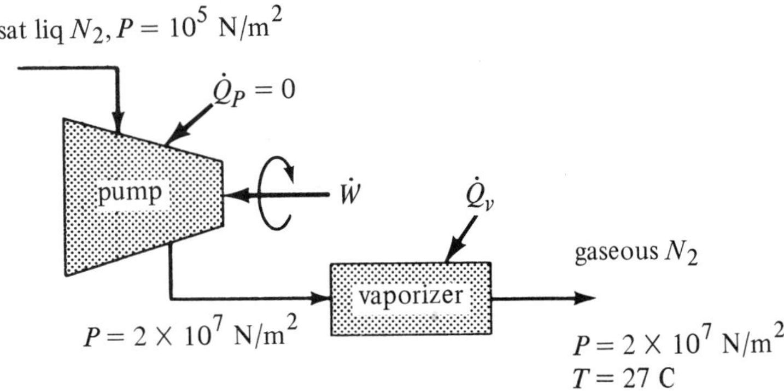

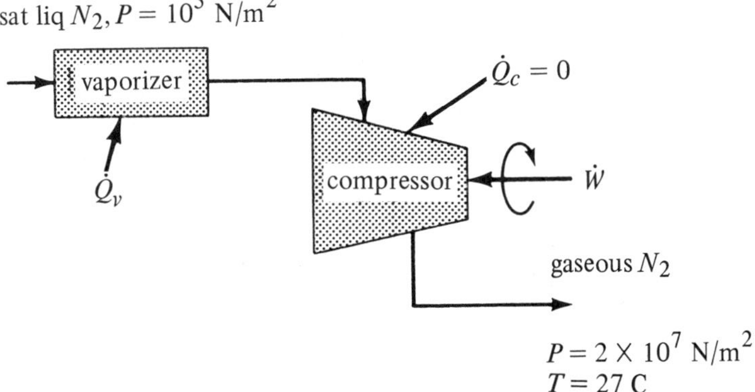

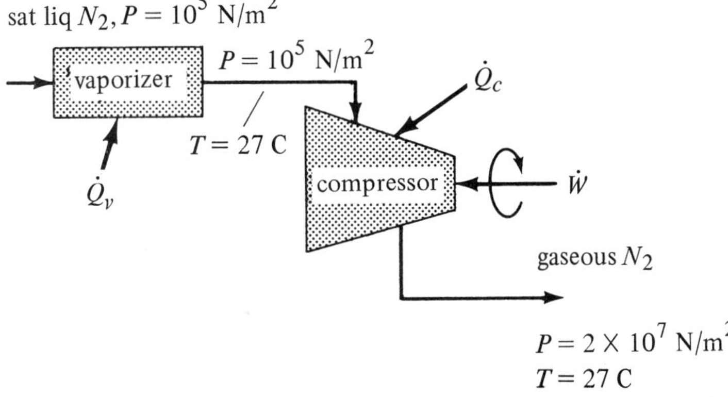

Figure 11P.4

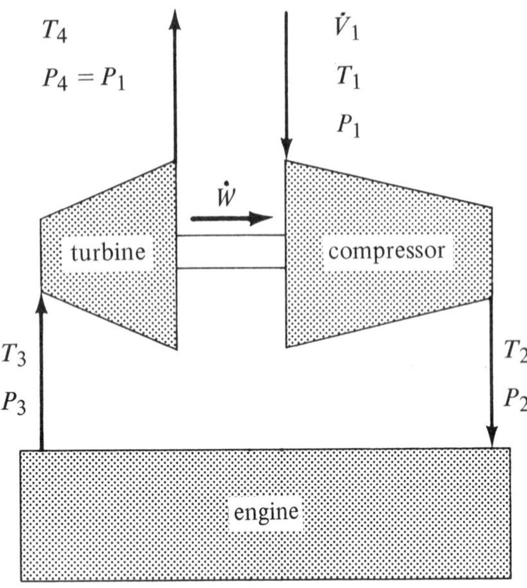

Figure 11P.5

10^5 N/m² and $T_1 = 288$ K. The compressor has a pressure ratio of 1.5 and an adiabatic shaft work efficiency of 80%. The compressor is driven directly by an exhaust turbine with an adiabatic shaft work efficiency of 70%. The turbo-charger and engine are arranged so that they work on the "constant pressure system" which means both turbine and compressor operate under conditions of steady flow. The turbine pressure ratio is the same as that for the compressor and the mass of engine fuel may be neglected.

Calculate the inlet temperature to the turbine and the power produced by the turbine. Assume air and exhaust gas are ideal with $\gamma = 1.4$ and $R = 287$ J/kgK.

11.6 A gas enters an insulated nozzle with a velocity ϑ_1 and is expanded adiabatically to an exit velocity ϑ_2. Show that if the expansion is also reversible and if the gas can be modeled as an ideal gas with constant specific heats, the exit velocity is given by

$$\vartheta_2 = \sqrt{\vartheta_1^2 + \frac{2RT}{\gamma - 1}\left[1 - \left(\frac{P_2}{P_1}\right)^{(\gamma-1)/\gamma}\right]}$$

where $\gamma = c_P/c_v$.

11.7 Steam at 6×10^5 N/m² and 220 C expands reversibly and adiabatically in a nozzle to a pressure of 10^5 N/m². The mass flow rate through the nozzle is 5 kg/s. The velocity of the H_2O at the inlet port is negligible compared to the velocity at the exit port.

(a) Determine the exit velocity and the required area at the nozzle outlet if the state at the exit is an equilibrium, two-phase state of vapor and liquid.

(b) Repeat (a) for the nozzle efficiency equal to 0.96 rather than 1.0.

11.8 Air at a pressure of 7×10^4 N/m² and a temperature of −40 C enters a diffuser at a velocity of 1000 m/s and leaves at a velocity of 100 m/s. If the process is reversible and adiabatic, what are the pressure and temperature of the leaving air? Assume that the air can be modeled as an ideal gas with $c_v = 715.9$ J/kgK and $R = 287$ J/kgK.

11.9 A horizontal adiabatic nozzle is attached to a large tank containing gas with a state designated by T_0, P_0, ρ_0. For a large tank, T_0, P_0 and ρ_0 may be taken as constant and the flow through the nozzle is steady. Since the fluid velocity in the tank is zero, the state in the tank is called the stagnation state for the flow in the nozzle and T_0, P_0, ρ_0 etc. are called the stagnation properties.

(a) Show that for an ideal gas with constant specific heats

$$\frac{T_0}{T} = 1 + \frac{\gamma - 1}{2} M^2$$

where γ is the ratio of the specific heats c_P/c_v and T is the temperature at that point in the flow for which the flow is characterized by the Mach number M. The Mach number is defined as the ratio of the local velocity to the velocity of sound "a" at the point in question where

$$a^2 = \left(\frac{\partial P}{\partial \rho}\right)_s$$

and ρ is the density.

(b) Show that if this expanding flow is also reversible

$$\frac{P_0}{P} = \left[1 + \frac{\gamma - 1}{2} M^2\right]^{\gamma/(\gamma-1)}$$

11.10 Saturated liquid carbon dioxide at 20 C flows at a steady rate through a throttle valve to a pressure of 1.005×10^6 N/m^2.

(a) Determine the exit temperature and the condition of the stream at a cross section downstream of the throttle valve where the kinetic energy is negligible.

(b) If the inlet and outlet cross section areas for the throttle valve are each 10^{-3} m^2 and the mass flow rate of the CO_2 is 0.5 kg/s, show that the assumption of negligible kinetic energy is justified.

11.11 A throttling calorimeter, like that in Figure 11P.11, is a device used to determine the quality of a liquid-vapor mixture. Its operation depends upon the evaporation of the liquid present in the mixture by expanding the mixture from a high pressure state to a low pressure state.

(a) If the pressure in the steam line is 3×10^6 N/m^2 and the pressure and temperature in the calorimeter expansion chamber are 10^5 N/m^2 and 150 C, respectively, what is the quality of the steam in the main steam line?

(b) Assuming steady flow through the sampling tube and no heat loss, determine the form of the first law applicable to this device. State all assumptions.

(c) Sketch this process on T-s and h-s diagrams and obtain a check on the quality using the T-s chart for H_2O.

11.12 Liquid water flows adiabatically in a constant diameter pipe past a throttling valve that is partially open. Let the subscripts 1 and 2 refer to conditions before and after the valve, respectively.

(a) Assuming the liquid can be modeled as an incompressible fluid, is:
$(h_2 - h_1)$ positive, negative, or zero?
$(P_2 - P_1)$ positive, negative, or zero?
$(u_2 - u_1)$ positive, negative, or zero?
$(T_2 - T_1)$ positive, negative, or zero?
$(s_2 - s_1)$ positive, negative, or zero?

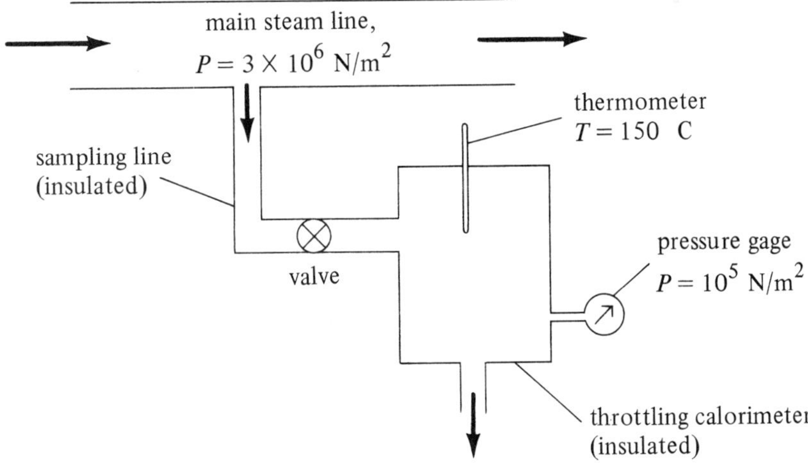

Figure 11P.11

(b) Assuming the liquid is saturated in state 1, repeat part (a). How do you account for any differences in the answers to (a) and (b)?

11.13 Steam enters a steam turbine at a pressure of 4×10^6 N/m^2 as shown in Figure 11P.13. The steam flows at a steady rate of 100 kg/hr through the turbine and exhausts to a heat exchanger at a pressure of 7×10^3 N/m^2 which is maintained inside the heat exchanger on the steam side. All of the steam is condensed to a saturated liquid at this pressure. The condensation of the steam is effected by a steady stream of cooling water flowing at a rate of 3110 kg/hr. The cooling water enters the heat exchanger at a temperature of 15 C and exits at a temperature of 33 C.

(a) If the expansion of the steam in the turbine is reversible isentropic and if the heat exchanger is adiabatic, determine the power output of the turbine. Assume that kinetic and potential energies are negligible and that the cooling water can be modeled as a pure thermal system with a specific heat of c = 4.187 kJ/kgK.

(b) Show the processes of the steam on a temperature-entropy diagram.

11.14 In steam turbines any liquid present in the high velocity stream can cause severe blade damage by erosion of the blade surface. After operating the turbine for any extended period in this fashion, the damage becomes extensive enough so that turbine performance suffers. Therefore, it is worthwhile to take steps to reduce the amount of liquid present in the turbines. In a typical turbine installation as shown in Figure 11P.14, steam enters the turbine at a pressure of 1.4×10^6 N/m^2 and a temperature of 500 C. The steam is expanded reversibly and adiabatically in the turbine to 10^4 N/m^2. The steam then enters the condenser in which it is condensed at constant pressure to a saturated liquid at 7×10^3 N/m^2. There is obviously a fair amount of liquid in the fluid stream. To reduce this amount of liquid, an engineer has suggested throttling the steam from 1.4×10^6 N/m^2 to 4×10^5 N/m^2 prior to the turbine. He then maintains that there is no change in turbine output but that the liquid present in the fluid stream has been reduced substantially. Further, he claims that since the turbine life is extended at no expense in turbine output, he has gotten something for nothing and has therefore beaten the second law of thermodynamics at its own game.

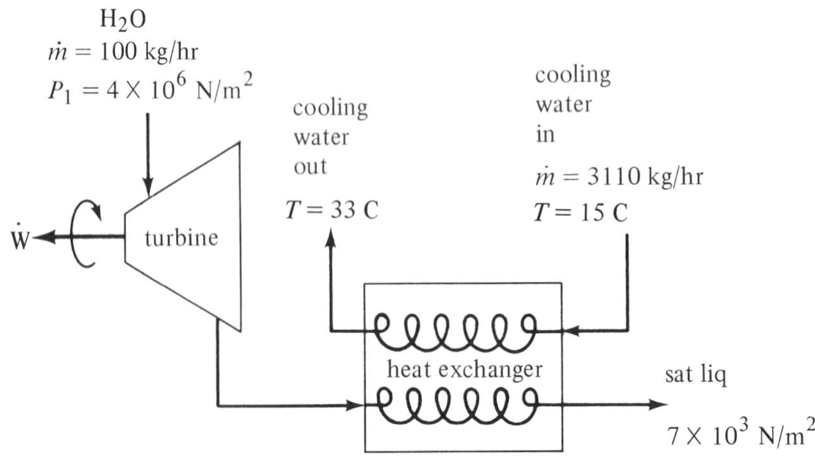

Figure 11P.13

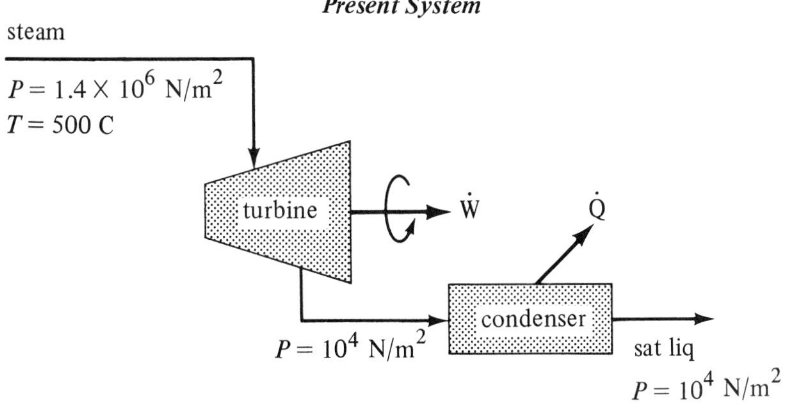

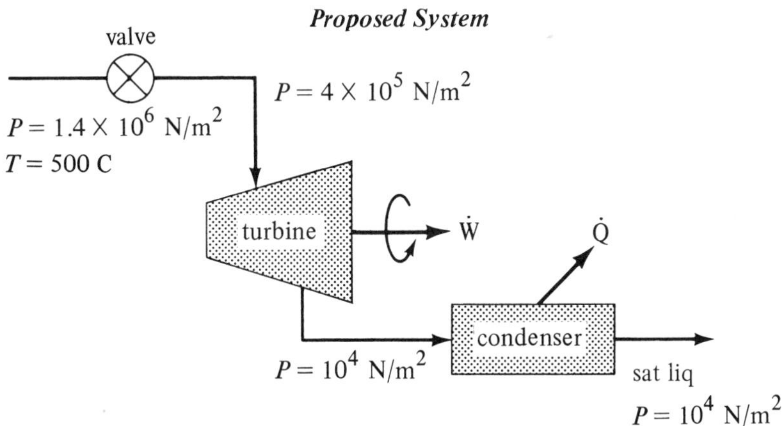

Figure 11P.14

(a) Show the two alternate systems on a *T-s* diagram.

(b) Determine the amount of liquid present in 1 kg of fluid for the two modes of operation.

(c) Determine the work transfer from the turbine per unit of mass of steam flow in the two cases.

(d) Do you agree that the engineer has gotten something for nothing? Why or why not?

11.15 A throttle valve is used to control the power output of a reversible adiabatic gas turbine as shown in Figure 11P.15.

The two states 1 and 3 are fixed as follows;

state 1: $P_1 = 4 \times 10^5$ N/m²
$T_1 = 1000$ K

state 3: $P_3 = 10^5$ N/m²
T_3 = varies depending upon load

The throttle operates by controlling the pressure at the inlet to the turbine (state 2). At full load, there is no pressure drop across the throttle valve and $P_2 = P_1$. At zero load, the throttle is closed and there is no flow through the turbine so that $P_2 = P_3$. For any intermediate load on the turbine, $P_1 > P_2 > P_3$. As-

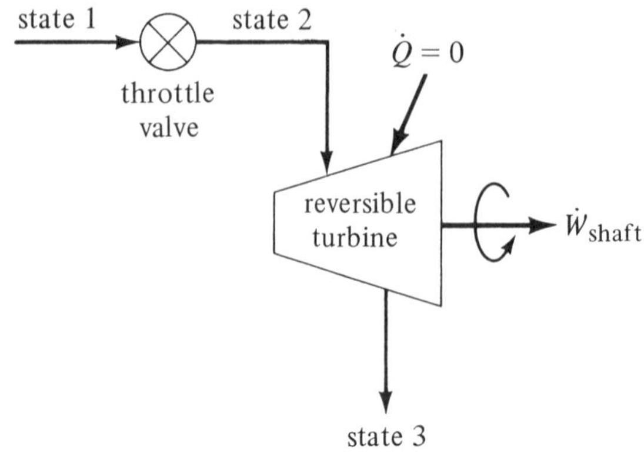

Figure 11P.15

sume that the fluid flowing can be modeled as an ideal gas (air) with $c_v = 715.9$ J/kgK and $R = 287$ J/kgK.

(a) What is the turbine shaft work transfer per unit mass of fluid flowing for full load output ($P_2 = P_1$)?

(b) Determine the pressure in state 2 for a shaft work transfer per unit mass equal to one-half the full load output.

(c) Sketch the processes of (a) and (b) on the T-s and on the h-s diagrams for air.

11.16 One method of producing liquid nitrogen is to use the system shown in Figure 11P.16. Nitrogen gas at a pressure of 10^7 N/m² and a temperature of 300 K flows at a rate of 10 m³/min (measured at 10^5 N/m²) through the heat exchanger, thereby decreasing in temperature. As it flows through the Joule-Thomson valve, its pressure is reduced from 10^7 N/m² to 10^5 N/m² and in the process, some liquid is formed. The gas which is not liquefied, but is reduced in temperature, flows out through the counterflow heat exchanger. The temperature of this discharge stream is 297 K. Assume that the heat exchanger is externally adiabatic.

(a) Determine the rate at which liquid nitrogen is delivered by this system.

(b) Plot the temperature difference in the heat exchanger versus the temperature of the low pressure stream.

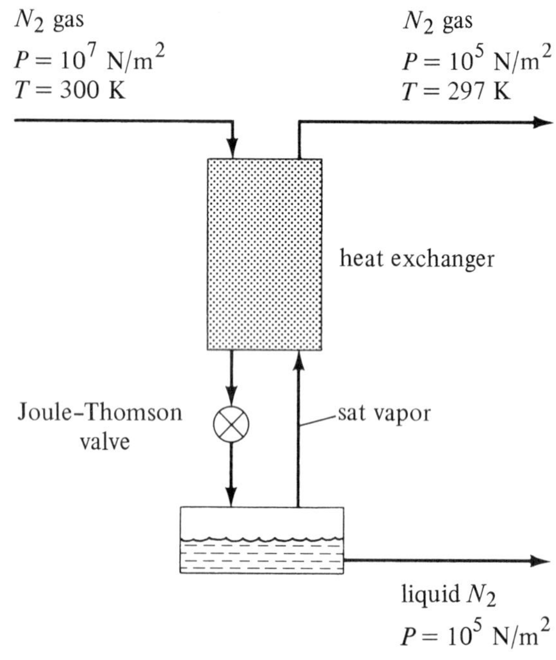

Figure 11P.16

(c) Is the heat transfer process in the heat exchanger reversible? If not, how would you try to make it more nearly reversible?

11.17 Liquid H_2O at 27 C and 7×10^6 N/m² enters a steady flow steam generator at a rate of 500,000 kg/hr. The superheated stream leaves the steam generator at 565 C and 7×10^6 N/m².

(a) Determine the rate at which heat is transferred to the H_2O in the steam generator.

(b) Estimate the fraction of this heat transfer which occurs at constant temperature during the phase change in the boiler.

11.18 A contact feedwater heater is used to preheat the water going into a boiler and operates on the principle of mixing steam and water. A typical feedwater heater of this type is shown in Figure 11P.18.

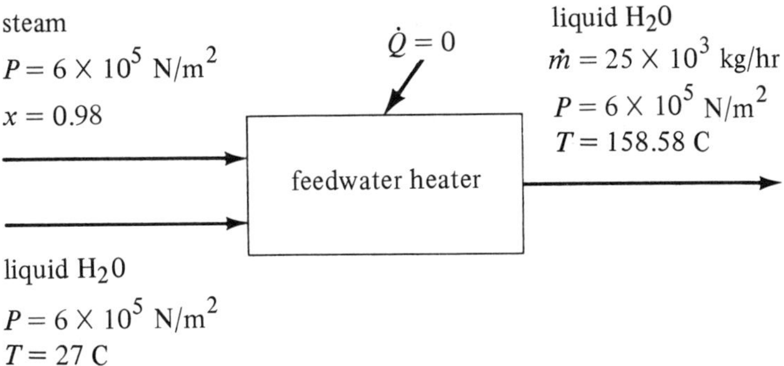

Figure 11P.18

(a) Determine the mass flow rates of the steam and water at entrance.
(b) Is this system reversible?
(c) If the system is not reversible, describe a modification to the system which would accomplish the mixing in a reversible manner and evaluate the performance of this new system.

11.19 A tube-in-shell heat exchanger is to be used to heat water as shown in Figure 11P.19. The water passes through the tubes, and the steam is in the shell around the tubes. Determine the required steam flow rate for the conditions shown in Figure 11P.19. State clearly any modeling which you use to solve the problem.

saturated steam
$P = 1.4 \times 10^5$ N/m^2

water
$P = 3 \times 10^5$ N/m^2
$T = 70$ C
$\dot{m} = 5000$ kg/hr

water
$T = 5$ C

condensate
$T = 38$ C

Figure 11P.19

11.20 A heat exchanger, Figure 11P.20, is being designed to heat a steady stream of liquid water with a steady stream of steam. The design requirements are as follows:

Liquid water flow rate, $\dot{m}_L = 5390$ kg/hr. Steam flow rate, $\dot{m}_s = 100$ kg/hr. Liquid water entrance temperature, $T_{L1} = 15$ C. Steam state at entrance is saturated vapor at $P_{s1} = 10^4$ N/m^2. Pressure in both fluid streams is constant.

Neglect changes in kinetic and potential energy.

(a) What is the maximum heat transfer rate that this heat exchanger can provide between these two fluid streams?
(b) Specify completely the state $s2$ of the discharge steam.

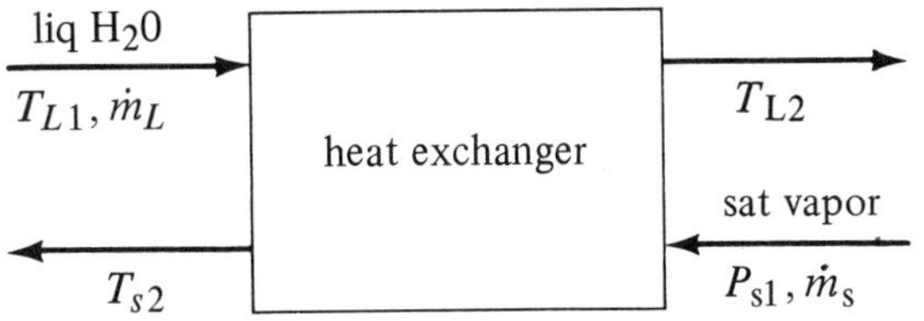

Figure 11P.20

(c) What is the temperature T_{L2} of the water at discharge?

(d) Describe what happens to the states $L2$ and $s2$ as the liquid water flow rate is reduced at constant inlet temperature, T_{L1}. The entrance state $s1$ of the steam and $\dot{m}_s$ remain fixed.

11.21 The adiabatic heat exchanger shown in Figure 11P.21 operates in steady flow with fixed conditions for stream a and an unspecified flow rate for stream b. The counter flowing streams a and b are incompressible fluids with

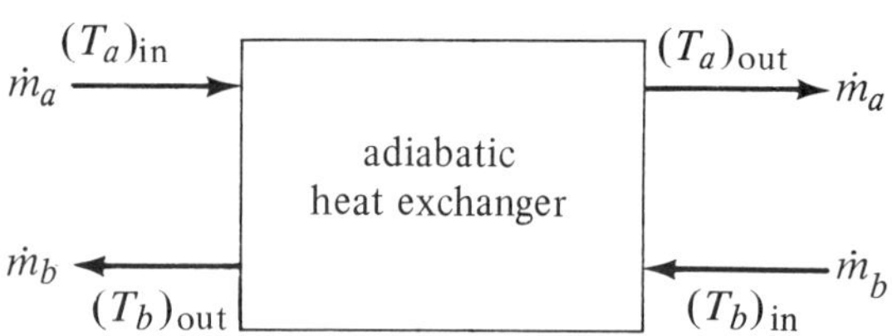

Figure 11P.21

$c_a = 4$ kJ/kgK and $c_b = 2$ kJ/kgK. The mass flow rate of stream a is $\dot{m}_a = 1$ kg/s. The temperatures for stream a are: $(T_a)_{in} = 400$ K, $(T_a)_{out} = 300$ K. The inlet temperature for stream b is $(T_b)_{in} = 300$ K, but the outlet temperature is unspecified. The pressure drops in the flows are zero. For the conditions specified

(a) Express $(T_b)_{out}$ as a function of $\dot{m}_b$.

(b) Express the entropy generated, $\sum \dot{m}s_{out} - \sum \dot{m}s_{in}$, as a function of $(T_b)_{out}$ and $\dot{m}_b$.

(c) Show that for $\dot{m}_b = 2$ kg/s, the entropy generation is zero.

(d) What values for $\dot{m}_b$ will violate the second law of thermodynamics for the specified conditions?

11.22 The long term storage of certain foods can be carried out most effectively if the food is stored at the normal boiling point of nitrogen. The apparatus shown in Figure 11P.22 is one possible method of preparing foods for storage under these conditions. The food, which is 50 percent H_2O and 50 per cent solid (by mass), enters the apparatus at a temperature of 27 C and a mass flow rate of 10 kg/min and leaves at a temperature equal to the saturation temperature of nitrogen at a pressure of 10^5 N/m^2. The nitrogen enters as a saturated liquid at a pressure 10^5 N/m^2. The nitrogen is sprayed onto the food thereby freezing it. The pressure throughout the apparatus is 10^5 N/m^2.

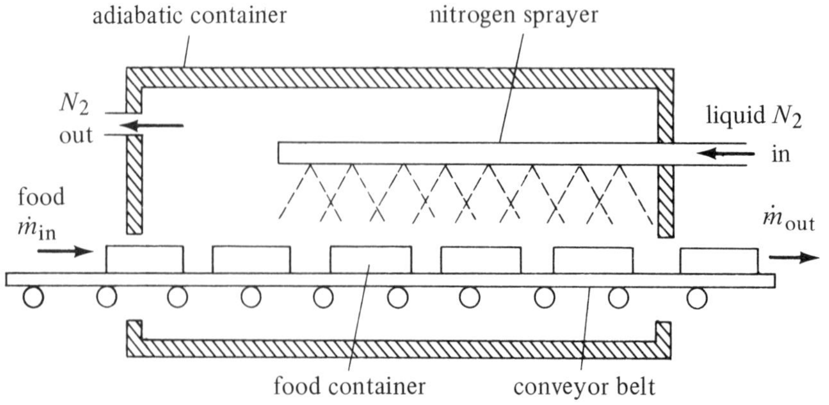

Figure 11P.22

Assume the following:

The ice and solid food component can be modeled as pure thermal systems with constant specific heats of 2.093 kJ/kgK and 0.837 kJ/kgK respectively. The liquid H_2O can be modeled as an incompressible fluid with a constant specific heat of 4.187 kJ/kgK. The latent heat of fusion for H_2O is 335 kJ/kg at 10^5 N/m^2 and 0 C.

Determine the minimum mass flow rate (kg/min) of nitrogen.

CHAPTER 12

Steady-Flow Thermodynamic Plants (Cycles and Processes)

In the previous chapter we described and analyzed the steady flow components that are most commonly used in steady-flow thermodynamic plants. In this chapter, we shall consider a number of important thermodynamic plants which can be modeled simply by interconnecting the components described previously. These plants fall naturally into two groups—the closed plant in which the same fluid is processed in a cycle and the open or process plant in which a stream of fluid is processed as it flows once through the plant. Many authors refer to the first group as steady-flow cycles and the second group as open cycles. The term open cycle is somewhat of a contradiction but still finds considerable usage.

12.1 Steam Power Plants

In general, a steam power plant is a complicated thermodynamic system of many components, all of which function together as a unit with one objective—to produce useful positive work transfer from an energy source of chemical or nuclear origin (fuel). As shown in Figure 12.1, we can schematically represent such a fossil fueled plant as a control volume consisting of three major subsystems: (1) the energy source, (2) the heat engine, and (3) the energy sink. Fuel and air enter the energy source control volume, undergo combustion, and exit as the products of combustion (stack gases). As a result of this process, there is a heat transfer interaction between the energy source and the heat engine. The heat engine experiences a positive heat transfer with the energy source, produces a positive work transfer, and experiences a negative heat transfer with the energy sink. Cooling water enters the energy-sink control volume, increases in temperature because of the heat transfer interaction with the heat engine, and then leaves the control volume.

Historically, water has been the most widely used working fluid in the heat engine. For this reason, this thermodynamic

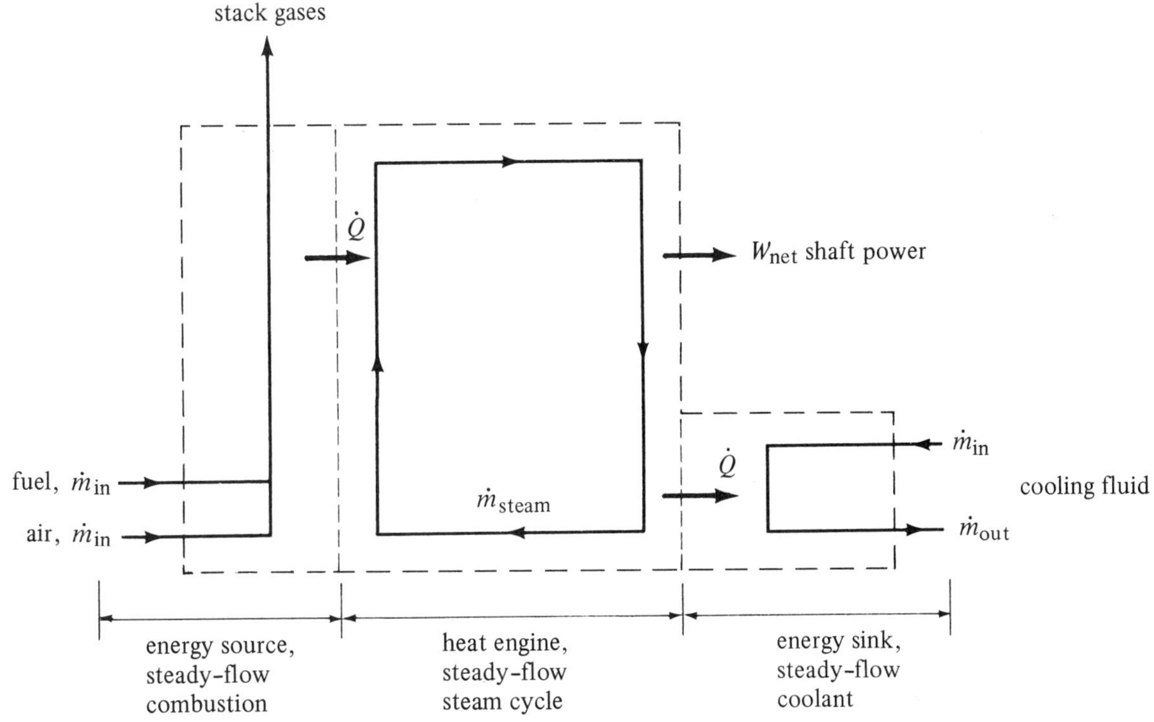

Figure 12.1 Control volume for complete power plant

plant is known as a steam power plant. Furthermore, since the water is processed in a cycle, the steam power plant is regarded as a closed plant even though the fuel, air, and condenser coolant flow only once through the plant. Since its inception, the heat engine has been the focus of attention in the thermodynamic analysis of steam power plants. In keeping with this tradition, our study of steam power plants will concentrate first on this subsystem and then briefly consider the energy source. The heat sink is simple (one passage of a heat exchanger) and will not be discussed here. Our discussion of the plant is greatly simplified if we adopt the historical approach, beginning with the simplest unit and following with the various modifications in order of increasing complexity.

12.1.1 Rankine-Cycle Steam Plant

This thermodynamic plant (Figure 12.2) is the oldest and, as expected, the simplest of all steam power plants. There are four major physical components of the plant: (1) boiler, (2) reciprocating steam engine, (3) condenser, and (4) boiler feed pump. As shown in Figure 12.2, the boundary between the energy source and the heat engine is the heat transfer surface between the combustion gases and the boiling water in the boiler. The boundary between the heat engine and the energy sink is the heat transfer surface between the condensing steam and the coolant in the condenser. In the heat engine, water at high pressure enters the boiler as a compressed liquid, experiences a positive heat transfer from the energy source in the boiler and exits as a saturated vapor. Upon leaving the boiler, the high pressure saturated vapor enters the reciprocating steam engine in which it produces a positive work transfer by expanding to a low pressure, liquid-vapor, two-phase state. The two-phase fluid exhausted from the engine then enters the condenser, experiences a negative heat transfer to the energy sink, and is discharged as a saturated liquid. This low pressure saturated liquid is then pumped to the boiler pressure by a negative work transfer in the boiler feed pump. Thus, the water executes a cycle. The cycle produces a net positive work transfer since the expansion of the high specific volume steam through the reciprocating engine pro-

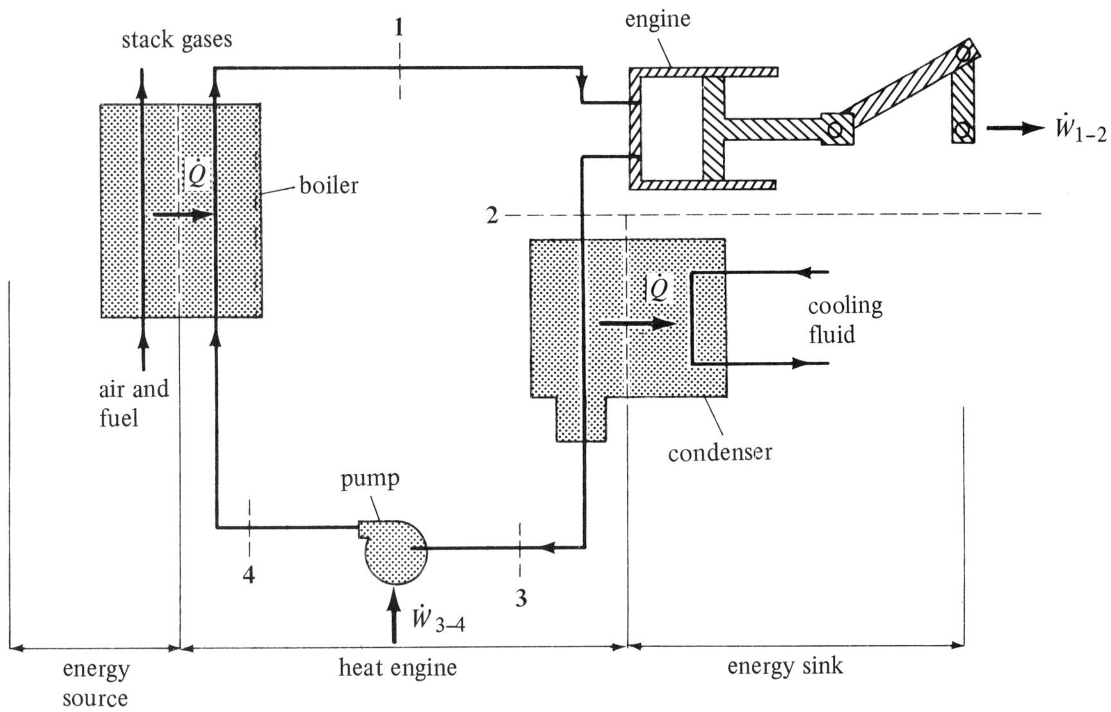

Figure 12.2(a) Rankine-cycle steam plant

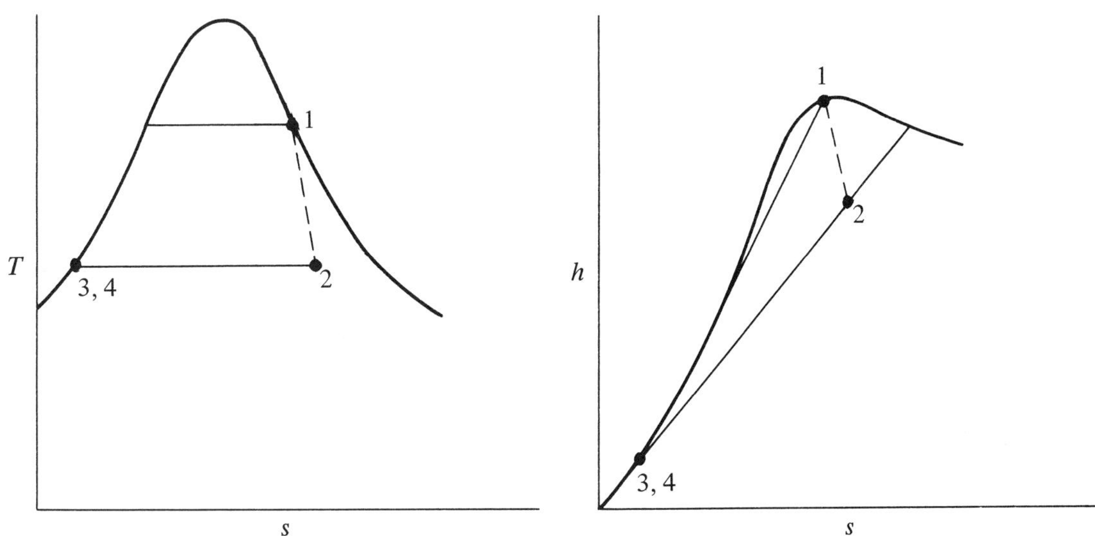

Figure 12.2(b) Thermodynamic states for the Rankine-cycle steam plant

duces more positive work transfer than is required to pump the low specific volume liquid into the boiler.

In formulating the simplest model of this plant, we model the boiler and condenser as constant pressure heat exchangers and the reciprocating steam engine and boiler feed pump as adiabatic shaft work machines (Figure 12.2). The resulting model for the cycle is known as the Rankine cycle while the plant itself is known as a Rankine-cycle plant. Since the steam discharged from the boiler is normally considered dry saturated steam and the pump is normally considered to draw saturated liquid from the hot well at the bottom of the condenser, the boiler pressure, the condenser pressure, the engine efficiency, the pump efficiency, and the mass flow rate of the water completely describe the cycle. With this information, all operating states and interactions can be evaluated, from the formulas shown in Table 12.1.

One significant characteristic of thermodynamic plants in general is the net work ratio, NWR. The NWR is defined as the ratio of the net shaft work transfer to the gross positive shaft work transfer. This parameter is a measure of the fraction of the gross positive shaft work transfer produced by the cycle which is available for useful purposes. Of course, the fraction which is not available is expended by negative shaft work transfer components of the plant. For the Rankine-cycle plant, the relevant

TABLE 12.1
Performance of Rankine-Cycle Steam Plant

Engine power:	$\dot{W}_{1-2} = \dot{m}(h_1 - h_2)$
Condenser load:	$\dot{Q}_{2-3} = \dot{m}(h_3 - h_2) = \dot{m}T_3(s_3 - s_2)$
Pump power:	$\dot{W}_{3-4} = \dot{m}(h_3 - h_4)$
Boiler load:	$\dot{Q}_{4-1} = \dot{m}(h_1 - h_4)$
Energy conversion efficiency:	$\eta = \dfrac{\dot{W}_{\text{net}}}{\dot{Q}_{\text{boiler}}} = \dfrac{(h_1 - h_2) + (h_3 - h_4)}{(h_1 - h_4)}$
	$\eta = 1 - \dfrac{(h_2 - h_3)}{(h_1 - h_4)}$
Net work ratio:	$\text{NWR} = \dfrac{\dot{W}_{\text{net}}}{\dot{W}_{\text{gross positive}}} = \dfrac{(h_1 - h_2) + (h_3 - h_4)}{(h_1 - h_2)}$
	$\text{NWR} = 1 - \dfrac{(h_4 - h_3)}{(h_1 - h_2)}$

expression for NWR is given in Table 12.1. From this expression it is apparent that the net work ratio for the Rankine cycle is very nearly unity. Thus, useful power can be generated even when the engine and pump both have very low efficiencies. In fact, it was the insensitivity of cycle performance to engine efficiency that made the Rankine-cycle plant so prominent in the early development of power generation.

In Chapter 7 we showed that the efficiency of a reversible heat engine increases as the average temperature of the heat transfer from the high temperature energy source is increased. To increase the heat addition temperature in the Rankine cycle, we must increase the steam pressure in the boiler since the H_2O is in a two-phase state. The cycle efficiency could also be improved by lowering condenser pressure. This increases the work transfer and reduces the heat transfer from the cycle for a fixed heat transfer into the cycle. The minimum condenser pressure is the saturation pressure corresponding to the temperature of the cooling fluid available to cool the condenser. Low heat rejection temperatures can only be obtained if essentially all air is excluded from the condenser.

As is shown in Figure 12.3, the Rankine cycle model for the steam power plant can be used to show clearly the influence of boiler pressure on the thermal efficiency. Early steam plants

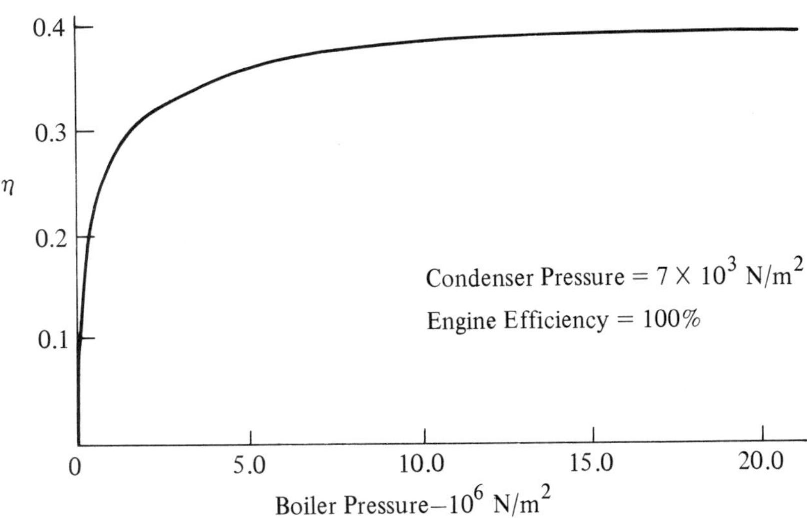

Figure 12.3 Influence of boiler pressure on Rankine cycle efficiency

did not achieve the efficiencies of Figure 12.3 because of nonthermodynamic limits. The first of these limits was the mechanical strength of the boiler and the second was an upper limit on steam temperatures imposed by the oil used to lubricate the engine cylinder. Temperatures above approximately 230 C decomposed the lubricating oil which was injected for lubrication.

The demand for more powerful and more efficient power plants to generate electric power and for ship propulsion lead to the development of steam turbines to replace reciprocating engines, and water tube boilers to replace pool-type boilers. The steam turbine eliminated the problem of high temperature lubrication and enabled the construction of single units several hundred times more powerful than practical with reciprocating engines. The water tube boiler, or steam generator, contains the high pressure steam in many small tubes which considerably simplifies the mechanical design of high pressure boilers and greatly reduces the probability of catastrophic boiler explosions.

12.1.2 Superheat-Rankine-Cycle Plant

With the advent of water tube boilers, it was a simple matter to continue the positive heat transfer process until the discharge was a superheated vapor state rather than a saturated vapor state. The simplest model of the superheat-Rankine-cycle plant is identical to that of the Rankine-cycle plant except that the constant pressure heat exchanger with the positive heat transfer now represents both the boiler and the superheater. The complete description of the superheat Rankine cycle is also the same as before except that the state of the steam leaving the superheater must be specified since it is no longer in a saturated vapor state. Figure 12.4 illustrates the influence of superheat on Rankine cycle efficiency.

In the turbine plant the thermal efficiency is limited by the design of the superheater tubes. The allowable operating stress for the superheater tubes is a strong function of the tube temperature. This limits the temperature of the steam leaving the superheater to about 560 C. For higher steam temperatures, the tubes reach a temperature at which the sulphur and other impurities in the flue gases corrode the tubes and cause early failure. The op-

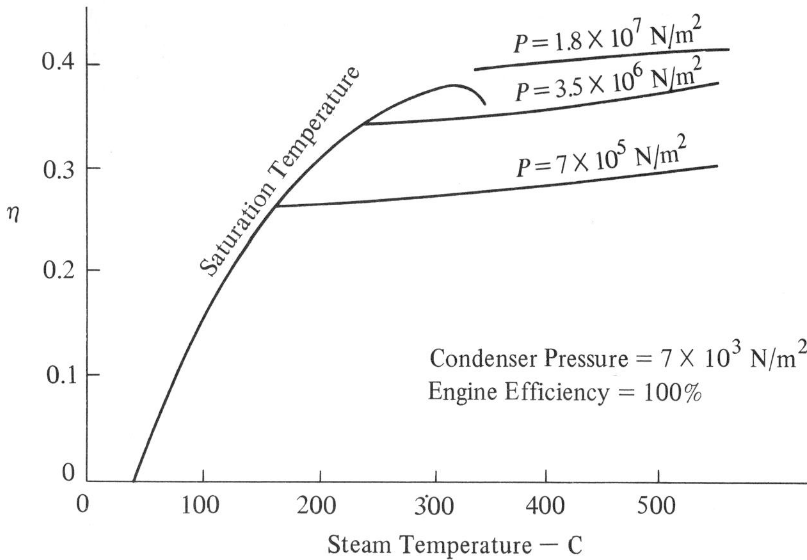

Figure 12.4 Influence of superheat on Rankine cycle efficiency

timum steam pressure is determined by balancing the increase in efficiency at high steam pressures against the increased cost of the boiler and boiler feed pump. The economic limit on steam pressure has increased continuously over the years and steam generators now operate at supercritical pressures.

12.1.3 Reheat-Rankine-Cycle Plant

The condensation of steam as it passes through the turbine is an important factor in the design of steam plants. As is shown in Figure 12.5, for fixed temperature at the discharge of the boiler, the quality of the steam leaving the turbine decreases as the steam pressure is increased. When a high performance turbine handles wet steam, the water droplets impact the blades with sufficient velocity to cause serious erosion of the blade surfaces. Because of this erosion, high pressure steam plants include a high pressure turbine, a low pressure turbine, and a re-superheater. When the superheated steam has expanded sufficiently in the high pressure turbine to become a two-phase fluid, or near the saturated state, the steam is returned to the re-

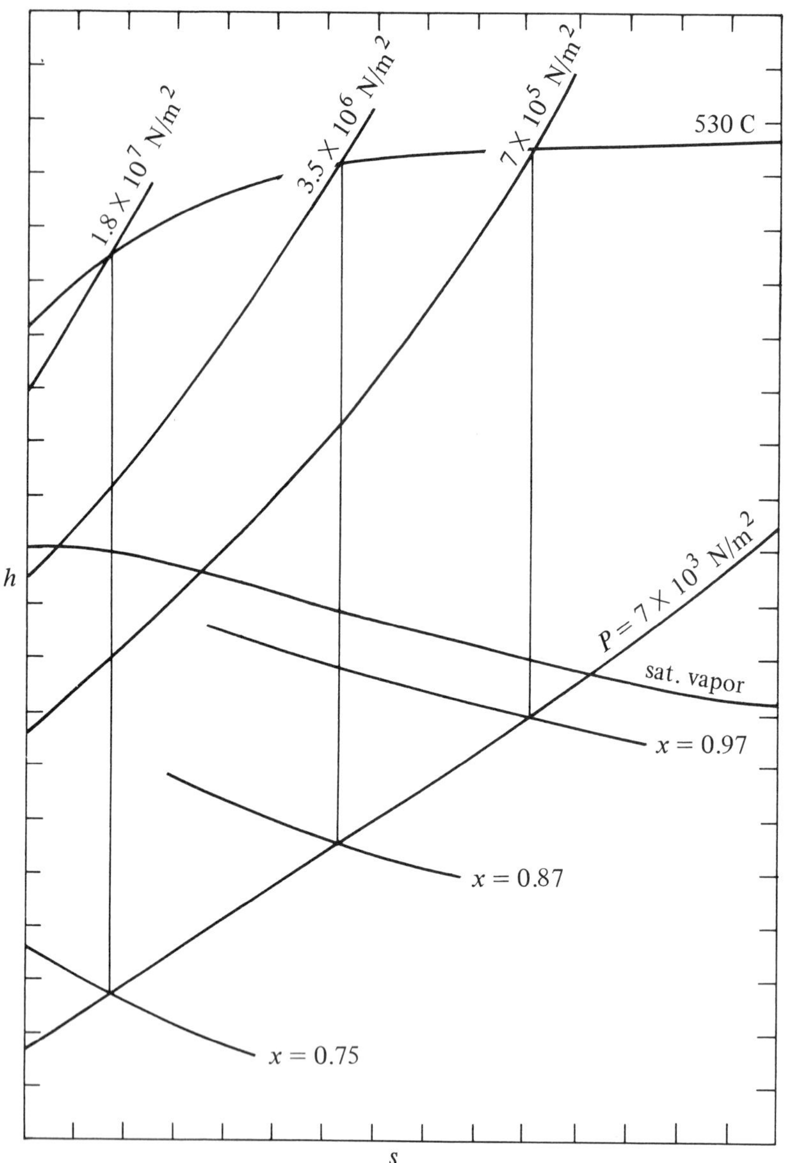

Figure 12.5 Influence of boiler pressure on quality of steam at turbine exit

superheater (reheater) in the steam generator. After being reheated, the now superheated steam returns to the low pressure turbine where it expands to condenser pressure as in Figure 12.6. The simplest model for the reheater is a constant pres-

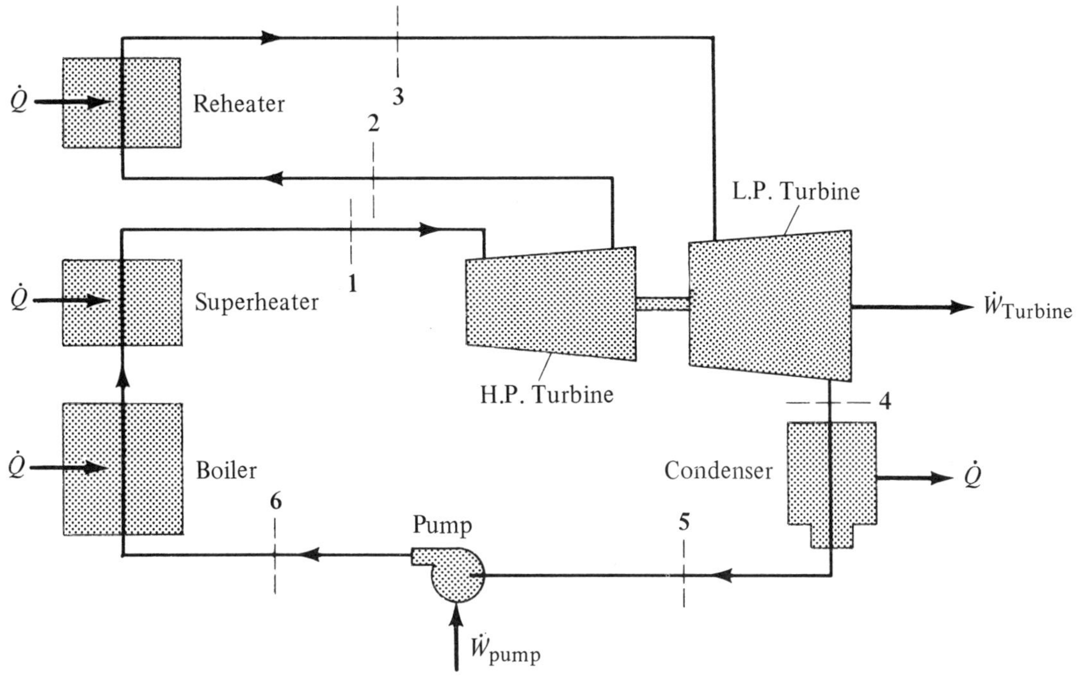

Figure 12.6(a) High pressure steam plant with reheat

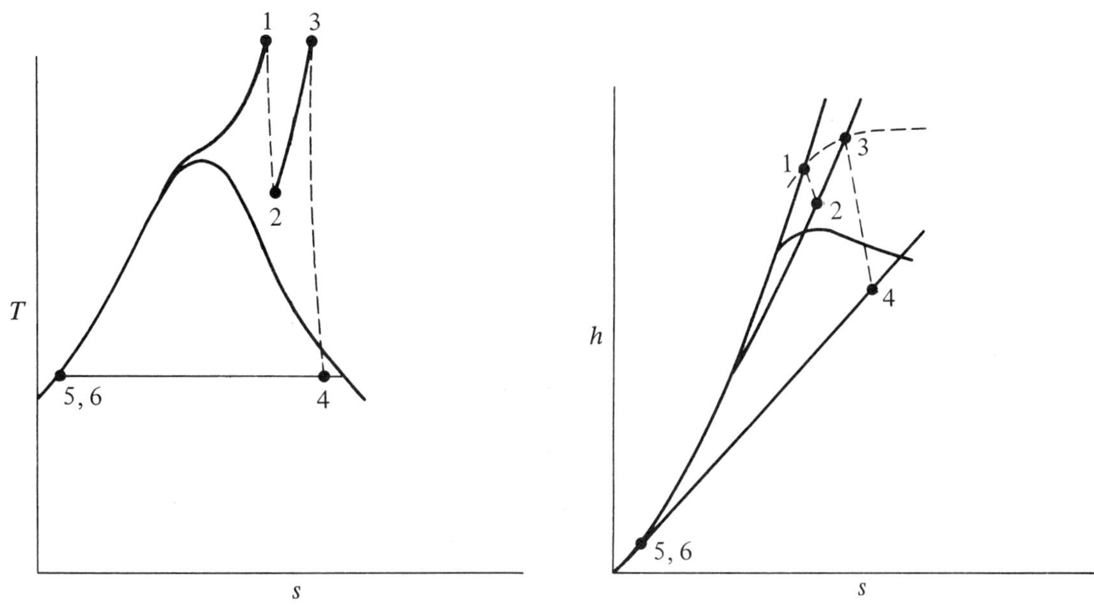

Figure 12.6(b) Thermodynamic states for the high pressure steam plant with reheat

TABLE 12.2
Performance of High Pressure Steam Plant with Reheat
(as shown in Figure 12.6(a))

Turbine power:	$\dot{W}_{\text{turbine}} = \dot{m}(h_1 - h_2) + \dot{m}(h_3 - h_4)$
Condenser load:	$\dot{Q}_{\text{cond}} = \dot{m}(h_5 - h_4) = \dot{m}T_5(s_5 - s_4)$
Pump power:	$\dot{W}_{\text{pump}} = \dot{m}(h_5 - h_6)$
Boiler and superheater load:	$\dot{Q}_{bs} = \dot{m}(h_1 - h_6)$
Reheater load:	$\dot{Q}_{\text{reheat}} = \dot{m}(h_3 - h_2)$
Energy conversion efficiency:	$\eta = \dfrac{(h_1 - h_2) + (h_3 - h_4) + (h_5 - h_6)}{(h_1 - h_6) + (h_3 - h_2)}$
	$\eta = 1 - \dfrac{(h_4 - h_5)}{(h_1 - h_6) + (h_3 - h_2)}$
Net work ratio:	$\text{NWR} = \dfrac{(h_1 - h_2) + (h_3 - h_4) + (h_5 - h_6)}{(h_1 - h_2) + (h_3 - h_4)}$
	$\text{NWR} = 1 - \dfrac{(h_6 - h_5)}{(h_1 - h_2) + (h_3 - h_4)}$

sure heat exchanger. The complete description of the reheat Rankine cycle includes the specification of the steam pressure in the reheater and the temperature of the superheated steam at the reheater exit in addition to the information that describes the superheat Rankine cycle. The formulas for the cycle performance are given in Table 12.2. The reheating of the steam not only reduces blade erosion by increasing the quality of the steam at the exit of the turbine, but also increases the thermal efficiency of the cycle as well.

12.1.4 Regenerative-Rankine-Cycle Plant

Another method for increasing the thermal efficiency of a steam power plant is known as regenerative feed water heating and is almost universally employed in steam plants. In the simplest terms, the improvement in efficiency is the result of preheating the cold boiler feed water with low pressure, low temperature steam, before the feed water enters the boiler and is heated by high temperature combustion gases. Regenerative feed water heating results in heat transfer to the cold feed water across a smaller temperature difference with attendant lower irreversibility and a high efficiency.

The steam for heating the feed water is extracted from the turbine at an intermediate pressure as is shown in Figure 12.7. The feed water heater may be "closed" so that the extraction pressure may be lower than the feed water pressure, or the feed water heater may be "open" so that the extraction steam simply mixes with the feed water which must be at the extraction pressure. In the closed heater, which is simply a two-passage heat exchanger, the extraction steam condenses and the liquid collects in the bottom of the heater. A steam trap (essentially a float valve) allows the condensate to leave the heater but blocks the flow of vapor. The heater drip normally flows back to the condenser.

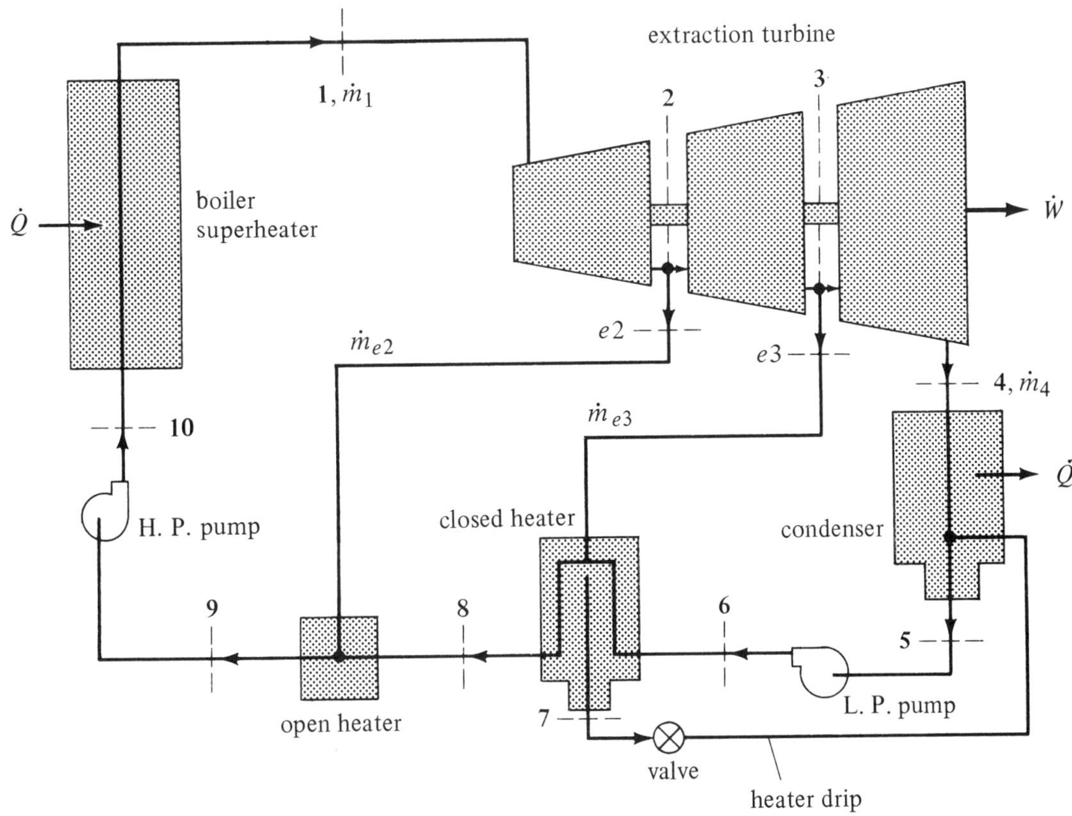

Figure 12.7 Steam plant with two regenerative feed water heaters

The closed heater may be modeled as having sufficient heat transfer surface so that the feed water leaves the heater at the saturation temperature corresponding to the extraction pressure. However, it is still subcooled since the feed water pressure is higher than the extraction pressure. The heater drip is taken as saturated liquid. The open heater is normally modeled as having ideal contact between liquid and vapor so that the feed water leaves as a saturated liquid.

The basic requirement of continuity for the plant of Figure 12.7 is

$$\dot{m}_1 = \dot{m}_{e2} + \dot{m}_{e3} + \dot{m}_4 \tag{12.1}$$

The specified steam state at 1 together with the specificed extraction pressures and the condenser pressure are sufficient to determine states 2, 3, and 4 if the efficiencies of each section of the turbine are specified. States 5, 7, and 9 are saturated liquid states. State 8 is compressed liquid at T_{sat} corresponding to P_3. The pump efficiencies fix states 6 and 10. With all the states specified, the extraction flow rates are fixed by the first law of thermodynamics for the feed water heat exchangers.

$$\dot{m}_{e3}(h_3 - h_7) = (\dot{m}_4 + \dot{m}_{e3})(h_8 - h_6) \tag{12.2}$$

$$\dot{m}_{e2}(h_2 - h_9) = (\dot{m}_1 - \dot{m}_{e2})(h_9 - h_8) \tag{12.3}$$

The work transfer rate for the turbine must be scaled to account for the reduced flow through the low pressure stages as follows:

$$\begin{aligned} \dot{W}_{turb} = {} & \dot{m}_1(h_1 - h_2) + (\dot{m}_1 - \dot{m}_{e2})(h_2 - h_3) \\ & + (\dot{m}_1 - \dot{m}_{e2} - \dot{m}_{e3})(h_3 - h_4) \end{aligned} \tag{12.4}$$

12.1.5 Energy Source

In the previous discussions we have considered only the heat engine portion of the steam power plant. The heat transfer to the water in the steam generator is from the high temperature combustion gases resulting from the combustion of fuel with air. The most elementary model for a combustion process is based on the limiting case of reaction in an infinitely dilute mixture of fuel in air in which the thermodynamic constitutive relations of the products of combustion are essentially identical to the thermodynamic constitutive relations of the reactants (fuel and air mixture prior to combustion) even though the chemistry of one is significantly different from the other. We account for the effect of the combustion process itself by means of the *heating value*, or *enthalpy of reaction*, of the fuel as illustrated in Figure 12.8. To obtain the total enthalpy of the reactants, we simply add to the enthalpy of the reactants prior to combustion an additional enthalpy equal to the heating value for each unit of mass of unburned fuel present in the reactants. Since the diluent of the dilute mixture is modeled as an ideal gas, the influence of temperature is taken into account by the specific heat of the air or other inert material comprising the diluent. In this manner, the heating value of the fuel is analogous to the latent heat associated with a constant temperature phase change in the pure substance model.

If we apply the first law of thermodynamics to the open-system, isothermal reaction chamber shown schematically in Figure 12.8, we find that the heating value, HV, of the fuel is equal to the amount of heat transfer *from* the fluid stream for each unit of mass of fuel burned. Thus,

$$\dot{Q} = -\dot{m}_f(HV) = \dot{m}_f h_{rpo} = \dot{m}_f(h_{\text{out}} - h_{\text{in}})_{\text{at same } T} \tag{12.5}$$

As is shown in equation (12.5), the enthalpy of reaction, h_{rpo}, is the negative of the heating value so that an exothermic chemical reaction is one in which the enthalpy of the products of the reaction is less than the enthalpy of the reactants (at the same temperature). The enthalpy of reaction gives the influence of the chemical composition on the enthalpy.

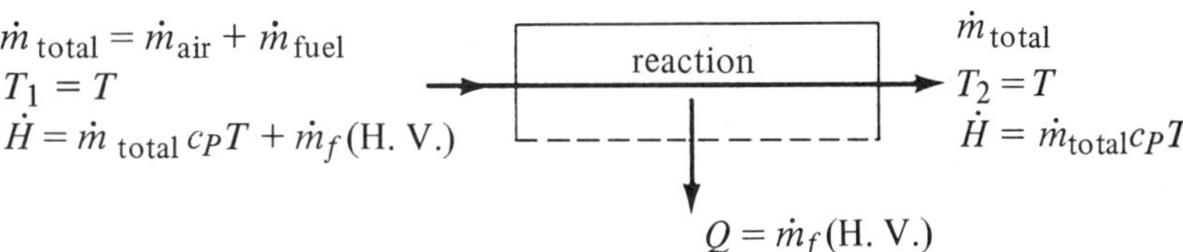

Figure 12.8 Definition of heating value of a fuel

In more advanced treatments[1] the influence of the change in chemical composition on the thermodynamic properties is considered in detail. The simple model is somewhat better than expected when used for the combustion of hydrocarbons in air because the properties of the CO_2 and H_2O produced by the combustion are not too different from the properties of the O_2 consumed in the combustion. In any case the products are diluted by the nitrogen which is 79 percent of the air on a mole basis.

Returning now to Figure 12.2, we find that the heat transfer rate to the steam would be $\dot{m}_f(HV)$ if the energy source (steam generator) were to exhaust the stack gases at the inlet air temperature. The actual heat transfer is less than this for a number of reasons so it is conventional to define the boiler (or steam generator) efficiency as

$$\eta_{\text{S.G.}} = \frac{\dot{Q}}{\dot{m}_f(HV)} \tag{12.6}$$

The common causes of loss of efficiency are: temperature of stack gases higher than inlet air temperature, incomplete combustion of the fuel, and heat transfer to the environment through the walls of the steam generator.

The overall efficiency of the power plant is the product of the boiler efficiency and the steam cycle efficiency.

$$\eta_{\text{overall}} = \frac{\dot{W}_{\text{net}}}{\dot{m}_f(HV)} \tag{12.7}$$

Figure 12.9 shows in a schematic way the arrangement of a steam generator which will give a minimum stack temperature and prevent the overheating and burnout of the water tubes.

The preheated air is mixed with pulverized or atomized fuel and burns in the open furnace volume. The furnace is completely surrounded by close-packed water tubes which form the furnace

[1] Denbigh, K., *The Principles of Chemical Equilibrium,* Cambridge University Press, London, 1968. Williams, F. A.; *Combustion Theory,* Addison-Wesley Publishing Co., Reading, Mass., 1965.

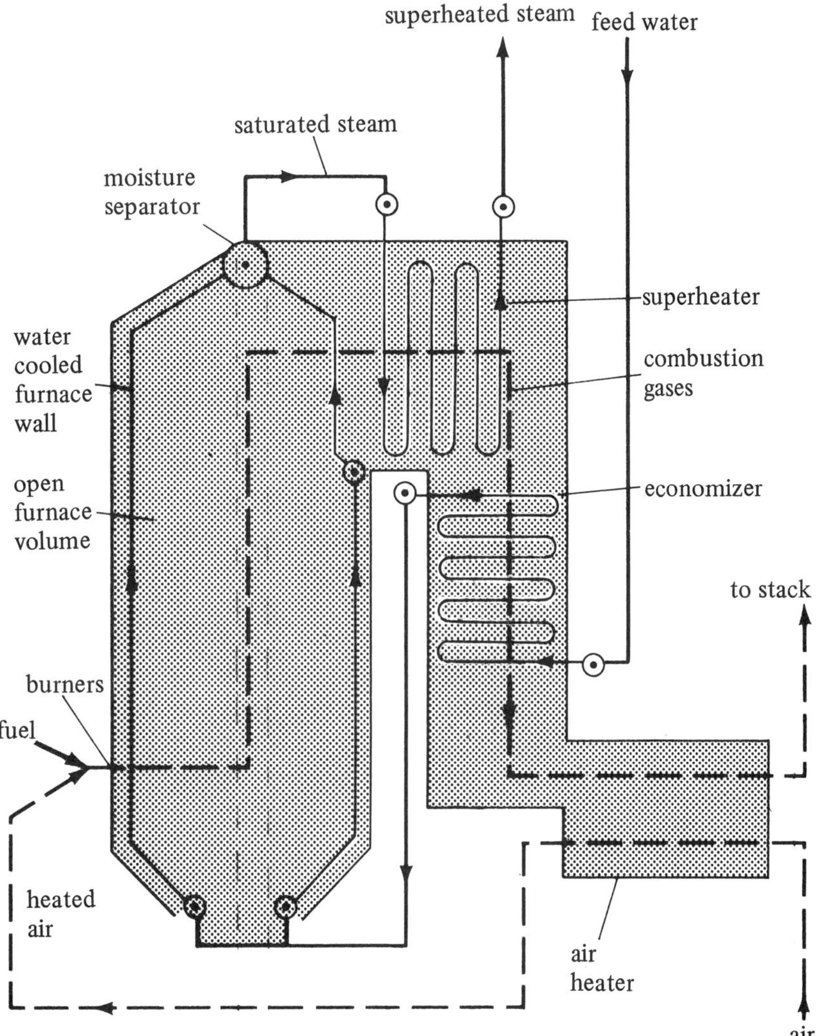

Figure 12.9 Schematic layout of a steam generator

walls. The high temperature flame heats the water in the tubes by thermal radiation. A high heat transfer coefficient is associated with boiling water so the tubes are kept from overheating. The furnace volume must be sufficient to allow complete combustion of the fuel. In smaller furnaces some of the wall surface is refractory material which becomes hot and reradiates to the flame to prevent cooling before combustion is complete. A major fraction of the heat transfer for the boiler is through the water tube walls of the furnace.

The combustion gases, which have been significantly cooled in the furnace, now pass over the superheater tubes. The gases are still hot enough to overheat the superheater tubes; however, the flow of steam at high velocity inside the tubes provides good heat transfer between the tube walls and the steam which keeps the tube wall temperature near the steam temperature. The steam temperature at the superheater outlet is controlled by by-passing some of the gases around the superheater.

After the superheater, the gases pass into the economizer section where the last stage of feed water heating is accomplished, and the temperature of the feed water is increased nearly to the boiling point. The combustion gases are finally cooled to stack temperature in the air heater which is used to preheat the incoming combustion air.

The minimum stack temperature is determined by the necessity to avoid condensation in the stack and air heater. If condensation occurs, it is very corrosive because of the sulphur in the fuel. The condensate also causes the ash dust to collect as a hard deposit. The minimum stack temperature is about 150 C depending upon the hydrogen and sulphur content of the fuel and the design of the air heater and stack.

12.1.6 Nuclear Powered Steam Plant

A nuclear powered steam plant is basically a conventional steam plant with a nuclear reactor replacing the combustion of fuel as the energy source. However, the design of the entire plant must be adjusted to the characteristics of the reactor. First, the requirements of nuclear safety limit the thermodynamic design of the plant, specifically the maximum steam temperature. The basic nuclear fuel costs are lower than conventional fuel costs; however, reactor capital costs are higher than conventional steam generator costs. As a result, the conventional components of a nuclear steam plant tend to sacrifice efficiency to reduce capital cost.

Two basic designs for nuclear power plants have been accepted by the power industry. The first is the pressurized water reactor (PWR). In this plant (Figure 12.10), the reactor is

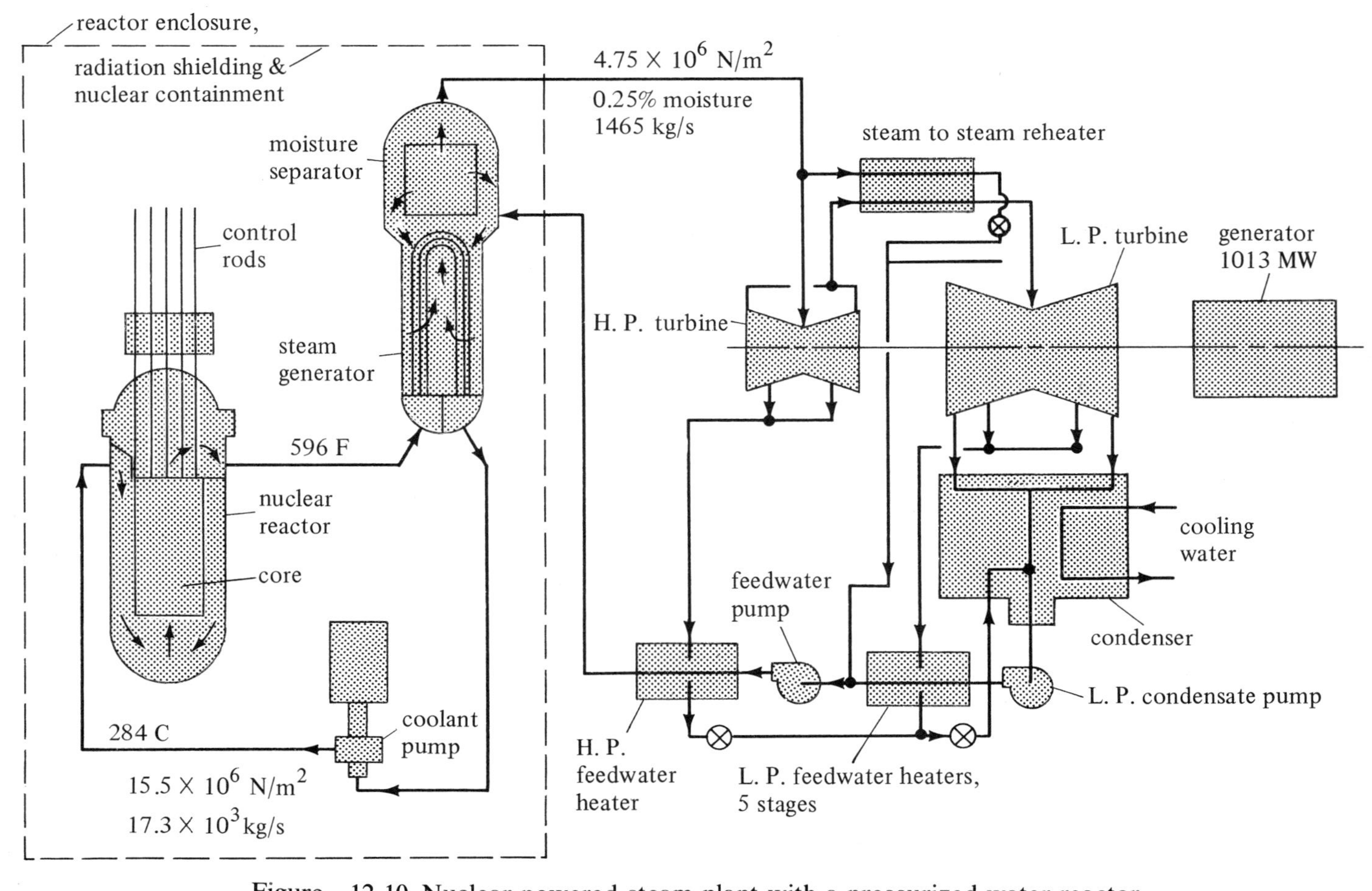

Figure 12.10 Nuclear powered steam plant with a pressurized water reactor

cooled by pressurized water circulated without boiling between the reactor and the steam generator. In the steam generator, heat transfer from the primary coolant (pressurized water) to the secondary circuit (low pressure water) generates steam which circulates through the turbine and condenser.

The turbine, condenser, feed water pumps, and regenerative feed water heater are similar to the same components in the fuel fired plant. The major difference here is that the turbine must expand wet steam because of the rather low maximum temperature in the reactor. In order to handle wet steam, the turbine has both internal and external separators to separate moisture drops from the steam flow. In addition, Figure 12.10 shows how steam directly from the steam generator can be used to reheat intermediate pressure steam and reduce the amount of moisture passing through the low pressure turbine.

The second accepted design for nuclear power plants is the boiling water reactor (BWR). In this design (Figure 12.11) the reactor is cooled directly by the H_2O which circulates through the turbine, condenser, and feed water heaters. A circulating pump is employed to circulate a large volume of saturated liquid H_2O through the reactor core. The vapor which is formed in the core cooling passages is separated from the liquid in the upper part of the reactor pressure vessel. The saturated vapor then passes directly to the turbine. The circulating pump is necessary to adequately and uniformly cool the reactor core since simple natural convection (used in smaller conventional fuel fired boilers) cannot provide adequate heat transfer to absorb the high heat transfer fluxes produced in the reactor core.

The small quantities of impurities in the steam (especially air) become radioactive in the reactor and are carried into the turbine. As a result, the conventional plant components contain low level radioactive material while in operation; however, this activity has a short half life so activity decreases rapidly on shut down.

12.2 Gas Turbine Power Plants

A gas turbine power plant is a complicated thermodynamic system which processes fuel and air to produce positive useful

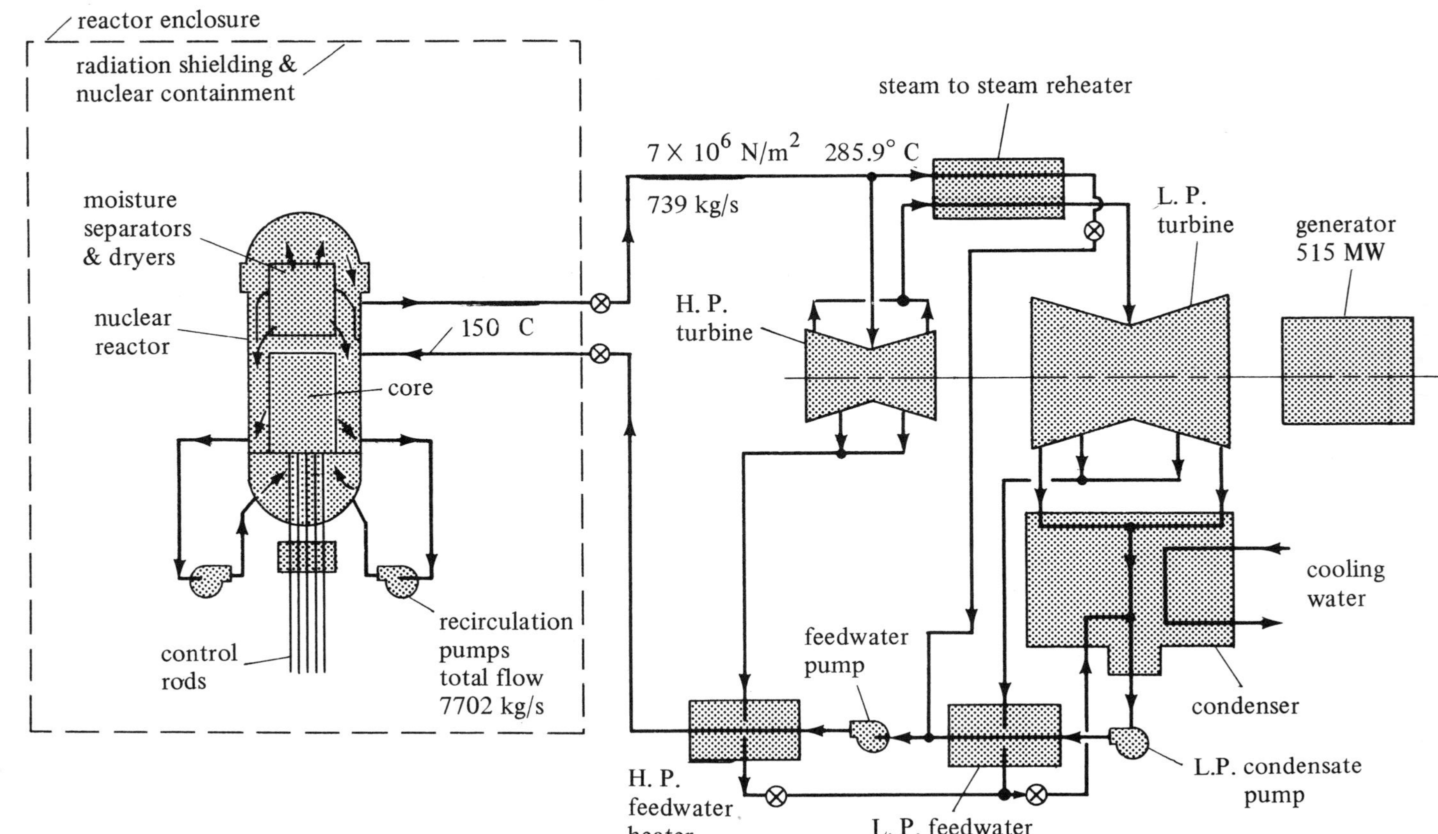

Figure 12.11 Nuclear powered steam plant with a boiling water reactor

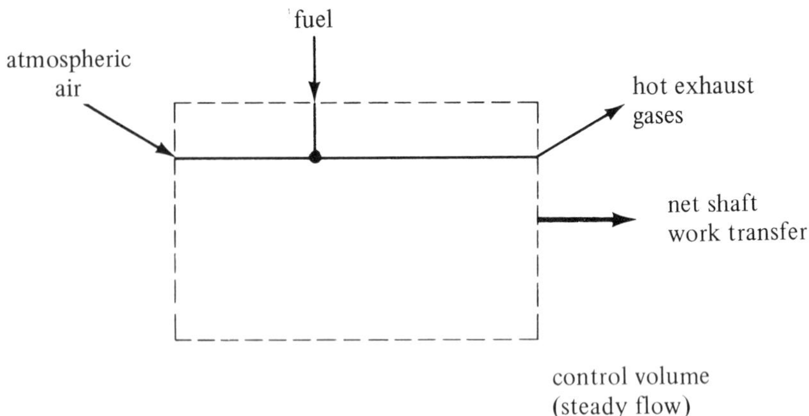

Figure 12.12 Control volume for gas turbine plant

work transfer. The overall plant can be represented by the control volume shown in Figure 12.12. From the overall point of view the gas turbine plant is the same as the steam plant—it processes air and fuel to products of combustion and discharges hot fluid to the environment. In contrast however, the gas turbine does not have an internal closed system (the heat engine) which executes a cycle. Therefore, the gas turbine plant is a process plant (open cycle) rather than a closed cycle plant. In addition, the plant has only one fluid circuit (rather than three) and no internal heat transfers.

The physical components of the gas turbine plant are a steady flow turbo-compressor, a burner, and a steady flow turbine expander arranged in series as shown in Figure 12.13. In operation, atmospheric air is compressed in the compressor by means of a negative work transfer. The high pressure air from the compressor is increased in temperature and specific volume by the combustion of fuel in the burner. The hot air is expanded through the turbine producing a positive shaft work transfer. Hot gases from the turbine are exhausted to the atmosphere. The plant produces a net positive shaft work transfer because the expansion of the hot (high specific volume) air through the turbine produces more shaft work transfer than is required to compress the cold (low specific volume) air in the compressor.

The basic model for a gas turbine power plant consists of a steady-flow adiabatic compressor, a constant pressure burner,

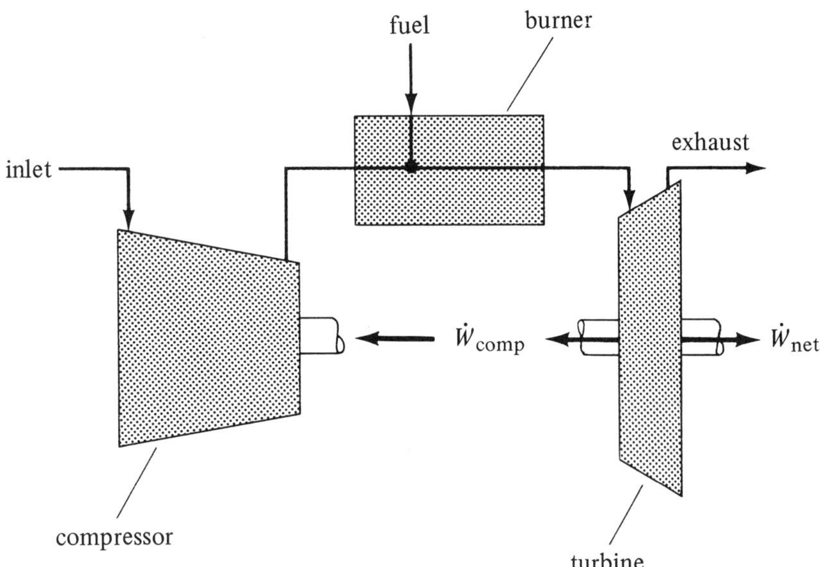

Figure 12.13 Basic gas turbine plant

and a steady-flow adiabatic turbine arranged in series as shown in Figure 12.14. The gas turbine plant, in contrast to the steam plant, utilizes the same atmospheric air for the work transfer components as for combustion of the fuel. As a result, the fuel must be free of corrosive and abrasive materials which could damage the work producing turbine.

The simplest model for the gas turbine plant considers the burner as an equivalent constant pressure heat transfer process. The equivalent heat transfer rate to the air is the product of the mass flow rate of fuel burned and the heating value of the fuel ($-h_{rpo}$). As before, this model neglects the increased mass flow because of fuel addition to the air stream. In addition, the change in chemical composition as a result of the combustion is neglected. This simple model for the burner allows the gas turbine plant to be modeled as an equivalent simple closed cycle operating with air as the working fluid. The hot exhaust air is returned to the inlet state by means of an equivalent constant pressure heat transfer process as shown in Figure 12.14. The closed cycle model for the gas turbine plant (Figure 12.14) is called the Brayton cycle or the air standard cycle. It consists of

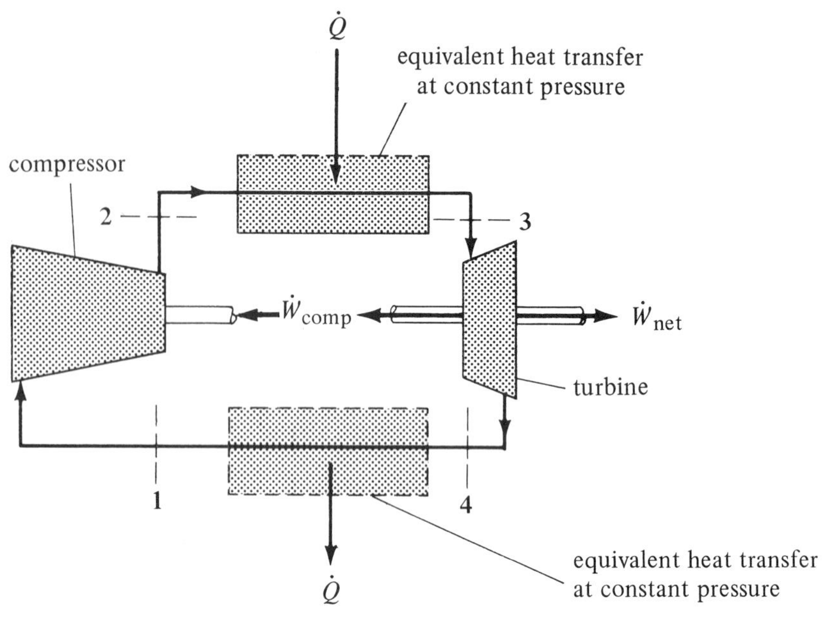

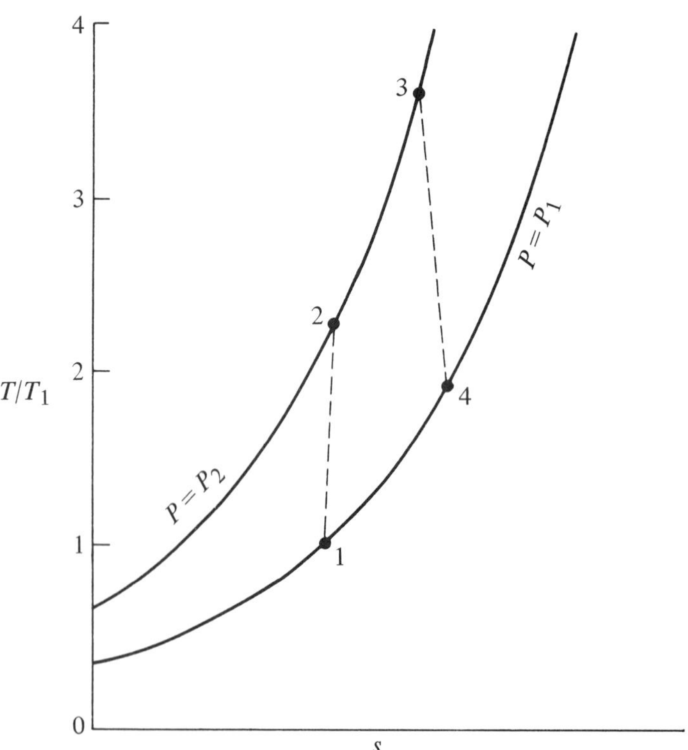

Figure 12.14 Brayton cycle model for a gas turbine plant

two adiabatic work transfer processes and two constant pressure heat transfer processes.

The operating conditions for the gas turbine plant are normally specified by the mass flow rate of air, the compressor inlet temperature, pressure (normally atmospheric), and the turbine inlet temperature and pressure. The compressor and turbine efficiencies are necessary to define the performance of the plant.

The closed cycle model for the gas turbine plant is useful to show the influence of the turbine inlet temperature (T_3), the pressure ratio (P_2/P_1) and the turbine and compressor efficiencies (η_T and η_C) on the performance of the plant. The overall performance is judged by the cycle efficiency which represents the fuel economy of the plant and by the net work transfer rate per unit mass flow which represents the power density or the relative size and weight of the plant.

As shown in Table 12.3 the performance of the cycle is derived directly from the equations for the components given in Chapter 11. The basic characteristics of the Brayton cycle can be shown simply by neglecting all changes in kinetic energy and assuming constant specific heats. With these simplifications, the expressions for the cycle efficiency and the net work transfer rate become

TABLE 12.3
Performance of Brayton Cycle Model for a Gas Turbine Plant

Compressor power:	$\dot{W}_{comp} = \dot{m}(h_1 - h_2) = \dot{m}c_P(T_1 - T_2)$
Equivalent heat addition:	$\dot{Q}_{add} = \dot{m}(h_3 - h_2) = \dot{m}c_P(T_3 - T_2)$
Turbine power:	$\dot{W}_{turb} = \dot{m}(h_3 - h_4) = \dot{m}c_P(T_3 - T_4)$
Equivalent heat rejection:	$\dot{Q}_{reject} = \dot{m}(h_1 - h_4) = \dot{m}c_P(T_1 - T_4)$
Energy conversion efficiency:	$\eta = \dfrac{(h_1 - h_2) + (h_3 - h_4)}{(h_3 - h_2)}$
	$\eta = 1 - \dfrac{(h_4 - h_1)}{(h_3 - h_2)} = 1 - \dfrac{(T_4 - T_1)}{(T_3 - T_2)}$
Net work ratio:	$\text{NWR} = \dfrac{(h_1 - h_2) + (h_3 - h_4)}{(h_3 - h_4)} = 1 - \dfrac{(h_2 - h_1)}{(h_3 - h_4)}$
	$\text{NWR} = 1 - \dfrac{(T_2 - T_1)}{(T_3 - T_4)}$

$$\eta = \frac{\frac{1}{\eta_C}(1 - \tau) + \eta_T\left(1 - \frac{1}{\tau}\right) T_3/T_1}{(T_3/T_1) - 1 - \frac{1}{\eta_C}(\tau - 1)} \qquad \textbf{(12.8)}$$

$$\frac{\dot{W}_{\text{net}}}{\dot{m}c_P T_1} = \frac{1}{\eta_C}(1 - \tau) + \eta_T (T_3/T_1)\left(1 - \frac{1}{\tau}\right) \qquad \textbf{(12.9)}$$

$$\frac{\dot{W}_{\text{net}}}{\dot{m}c_P T_1} = (\tau - 1)\left\{\frac{\eta_T(T_3/T_1)}{\tau} - \frac{1}{\eta_C}\right\} \qquad \textbf{(12.10)}$$

Here τ is the isentropic temperature ratio which is given by

$$\tau = (P_2/P_1)^{(\gamma-1)/\gamma} = (P_3/P_4)^{(\gamma-1)/\gamma}$$

The use of dimensionless variables in equations (12.8) and (12.10) reduces the number of independent variables necessary to specify the efficiency and the net work transfer rate.

Equations (12.8) and (12.10) are shown in Figure 12.15 for fixed T_3/T_1 and various values of η_C and η_T. The figure clearly shows the pressure ratio for maximum power and a different pressure ratio for maximum efficiency for each combination of η_T and η_C. This illustrates in a simple manner the normal compromise required in the design of thermal power systems. The design for a light, inexpensive, and powerful plant usually results in an efficiency lower than the maximum obtainable with a larger and more expensive plant. The best design for a given application depends upon the relative importance of fuel economy and size and cost.

Figure 12.15 also shows the sensitivity of the gas turbine to the compressor and turbine efficiency. With compressor and turbine efficiencies of 0.85 the maximum power is only 49 percent of the power which could be obtained with perfect components. This sensitivity to losses is the result of the relatively low net work ratio. For example, the maximum power design with $\eta_T = \eta_C = 0.85$ has a net work ratio of approximately 0.26; thus, nearly three-fourths of the turbine power is used to power the compressor. A one percent loss in the turbine therefore represents a 4 percent loss in net power.

Figure 12.16 shows how the maximum efficiency and the maximum power (shown in Figure 12.15) individually vary with the overall temperature ratio T_3/T_1 for fixed compressor

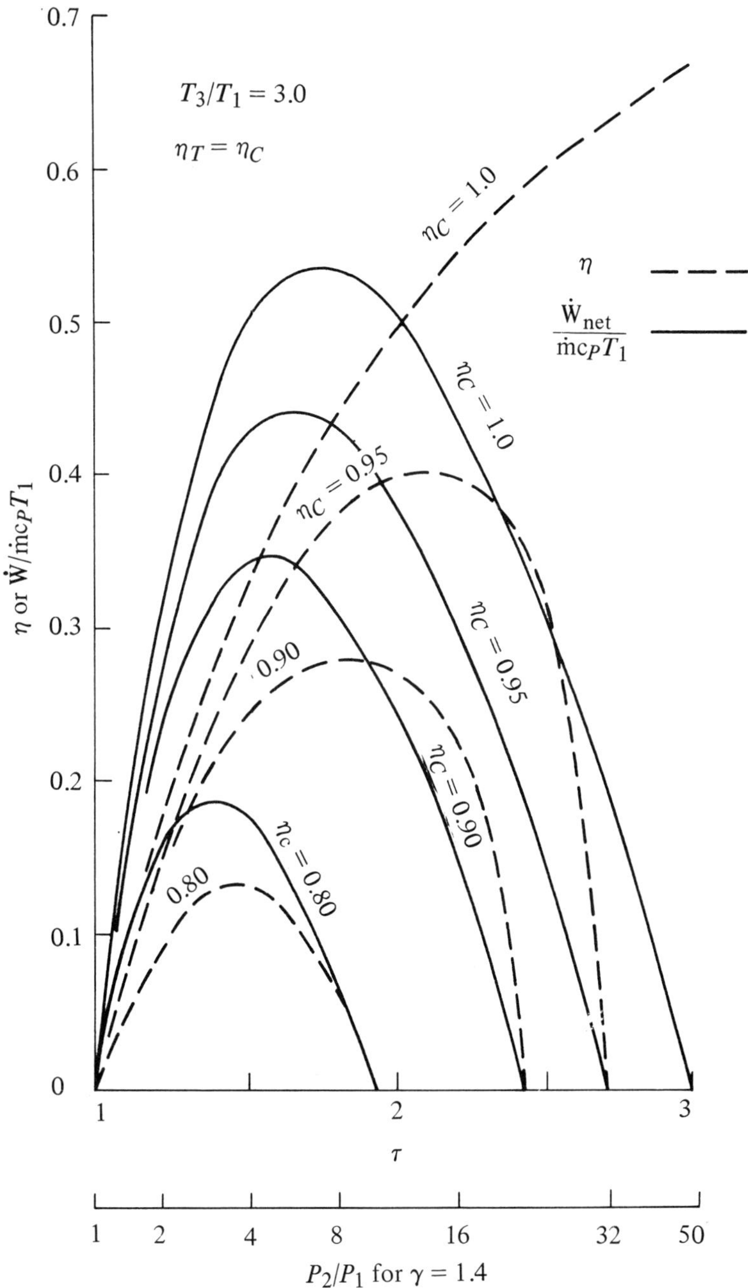

Figure 12.15 Influence of pressure ratio and component efficiency on the Brayton cycle

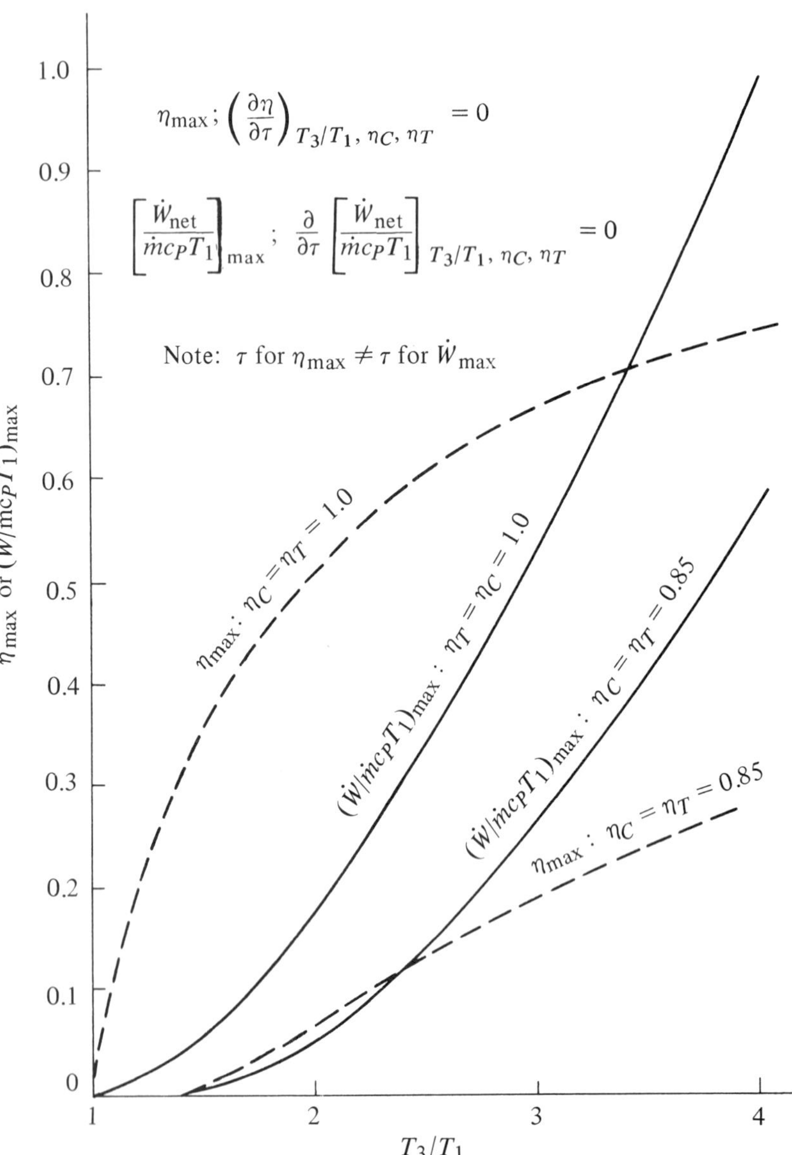

Figure 12.16 Influence of overall temperature ratio on maximum performance of Brayton cycle

and turbine efficiencies. Figure 12.16 also shows clearly that a high turbine inlet temperature is necessary for a high performance (high efficiency and high power) gas turbine plant. Unfortunately, in practice the turbine inlet temperature is limited by the mechanical strength of the turbine blades which operate essentially in thermal equilibrium with the hot gases passing through the turbine while being subjected to high centrifugal stresses. This need for high strength, high temperature materials for turbine blades has motivated considerable metallurgical research, and gas turbines now operate with turbine inlet temperatures in the range of 800 to 1100 C. The most recent development has been gas turbines with internally cooled turbine blades which operate at a temperature below the gas temperature. Historically the thermodynamics of the gas turbine were well understood before the availability of high temperature materials allowed the construction of practical plants. The development of efficient axial-flow compressors was also a significant factor in the realization of practical gas turbines. Currently, compressor efficiencies in the range 0.85–0.90 and turbine efficiencies in the range 0.90 – 0.95 are commonly achieved.

The gas turbine is most commonly employed for aircraft propulsion because the plant can develop a high power per unit mass of equipment and per unit volume of the plant. (The engine also has a long cylindrical shape which fits well into the aircraft structure.) These high power densities are possible because the engine does not depend upon heat transfer processes for its thermodynamic operation. The size of the gas turbine is determined by pressure equalization rates and chemical reaction rates in the burner, rather than by slower heat transfer rates. Specifically, it is not necessary to have heat transfer between a fixed wall and the thermodynamic operating fluid. The temperature rise in the burner is the result of the chemical reaction within the fluid and the walls of the burner are below the temperature of the flame. A heat exchanger to provide the heat transfer equivalent to the process in the burner would be several times larger than turbine and compressor.

The gas turbine also does not need a heat exchanger for heat transfer to the environment. This heat transfer is effectively accomplished by exhausting hot gases to the atmosphere. In contrast, the steam plant requires large heat exchangers for both positive heat transfer and the negative heat transfer (boiler-superheater and condenser, respectively). The steam plant is most appropriate when high efficiency and the use of unrefined fuel are more important than the large size and complexity. The gas turbine plant is most appropriate when light weight and high power are more important than efficiency and fuel costs.

In the application of gas turbines to aircraft propulsion, the power plant may be separate from or combined with the propulsive mechanism. The functions are completely separate in the turbo-prop engine in which the gas turbine drives a propeller which changes the momentum of the propeller slip stream. This change in momentum gives rise to the force (thrust) that propels the aircraft. The two functions are completely combined in the pure turbojet shown in Figure 12.17. The equivalent-heat-addition model for the turbojet propulsion plant is shown in the *T-s* diagram of Figure 12.17. In this case, the pressure ratio for the turbine is lower than that for the compressor so no net shaft power is produced. In effect, P_4 is fixed by the requirement $\dot{W}_{\text{turb}} = \dot{W}_{\text{comp}}$. The pressure ratio for the nozzle is therefore greater than that for the diffuser. This causes velocity in the exhaust jet, ϑ_j, to be larger than the flight speed, ϑ_∞. The change in momentum of the air processed through the turbojet produced the thrust. The pressure at the exit of the nozzle is the environmental pressure, P_∞, when the jet exhaust velocity is subsonic. The performance of supersonic diffusers and nozzles is considerably more complex.[2] The inlet diffuser is necessary because the flight speed is normally higher than the inlet velocity required for efficient operation of the compressor.

The propjet engine and the fanjet engine (or bypass jet) are combined plants which derive thrust from the change of momentum of the stream processed by the gas turbine as well as from the slip stream of the propeller or fan. In the fanjet, the fan is essentially an enclosed propeller operating down stream of the inlet diffuser. Thus, the fanjet can operate at flight speeds which are excessive for efficient operation of propellers.

---

[2] For further details see Hill, P. G. and Peterson, C. R., *Mechanics and Thermodynamics of Propulsion,* Addison-Wesley Publishing Co., Reading, Mass. 1965.

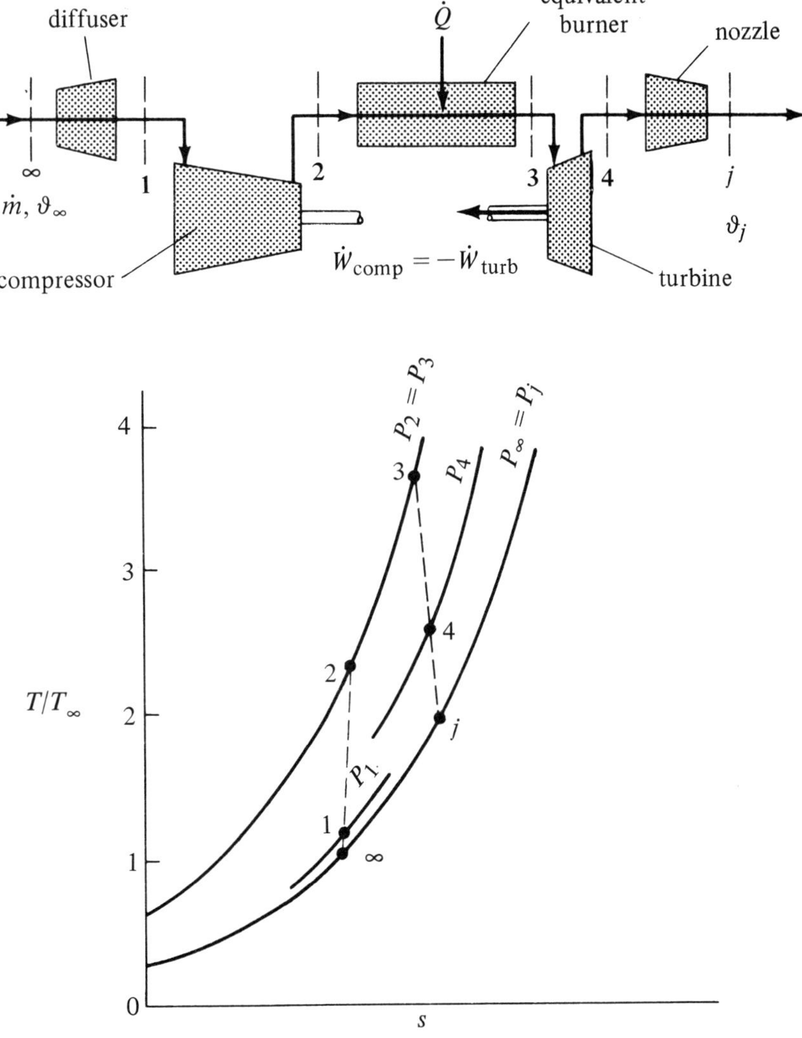

Figure 12.17 Model for a turbojet plant

Regenerative heat transfer can be applied to the gas turbine plant in much the same way that it is employed for feed water heating in the steam plant. Figure 12.18 shows a gas turbine plant employing a counter-flow heat exchanger to preheat the compressed air with the hot turbine exhaust stream. Figure 12.18 also shows the *T-s* diagram for the equivalent-heat-addition model for the regenerative gas turbine. The compressed

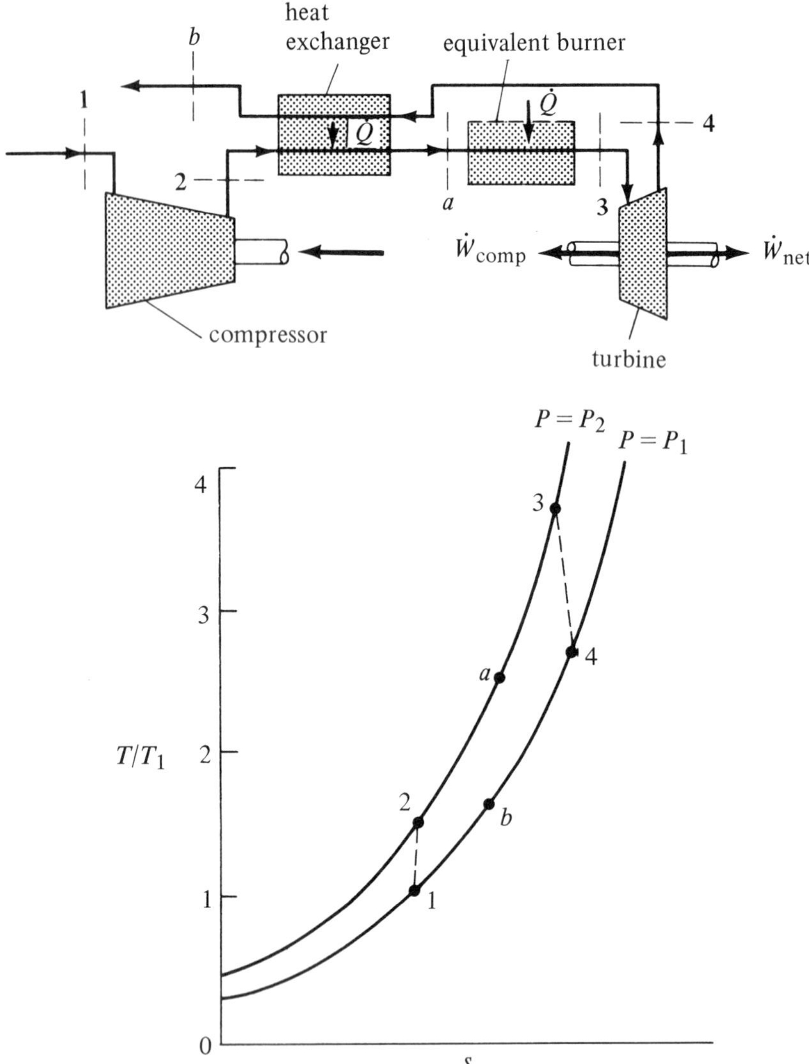

Figure 12.18 Model for a regenerative gas turbine plant

air is heated from state 2 to state a by the exhaust gases which are cooled from state 4 to state b before being exhausted. If the specific heats of the two streams are assumed constant and equal, the first law of thermodynamics for the heat exchanger reduces to

$$T_a - T_2 = T_4 - T_b \quad \textbf{(12.11)}$$

The heat transfer process equivalent to the process occurring in the burner increases the temperature of the gas from state a to 3. The net work transfer is not changed if the heat exchanger is modeled as constant pressure as shown in Figure 12.18.

The major effect of the regenerative heat exchanger is to decrease the pressure ratio for maximum efficiency. In the regenerative cycle the pressure ratio for maximum efficiency is less than that for maximum net work transfer (Figure 12.19). In fact, the

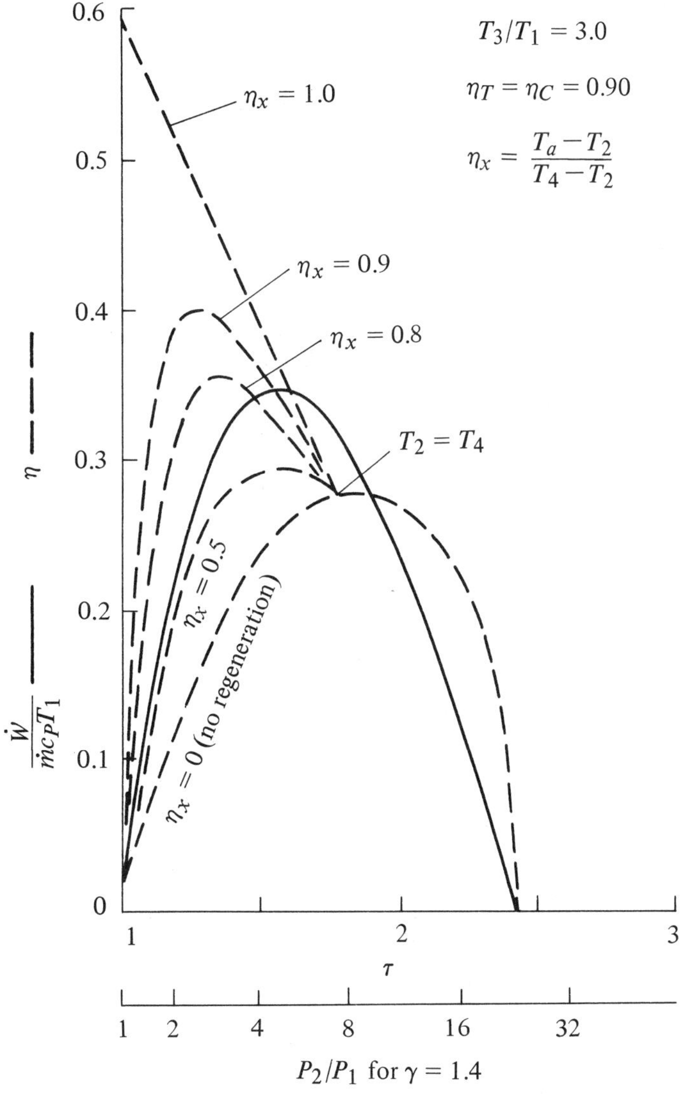

Figure 12.19 Performance of a regenerative gas turbine

maximum efficiency occurs at a pressure ratio of 1 for a regenerative cycle with reversible components. At high pressure ratios, the regenerative heat exchanger is of no use since T_2 is greater than T_4.

The use of the regenerative heat exchanger increases the maximum efficiency that can be obtained, but this increase is at the expense of a significant increase in the size and weight of the gas turbine plant. The regenerative heat exchanger is too large and heavy for aircraft applications, and does not increase the efficiency sufficiently to make the gas turbine competitive with central station steam plants. Thus, the regenerative gas turbine has not found extensive application.

12.3 Reciprocating Internal Combustion Engines

A reciprocating internal combustion (I.C.) engine is a thermodynamic power plant which processes fuel and air to produce positive useful work transfer. From the overall point of view, the reciprocating I.C. engine process is essentially the same as the gas turbine process, as is shown in Figure 12.20. The major difference is that a significant cooling is required for the internal mechanism of the I.C. engine. The I.C. engine does not have an internal closed thermodynamic system and is thus termed a process plant or an open cycle.

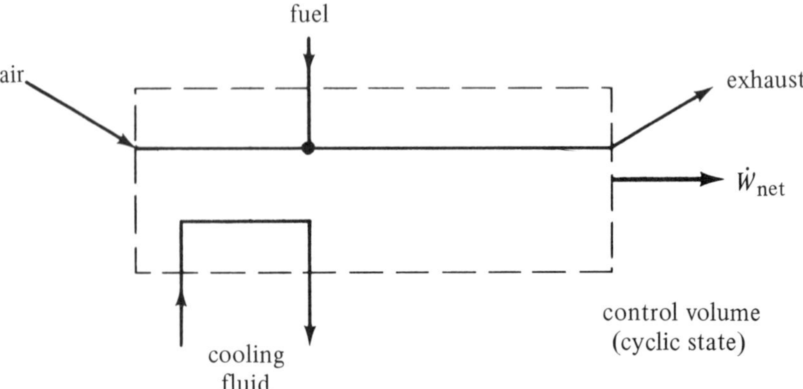

Figure 12.20 Control volume for reciprocating internal combustion engine plant

In contrast to the gas turbine plant, the I.C. engine is not composed of interconnected components, but rather the entire series of thermodynamic processes occur within the same physical apparatus. Since we have not previously analyzed this type of apparatus, we will consider the thermodynamics of the I.C. engine in more detail.

The reciprocating internal combustion engine (Figure 12.21) consists essentially of a reciprocating piston inside a cylinder fitted with two valves. The piston is connected by a crank mechanism to the shaft which delivers the useful work transfer. The engine operates as a periodic-flow open system since the fuel and air that enter the cylinder are processed and then discharged to the atmosphere. The states of the engine (control volume) are cyclic, but a given mass of fuel and air undergoes a process rather than a cycle. In operation, the cylinder

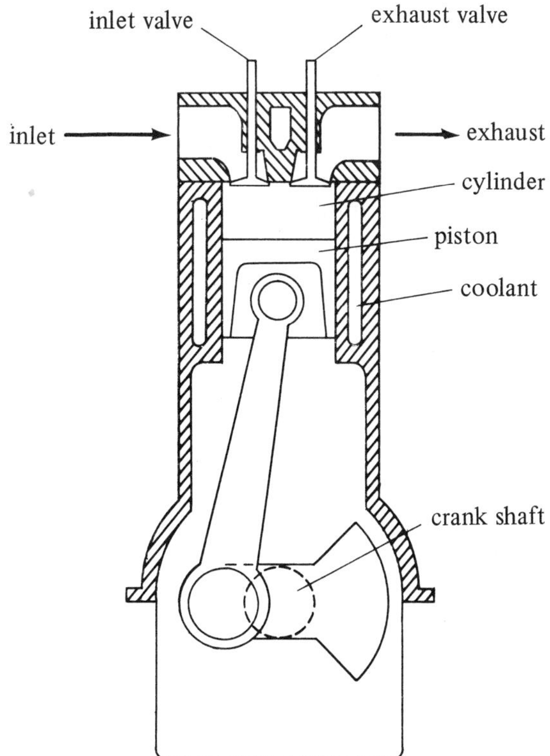

Figure 12.21 Reciprocating internal combustion engine

is filled with a fresh charge of fuel and air, both valves are closed, and the piston compresses the charge to a smaller volume by means of a negative work transfer. A combustion process is initiated within the charge causing the pressure and temperature within the cylinder to increase significantly. This high pressure, high temperature charge is then expanded to produce a positive work transfer. The net work transfer is the excess of this expansion work transfer over the work transfer required for compression. The burned charge leaves the engine through the exhaust valve, and the fresh charge enters through the inlet valve.

The engine is termed a four-stroke engine if a compression stroke is used to exhaust the burned charge and an expansion stroke is used to introduce the fresh charge. The engine is termed a two-stroke engine if the fresh charge merely displaces the burned charge while the piston is essentially at the maximum volume position. The engine is a premix, spark ignition, or Otto engine, if the fresh charge is a mixture of air and fuel vapor. The engine is a compression ignition, or diesel engine, if the fresh charge is only air and the combustion is initiated by injection of fuel into the hot compressed air.

A simple model for the I.C. engine is formulated by replacing the combustion process by a simple positive heat transfer process and by neglecting the heat transfer between the operating fluid and the mechanical parts of the engine. Two very simple models are widely used, one for the spark ignition engine and one for the compression ignition engine. Although the models are too simple to provide useful quantitative results, they do show qualitatively the influence of compression ratio and illustrate the limits on engine performance.

12.3.1 Otto Engine

The simplest model for the Otto or spark ignition engine is best described in terms of the diagram of pressure versus cylinder volume shown in Figure 12.22. The fresh charge of premixed air and fuel vapor at state 1 is compressed reversibly and adiabatically to the minimum volume of state 2. At state 2 the spark initiates the combustion process which is modeled as an equivalent constant volume heat addition process to state 3. The hot gas is

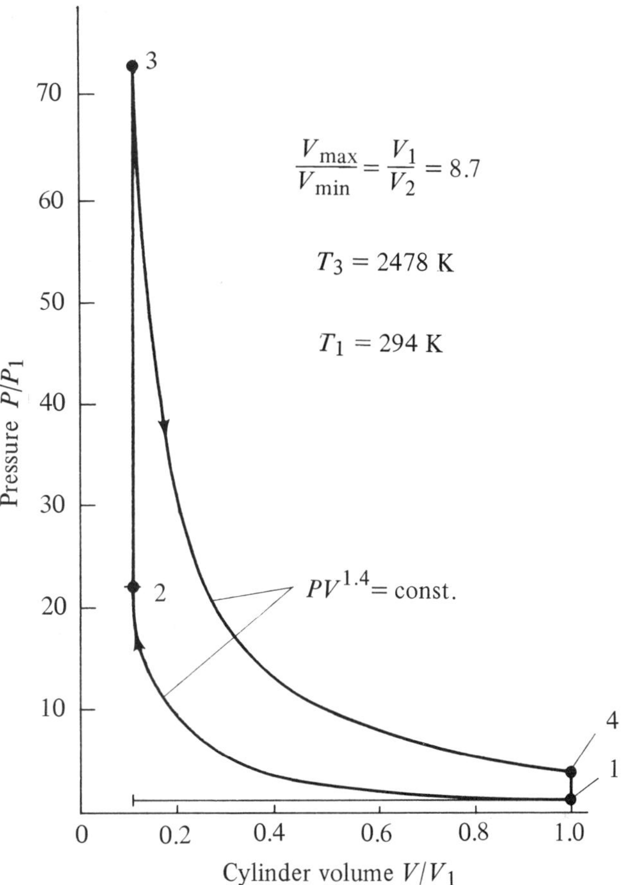

Figure 12.22 Pressure versus cylinder volume for the Otto cycle

then expanded to maximum volume, state 4, in a reversible adiabatic process. At state 4 the exhaust valve opens and the hot gas expands down to exhaust pressure. The remaining exhaust gases are then replaced with a fresh charge, and the cycle of the control volume is completed with return to state 1. Since we have neglected the change in composition caused by the combustion, the mass within the cylinder at state 4 is the same as at state 1. In addition, the volume of the cylinder is the same at 4 and 1; therefore, the pressure release, the exhaust of the old charge, and the intake of the fresh charge may be modeled as an equivalent constant volume heat transfer process for a fixed charge from state 4 to state 1. As with the gas turbine, we have now modeled the cyclic states within the I.C. engine cylinder with a closed cycle. In contrast to the model for the gas turbine, the cycle is executed by a closed system consisting of the mass of the charge in the cylinder.

We can now evaluate the net work transfer for the cycle. It is common with this model to assume that the specific heats of the air are constant. Although the temperature range of the cycle is large enough so that considerable changes do occur, the errors do not change the qualitative results provided by this simple model.

The net work transferred from the gas to the piston per cycle is

$$W_{net} = W_{1-2} + W_{3-4} \tag{12.12}$$

since the work transfer for processes 2-3 and 4-1 are each zero. Processes 1-2 and 3-4 are adiabatic; therefore, the work transfer is the negative of the internal energy change.

$$W_{net} = mc_v(T_1 - T_2) + mc_v(T_3 - T_4) \tag{12.13}$$

Since process 2-3 is constant volume the heat transfer is

$$Q_{2-3} = mc_v(T_3 - T_2) \tag{12.14}$$

Then the thermal efficiency is

$$\begin{aligned} \eta &= \frac{W_{net}}{Q_{2-3}} = \frac{T_1 - T_2 + T_3 - T_4}{T_3 - T_2} = 1 - \frac{T_4 - T_1}{T_3 - T_2} \\ &= 1 - \frac{T_1}{T_2}\left[\frac{T_4/T_1 - 1}{T_3/T_2 - 1}\right] \end{aligned} \tag{12.15}$$

Processes 3-4 and 1-2 are reversible adiabatic with $V_1 = V_4$ and $V_2 = V_3$; therefore, $T_2/T_1 = T_3/T_4$ or $T_4/T_1 = T_3/T_2$. The efficiency then becomes

$$\eta = 1 - \frac{T_1}{T_2} \tag{12.16}$$

For process 1-2,

$$\frac{T_1}{T_2} = \left(\frac{V_2}{V_1}\right)^{(\gamma-1)} = \frac{1}{r^{(\gamma-1)}} \tag{12.17}$$

where the ratio $r = V_1/V_2$ is called the compression ratio for the cycle. The efficiency then depends only on the compression ratio in this simple model of the Otto engine.

$$\eta = 1 - \frac{1}{r^{(\gamma-1)}} \tag{12.18}$$

Equation 12.18 is plotted in Figure 12.23.

The net work per unit mass of working fluid is also important for showing the relative size and weight of the engine for a given power.

$$\frac{W_{\text{net}}}{mRT_1} = \frac{W_{\text{net}}}{P_1V_1} = \frac{\eta Q_{2-3}}{mRT_1} \tag{12.19}$$

But the heat transfer into the engine per cycle is given by

$$Q_{2-3} = m_f(HV) \tag{12.20}$$

where m_f is the mass of fuel burned per cycle. Thus,

$$\frac{W_{\text{net}}}{P_1V_1} = \left(\frac{m_f}{m}\right)\left(\frac{HV}{RT_1}\right)\left[1 - \frac{1}{r^{(\gamma-1)}}\right] \tag{12.21}$$

In reciprocating machines, it is common practice to define the mean effective pressure which can be used to show the relative size and weight of machines.

$$P_{\text{mean effective}} = \frac{W_{\text{net per cycle}}}{V_{\text{displaced per cycle}}} \tag{12.22}$$

For our model of the Otto engine

$$P_{m.e.} = \frac{W_{\text{net}}}{V_1 - V_2} \tag{12.23}$$

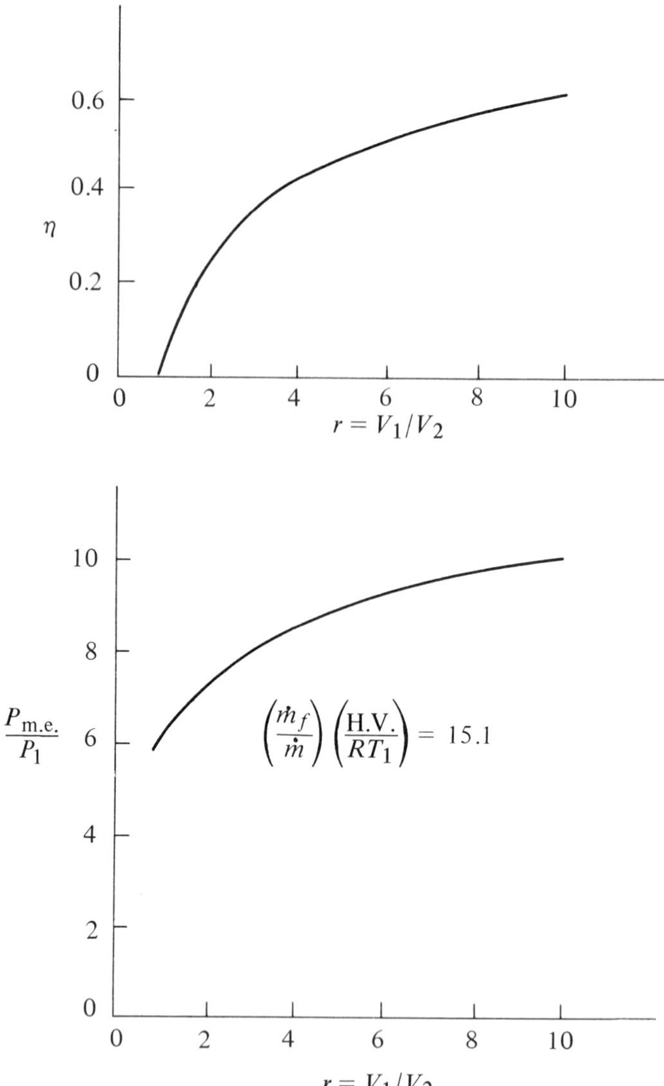

Figure 12.23 Influence of compression ratio on Otto cycle

If we combine this definition with equation (12.21), we have

$$\frac{P_{m.e.}}{P_1} = \left(\frac{m_f}{m}\right)\left(\frac{HV}{RT_1}\right)\left(\frac{r}{r-1}\right)\left(1 - \frac{1}{r^{(\gamma-1)}}\right) \qquad \textbf{(12.24)}$$

The influence of r on $P_{m.e.}$ is shown in Figure 12.23. In addition, the mean effective pressure varies directly with the inlet pressure, P_1, and with the product $(m_f/m)(HV/RT_1)$. The heating value and ratio of mass of fuel to mass of air are determined by the fuel. The mass of fuel is that which reacts completely with the oxygen in the air. Our simple adiabatic model overestimates the power by a factor of about 2 because it neglects heat transfer to the cylinder wall, assumes complete burning of the fuel, and assumes constant specific heats.

Equation (12.18) for the efficiency and equation (12.21) show that both the efficiency and the power of an Otto engine increase with compression ratio. The compression ratio is, however, limited by the preignition or detonation limit of the fuel which cannot be accounted for directly by the model. The temperature prior to ignition is T_2 which increases with compression ratio for fixed T_1.

$$\frac{T_2}{T_1} = r^{(\gamma-1)} \tag{12.25}$$

If the compression ratio is too high the temperature T_2 will be high enough to cause autoignition, and the charge will preignite spontaneously before the piston reaches the minimum volume. Usually the detonation limit is reached at lower compression ratios. When the engine knocks or detonates, the combustion occurs in a detonation wave rather than by the uniform travel of a diffusive flame front or deflagration wave. In a detonation wave the chemical reaction is initiated by the temperature rise associated with the sudden pressure increase of a shock wave. In a flame or deflagration wave the reaction is initiated by thermal conduction into the unburned gases. The higher T_2, the smaller the additional pressure increase needed to initiate reaction and the easier to start a detonation wave. The detonation limited compression ratio is increased by adding heavy molecules of tetra-ethyl-lead or similar compounds to the fuel to change the chemical reaction kinetics. More recently, careful shaping of the combustion space has been used to eliminate the focusing of small pressure waves into waves strong enough to start detonation.

In contrast to the steam plant and the gas turbine, the performance of the Otto engine is not limited by the maximum fluid temperature imposed by a structural limitation. In the Otto engine, there is no mechanical structure in thermal equilibrium with the hot gases. The maximum gas temperature is on the order of 2200 C so that the cylinder must be well cooled to prevent mechanical failure. This large temperature difference between the gas and the cylinder produces a significant heat transfer which reduces the performance of the engine considerably below that predicted by the adiabatic model. In addition, the temperature T_3 is so high that at equilibrium the combustion is incomplete. That is, the products of combustion include some CO and other compounds as well as CO_2 and H_2O. The rapid expansion 3-4 freezes the incomplete reaction producing CO and other nonequilibrium compounds in the exhaust gases.

The power delivered by the engine is the net work per cycle times the cyclic speed N of the engine.

$$\dot{W} = NW_{\text{net}} \tag{12.26}$$

In the Otto engine the cyclic speed is normally limited by the mechanical valve mechanism. The cyclic speed tends to vary inversely with the linear dimension of the engine so that the average piston velocity remains relatively constant with the size of the engine for the same service. Thus

$$\dot{W} = A_{\text{piston}} \vartheta_{\text{piston}} P_{m.e.} \tag{12.27}$$

so that the power varies as $P_{m.e.}$ for engines of the same size and the same piston velocity.

The proportionality between the work transfer per cycle and the inlet pressure P_1 is the basis for the usual power control for the engine. For low power levels the inlet air is throttled to reduce P_1. This throttling does not influence the efficiency as predicted by our simple model which neglects all mechanical friction. Supercharged engines use an air compressor to increase P_1 thereby increasing the power from the engine.

12.3.2 Diesel Engine

The compression ignition or diesel engine is similar to the spark-ignition engine, except that the cylinder is charged with air. The air is compressed more than in the spark-ignition engine so at minimum cylinder volume, the air temperature is well above the ignition temperature of the fuel. Fuel injectors spray the liquid fuel into the cylinder when the piston is near the minimum volume position. The fuel spray evaporates, mixes with the air, ignites and burns. Since a heat transfer is required to evaporate and heat the fuel to ignition, the combustion process in the diesel engine is not as rapid as in the Otto engine. The slower combustion process in the diesel engine is often modeled as a constant pressure combustion as is shown, Figure 12.24, although in reality the actual combustion process is quite complex.

The simplest model for the diesel engine, Figure 12.24 consists of an adiabatic compression, 1-2, an equivalent constant pressure heat addition, 2-3, an adiabatic expansion, 3-4, and an equivalent constant volume heat rejection, 4-1. The thermodynamic analysis of the model processes shown in Figure 12.24 is similar to the analysis used in the case of the Otto engine and the results are also shown in Table 12.4. The expression for the effi-

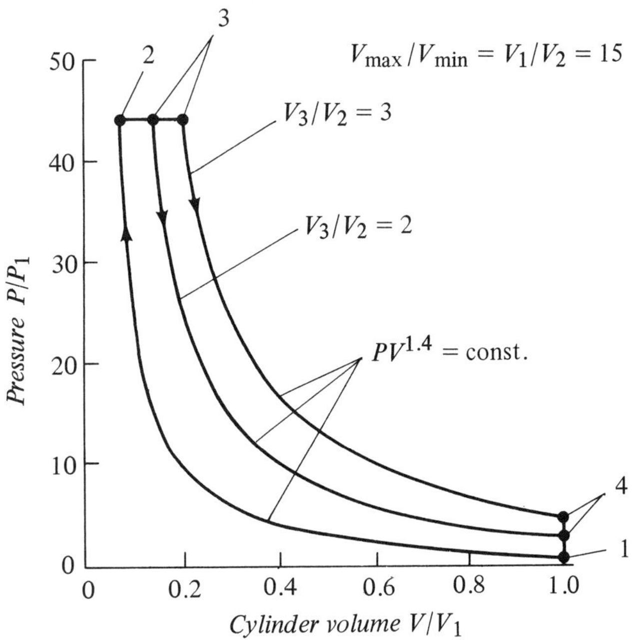

Figure 12.24 Pressure versus cylinder volume for the diesel cycle

TABLE 12.4
Performance of a Simple Diesel Cycle Model

Work of compression:	$W_{1-2} = mc_v(T_1 - T_2) = mc_vT_1(1 - r^{\gamma-1})$
Work of constant pressure expansion:	$W_{2-3} = mR(T_3 - T_2) = mc_vT_1(\gamma - 1)(r^{\gamma-1})(r_c - 1)$
Work of adiabatic expansion:	$W_{3-4} = mc_v(T_3 - T_4) = mc_vT_1r_c^{\gamma}[r^{\gamma-1}r_c^{1-\gamma} - 1)$
Equivalent heat addition:	$Q_{2-3} = mc_P(T_3 - T_2) = mc_vT_1\gamma(r^{\gamma-1})(r_c - 1)$
Equivalent heat rejection:	$Q_{4-1} = mc_v(T_1 - T_4) = mc_vT_1(1 - r_c^{\gamma})$
Energy conversion efficiency:	$\eta = 1 + \dfrac{Q_{4-1}}{Q_{2-3}} = 1 - \dfrac{1}{r^{\gamma-1}}\dfrac{r_c^{\gamma} - 1}{\gamma(r_c - 1)}$
where	$r = \dfrac{V_1}{V_2}$ and $r_c = \dfrac{V_3}{V_2} = 1 + \dfrac{m_f}{m}\left[\dfrac{-HV}{c_PT_1}\right]\left[\dfrac{1}{r^{\gamma-1}}\right]$
Net work per unit mass:	$\dfrac{W_{\text{net}}}{P_1V_1} = \dfrac{\gamma}{\gamma - 1}(r^{\gamma-1})(r_c - 1) - \dfrac{1}{\gamma - 1}(r_c^{\gamma} - 1)$
Mean effective pressure:	$\dfrac{P_{\text{mean}}}{P_1} = \dfrac{r}{r - 1}\left[\dfrac{\gamma}{\gamma - 1}\right]\left[(r^{\gamma-1})(r_c - 1) - \dfrac{1}{\gamma}(r_c^{\gamma} - 1)\right]$

ciency shows that at the same compression ratio, the diesel engine model is less efficient than the Otto engine model. In practice, however, diesel engines have higher efficiencies than Otto engines because the diesel engine compression ratio is not limited by detonation of the air-fuel mixtures. Diesel engines operate with compression ratios in the range of 15 to 18. Higher compression ratios tend to require heavier mechanical components without a corresponding increase in engine power.

In the Otto engine, the fuel-air ratio, which is fixed by the requirements for rapid burning of the premixed charge, determines the positive heat transfer in the model. Thus, changing the inlet pressure P_1 is the proper method for control of engine power. In contrast, in the diesel engine small quantities of fuel will burn satisfactorily when injected as a fine spray into the hot air in the cylinder. Thus the power of the diesel engine is normally controlled by adjusting the amount of fuel injected while the inlet pressure remains constant. The ratio, r_c, of the cylinder volume after constant pressure combustion to the volume prior to combustion varies with adjustments in the mass of fuel injected. Figure 12.24 shows the model for two different values of r_c. Note the difference in the area enclosed by these two cycles on the P-V plane. Figure 12.25 shows how the efficiency

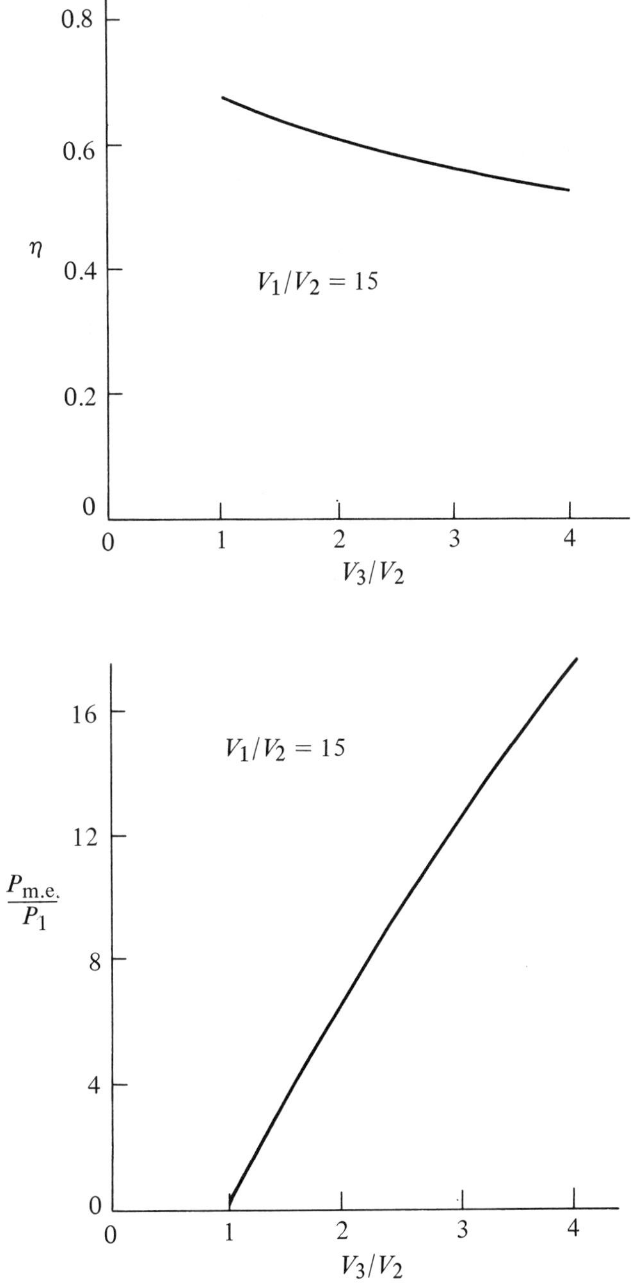

Figure 12.25 Influence of fuel cutoff ratio on the diesel cycle

for the model increases as r_c decreases, and Figure 12.25 shows how the power (at constant speed) increases with r_c.

12.4 Refrigeration Plants

The purpose of a refrigeration plant is to receive a positive heat transfer (the refrigeration effect) from a low temperature fluid stream and with the aid of a work transfer into the plant, to transfer heat to an environmental fluid (air or water) at a higher temperature. An overall control volume for the plant, as shown in Figure 12.26, consists of a refrigeration load, a closed refrigeration cycle, and an energy sink.

The vapor compression refrigeration plant is most common. It consists of four steady flow components: a compressor, a condenser (heat exchanger), a throttle valve, and an evaporator (heat exchanger), as shown in Figure 12.27. In operation, the compressor takes low pressure vapor from the evaporator and compresses it to a high enough pressure so that it condenses in the condenser. The working fluid then experiences a negative

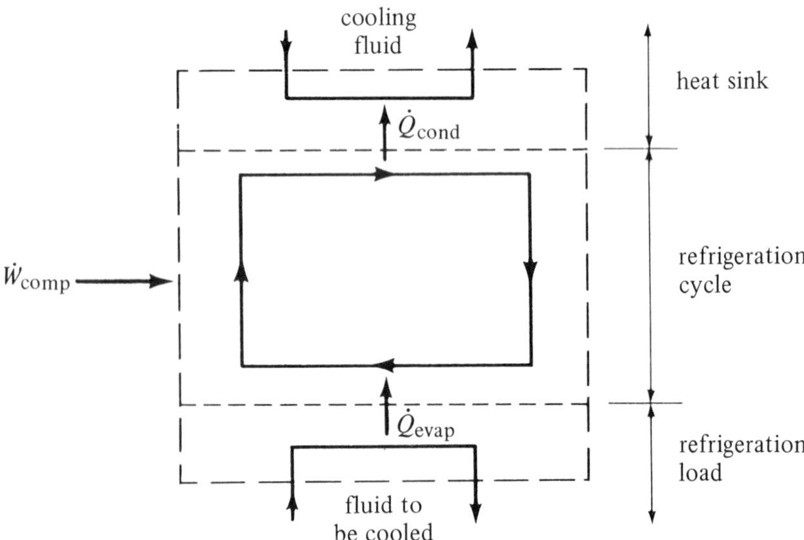

Figure 12.26 Overall control volume for a refrigeration plant

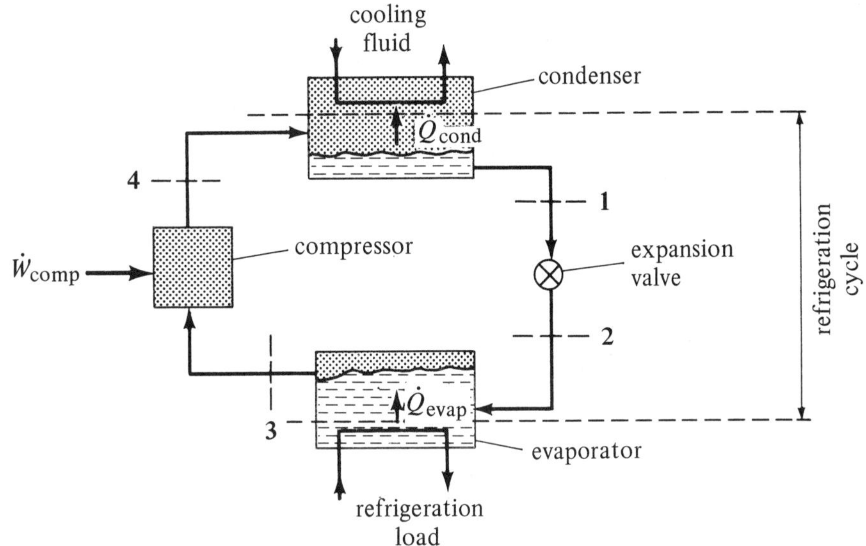

Figure 12.27 Vapor compression refrigeration plant

heat transfer to the energy sink as the high pressure vapor condenses. Liquid from the condenser passes through the throttle valve. As the pressure decreases, a fraction of the liquid evaporates and cools the remaining liquid to the saturation temperature corresponding to evaporator pressure. The working fluid then experiences a positive heat transfer from the refrigeration load as the liquid evaporates in the evaporator.

The plant is designed so that saturated liquid leaves the bottom of the condenser at 1 and saturated vapor leaves the top of the evaporator at 3. The condenser is normally modeled as a constant pressure heat exchanger so that the condenser pressure, for states 4 and 1, is the saturation pressure corresponding to the temperature of the environmental heat sink. The evaporator is also modeled as a constant pressure heat exchanger so that the evaporator pressure for states 2 and 3 is the saturation pressure corresponding to the temperature of the refrigeration load. The states for the refrigeration cycle are shown on the temperature-entropy diagram in Figure 12.28, and the performance equations are given in Table 12.5.

The working fluid or refrigerant for a refrigeration plant must have a reasonable saturation pressure at ambient tempera-

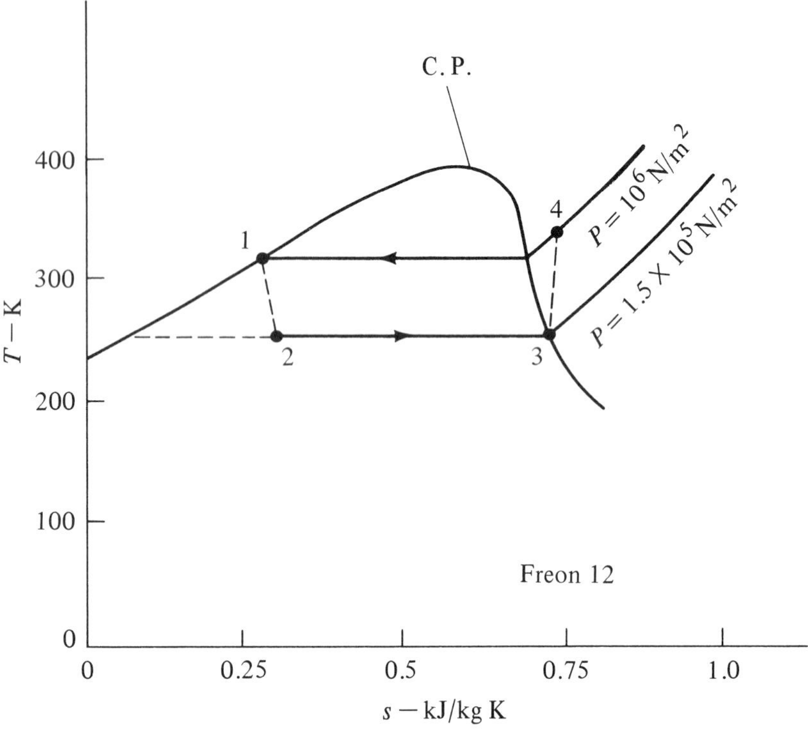

Figure 12.28 Vapor compression refrigeration cycle

tures and must have a triple point below the desired refrigeration temperature. In addition, the refrigerant should have a reasonable saturated vapor density at the refrigeration temperature. Compressors to handle very large volumes of low density gas are either very large or if more reasonable in size, very inefficient. This vapor-density limit determines the lowest practical refrigeration temperature for a number of the low boiling point refrigerants.

The most commonly used refrigerants are ammonia, refrigerant 12 (dichlorodifluoromethane, CCl_2F_2, Freon[3]-12) and refrigerant 22 (chlorodifluoromethane, $CHClF_2$, Freon[3]-22). Ammonia is most frequently used for large industrial refrigeration and air conditioning plants, while the Freons are almost exclusively used for household refrigeration and air condi-

[3] Dupont Co. trademark.

TABLE 12.5
Performance of Vapor Compression Refrigeration Cycle

Refrigeration load:	$\dot{Q}_{evap} = \dot{m}(h_3 - h_2) = \dot{m}T_3(s_3 - s_2)$
Compressor power:	$\dot{W}_{comp} = \dot{m}(h_3 - h_4)$
Condenser load:	$\dot{Q}_{cond} = \dot{m}(h_1 - h_4)$
Coefficient of performance:	$\text{COP} = \dfrac{h_3 - h_2}{-(h_3 - h_4)}$

tioning. Sulphur dioxide, carbon dioxide, and ethylene as well as water have also been used as refrigerants in vapor compression cycles. Many other light hydrocarbons and halogenated hydrocarbons are also used as refrigerants.

The practical vapor compression refrigeration plant has irreversible processes which reduce the COP of the refrigeration cycle. For example, the irreversible throttle valve could be replaced by an expansion machine, and the compressor could compress a two-phase mixture and avoid superheating the vapor that goes to the condenser. The cycle for this reversible refrigeration plant would then be as is shown in Figure 12.29 where

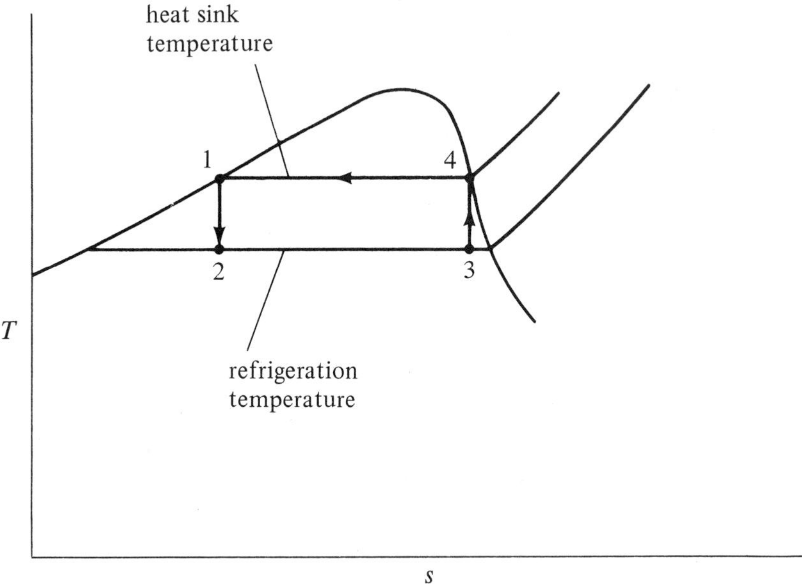

Figure 12.29 Reversible refrigeration cycle

reversible adiabatic compression and expansion processes have been assumed. The condenser and the evaporator have large heat transfer surface areas so that the refrigerant condenses at the sink temperature and evaporates at the refrigeration load temperature; thus, the refrigeration cycle of Figure 12.29 operates reversibly between these two temperatures. If we attempt to construct a refrigeration plant which closely approaches this reversible plant, the results are not very attractive from an economic point of view. The improvement in COP resulting from the use of an expansion machine in place of the throttle valve is not worth the increased cost and complexity. The increase in COP resulting in the elimination of the large heat transfer temperature difference associated with cooling the compressor outlet vapor to saturation does not justify the metering system required to feed the compressor with the proper two-phase mixture. The heat transfer temperature differences for the condenser and the evaporator are selected by balancing the increase in COP against the increased size and cost of the heat exchangers required for a smaller temperature difference.

12.5 General Considerations in the Design of Thermodynamic Plants

In the preceding sections of this chapter, we discussed several important thermodynamic plants and have pointed out some of the reasons for these particular plant designs. In this section, certain features are singled out for consideration because they are common to all thermodynamic plants. We shall show in a general way why plants must have these features in order to be practical.

From a practical point of view, the design of a thermodynamic plant requires a compromise between the need for high efficiency achieved by the minimization of irreversibilities and the need for a light weight, and hence low cost, plant which processes energy at a high rate. Unfortunately, these two objectives are in direct conflict since reversibility can be achieved only at vanishingly small rates of energy transfer. Confronted with this obstacle, the plant designer then devises equipment, evaluates working fluids, and develops thermodynamic cycles that will provide high rates of energy processing at levels of irreversibility which are considered reasonable in the light of other economic and performance constraints. This procedure requires detailed consideration of disciplines other than thermodynamics, especially heat transfer and fluid mechanics. However, without getting involved in the details of these other disciplines, we can point out several thermodynamic facts which must be considered in the design of a plant, especially with regard to energy sources and cycle design.

12.6 Energy Sources

In each of the thermodynamic plants previously discussed, we ultimately attributed the power produced to an energy source consisting of a thermodynamically metastable system that has come into thermodynamic equilibrium with the environment. In most cases, these energy sources have exhibited coupling of one form or another, but thermodynamic coupling is not a necessary characteristic of an energy source.

Consider some typical energy sources. In fossil-fuel fired thermodynamic plants, the metastable fuel and oxygen mixture experiences a chemical reaction to form thermodynamically stable combustion products. The maximum potential work transfer from the fuel-oxygen system is the difference between the Gibbs free energy of the fuel-oxygen mixture at atmospheric temperature and pressure and the Gibbs free energy of the combustion products at atmospheric temperature and pressure. (See Chapter 7.) However, this maximum work transfer can be obtained only if the chemical reaction is reversible. Unfortunately, the pressure and temperature required to hold the chemical reaction in equilibrium are well above those which can be achieved practically. In fact, the temperature is the same as that required to dissociate the combustion products; consequently, the practical power plant employs an irreversible chemical reaction. By virtue of the coupling between the chemical and thermal aspects of the reactants and products, the chemical metastability of the fuel-oxygen system is converted to a temperature difference relative to the atmosphere. This temperature difference is then used to drive a heat engine such as we have described for the steam plant or the gas turbine plant. In the reciprocating internal combustion engine, the chemical metastability of the fuel-oxygen system is converted by the combustion process to a pressure rise as well as a temperature rise. The difference in pressure between the combustion products and the environment can be used directly to produce a work transfer. Pressure differences produced by chemical reactions in systems which exhibit coupling between their chemical and thermodynamic aspects are also used directly in gun powder propellants and in solid-fuel rocket motors.

Nuclear reactors used as energy sources perform in much the same way as the fossil fuel-oxygen systems. The heavy uranium nucleus undergoes fission to form more stable light elements. The fission reaction occurs in a controlled but irreversible manner. By virtue of the coupling between the thermal and nuclear aspects of the nuclear fuel, the reactor produces a temperature difference relative to the environment. This temperature difference is then exploited to produce power in a heat engine.

Electro-chemical energy sources have the distinguishing characteristic that the electro-chemical coupling of properties allows the chemical metastability to be converted directly to an electric potential without changes or differences in the temperature or pressure of the system. The dry cell battery is such a chemically metastable system. In this system, the microscopic electric charge transfer resulting from the chemical reaction in the battery is harnessed to produce a macroscopic charge transfer between the terminals of the battery. In a storage battery, the chemical reaction is essentially held in equilibrium by the electric potential so that the battery may be charged and discharged. Electro-chemical systems are considered in detail in physical chemistry or chemical thermodynamics.[4]

In recent years renewed effort has been devoted to the old idea of harnessing the microscopic charge transfer involved in the oxidation of fossil fuels in devices called fuel cells. Unfortunately, the problem of producing an inexpensive and effective

---

[4] Denbigh, K., *The Principles of Chemical Equilibrium,* Cambridge Univ. Press, London, 1968.

catalyst for the charge transfer reactions at the electrodes of the fuel cell has prevented them from becoming competitive with the more common methods of power production.

Energy sources can also utilize the uncoupled mechanical properties of a system which is in a metastable mechanical state for power production. These uncoupled systems are mechanical rather than thermodynamic. The gravitational potential of a stream of water (hydropower) is the only uncoupled mechanical aspect which is normally used for large scale power production. On a small scale, springs, weights, and environmental kinetic energy (winds, tides and waves) are commonly used for power production. Energy storage in capacitors, inductors, and flywheels is also commonly utilized. Geophysical and atmospheric temperature differences have been proposed for power production and in a few cases have actually been utilized.

In considering the various types of energy sources for power production, the availability or potential for work transfer per unit mass of the metastable system is an important characteristic since it determines the specific size and weight of the power plant. The class of systems which utilize some sort of chemical coupling has a broad spectrum of specific availabilities. The systems with high specific availability which are normally used for fuels are difficult to react reversibly because of the kinetics of the reaction. However, the rapid reaction and high temperatures result in plants with favorable specific size and weight. The systems with lower specific availability which are normally used in storage batteries are easier to hold in equilibrium and react nearly reversibly, but are heavy and bulky. A reversible nuclear reaction is out of the question except on an astronomical scale. In contrast, systems with uncoupled mechanical properties have a small specific availability and are therefore used primarily for very small scale power plants. An exception is the gravitational energy utilized in hydroelectric plants where large quantities of water are available and can be handled economically.

In summary, the considerations of a high availability per unit mass for an energy source dictate the use of energetic fuels which react rapidly and irreversibly. The resulting high temperatures are utilized to power heat engines which use the atmosphere as a low temperature heat reservoir.

12.7 Thermodynamic Plant Cycles

Each of the thermodynamic plants which we have studied has had the characteristic of processing a fluid with strong thermodynamic coupling through an apparatus in which heat transfers (actual, not equivalent to combustion) were separated from work transfers. This design feature insures that only the state of the strongly coupled fluid (such as a gas) is cycled while the thermal state of the essentially uncoupled parts of the apparatus remain virtually unchanged. Heat transfer to the fluid is effected by moving the fluid to a region of higher wall temperatures rather than increasing the temperature of the wall confining the gas. If the thermal state of the walls confining a gas is cycled in order to effect heat interactions with the gas, the energy changes for the walls are larger than the energy changes for the gas (for normal materials over the range of states of practical importance). Since every energy transfer interaction is irreversible, especially if it is at a practical rate, the thermal cycling of any structural members or confining walls will seriously increase the irreversibilities of any thermodynamic plant.

These facts are two of the basic reasons that all practical thermo-mechanical energy conversion plants employ flow systems. Although we cannot express these generalities quantitatively, we can demonstrate their significance with a specific example. For the purpose of illustrating the limitations of a non-flow thermodynamic plant, we will consider in some detail an ideal gas Carnot cycle operating entirely within a single cylinder. In our analysis, we shall consider two models: first a model for the plant which ignores the thermal aspects of the cylinder and second a model which includes the heat capacity of the cylinder. In the entire discussion we shall be considering only an ideal gas Carnot cycle plant which operates reversibly between a reservoir at T_H and a reservoir at T_L.

Consider the performance of our first model. As we have previously discussed in Chapter 6, one useful measure of performance is the thermodynamic energy conversion efficiency given by

$$\eta = \frac{W_{\text{net}}}{Q_{\text{add}}} = 1 - \frac{T_L}{T_H} \qquad \textbf{(12.28)}$$

Since we are considering only reversible cycles in this discussion, each of the models will have this reversible efficiency. Absent from our previous discussion of ideal gas Carnot cycles were two characteristics which are important considerations for a practical plant: first, the mean effective pressure which is a measure of the work per cycle and second, the net work ratio, NWR, which is an indicator of the influence of irreversibilities on the cycle. Recall from Chapter 6 that the work transfers for the four processes of the cycle are:

(1) adiabatic expansion

$$W_{1-2} = mc_v(T_H - T_L) \tag{12.29}$$

(2) isothermal compression

$$W_{2-3} = mRT_L \ln (V_3/V_2) = -mRT_L \ln (V_1/V_4) \tag{12.30}$$

(3) adiabatic compression

$$W_{3-4} = -mc_v(T_H - T_L) \tag{12.31}$$

(4) isothermal expansion

$$W_{4-1} = mRT_H \ln (V_1/V_4) \tag{12.32}$$

From our definition of the mean effective pressure given in equation (12.22), we obtain

$$P_{m.e.} = \frac{mR(T_H - T_L) \ln (V_1/V_4)}{V_2 - V_4} \tag{12.33}$$

As is shown by equation (12.33), the mean effective pressure varies directly with the mass of the charge in the cylinder. However, as this mass is increased, the maximum pressure of the cycle also increases. This increase in pressure in turn requires increased strength and weight of the mechanism. As a first order approximation, we assume that the mass of the apparatus varies directly with the maximum pressure of the cycle; therefore, a reasonable measure of the specific work of the machine is the ratio $P_{m.e.}/P_4$. Thus, substituting the ideal gas constitutive relation into equation (12.33), we obtain

$$\frac{P_{m.e.}}{P_4} = \frac{[1 - (T_L/T_H)]}{(V_2/V_1)(V_1/V_4) - 1} \ln (V_1/V_4) \tag{12.34}$$

Equation (12.34) shows that for a fixed temperature ratio T_H/T_L, the ratio $P_{m.e.}/P_4$ depends only on the volume ratio during the isothermal expansion, V_1/V_4, since V_2/V_1 is fixed by the temperature ratio of the isentropic processes which are utilized to change the temperature in the cycle. As is shown in Figure 12.30, the ratio $P_{m.e.}/P_4$ reaches a maximum at $V_1/V_2 = 2.57$. In geometric terms this value gives a pressure volume diagram for the cycle which has the maximum pressure difference (between expansion and compression) for its length. That is, it is the ''fat-

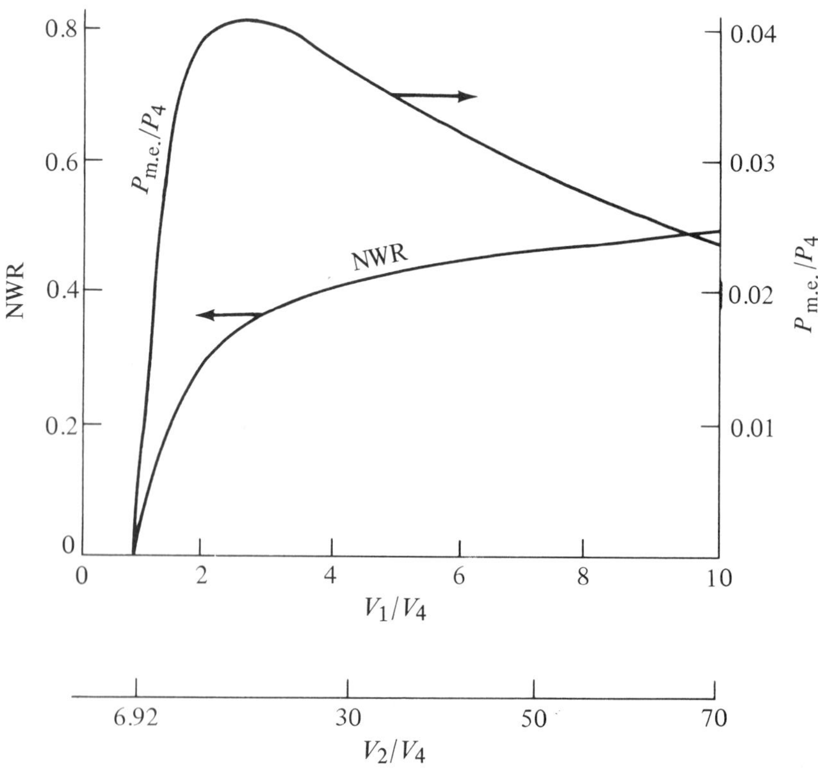

Figure 12.30 Carnot cycle performance

test'' diagram for the cycle. However, even at this maximum pressure difference, the mean effective pressure is only slightly over 4 percent of the maximum pressure. This very low utilization of the displaced volume is the result of the very sharp pressure rise up to state 4 as was shown previously in Chapter 6, Figure 6.1.

The net work ratio for the cycle (as defined in Section 12.1.1) is formed directly from equations (12.29) through (12.32). Thus

$$\text{NWR} = \frac{R(T_H - T_L)\ln(V_1/V_4)}{RT_H \ln(V_1/V_4) + c_v(T_H - T_L)} \quad \textbf{(12.35)}$$

$$\text{NWR} = \eta\left[\frac{1}{1 + \dfrac{1}{\gamma - 1}\dfrac{\eta}{\ln(V_1/V_2)}}\right] \quad \textbf{(12.36)}$$

As is shown in Figure 12.30, the net work ratio increases rapidly with the volume ratio V_1/V_4. At the maximum mean effective pressure, the net work ratio is about 0.34, which is a bit less than half of the asymptotic value 0.725. Thus, some compromise must be reached between the size of the system and the effect of irreversibilities.

Our second model illustrates the most serious practical difficulty with the closed system Carnot cycle—namely, that the gas must be both heated and cooled by heat transfers from the same cylinder walls. Since the gas must have positive Q when its temperature is high and must have negative Q when its temperature is low, the cylinder walls must be made to change in temperature over the same range as the gas. As we shall see, the heat transfers required to change the temperature of the cylinder walls are considerably larger than those required to change the temperature of the gas.

Since the cycle must operate reversibly while in communication with only two heat reservoirs, the system which undergoes the cycle must be the gas plus the piston-cylinder apparatus which confines the gas. For reversibility, the gas and the piston-cylinder must be in equilibrium at all times. The inclusion of the structural parts in the system does not change the pressure-volume-temperature relation.

$$PV = m_g RT \tag{12.37}$$

Here m_g is the mass of the gas in the cylinder. However, the energy of the system must include the thermal energy of the structural parts, which are modeled as pure thermal system elements and rigid nonelastic mechanical elements, as well as the internal energy of the gas. Then

$$U = m_g \left[c_v + \frac{m_s}{m_g} c \right] T \tag{12.38}$$

where m_s is the mass of the structural parts and c is their specific heat. The system thus behaves as a gas with the gas constant R of the gas alone, but with an augmented specific heat, $(c_v)_{\text{eff}}$.

$$(c_v)_{\text{eff}} = \left[c_v + \frac{m_s}{m_g} c \right] \tag{12.39}$$

The effective γ for the system is

$$\gamma_{\text{eff}} = \frac{R}{(c_v)_{\text{eff}}} + 1 = \frac{1}{(c_v/R) + (m_s/m_g)(c/R)} + 1 \tag{12.40}$$

and the results of our analysis for the first model can be used directly.

We can now show that the term $(m_s/m_g)(c/R)$ is a number large relative to c_v/R so that γ_{eff} is only slightly greater than unity. As γ_{eff} approaches unity, the system approaches uncoupled behavior and the NWR and $P_{m.e.}/P_4$ both approach zero. A simple first order estimate of the ratio m_s/m_g can be obtained by considering the cylinder wall as the only contribution to the mass of the structure. The mass of the structure is then limited by the maximum stress σ which the cylinder wall material can withstand. The hoop stress in the cylinder is

$$\sigma = P \frac{D}{2t} \tag{12.41}$$

where the symbols are defined in Figure 12.31. The volume V_s of structural materials is

$$V_s = DtL = \frac{D^2 LP}{2\sigma} \tag{12.42}$$

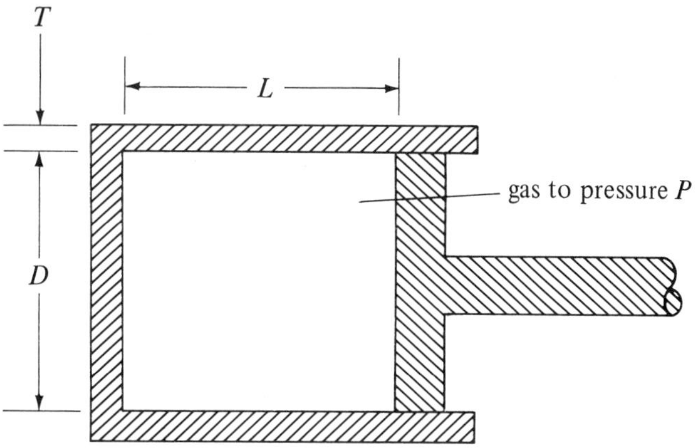

Figure 12.31 Piston and cylinder

Thus, the mass of the structure is

$$m_s = \frac{D^2 L P \rho_s}{2} \tag{12.43}$$

where ρ_s is the density of the cylinder material. The mass of gas in the cylinder is from the ideal gas constitutive relation

$$m_g = \frac{D^2}{4} L \frac{P}{RT} \tag{12.44}$$

so that the ratio m_s/m_g becomes

$$\frac{m_s}{m_g} = \frac{2\rho_s RT}{\sigma} \tag{12.45}$$

The limiting value of this ratio will occur at the highest temperature at which the allowable stress is lowest. As a typical value for this ratio consider a steel cylinder and helium gas for which the relevant properties are:

$$\begin{aligned} \rho_s &= 7817 \text{ kg/m}^3 \\ \sigma &= 137 \text{ MPa} \\ c &= 472 \text{ J/kgK} \\ R &= 2073 \text{ J/kgK} \\ c_v &= 1.5 \text{ R} \end{aligned}$$

For these values and a temperature $T = 1200$ K, the ratio m_s/m_g has the value

$$\frac{m_s}{m_g} = \frac{2(7817)(2073)(1200)}{137 \times 10^6} = 284 \qquad \textbf{(12.46)}$$

The effective specific heat ratio is then

$$\gamma_{\text{eff}} = 1 + \frac{1}{1.5 + 284(472/2073)} = 1.0151 \qquad \textbf{(12.47)}$$

From the expression for the path of an isentropic process (PV^γ = constant), we find that with this very low specific heat ratio, a volume ratio V_2/V_1 of about 5.8×10^{39} is required for $T_H = 1200$ K and $T_L = 300$ K: Then the ratio $P_{m.e.}/P_4$ is reduced to about 4.7×10^{-41} at $V_1/V_4 = 2.72$ and the NWR at this same value of V_1/V_4 is reduced to 1.1 percent. These values show that thermally cycling the mechanical structure of the Carnot engine cylinder causes the closed system Carnot engine to be completely impractical. Even though the engine theoretically still has the reversible efficiency, the very small NWR indicates that a very small irreversibility will drastically reduce the thermal efficiency.

The two examples presented here show that the impracticability of a thermodynamic plant can be deduced from simple calculations based on reversibility without detailed consideration of the rate processes. In the first model we saw that the low mean effective pressure for the Carnot cycle renders that cycle completely impracticable in spite of its relatively high thermal efficiency. In the second model we saw that cycling of the cylinder walls not only further reduced the mean effective pressure but also reduced the net work ratio to an unacceptable level.

Problems

12.1 A steam plant operates on a simple Rankine cycle as shown in Figure 12.2. The pressure in the boiler is 6×10^6 N/m^2 while the pressure in the condenser is 10^4 N/m^2. The engine and pump efficiencies are 1.0.

(a) Calculate the engine power output per unit flow rate of steam.

(b) Calculate pump power input per unit flow rate of steam.

(c) What is the net power output of the cycle per unit flow rate of steam?

(d) Calculate the energy conversion efficiency of the cycle.

(e) Repeat parts (a) through (d) above for a

Carnot cycle (with H_2O as the working fluid) operating between the maximum boiler temperature and the condenser temperature. The heat transfer at the high temperature changes the state of the steam from a saturated liquid to a saturated vapor.

(f) Compare the ratio of the net power to the expansion power (NWR) for the two cycles.

12.2 A superheat-Rankine-cycle plant operates with a boiler pressure of 6×10^6 N/m² and a condenser pressure of 7.384×10^3 N/m². The temperature of the steam at the discharge from the superheater is 500 C. The turbine and pump efficiencies are both 1.0.

(a) Calculate the turbine power output per unit flow rate of steam.

(b) Calculate the pump power output per unit flow rate of steam.

(c) What is the net power output of the cycle per unit flow rate of steam?

(d) Calculate the energy conversion efficiency of the cycle.

(e) Compare the performance of this cycle with the simple Rankine cycle of problem 12.1. How do you account for the difference in performance?

12.3 A steam plant operates on a reheat cycle with steam entering the high pressure turbine at a pressure of 1.2×10^7 N/m² and a temperature of 540 C. The high pressure turbine ($\eta_T = 0.95$) exhausts into a reheater (modeled as a constant pressure heat exchanger) at a pressure of 2×10^6 N/m². The steam is reheated to a temperature of 540 C before entering the low pressure turbine ($\eta_T = 0.90$). The condenser pressure is 10^4 N/m².

(a) Calculate the total turbine output per unit mass flow rate of steam.

(b) What is the energy conversion efficiency of this plant?

(c) What is the net work ratio for this plant if the feed water pump is reversible and adiabatic?

12.4 It is proposed to operate a steam plant on the reheat-regenerative Rankine cycle with one stage of reheat and two stages of feed water heating. (The feed water heaters are connected as shown in Figure 12.7.) Steam is supplied to the high pressure turbine ($\eta_T = 1.0$) at a pressure of 1.4×10^7 N/m² and a temperature of 540 C. The high pressure turbine discharges into the reheater and the second stage of feed water heating at a pressure of 2×10^6 N/m². The steam to the intermediate pressure turbine passes through the reheater (not shown in Figure 12.7). Steam leaves the reheater (modeled as a constant pressure heat exchanger) at a temperature of 480 C. The first stage of feed water heating bleeds off at a pressure of 1.4×10^5 N/m² ($\eta = 1.0$ for intermediate pressure turbine). The condenser pressure is at 10^4 N/m². The low pressure turbine has $\eta_T = 1.0$. Assume ideal components where performance is not specified.

(a) Show the locus of states on a *T-s* diagram.

(b) Calculate the fraction of boiler steam flow that is bled off at each feed water heating stage.

(c) Calculate the total turbine work transfer per unit mass flow rate of steam through the boiler.

(d) For the reheater calculate the heat transfer per unit mass flow rate of boiler steam flow.

(e) What is the energy conversion efficiency of the plant?

12.5 Consider the boiling water reactor of Figure 12.11, with the operating conditions shown in the figure.

(a) Calculate the energy conversion efficiency of this plant assuming the reactor to be a constant pressure heat exchanger and the generator to have no losses. Neglect pump work transfers. Assume the steam leaving the reactor to be saturated vapor.

(b) Determine the required reactor steam

flow to the steam-to-steam reheater as a function of the high pressure turbine outlet pressure for a constant total reactor steam flow of 739 kg/s. The high pressure turbine efficiency is $\eta = 0.9$. Model the reheater as a constant pressure heat exchanger with the reheated steam leaving at 280 C and the condensed reactor steam leaving as saturated liquid at reactor pressure. Neglect the steam flow from the high pressure turbine to the high pressure feed water heater.

12.6 The steam power plant shown in Figure 12P.6 must supply shaft power and 6×10^5 N/m² steam for process heating. The power requirement is 1000 kW and the heat transfer requirement for the process heater is 1400 kW. The thermodynamic states are shown in Figure 12P.6. The pumps and turbines are modeled as reversible and adiabatic. The heat exchangers are modeled as constant pressure.

(a) Determine the mass flow rate to the process heater, m_1.
(b) Determine the mass flow rate at the boiler exit, m_{total}.
(c) What is the condenser heat transfer requirement?
(d) What is the energy conversion efficiency of this system?

12.7 The energy conversion efficiency of a simple Rankine cycle is influenced by the condenser pressure.

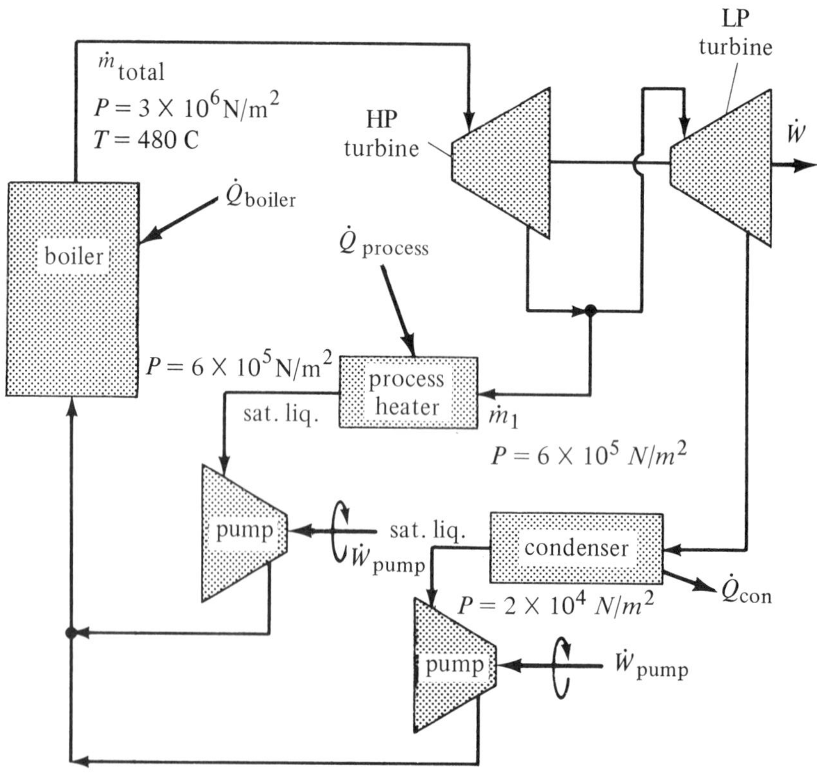

Figure 12P.6

(a) Show this effect by plotting efficiency versus condenser pressure. Assume the boiler discharge state to be saturated vapor at 3×10^6 N/m^2, and the engine efficiency to be 1.0.

(b) Show this effect for a Rankine cycle with superheat (boiler discharge state 3×10^6 N/m^2 and 540 C) by plotting efficiency versus condenser pressure. Assume the turbine efficiency to be 1.0.

(c) What conclusions can you draw from a comparison of your results for parts (a) and (b) above?

12.8 A gas turbine engine is to be modeled by a Brayton cycle employing air with constant specific heats (c_P = 1.003 kJ/kgK, c_v = 0.716 kJ/kgK) as the working fluid. The compressor and the turbine are modeled as reversible adiabatic ($\eta_C = \eta_T = 1.0$). Starting with the basic equations (Table 12.3):

(a) Show that the net shaft power and the ratio $\dot{W}_{net}/\dot{Q}_{added}$ are:

$$\dot{W}_{net}/\dot{m}c_P = (T_3 - T_2)\left[1 - \left[\frac{P_{atm}}{P_{max}}\right]^{(\gamma-1)/\gamma}\right]$$

$$\dot{W}_{net}/\dot{Q}_{add} = 1 - \left[\frac{P_{atm}}{P_{max}}\right]^{(\gamma-1)/\gamma}$$

(b) Determine the values for $\dot{W}_{net}/\dot{m}c_P$ and $\dot{W}_{net}/\dot{Q}_{added}$ for a gas turbine with T_1 = 295 K, T_3 = 950 K, $P_{max}/P_{min} = 5$.

(c) If the compressor and the turbine each have an efficiency of 0.8, and the other conditions are the same as in part (b), determine the values for $\dot{W}_{net}/\dot{m}c_P$ and $\dot{W}_{net}/\dot{Q}_{added}$.

12.9 Figure 12.15 shows the influence of the compressor efficiency and turbine efficiency on the net work transfer and the cycle efficiency for the Brayton cycle model of a gas turbine plant.

(a) For the conditions used for Figure 12.15 ($T_3/T_1 = 3.0$, $\gamma = 1.4$, and $\eta_C = \eta_T$) find the value for η_C which reduces the maximum point of the net work transfer curve down to zero.

(b) What is the cycle efficiency at this condition?

(c) What is the pressure ratio at this condition?

12.10 It has been proposed to use a gas turbine plant for automotive propulsion. The plan is to use the Brayton cycle model (shown in Figure 12.14) with one modification. Two turbines will be employed (a split shaft plant). One turbine will be used to drive the compressor while the other turbine will serve as the power turbine to drive the rear wheels of the vehicle. For purposes of control, the turbines will be interconnected with a throttle valve as shown in Figure 12P.10. Assume that the mass flow rate of fuel added in the burner is negligible compared to the mass flow rate of air. Model the burner as a constant pressure heat exchanger and the compressor and turbines as adiabatic shaft work machines. The efficiency of the compressor is $\eta_C = 0.75$ while the turbines have identical efficiencies of $\eta_T = 0.80$. Assume the air can be modeled as an ideal gas with constant specific heat c_v = 715.9 J/kgK and R = 287.0 J/kgK.

(a) Determine the pressures and temperatures of states 5 and 6 for the case in which the power turbine operates at full load (throttle valve wide open, $P_4 = P_5$) and at zero load (throttle valve partially open, $P_5 = P_6$). Show both of these operating conditions on a T-s diagram.

(b) Calculate the compressor work transfer per unit mass flow rate of air for both conditions in part (a).

(c) Calculate the turbine work transfer per unit mass flow rate of air for both turbines under both load conditions in part (a).

(d) Calculate the heat transfer per unit mass flow rate of air in the burner for both load conditions in part (a).

(e) Calculate the energy conversion efficiency of the plant at full load.

(f) If 250 kW are required for a specific

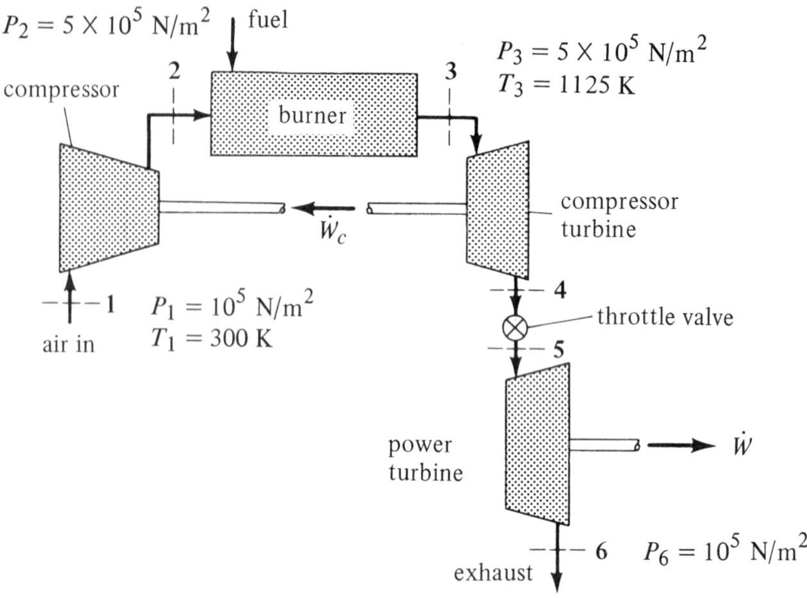

Figure 12P.10

vehicle application, what is the necessary air flow rate?

(g) If the heating value of the fuel is 4.4×10^7 J/kg, calculate the fuel flow rate and the specific fuel consumption (kg/kWh) at full load.

12.11 A reciprocating spark ignition engine is to be modeled as an Otto cycle using air with $c_P = 1.003$ kJ/kgK and $R = 0.287$ kJ/kgK as the working fluid. The cycle has a compression ratio of 10.7 and an inlet temperature of 20 C.

(a) What is the temperature at the end of the compression stroke?

(b) If the temperature at the end of combustion is 1000 C what is the cycle efficiency and the work per cycle per unit mass of air in the cylinder?

(c) Repeat part (b) with the temperature at the end of combustion equal to 2000 C.

(d) What is the temperature at the start of the equivalent exhaust process for parts (b) and (c)?

12.12 The spark ignition internal combustion engine is limited in compression ratio by the temperature of the pre-mixed charge at the end of the compression stroke. For $T_1 = 300$ K and $\gamma = 1.4$ determine this temperature T_2 (Figure 12.22) as a function of compression ratio.

12.13 A truck engine is to be modeled with the diesel cycle shown in Figure 12.24. The compression ratio is 15 and the fuel cutoff ratio is 2. The air at the beginning of compression has a temperature of 15 C and a pressure of 1.013×10^5 N/m^2. Assume that the air can be modeled as an ideal gas with constant specific heat $c_v = 715.9$ J/kgK and $R = 287$ J/kgK.

(a) Calculate the mean effective pressure of the engine.

(b) Calculate the energy conversion efficiency of this engine.

(c) What is the temperature of the air at the end of the compression stroke (prior to injection in the actual engine)?

(d) If the engine must deliver 186.4 kW at 2000 rpm, what is the required piston displacement for a four-stroke engine?

12.14 A simple *open* system model for the four-stroke, spark ignition, reciprocating, internal combustion engine employs air with constant specific heats ($c_P = 1.005$ kJ/kgK, $c_v = 715.9$ J/kgK) as the working fluid and replaces the combustion process with an equivalent heat addition. The detail processes as shown on the pressure versus cylinder volume diagram and the pressure versus specific volume diagram of Figure 12P.14 are:

1-2 closed system, $m = m_1$, reversible adiabatic compression from V_{max} to V_{min}.

2-3 closed system, $m = m_1$, equivalent heat addition at V = constant = V_{min}.

3-4 closed system, $m = m_1$, reversible adiabatic expansion from V_{min} to V_{max}.

4-5 open system, adiabatic, cylinder volume = const = V_{max}, flow out of exhaust valve until $m = m_5$, gas m_5 remaining within cylinder expands reversibly and adiabatically to $P_{exhaust} = P_1 = P_{inlet}$.

5-6 open system, P = constant = $P_{exhaust}$, adiabatic, volume in cylinder decreases from V_{max} to V_{min}, mass flows out exhaust valve as m decreases to m_6.

6-1 open system, P = constant = P_{inlet}, adiabatic, volume in cylinder increases from V_{min} to V_{max}, mass flows in through inlet valve as m increases to m_1. The temperature decreases from T_6 to T_1 as cold air at T_{inlet} enters the cylinder.

(a) Show that the net work per cycle of the piston is

$$W_{net} = m_1 c_v [T_3 - T_2]\left[1 - \left[\frac{V_{min}}{V_{max}}\right]^{\gamma-1}\right]$$

and the ratio

$$\frac{W_{net}}{Q_{add}} = 1 - \left[\frac{V_{min}}{V_{max}}\right]^{\gamma-1}$$

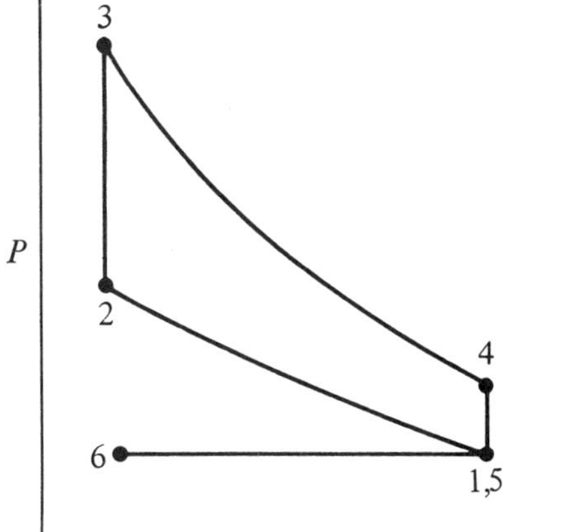

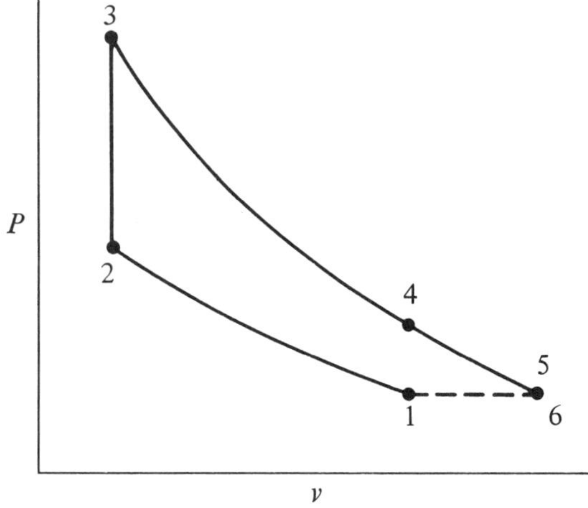

Figure 12P.14

Show that these expressions, where V_{max}/V_{min} is the compression ratio for the engine, are the same as for the completely closed system model.

(b) If $T_1 = 330$ K, $P_1 = 1$ atm, $T_3 = 1650$ K and $V_{max}/V_{min} = 7$, find T_2, P_4, T_5 and T_{inlet}. Also determine the average temperature of the gas leaving the engine exhaust.

12.15 The Ericson cycle is a reversible model for a heat engine which operates with a closed system of m kg of an ideal gas with constant specific heats as the working fluid. The cycle is composed of two isothermal processes and two constant pressure processes, and operates as follows:

Process I) The gas expands reversibly and isothermally from volume V_1 to volume V_2 while in communication with a heat reservoir at the temperature T_H.

Process II) The gas is then compressed reversibly and at constant pressure P_L from volume V_2 to volume V_3 while passing through an energy storage device called a thermal regenerator. During this process, the gas rejects heat to the regenerator thereby increasing the stored thermal energy of the regenerator. The gas enters the regenerator at the temperature T_H and leaves at the temperature T_L.

Process III) The gas is then compressed reversibly and isothermally from volume V_3 to volume V_4 while in communication with a heat reservoir at the temperature T_L.

Process IV) Finally the gas is expanded reversibly from volume V_4 to volume V_1 at constant pressure P_L while passing through the regenerator. During this process, the gas which enters the regenerator at the temperature T_L and leaves at the temperature T_H experiences a heat transfer which reduces the stored thermal energy of the regenerator to its original level.

(a) Show the cycle on T-s and P-V diagrams.
(b) Determine the net work transfer for the cycle in terms of the pressures P_H and P_L, the temperatures T_H and T_L, and the properties of the gas.
(c) Determine the net heat transfer for the cycle in terms of variables listed above.
(d) Evaluate the energy conversion efficiency for the engine and compare it with the energy conversion efficiency of a Carnot cycle operating between the same two temperatures.

12.16 A CCl_2F_2 (Freon-12) refrigeration system is shown schematically in Figure 12P.16. Model both the condenser and the evaporator as constant pressure heat exchangers. Model the compressor as an adiabatic, reversible compressor. Assume that all components operate in a steady manner with negligible changes in the kinetic and gravitational potential energy of the working fluid. The states are as follows:

state 1: saturated vapor at $T_1 = -25$ C
state 2: pressure $P_2 = 6.865 \times 10^5$ N/m^2
state 3: saturated liquid at $P_2 = 6.865 \times 10^5$ N/m^2 (27 C)

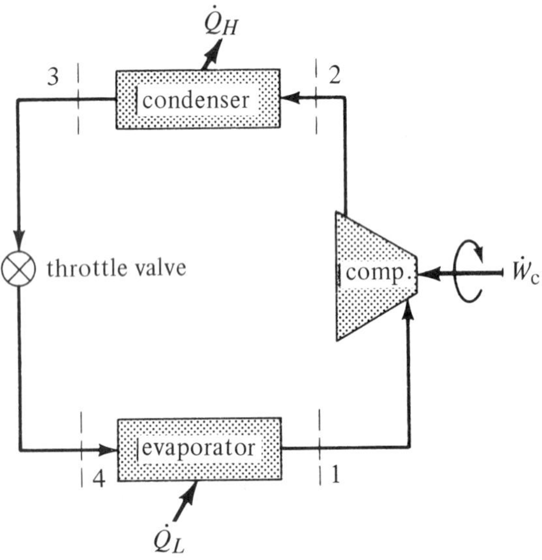

Figure 12P.16

(a) Show the states of the working fluid on a T-s diagram and on the h-P diagram.
(b) Determine the refrigeration rate per unit mass of refrigerant circulated, $\dot{Q}_1/\dot{m}$.
(c) Determine the necessary compressor power per unit mass of refrigerant circulated.
(d) Is this a reversible cycle?

12.17 In the idealized vapor compression refrigeration plant shown in Figure 12P.17, the fluid flows steadily into the compressor at a temperature of − 17 C and is compressed reversibly and adiabatically to the saturated vapor state at a temperature of 27 C. The fluid is condensed at constant pressure in a heat exchanger until it attains the saturated liquid state. It is then throttled adiabatically to a two-phase state at a temperature of − 17 C. The two-phase fluid then enters the evaporator where it is partially evaporated. The discharge from the evaporator enters the compressor completing the cycle.

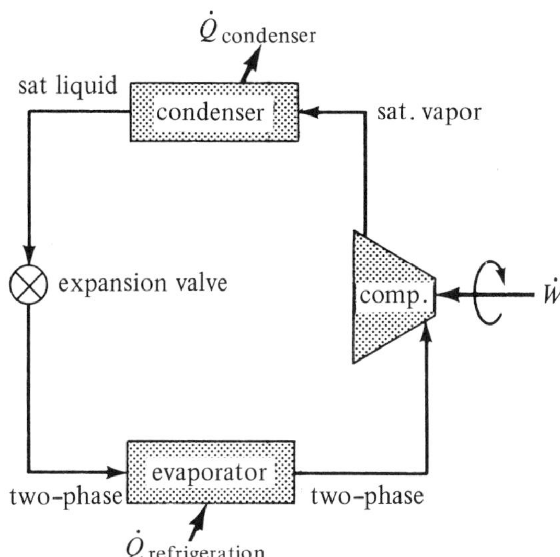

Figure 12P.17

A partial tabulation of properties is given below.

T(C)	P(N/m²)	h_f(kJ/kg)	s_f(kJ/kgK)	s_g(kJ/kgK)
− 17	2.7×10^5	24.73	0.1004	unknown
27	1.1×10^6	79.72	0.2965	0.9007

(a) Show the cycle on a T-s diagram.
(b) Determine the specific enthalpy at inlet to the compressor.
(c) Determine the refrigeration load (per unit mass of refrigerant).
(d) Determine the work transfer (per unit mass of refrigerant) to the fluid in the compressor.
(e) Determine the COP and compare it with a Carnot refrigerator operating between the same two temperatures.

12.18 As illustrated in Figure 12P.18 one method of heating a home during the winter is to circulate warm air through the house. The cool discharge air returns to a heat exchanger where it is heated by means of a heat pump.

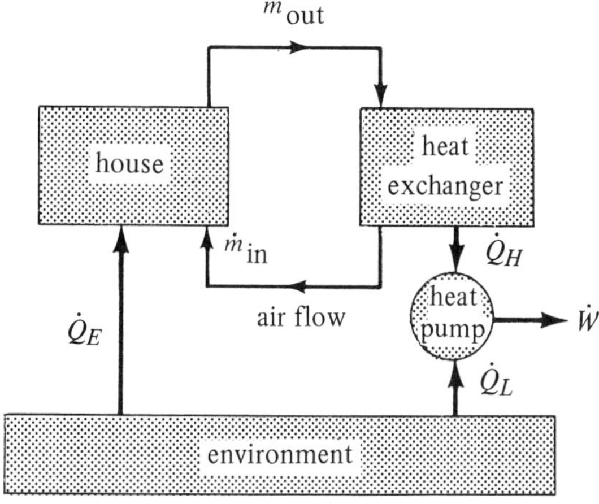

Figure 12P.18

This heat pump is also in communication with the environment.

The steady operating conditions are: The house experiences heat transfer with the environment at a rate of $\dot{Q}_E = -27.77$ kW. The air enters the house at a temperature of 27 C and leaves at a temperature of 21 C. The temperature of the environment is -17 C. The heat pump executes an integral number of cycles.

Assume the following models are valid: The enviornment can be modeled as a heat reservoir. The air in the house can be modeled as an ideal gas with $c_v = 715.9$ J/kgK and $R = 287$ J/kgK. The pressure of the air everywhere is 1 atm (1.013×10^5 N/m^2). Changes in kinetic and potential energies are negligible.

(a) What is the mass flow rate of the air through the heat exchanger?

(b) What is the heat transfer rate, $\dot{Q}_H$, between the heat pump and the heat exchanger?

(c) What is the minimum power required by the heat pump?

APPENDIX

Thermodynamic Properties in SI Units

Properties of Saturated H_2O

t C	T K	P N/m^2	v_f m^3/kg	v_{fg} m^3/kg	v_g m^3/kg	u_f kJ/kg	u_{fg} kJ/kg	u_g kJ/kg	h_f kJ/kg	h_{fg} kJ/kg	h_g kJ/kg	s_f kJ/kgK	s_{fg} kJ/kgK	s_g kJ/kgK
.01	273.16	611.3	1.0002-3	206.14	206.14	0	2375.3	2375.3	0.01	2501.3	2501.4	0	9.1562	9.1562
5	278.15	872.1	1.0001-3	147.12	147.12	20.97	2361.3	2382.3	20.98	2489.6	2510.6	0.0761	8.9496	9.0257
10	283.15	1.2276+3	1.0004-3	106.38	106.38	43.00	2347.2	2389.2	42.01	2477.7	2519.8	0.1510	8.7498	8.9008
15	288.15	1.7051+3	1.0009-3	77.925	77.926	62.99	2333.1	2396.1	62.99	2465.9	2528.9	0.2245	8.5569	8.7814
20	293.15	2.339+3	1.0018-3	57.790	57.791	83.95	2319.0	2402.9	83.96	2454.1	2538.1	0.2966	8.3706	8.6672
25	298.15	3.169+3	1.0029-3	43.359	43.360	104.88	2304.9	2409.8	104.89	2442.3	2547.2	0.3674	8.1905	8.5580
30	303.15	4.246+3	1.0043-3	32.893	32.894	125.78	2290.8	2416.6	125.79	2430.5	2556.3	0.4369	8.0164	8.4533
35	308.15	5.628+3	1.0060-3	25.215	25.216	146.67	2276.7	2423.4	146.68	2418.6	2565.3	0.5053	7.8478	8.3531
40	313.15	7.384+3	1.0078-3	19.522	19.523	167.56	2262.6	2430.1	167.57	2406.7	2574.3	0.5725	7.6845	8.2570
45	318.15	9.593+3	1.0099-3	15.257	15.258	188.44	2248.4	2436.8	188.45	2394.8	2583.2	0.6387	7.5261	8.1648
50	323.15	12.349+3	1.0121-3	12.031	12.032	209.32	2234.2	2443.5	209.33	2382.7	2592.1	0.7238	7.3725	8.0763
55	328.15	15.758+3	1.0146-3	9.567	9.568	230.21	2219.9	2450.1	230.23	2370.7	2600.9	0.7679	7.2234	7.9913
60	333.15	19.940+3	1.0172-3	7.670	7.671	251.11	2205.5	2456.6	251.13	2358.5	2609.6	0.8312	7.0784	7.9096
65	338.15	25.03+3	1.0199-3	6.196	6.197	272.02	2191.1	2463.1	272.06	2346.2	2618.3	0.8935	6.9375	7.8310
70	343.15	31.19+3	1.0228-3	5.041	5.042	292.95	2176.6	2469.6	292.98	2333.8	2626.8	0.9549	6.8004	7.7553
75	348.15	38.58+3	1.0259-3	4.130	4.131	313.90	2162.0	2475.9	313.93	2321.4	2635.3	1.0155	6.6669	7.6824
80	353.15	47.39+3	1.0291-3	3.406	3.407	334.86	2147.4	2482.2	334.91	2308.8	2643.7	1.0753	6.5369	7.6122
85	358.15	57.83+3	1.0325-3	2.827	2.828	355.84	2132.6	2488.4	355.90	2296.0	2651.9	1.1343	6.4102	7.5445
90	363.15	70.14+3	1.0360-3	2.360	2.361	376.85	2117.7	2494.5	376.92	2283.2	2660.1	1.1925	6.2866	7.4791
95	368.15	84.55+3	1.0397-3	1.9809	1.9819	397.88	2102.7	2500.6	397.96	2270.2	2668.1	1.2500	6.1659	7.4159
100	373.15	101.35+3	1.0435-3	1.6719	1.6729	418.94	2087.6	2506.5	419.04	2257.0	2676.1	1.3069	6.0480	7.3549
105	378.15	120.82+3	1.0475-3	1.4184	1.4194	440.02	2072.3	2512.4	440.15	2243.7	2683.8	1.3630	5.9328	7.2958
110	383.15	143.27+3	1.0516-3	1.2091	1.2102	461.14	2057.0	2518.1	461.30	2230.2	2691.5	1.4185	5.8202	7.2387
115	388.15	169.06+3	1.0559-3	1.0355	1.0366	482.30	2041.4	2523.7	482.48	2216.5	2699.0	1.4734	5.7100	7.1833
120	393.15	198.53+3	1.0603-3	0.8908	0.8919	503.50	2025.8	2529.3	503.71	2202.6	2706.3	1.5276	5.6020	7.1296
125	398.15	232.1+3	1.0649-3	0.7695	0.7706	524.74	2009.9	2534.6	524.99	2188.5	2713.5	1.5813	5.4962	7.0775
130	403.15	270.1+3	1.0697-3	0.6674	0.6685	546.02	1993.9	2539.9	546.31	2174.2	2720.5	1.6344	5.3925	7.0269
135	408.15	313.0+3	1.0746-3	0.5811	0.5822	567.35	1977.7	2545.0	567.69	2159.6	2727.3	1.6870	5.2907	6.9777
140	413.15	361.3+3	1.0797-3	0.5078	0.5089	588.74	1961.3	2550.0	589.13	2144.7	2733.9	1.7391	5.1908	6.9299
145	418.15	415.4+3	1.0850-3	0.4452	0.4463	610.18	1944.7	2554.9	610.63	2129.6	2740.3	1.7907	5.0926	6.8833
150	423.15	475.8+3	1.0905-3	0.3917	0.3928	631.68	1927.9	2559.5	632.20	2114.3	2746.5	1.8418	4.9960	6.8379
155	428.15	543.1+3	1.0961-3	0.3457	0.3468	653.24	1910.8	2564.1	653.84	2098.6	2752.4	1.8925	4.9010	6.7935
160	433.15	617.8+3	1.1020-3	0.3060	0.3071	674.87	1893.5	2568.4	675.55	2082.6	2758.1	1.9427	4.8075	6.7502
165	438.15	700.5+3	1.1080-3	0.2716	0.2727	696.56	1876.0	2572.5	697.34	2066.2	2763.5	1.9925	4.7153	6.7078
170	443.15	791.7+3	1.1143-3	0.2417	0.2428	718.33	1858.1	2576.5	719.21	2049.5	2768.7	2.0419	4.6244	6.6663
175	448.15	892.0+3	1.1207-3	0.2157	0.2168	740.17	1840.0	2580.2	741.17	2032.4	2773.6	2.0909	4.5347	6.6256
180	453.15	1.0021+6	1.1274-3	0.19292	0.19405	762.09	1821.6	2583.7	763.22	2015.0	2778.2	2.1396	4.4461	6.5857
185	458.15	1.1227+6	1.1343-3	0.17296	0.17409	784.10	1802.9	2587.0	785.37	1997.1	2782.4	2.1879	4.3586	6.5465
190	463.15	1.2544+6	1.1414-3	0.15399	0.15654	806.19	1783.8	2590.0	807.62	1978.8	2786.4	2.2359	4.2720	6.5079
195	468.15	1.3978+6	1.1488-3	0.13990	0.14105	828.37	1764.4	2592.8	829.98	1960.0	2790.0	2.2835	4.1863	6.4698
200	473.15	1.5538+6	1.1565-3	0.12620	0.12736	850.65	1744.7	2595.3	852.45	1940.7	2793.2	2.3309	4.1014	6.4323
205	478.15	1.7230+6	1.1644-3	0.11405	0.11521	873.04	1724.5	2597.5	875.04	1921.0	2796.0	2.3780	4.0172	6.3952
210	483.15	1.9062+6	1.1726-3	0.10324	0.10441	895.53	1703.9	2599.5	897.76	1900.7	2798.5	2.4248	3.9337	6.3585
215	488.15	2.104+6	1.1812-3	93.61-3	94.79-3	918.14	1682.9	2601.1	920.62	1879.9	2800.5	2.4714	3.8507	6.3221
220	493.15	2.318+6	1.1900-3	85.00-3	86.19-3	940.87	1661.5	2602.4	943.62	1858.5	2802.1	2.5178	3.7683	6.2861
225	498.15	2.548+6	1.1992-3	77.29-3	78.49-3	963.73	1639.6	2603.3	966.78	1836.5	2803.3	2.5639	3.6863	6.2503
230	503.15	2.795+6	1.2088-3	70.37-3	71.58-3	986.74	1617.2	2603.9	990.12	1813.8	2804.0	2.6099	3.6047	6.2146
235	508.15	3.060+6	1.2187-3	64.15-3	65.37-3	1009.89	1594.2	2604.1	1013.62	1790.5	2804.2	2.6558	3.5233	6.1791
240	513.15	3.344+6	1.2291-3	58.53-3	59.76-3	1033.21	1570.8	2604.0	1037.32	1766.5	2803.8	2.7015	3.4422	6.1437
245	518.15	3.648+6	1.2399-3	53.47-3	54.71-3	1056.71	1546.7	2603.4	1061.23	1741.7	2803.0	2.7472	3.3612	6.1083
250	523.15	3.973+6	1.2512-3	48.88-3	50.13-3	1080.39	1522.0	2602.4	1085.36	1716.2	2801.5	2.7927	3.2802	6.0730
255	528.15	4.319+6	1.2631-3	44.72-3	45.98-3	1104.28	1496.7	2600.9	1109.73	1689.8	2799.5	2.8383	3.1992	6.0375
260	533.15	4.688+6	1.2755-3	40.93-3	42.21-3	1128.39	1470.6	2599.0	1134.37	1662.5	2796.9	2.8838	3.1181	6.0019
265	538.15	5.081+6	1.2886-3	37.48-3	38.77-3	1152.74	1443.9	2596.6	1159.28	1634.4	2793.6	2.9294	3.0368	5.9662
270	543.15	5.499+6	1.3023-3	34.34-3	35.64-3	1177.36	1416.3	2593.7	1184.51	1605.2	2789.7	2.9751	2.9551	5.9301
275	548.15	5.942+6	1.3168-3	31.47-3	32.79-3	1202.25	1387.9	2590.2	1210.07	1574.9	2785.0	3.0208	2.8730	5.8938
280	553.15	6.412+6	1.3321-3	28.84-3	30.17-3	1227.46	1358.7	2586.1	1235.99	1543.6	2779.6	3.0668	2.7903	5.8571
285	558.15	6.909+6	1.3483-3	26.42-3	27.77-3	1253.00	1328.4	2581.4	1263.31	1511.0	2773.3	3.1130	2.7070	5.8199
290	563.15	7.436+6	1.3656-3	24.20-3	25.57-3	1278.92	1297.1	2576.0	1289.07	1477.1	2766.2	3.1594	2.6227	5.7821
295	568.15	7.993+6	1.3839-3	22.16-3	23.54-3	1305.2	1264.7	2569.9	1316.3	1441.8	2758.1	3.2062	2.5375	5.7437
300	573.15	8.581+6	1.4036-3	20.27-3	21.67-3	1332.0	1231.0	2563.0	1344.0	1404.9	2749.0	3.2534	2.4511	5.7045
305	578.15	9.202+6	1.4247-3	18.52-3	19.948-3	1359.3	1195.9	2555.2	1372.4	1366.4	2738.7	3.3010	2.3633	5.6643
310	583.15	9.856+6	1.4474-3	16.90-3	18.350-3	1387.1	1159.4	2546.4	1401.3	1326.0	2727.3	3.3493	2.2737	5.6230
320	593.15	11.274+6	1.4988-3	13.99-3	15.488-3	1444.6	1080.9	2525.5	1461.5	1238.6	2700.1	3.4480	2.0882	5.5362
330	603.15	12.845+6	1.5607-3	11.44-3	12.996-3	1505.3	993.7	2498.9	1525.3	1140.6	2665.9	3.5507	1.8909	5.4417
340	613.15	14.586+6	1.6379-3	9.16-3	10.797-3	1570.3	894.3	2464.6	1594.2	1027.9	2622.0	3.6594	1.6763	5.3357
350	623.15	16.513+6	1.7403-3	7.07-3	8.813-3	1641.9	776.6	2418.4	1670.6	893.4	2563.9	3.7777	1.4335	5.2112
360	633.15	18.651+6	1.8925-3	5.05-3	6.945-3	1725.2	626.3	2351.5	1760.5	720.5	2481.0	3.9147	1.1379	5.0526
370	643.15	21.03+6	2.213-3	2.71-3	4.925-3	1844.0	384.5	2228.5	1890.5	441.6	2332.1	4.1106	0.6865	4.7971
374.14	647.29	22.09+6	3.155-3	0	3.155-3	2029.6	0	2029.6	2099.3	0	2099.3	4.4298	0	4.4298

Data from *Steam Tables* by Joseph H. Keenan, Frederick G. Keyes, Philip G. Hill, and Joan G. Moore.

Properties of H_2O

$P(t_{sat})$ Temp C	1 x 10^3 N/m² (6.98 C) v m³/kg	u kJ/kg	h kJ/kg	s kJ/kgK
Sat liq	1.0002-3	29.30	29.30	0.1059
Evap	129.21	2355.70	2484.90	8.8697
Sat vap	129.21	2385.0	2514.2	8.9756
0	1.0002-3	-0.03	-0.03	-0.0001
10	130.60	2389.2	2519.8	8.9956
20	135.23	2403.2	2538.5	9.0603
30	139.85	2417.3	2557.1	9.1228
40	144.47	2431.3	2575.8	9.1835
50	149.09	2445.4	2594.5	9.2423
60	153.71	2459.5	2613.2	9.2994
70	158.33	2473.7	2632.0	9.3550
80	162.95	2487.9	2650.8	9.4090
90	167.57	2502.1	2669.7	9.4616
100	172.19	2516.4	2688.6	9.5129
120	181.42	2545.0	2726.5	9.6119
140	190.65	2573.9	2764.5	9.7063
160	199.89	2602.9	2802.8	9.7968
180	209.12	2632.2	2841.3	9.8836
200	218.35	2661.6	2880.0	9.9671
220	227.58	2691.3	2918.8	10.0476
240	236.82	2721.2	2958.0	10.1254
260	246.05	2751.3	2997.3	10.2006
280	255.28	2781.6	3036.9	10.2735
300	264.51	2812.2	3076.8	10.3443
320	273.74	2843.1	3116.8	10.4130
340	282.97	2874.2	3157.2	10.4799
360	292.20	2905.6	3197.8	10.5450
380	301.43	2937.2	3238.6	10.6085
400	310.66	2969.0	3279.7	10.6705
420	319.89	3001.2	3321.1	10.7310
440	329.12	3033.6	3362.7	10.7902
460	338.35	3066.2	3404.6	10.8482
480	347.58	3099.2	3446.7	10.9049
500	356.81	3132.4	3489.2	10.9605
550	379.89	3216.6	3596.5	11.0950
600	402.97	3302.5	3705.5	11.2235
650	426.04	3390.2	3816.2	11.3469
700	449.12	3479.6	3928.7	11.4655
750	472.19	3570.8	4043.0	11.5800
800	495.27	3663.8	4159.1	11.6908
850	518.34	3758.6	4276.9	11.7981
900	541.42	3855.1	4396.5	11.9023
950	564.49	3953.2	4517.7	12.0035
1000	587.57	4053.0	4640.6	12.1019
1100	633.72	4257.5	4891.2	12.2914
1200	679.87	4467.9	5147.8	12.4718
1300	726.02	4683.7	5409.7	12.6438

$P(t_{sat})$ Temp C	10 x 10^3 N/m² (45.81 C) v m³/kg	u kJ/kg	h kJ/kg	s kJ/kgK
Sat liq	1.0102-3	191.82	191.83	0.6493
Evap	14.673	2246.08	2392.87	7.5009
Sat vap	14.674	2437.9	2584.7	8.1502
0	1.0002-3	-0.03	-0.02	-0.0001
10	1.0010-3	41.91	41.92	0.1483
20	1.0018-3	83.85	83.86	0.2966
30	1.0048-3	125.70	125.71	0.4345
40	1.0098-3	167.56	167.57	0.5725
50	14.869	2443.9	2592.6	8.1749
60	15.336	2458.2	2611.5	8.2327
70	15.801	2472.5	2630.5	8.2887
80	16.267	2486.8	2649.5	8.3432
90	16.731	2501.1	2668.5	8.3963
100	17.196	2515.5	2687.5	8.4479
120	18.123	2544.4	2725.6	8.5474
140	19.050	2573.3	2763.8	8.6423
160	19.975	2602.5	2802.2	8.7331
180	20.900	2631.8	2840.8	8.8201
200	21.825	2661.3	2879.5	8.9038
220	22.749	2691.0	2918.5	8.9844
240	23.674	2720.9	2957.7	9.0623
260	24.598	2751.1	2997.0	9.1376
280	25.521	2781.5	3036.7	9.2105
300	26.445	2812.1	3076.5	9.2813
320	27.369	2842.9	3116.6	9.3501
340	28.292	2874.1	3157.0	9.4170
360	29.216	2905.4	3197.6	9.4821
380	30.139	2937.0	3238.4	9.5457
400	31.063	2968.9	3279.6	9.6077
420	31.986	3001.1	3320.9	9.6682
440	32.909	3033.5	3362.6	9.7275
460	33.832	3066.1	3404.5	9.7854
480	34.756	3099.1	3446.6	9.8422
500	35.679	3132.3	3489.1	9.8978
550	37.987	3216.5	3596.4	10.0323
600	40.295	3302.5	3705.4	10.1608
650	42.603	3390.1	3816.2	10.2842
700	44.911	3479.6	3928.7	10.4028
750	47.218	3570.8	4043.0	10.5174
800	49.526	3663.8	4159.0	10.6281
850	51.834	3758.6	4276.9	10.7354
900	54.141	3855.0	4396.4	10.8396
950	56.449	3953.2	4517.7	10.9408
1000	58.757	4053.0	4640.6	11.0393
1100	63.372	4257.5	4891.2	11.2287
1200	67.987	4467.9	5147.8	11.4091
1300	72.602	4683.7	5409.7	11.5811

$P(t_{sat})$ Temp C	20 x 10^3 N/m² (60.06 C) v m³/kg	u kJ/kg	h kJ/kg	s kJ/kgK
Sat liq	1.0172-3	251.38	251.40	0.8320
Evap	7.6473	2205.32	2358.30	7.0765
Sat vap	7.649	2456.7	2609.7	7.9085
0	1.0002-3	-0.03	-0.01	-0.0001
10	1.0010-3	41.96	41.98	0.1483
20	1.0018-3	83.95	83.97	0.2966
30	1.0048-3	125.75	125.77	0.4345
40	1.0078-3	167.56	167.58	0.5725
50	1.0125-3	209.34	209.35	0.7018
60	1.0172-3	251.12	251.13	0.8312
70	7.883	2471.1	2628.8	7.9649
80	8.117	2485.6	2647.9	8.0199
90	8.351	2500.1	2667.1	8.0734
100	8.585	2514.6	2686.2	8.1255
120	9.051	2543.6	2724.6	8.2256
140	9.516	2572.7	2763.0	8.3209
160	9.980	2602.0	2801.6	8.4120
180	10.444	2631.4	2840.2	8.4993
200	10.907	2660.9	2879.1	8.5831
220	11.370	2690.7	2918.1	8.6639
240	11.832	2720.7	2957.3	8.7418
260	12.295	2750.8	2996.7	8.8172
280	12.757	2781.3	3036.4	8.8903
300	13.219	2811.9	3076.3	8.9611
320	13.681	2842.8	3116.4	9.0299
340	14.143	2873.9	3156.8	9.0968
360	14.605	2905.3	3197.4	9.1620
380	15.067	2936.9	3238.3	9.2256
400	15.529	2968.8	3279.4	9.2876
420	15.991	3001.0	3320.8	9.3482
440	16.453	3033.4	3362.4	9.4074
460	16.915	3066.1	3404.3	9.4654
480	17.376	3099.0	3446.5	9.5222
500	17.838	3132.2	3489.0	9.5778
550	18.992	3216.4	3596.3	9.7123
600	20.147	3302.4	3705.3	9.8409
650	21.301	3390.1	3816.1	9.9642
700	22.455	3479.5	3928.6	10.0829
750	23.609	3570.8	4043.0	10.1975
800	24.763	3663.8	4159.1	10.3082
850	25.916	3758.5	4276.9	10.4155
900	27.070	3855.0	4396.4	10.5197
950	28.224	3953.2	4517.7	10.6209
1000	29.378	4053.0	4640.6	10.7193
1100	31.686	4257.5	4891.2	10.9088
1200	33.994	4467.9	5147.8	11.0892
1300	36.301	4683.7	5409.7	11.2612

$P(t_{sat})$ Temp C	30 x 10^3 N/m² (69.10 C) v m³/kg	u kJ/kg	h kJ/kg	s kJ/kgK
Sat liq	1.0223-3	289.20	289.23	0.9349
Evap	5.2280	2179.20	2336.07	6.8337
Sat vap	5.229	2468.4	2625.3	7.7686
0	1.0002-3	-0.03	0.00	-0.0001
10	1.0010-3	41.96	41.99	0.1483
20	1.0018-3	83.95	83.98	0.2966
30	1.0048-3	125.75	125.78	0.4345
40	1.0078-3	167.56	167.59	0.5725
50	1.0125-3	209.34	209.36	0.7018
60	1.0172-3	251.12	251.14	0.8312
70	5.243	2469.7	2627.0	7.7737
80	5.401	2484.4	2646.4	7.8293
90	5.558	2499.0	2665.7	7.8833
100	5.715	2513.6	2685.0	7.9357
120	6.027	2542.8	2723.6	8.0365
140	6.338	2572.1	2762.2	8.1323
160	6.648	2601.4	2800.9	8.2237
180	6.958	2630.9	2839.7	8.3112
200	7.267	2660.6	2878.6	8.3952
220	7.577	2690.4	2917.7	8.4762
240	7.885	2720.4	2957.0	8.5542
260	8.194	2750.6	2996.4	8.6297
280	8.502	2781.0	3036.1	8.7028
300	8.811	2811.7	3076.0	8.7736
320	9.119	2842.6	3116.2	8.8425
340	9.427	2873.8	3156.6	8.9095
360	9.735	2905.2	3197.2	8.9747
380	10.043	2936.8	3238.1	9.0383
400	10.351	2968.7	3279.2	9.1003
420	10.659	3000.9	3320.6	9.1609
440	10.967	3033.3	3362.3	9.2202
460	11.275	3066.0	3404.2	9.2781
480	11.583	3098.9	3446.4	9.3349
500	11.891	3132.1	3488.9	9.3906
550	12.661	3216.4	3596.2	9.5251
600	13.430	3302.3	3705.3	9.6537
650	14.200	3390.0	3816.0	9.7770
700	14.969	3479.5	3928.6	9.8957
750	15.739	3570.7	4042.9	10.0103
800	16.508	3663.7	4158.9	10.1210
850	17.277	3758.5	4276.8	10.2284
900	18.047	3855.0	4396.4	10.3325
950	18.816	3953.1	4517.6	10.4337
1000	19.585	4053.0	4640.5	10.5322
1100	21.124	4257.4	4891.1	10.7217
1200	22.662	4467.9	5147.7	10.9020
1300	24.201	4683.6	5409.7	11.0740

$P(t_{sat})$ Temp C	50 x 10^3 N/m² (81.33 C) v m³/kg	u kJ/kg	h kJ/kg	s kJ/kgK
Sat liq	1.0300-3	340.44	340.49	1.0910
Evap	3.2390	2143.46	2305.41	6.5029
Sat vap	3.240	2483.9	2645.9	7.5939
0	1.0002-3	-0.03	0.02	-0.0001
10	1.0010-3	41.96	42.01	0.1483
20	1.0018-3	83.95	84.00	0.2966
30	1.0048-3	125.75	125.80	0.4345
40	1.0078-3	167.56	167.60	0.5725
50	1.0125-3	209.34	209.38	0.7018
60	1.0172-3	251.12	251.16	0.8312
70	1.0231-3	292.99	293.02	0.9532
80	1.0291-3	334.87	334.89	1.0753
90	3.323	2496.8	2662.9	7.6414
100	3.418	2511.6	2682.5	7.6947
120	3.608	2541.3	2721.6	7.7968
140	3.796	2570.8	2760.6	7.8935
160	3.983	2600.4	2799.6	7.9856
180	4.170	2630.1	2838.6	8.0736
200	4.356	2659.9	2877.7	8.1580
220	4.542	2689.8	2916.9	8.2392
240	4.728	2719.9	2956.2	8.3174
260	4.913	2750.2	2995.8	8.3931
280	5.099	2780.6	3035.6	8.4663
300	5.284	2811.3	3075.5	8.5373
320	5.469	2842.3	3115.7	8.6062
340	5.654	2873.5	3156.2	8.6732
360	5.839	2904.9	3196.8	8.7385
380	6.024	2936.5	3237.8	8.8021
400	6.209	2968.5	3278.9	8.8642
420	6.394	3000.6	3320.4	8.9249
440	6.579	3033.1	3362.0	8.9841
460	6.764	3065.8	3404.0	9.0421
480	6.949	3098.7	3446.2	9.0989
500	7.134	3132.0	3488.7	9.1546
550	7.596	3216.2	3596.0	9.2891
600	8.057	3302.2	3705.1	9.4178
650	8.519	3389.9	3815.9	9.5412
700	8.981	3479.4	3928.5	9.6599
750	9.443	3570.6	4042.8	9.7744
800	9.904	3663.6	4158.9	9.8852
850	10.366	3758.4	4276.7	9.9926
900	10.828	3854.9	4396.3	10.0967
950	11.289	3953.1	4517.6	10.1979
1000	11.751	4052.9	4640.5	10.2964
1100	12.674	4257.4	4891.1	10.4859
1200	13.597	4467.8	5147.7	10.6662
1300	14.521	4683.6	5409.6	10.8382

Data from *Steam Tables* by Joseph H. Keenan, Frederick G. Keyes, Philip G. Hill, and Joan G. Moore.

Properties of H_2O

$P(t_{sat})$	70 x 10^3 N/m^2 (89.95 C)				100 x 10^3 N/m^2 (99.63 C)				120 x 10^3 N/m^2 (104.80 C)				140 x 10^3 N/m^2 (109.31 C)				160 x 10^3 N/m^2 (113.32 C)				
Temp C	v m^3/kg	u kJ/kg	h kJ/kg	s kJ/kgK	v m^3/kg	u kJ/kg	h kJ/kg	s kJ/kgK	v m^3/kg	u kJ/kg	h kJ/kg	s kJ/kgK	v m^3/kg	u kJ/kg	h kJ/kg	s kJ/kgK	v m^3/kg	u kJ/kg	h kJ/kg	s kJ/kgK	Temp C
Sat liq	1.0360-3	376.63	376.70	1.1919	1.0432-3	417.36	417.46	1.3026	1.0473-3	439.20	439.32	1.3608	1.0510-3	458.24	458.39	1.4109	1.0544-3	475.19	475.36	1.4550	Sat liq
Evap	2.364	2117.9	2283.3	6.2878	1.6930	2088.7	2258.0	6.0568	1.4274	2072.9	2244.2	5.9373	1.2355	2059.1	2232.0	5.8355	1.0903	2046.7	2221.1	5.7467	Evap
Sat vap	2.365	2494.5	2660.0	7.4797	1.6940	2506.1	2675.5	7.3594	1.4284	2512.1	2683.5	7.2981	1.2366	2517.3	2690.4	7.2464	1.0914	2521.9	2696.5	7.2017	Sat vap
0	1.0002-3	-0.03	0.04	-0.0001	1.0002-3	-0.03	0.07	-0.0001	1.0001-3	-0.02	0.09	-0.0001	1.0001-3	-0.03	0.11	-0.0001	1.0001-3	-0.03	0.13	-0.0001	0
10	1.0010-3	41.96	42.03	0.1483	1.0010-3	41.95	42.05	0.1483	1.0011-3	41.95	42.07	0.1483	1.0011-3	41.95	42.09	0.1483	1.0011-3	41.95	42.11	0.1483	10
20	1.0018-3	83.95	84.02	0.2966	1.0018-3	83.94	84.04	0.2966	1.0017-3	83.94	84.06	0.2966	1.0017-3	83.94	84.08	0.2966	1.0017-3	83.94	84.10	0.2966	20
30	1.0048-3	125.75	125.82	0.4345	1.0048-3	125.75	125.84	0.4345	1.0047-3	125.74	125.86	0.4345	1.0047-3	125.74	125.88	0.4345	1.0047-3	125.74	125.90	0.4344	30
40	1.0078-3	167.55	167.62	0.5725	1.0078-3	167.55	167.65	0.5725	1.0077-3	167.55	167.66	0.5725	1.0077-3	167.54	167.68	0.5725	1.0077-3	167.54	167.70	0.5725	40
50	1.0125-3	209.33	209.40	0.7018	1.0125-3	209.32	209.42	0.7018	1.0124-3	209.32	209.44	0.7018	1.0124-3	209.32	209.46	0.7018	1.0124-3	209.31	209.48	0.7018	50
60	1.0172-3	251.11	251.18	0.8312	1.0172-3	251.10	251.20	0.8312	1.0171-3	251.09	251.22	0.8311	1.0171-3	251.09	251.24	0.8311	1.0171-3	251.08	251.25	0.8311	60
70	1.0231-3	292.98	293.06	0.9532	1.0231-2	292.98	293.08	0.9532	1.0230-3	292.97	293.09	0.9531	1.0230-3	292.97	293.11	0.9531	1.0230-3	292.96	293.12	0.9531	70
80	1.0291-3	334.85	334.93	1.0753	1.0291-3	334.85	334.95	1.0753	1.0290-3	334.84	334.97	1.0752	1.0290-3	334.84	334.98	1.0752	1.0290-3	334.83	335.00	1.0752	80
90	2.365	2494.6	2660.1	7.4800	1.0363-3	376.89	377.00	1.1911	1.0363-3	376.88	377.01	1.1910	1.0363-3	376.88	377.02	1.1910	1.0363-3	376.87	377.04	1.1910	90
100	2.434	2509.7	2680.0	7.5341	1.6958	2506.7	2676.2	7.3614	1.0436-3	418.93	419.05	1.3068	1.0436-3	418.92	419.06	1.3068	1.0436-3	418.91	419.08	1.3068	100
120	2.571	2539.7	2719.6	7.6375	1.7929	2537.3	2716.6	7.4668	1.4904	2535.7	2714.5	7.3785	1.2743	2534.1	2712.5	7.3032	1.1122	2532.4	2710.4	7.2374	120
140	2.706	2569.6	2759.0	7.7351	1.8887	2567.7	2756.5	7.5659	1.5709	2566.4	2754.9	7.4786	1.3438	2565.1	2753.2	7.4043	1.1735	2563.8	2751.5	7.3395	140
160	2.841	2599.4	2798.2	7.8279	1.9838	2597.8	2796.2	7.6597	1.6506	2596.8	2794.9	7.5731	1.4126	2595.7	2793.5	7.4995	1.2340	2594.7	2792.1	7.4354	160
180	2.975	2629.2	2837.5	7.9164	2.0782	2627.9	2835.8	7.7489	1.7297	2627.1	2834.6	7.6629	1.4807	2626.2	2833.5	7.5898	1.2939	2625.3	2832.4	7.5263	180
200	3.108	2659.1	2876.7	8.0012	2.172	2658.1	2875.3	7.8343	1.8083	2657.3	2874.3	7.7486	1.5483	2656.6	2873.4	7.6759	1.3533	2655.9	2872.4	7.6127	200
220	3.241	2689.2	2916.1	8.0826	2.266	2688.2	2914.8	7.9162	1.8866	2687.6	2914.0	7.8308	1.6157	2687.0	2913.2	7.7584	1.4124	2686.4	2912.4	7.6955	220
240	3.374	2719.3	2955.5	8.1611	2.359	2718.5	2954.5	7.9949	1.9647	2718.0	2953.8	7.9098	1.6827	2717.5	2953.1	7.8376	1.4713	2716.9	2952.3	7.7749	240
260	3.507	2749.7	2995.2	8.2369	2,453	2749.0	2994.3	8.0710	2.043	2748.5	2993.6	7.9860	1.7496	2748.1	2993.0	7.9139	1.5299	2747.6	2992.4	7.8514	260
280	3.640	2780.2	3035.0	8.3102	2.546	2779.6	3034.2	8.1445	2.120	2779.2	3033.6	8.0596	1.8164	2778.8	3033.1	7.9877	1.5844	2778.4	3032.5	7.9254	280
300	3.772	2811.0	3075.0	8.3813	2.639	2810.4	3074.3	8.2158	2.198	2810.1	3073.8	8.1310	1.8830	2809.7	3073.3	8.0592	1.6468	2809.3	3072.8	7.9969	300
320	3.905	2842.0	3115.3	8.4504	2.732	2841.5	3114.6	8.2849	2.275	2841.1	3114.2	8.2002	1.9495	2840.8	3113.7	8.1285	1.7051	2840.5	3113.3	8.0663	320
340	4.037	2873.2	3155.8	8.5175	2.824	2872.7	3155.2	8.3521	2.353	2872.4	3154.8	8.2675	2.0160	2872.1	3154.3	8.1958	1.7633	2871.8	3153.9	8.1337	340
360	4.170	2904.6	3196.5	8.5828	2.917	2904.2	3195.9	8.4175	2.430	2903.9	3195.5	8.3330	2.0824	2903.6	3195.2	8.2614	1.8215	2903.4	3194.8	8.1993	360
380	4.302	2936.3	3237.4	8.6465	3.010	2935.9	3236.9	8.4813	2.508	2935.7	3236.6	8.3968	2.1487	2935.4	3236.2	8.3252	1.8796	2935.2	3235.9	8.2632	380
400	4.434	2968.2	3278.6	8.7086	3.103	2967.9	3278.2	8.5435	2.585	2967.7	3277.8	8.4590	2.215	2967.4	3277.5	8.3875	1.9376	2967.2	3277.2	8.3255	400
420	4.566	3000.4	3320.1	8.7693	3.195	3000.1	3319.6	8.6042	2.662	2999.9	3319.3	8.5197	2.281	2999.7	3319.1	8.4483	1.9956	2999.5	3318.8	8.3864	420
440	4.698	3032.9	3361.8	8.8286	3.288	3032.6	3361.4	8.6636	2.739	3032.4	3361.1	8.5791	2.347	3032.2	3360.8	8.5077	2.0536	3032.0	3360.6	8.4458	440
460	4.831	3065.6	3403.7	8.8866	3.380	3065.3	3403.4	8.7216	2.816	3065.1	3403.1	8.6372	2.414	3065.0	3402.9	8.5658	2.1116	3064.8	3402.6	8.5040	460
480	4.963	3098.6	3446.0	8.9434	3.473	3098.3	3445.6	8.7785	2.894	3098.1	3445.4	8.6941	2.480	3098.0	3445.2	8.6227	2.1695	3097.8	3444.9	8.5609	480
500	5.095	3131.8	3488.5	8.9991	3.565	3131.6	3488.1	8.8342	2.971	3131.4	3487.9	8.7498	2.546	3131.3	3487.7	8.6785	2.227	3131.1	3487.5	8.6167	500
540	5.359	3199.1	3574.2	9.1073	3.750	3198.9	3574.0	8.9424	3.125	3198.8	3573.8	8.8581	2.678	3198.6	3573.6	8.7868	2.343	3198.5	3573.4	8.7250	540
580	5.623	3267.5	3661.1	9.2116	3.935	3267.3	3660.9	9.0467	3.279	3267.2	3660.7	8.9625	2.810	3267.1	3660.5	8.8912	2.459	3267.0	3660.4	8.8294	580
620	5.887	3337.0	3749.1	9.3123	4.120	3336.8	3748.9	9.1475	3.433	3336.7	3748.7	9.0633	2.943	3336.6	3748.6	8.9920	2.574	3336.5	3748.4	8.9303	620
660	6.151	3407.6	3838.1	9.4099	4.305	3407.4	3838.0	9.2451	3.587	3407.4	3837.8	9.1609	3.075	3407.3	3837.7	9.0896	2.690	3407.2	3837.6	9.0279	660
700	6.415	3479.3	3928.3	9.3045	4.490	3479.2	3928.2	9.3398	3.741	3479.1	3928.1	9.2555	3.207	3479.0	3927.9	9.1843	2.806	3478.9	3927.8	9.1226	700
750	6.744	3570.6	4042.7	9.6191	4.721	3570.4	4042.5	9.4544	3.934	3570.4	4042.4	9.3701	3.372	3570.3	4042.3	9.2989	2.950	3570.2	4042.2	9.2372	750
800	7.074	3663.6	4158.8	9.7299	4.952	3663.5	4158.6	9.5652	4.126	3663.4	4158.6	9.4810	3.537	3663.3	4158.5	9.4097	3.094	3663.3	4158.4	9.3481	800
850	7.404	3758.4	4276.7	9.8372	5.183	3758.3	4276.5	9.6725	4.319	3758.2	4276.5	9.5883	3.702	3758.2	4276.4	9.5171	3.239	3758.1	4276.3	9.4555	850
900	7.734	3854.9	4396.2	9.9414	5.414	3854.8	4396.1	9.7767	4.511	3854.7	4396.1	9.6925	3.867	3854.7	4396.0	9.6213	3.383	3854.6	4395.9	9.5596	900
950	8.064	3953.0	4517.5	10.0426	5.644	3953.0	4517.4	9.8779	4.704	3952.9	4517.3	9.7937	4.032	3952.9	4517.3	9.7225	3.528	3952.8	4517.2	9.6609	950
1000	8.393	4052.9	4640.4	10.1411	5.875	4052.8	4640.3	9.9764	4.896	4052.7	4640.3	9.8922	4.196	4052.7	4640.2	9.8210	3.672	4052.6	4640.1	9.7594	1000
1100	9.053	4257.3	4891.0	10.3306	6.337	4257.3	4891.0	10.1659	5.281	4257.2	4890.9	10.0817	4.526	4257.2	4890.8	10.0105	3.960	4257.1	4890.8	9.9489	1100
1200	9.712	4467.8	5147.6	10.5109	6.799	4467.7	5147.6	10.3463	5.666	4467.7	5147.5	10.2621	4.856	4467.6	5147.5	10.1909	4.249	4467.6	5147.4	10.1293	1200
1300	10.372	4683.5	5409.6	10.6829	7.260	4683.5	5409.5	10.5183	6.050	4683.4	5409.5	10.4341	5.186	4683.4	5409.4	10.3629	4.538	4683.3	5409.4	10.3013	1300

Properties of H_2O

P(t_{sat}) Temp C	200 x 10^3 N/m^2 (120.23 C) v m^3/kg	u kJ/kg	h kJ/kg	s kJ/kgK
Sat liq	1.0605-3	504.49	504.70	1.5301
Evap	0.8846	2025.0	2202.0	5.5970
Sat vap	0.8857	2529.5	2706.7	7.1271
0	1.0001-3	-0.03	0.17	-0.0001
20	1.0017-3	83.94	84.13	0.2966
40	1.0077-3	167.54	167.74	0.5724
60	1.0171-3	251.08	251.28	0.8311
80	1.0290-3	334.82	335.03	1.0752
100	1.0435-3	418.90	419.11	1.3067
120	1.0603-3	503.50	503.71	1.5275
140	0.9350	2561.1	2748.1	7.2300
160	0.9841	2592.5	2789.3	7.3275
180	1.0324	2623.6	2830.1	7.4194
200	1.0803	2654.4	2870.5	7.5066
220	1.1279	2685.1	2910.7	7.5899
240	1.1752	2715.9	2950.9	7.6698
260	1.2223	2746.6	2991.1	7.7467
280	1.2693	2777.5	3031.4	7.8209
300	1.3162	2808.6	3071.8	7.8926
320	1.3629	2839.8	3112.4	7.9622
340	1.4096	2871.2	3153.1	8.0298
360	1.4562	2902.8	3194.1	8.0955
380	1.5028	2934.7	3235.2	8.1595
400	1.5493	2966.7	3276.6	8.2218
420	1.5958	2999.0	3318.2	8.2828
440	1.6422	3031.6	3360.0	8.3423
460	1.6886	3064.4	3402.1	8.4005
480	1.7350	3097.5	3444.5	8.4575
500	1.7814	3130.8	3487.1	8.5133
540	1.8741	3198.2	3573.0	8.6217
580	1.9667	3266.7	3660.0	8.7261
620	2.059	3336.3	3748.1	8.8270
660	2.152	3407.0	3837.3	8.9247
700	2.244	3478.8	3927.6	9.0194
750	2.360	3570.1	4042.0	9.1341
800	2.475	3663.1	4158.2	9.2449
850	2.591	3758.0	4276.2	9.3524
900	2.706	3854.5	4395.8	9.4566
950	2.822	3952.7	4517.1	9.5578
1000	2.937	4052.5	4640.0	9.6563
1100	3.168	4257.0	4890.7	9.8458
1200	3.399	4467.5	5147.3	10.0262
1300	3.630	4683.2	5409.3	10.1982

Temp C	300 x 10^3 N/m^2 (133.55 C) v m^3/kg	u kJ/kg	h kJ/kg	s kJ/kgK
Sat liq	1.0732-3	561.15	561.47	1.6718
Evap	0.6047	1982.5	2163.8	5.3201
Sat vap	0.6058	2543.6	2725.3	6.9919
0	1.0001-3	-0.03	0.27	-0.0001
20	1.0017-3	83.93	84.23	0.2965
40	1.0077-3	167.52	167.82	0.5724
60	1.0171-3	251.06	251.37	0.8310
80	1.0290-3	334.80	335.11	1.0751
100	1.0435-3	418.87	419.19	1.3066
120	1.0602-3	503.47	503.78	1.5275
140	0.6169	2554.3	2739.4	7.0263
160	0.6506	2587.1	2782.3	7.1276
180	0.6837	2619.1	2824.2	7.2223
200	0.7163	2650.7	2865.6	7.3115
220	0.7485	2682.0	2906.5	7.3963
240	0.7805	2713.1	2947.3	7.4774
260	0.8122	2744.3	2987.9	7.5551
280	0.8438	2775.4	3028.6	7.6299
300	0.8753	2806.7	3069.3	7.7022
320	0.9067	2838.1	3110.1	7.7722
340	0.9380	2869.7	3151.1	7.8401
360	0.9692	2901.4	3192.2	7.9061
380	1.0004	2933.4	3233.5	7.9704
400	1.0315	2965.6	3275.0	8.0330
420	1.0626	2998.0	3316.7	8.0941
440	1.0937	3030.6	3358.7	8.1538
460	1.1247	3063.5	3400.9	8.2121
480	1.1557	3096.6	3443.3	8.2692
500	1.1867	3130.0	3486.0	8.3251
540	1.2486	3197.5	3572.1	8.4337
580	1.3105	3266.1	3659.2	8.5383
620	1.3722	3335.8	3747.4	8.6393
660	1.4340	3406.5	3836.7	8.7371
700	1.4957	3478.4	3927.1	8.8319
750	1.5729	3569.8	4041.6	8.9467
800	1.6499	3662.9	4157.8	9.0576
850	1.7270	3757.7	4275.8	9.1650
900	1.8041	3854.2	4395.4	9.2692
950	1.8811	3952.4	4516.7	9.3705
1000	1.9581	4052.3	4639.7	9.4690
1100	2.1121	4256.8	4890.4	9.6585
1200	2.2661	4467.2	5147.1	9.8389
1300	2.4201	4683.0	5409.0	10.0110

Temp C	400 x 10^3 N/m^2 (143.64 C) v m^3/kg	u kJ/kg	h kJ/kg	s kJ/kgK
Sat liq	1.0836-3	604.31	604.74	1.7766
Evap	0.4614	1949.3	2133.9	5.1193
Sat vap	0.4625	2553.6	2738.6	6.8959
0	1.0000-3	-0.03	0.37	-0.0001
20	1.0016-3	83.93	84.32	0.2965
40	1.0076-3	167.51	167.91	0.5724
60	1.0170-3	251.03	251.45	0.8310
80	1.0289-3	334.78	335.19	1.0750
100	1.0434-3	418.84	419.26	1.3066
120	1.0602-3	503.43	503.85	1.5274
140	1.0797-3	558.72	589.15	1.7389
160	0.4838	2581.4	2774.9	6.9815
180	0.5093	2614.5	2818.2	7.0792
200	0.5342	2646.8	2860.5	7.1706
220	0.5588	2678.7	2902.3	7.2570
240	0.5831	2710.4	2943.6	7.3392
260	0.6071	2741.9	2984.7	7.4178
280	0.6310	2773.3	3025.7	7.4933
300	0.6548	2804.8	3066.8	7.5662
320	0.6785	2836.4	3107.8	7.6366
340	0.7021	2868.2	3149.0	7.7049
360	0.7257	2900.0	3190.3	7.7712
380	0.7492	2932.1	3231.8	7.8357
400	0.7726	2964.4	3273.4	7.8985
420	0.7960	2996.9	3315.3	7.9598
440	0.8194	3029.6	3357.4	8.0196
460	0.8427	3062.6	3399.7	8.0781
480	0.8661	3095.8	3442.2	8.1353
500	0.8893	3129.2	3484.9	8.1913
540	0.9359	3196.8	3571.2	8.3001
580	0.9823	3265.5	3658.4	8.4048
620	1.0287	3335.2	3746.7	8.5059
660	1.0751	3406.0	3836.1	8.6038
700	1.1215	3477.9	3926.5	8.6987
750	1.1794	3569.3	4041.0	8.8134
800	1.2372	3662.4	4157.3	8.9244
850	1.2951	3757.4	4275.3	9.0319
900	1.3529	3853.9	4395.1	9.1362
950	1.4107	3952.2	4516.4	9.2375
1000	1.4685	4052.0	4639.4	9.3360
1100	1.5840	4256.5	4890.2	9.5256
1200	1.6996	4467.0	5146.8	9.7060
1300	1.8151	4682.8	5408.8	9.8780

Temp C	500 x 10^3 N/m^2 (151.86 C) v m^3/kg	u kJ/kg	h kJ/kg	s kJ/kgK
Sat liq	1.0926-3	639.68	640.23	1.8607
Evap	0.3738	1921.5	2108.5	4.9606
Sat vap	0.3749	2561.2	2748.7	6.8213
0	1.0000-3	-0.02	0.47	-0.0001
20	1.0016-3	83.92	84.42	0.2965
40	1.0076-3	167.50	168.00	0.5723
60	1.0170-3	251.01	251.54	0.8309
80	1.0289-3	334.75	335.27	1.0750
100	1.0434-3	418.82	419.34	1.3065
120	1.0602-3	503.39	503.92	1.5273
140	1.0797-3	588.68	589.22	1.7388
160	0.3836	2575.6	2767.4	6.8648
180	0.4045	2609.7	2812.0	6.9656
200	0.4249	2642.9	2855.4	7.0592
220	0.4449	2675.5	2897.9	7.1473
240	0.4646	2707.6	2939.9	7.2307
260	0.4841	2739.4	2981.5	7.3102
280	0.5034	2771.2	3022.9	7.3865
300	0.5226	2802.9	3064.2	7.4599
320	0.5416	2834.7	3105.6	7.5308
340	0.5606	2866.6	3146.9	7.5994
360	0.5796	2898.7	3188.4	7.6660
380	0.5985	2930.8	3230.1	7.7307
400	0.6173	2963.2	3271.9	7.7938
420	0.6361	2995.8	3313.8	7.8552
440	0.6548	3028.6	3356.0	7.9152
460	0.6736	3061.6	3398.4	7.9738
480	0.6923	3094.9	3441.0	8.0312
500	0.7109	3128.4	3483.9	8.0873
540	0.7482	3196.1	3570.2	8.1962
580	0.7855	3264.9	3657.6	8.3011
620	0.8227	3334.7	3746.0	8.4023
660	0.8598	3405.5	3835.4	8.5003
700	0.8969	3477.5	3925.9	8.5952
750	0.9433	3568.9	4040.5	8.7101
800	0.9896	3662.1	4156.9	8.8211
850	1.0359	3757.0	4275.0	8.9287
900	1.0822	3853.6	4394.7	9.0329
950	1.1284	3951.9	4516.1	9.1343
1000	1.1747	4051.8	4639.1	9.2328
1100	1.2672	4256.3	4889.9	9.4224
1200	1.3596	4466.8	5146.6	9.6029
1300	1.4521	4682.5	5408.6	9.7749

Temp C	600 x 10^3 N/m^2 (158.85 C) v m^3/kg	u kJ/kg	h kJ/kg	s kJ/kgK	Temp C
Sat liq	1.1006-3	669.90	670.56	1.9312	Sat liq
Evap	0.3146	1897.5	2086.2	4.8288	Evap
Sat vap	0.3157	2567.4	2756.8	6.7600	Sat vap
0	1.0000-3	-0.02	0.57	-0.0001	0
20	1.0016-3	83.91	84.51	0.2965	20
40	1.0076-3	167.49	168.08	0.5723	40
60	1.0070-3	250.99	251.62	0.8309	60
80	1.0289-3	334.73	335.35	1.0749	80
100	1.0434-3	418.79	419.41	1.3064	100
120	1.0602-3	503.36	503.99	1.5272	120
140	1.0796-3	588.63	589.28	1.7386	140
160	0.3167	2569.5	2759.5	6.7663	160
180	0.3347	2604.9	2805.7	6.8705	180
200	0.3520	2638.9	2850.1	6.9665	200
220	0.3690	2672.1	2893.5	7.0562	220
240	0.3856	2704.7	2936.1	7.1409	240
260	0.4020	2737.0	2978.2	7.2214	260
280	0.4183	2769.0	3020.0	7.2984	280
300	0.4344	2801.0	3061.6	7.3724	300
320	0.4504	2833.0	3103.2	7.4437	320
340	0.4663	2865.1	3144.9	7.5127	340
360	0.4822	2897.3	3186.6	7.5796	360
380	0.4980	2929.6	3228.4	7.6446	380
400	0.5137	2962.1	3270.3	7.7079	400
420	0.5294	2994.7	3312.4	7.7695	420
440	0.5451	3027.6	3354.7	7.8297	440
460	0.5608	3060.7	3397.2	7.8884	460
480	0.5764	3094.0	3439.9	7.9459	480
500	0.5920	3127.6	3482.8	8.0021	500
540	0.6231	3195.4	3569.3	8.1112	540
580	0.6542	3264.3	3656.8	8.2162	580
620	0.6853	3334.1	3745.3	8.3176	620
660	0.7162	3405.0	3834.8	8.4156	660
700	0.7472	3477.0	3925.3	8.5107	700
750	0.7859	3568.5	4040.1	8.6256	750
800	0.8245	3661.8	4156.5	8.7367	800
850	0.8631	3756.7	4274.6	8.8442	850
900	0.9017	3853.4	4394.4	8.9486	900
950	0.9403	3951.6	4515.8	9.0499	950
1000	0.9788	4051.5	4638.8	9.1485	1000
1100	1.0559	4256.1	4889.6	9.3381	1100
1200	1.1330	4466.5	5146.3	9.5185	1200
1300	1.2101	4682.3	5408.3	9.6906	1300

Properties of H_2O

$P(t_{sat})$	800 x 10^3 N/m^2 (170.43 C)				1 x 10^6 N/m^2 (179.91 C)				1.2 x 10^6 N/m^2 (187.99 C)				1.4 x 10^6 N/m^2 (195.07 C)				1.6 x 10^6 N/m^2 (201.41 C)				
Temp C	v m^3/kg	u kJ/kg	h kJ/kg	s kJ/kgK	v m^3/kg	u kJ/kg	h kJ/kg	s kJ/kgK	v m^3/kg	u kJ/kg	h kJ/kg	s kJ/kgK	v m^3/kg	u kJ/kg	h kJ/kg	s kJ/kgK	v m^3/kg	u kJ/kg	h kJ/kg	s kJ/kgK	Temp C
Sat liq	1.1148-3	720.22	721.11	2.0462	1.1273-3	761.68	762.81	2.1387	1.1385-3	797.29	798.65	2.2166	1.1489-3	828.70	830.30	2.2842	1.1587-3	856.94	858.79	2.3442	Sat liq
Evap	0.2393	1856.6	2048.0	4.6166	0.19331	1821.9	2015.3	4.4478	0.16219	1791.5	1986.2	4.3067	0.13969	1764.1	1959.7	4.1851	0.12264	1739.1	1935.2	4.0776	Evap
Sat vap	0.2404	2576.8	2769.1	6.6628	0.19444	2583.6	2778.1	6.5865	0.16333	2588.8	2784.8	6.5233	0.14084	2592.8	2790.0	6.4693	0.12380	2596.0	2794.0	6.4218	Sat vap
0	0.9999-3	-0.02	0.78	-0.0001	0.9997-3	-0.02	0.98	-0.0001	0.9996-3	-0.02	1.18	-0.0001	0.9995-3	-0.01	1.39	-0.0000	0.9994-3	-0.01	1.59	-0.0000	0
40	1.0075-3	167.46	168.27	0.5722	1.0074-3	167.44	168.44	0.5720	1.0073-3	167.41	168.62	0.5720	1.0072-3	167.39	168.80	0.5719	1.0071-3	167.36	168.97	0.5719	40
80	1.0287-3	334.69	335.51	1.0748	1.0287-3	334.64	335.67	1.0747	1.0286-3	334.59	335.82	1.0745	1.0285-3	334.55	335.98	1.0744	1.0284-3	334.50	336.14	1.0743	80
120	1.0600-3	503.29	504.13	1.5271	1.0597-3	503.21	504.27	1.5269	1.0597-3	503.14	504.41	1.5267	1.0596-3	503.07	504.56	1.5265	1.0595-3	503.00	504.70	1.5263	120
160	1.1018-3	674.78	675.66	1.9424	1.1017-3	674.68	675.77	1.9422	1.1015-3	674.57	675.89	1.9420	1.1014-3	674.47	676.01	1.9417	1.1012-3	674.36	676.12	1.9415	160
170	1.1147-3	718.50	719.39	2.0411	1.1146-3	718.39	719.54	2.0409	1.1144-3	718.27	719.60	2.0406	1.1142-3	718.16	719.71	2.0403	1.1140-3	718.04	719.82	2.0401	170
180	0.2472	2594.7	2792.4	6.7146	0.19450	2583.8	2778.3	6.5870	1.1273-3	761.97	763.32	2.1393	1.1271-3	761.85	763.42	2.1390	1.1269-3	761.73	763.52	2.1388	180
190	0.2540	2612.9	2816.1	6.7663	0.20030	2603.2	2803.5	6.6419	0.16435	2592.9	2790.1	6.5348	1.1421-3	806.27	807.91	2.2348	1.1416-3	806.16	807.96	2.2348	190
200	0.2608	2630.6	2839.3	6.8158	0.2060	2621.9	2827.9	6.6940	0.16930	2612.8	2815.9	6.5898	0.14302	2603.1	2803.3	6.4975	1.1564-3	850.6	852.4	2.3308	200
210	0.2674	2648.1	2862.0	6.8634	0.2115	2640.2	2851.7	6.7437	0.17411	2631.9	2840.9	6.6420	0.14734	2623.3	2829.6	6.5524	0.12719	2614.2	2817.7	6.4713	210
220	0.2740	2665.2	2884.4	6.9093	0.2169	2658.0	2874.9	6.7914	0.17880	2650.6	2865.1	6.6918	0.15152	2642.8	2854.9	6.6044	0.13101	2634.7	2844.3	6.5257	220
230	0.2804	2682.1	2906.5	6.9536	0.2222	2675.6	2897.8	6.8374	0.18339	2668.8	2888.9	6.7394	0.15560	2661.8	2879.6	6.6539	0.13471	2654.4	2870.0	6.5773	230
240	0.2868	2698.9	2928.3	6.9966	0.2275	2692.9	2920.4	6.8817	0.18790	2686.6	2912.1	6.7852	0.15959	2680.2	2903.7	6.7012	0.13832	2673.6	2894.9	6.6263	240
250	0.2931	2715.5	2950.0	7.0384	0.2327	2709.9	2942.6	6.9247	0.19234	2704.2	2935.0	6.8294	0.16350	2698.3	2927.2	6.7467	0.14184	2692.3	2919.2	6.6732	250
260	0.2994	2732.0	2971.5	7.0791	0.2378	2726.8	2964.6	6.9664	0.19673	2721.5	2957.6	6.8721	0.16735	2716.1	2950.4	6.7906	0.14530	2710.6	2943.0	6.7183	260
270	0.3056	2748.3	2992.9	7.1188	0.2429	2743.6	2986.5	7.0069	0.20106	2738.7	2979.9	6.9136	0.17115	2733.7	2973.3	6.8331	0.14869	2728.5	2966.4	6.7618	270
280	0.3118	2764.7	3014.1	7.1577	0.2480	2760.2	3008.2	7.0465	0.20535	2755.6	3002.1	6.9540	0.17490	2751.0	2995.8	6.8743	0.15204	2746.2	2989.5	6.8039	280
290	0.3180	2780.9	3035.3	7.1956	0.2530	2776.7	3029.7	7.0851	0.20960	2772.5	3024.0	6.9933	0.17861	2768.1	3018.2	6.9144	0.15535	2763.7	3012.3	6.8447	290
300	0.3241	2797.2	3056.5	7.2328	0.2579	2793.2	3051.2	7.1229	0.2138	2789.2	3045.8	7.0317	0.12828	2785.2	3040.4	6.9534	0.15862	2781.1	3034.8	6.8844	300
320	0.3363	2829.6	3098.6	7.3051	0.2678	2826.1	3093.9	7.1962	0.2222	2822.5	3089.1	7.1060	0.18954	2818.9	3084.3	7.0287	0.16507	2815.3	3079.4	6.9609	320
340	0.3484	2862.0	3140.7	7.3749	0.2776	2858.8	3136.4	7.2667	0.2304	2855.7	3132.2	7.1774	0.19671	2852.5	3127.8	7.1009	0.17142	2849.2	3123.5	7.0340	340
360	0.3604	2894.4	3182.7	7.4424	0.2873	2891.6	3178.9	7.3349	0.2386	2888.7	3175.1	7.2462	0.2038	2885.8	3171.2	7.1704	0.17769	2882.9	3167.2	7.1042	360
380	0.3724	2927.0	3224.9	7.5079	0.2970	2924.4	3221.4	7.4009	0.2467	2921.8	3217.9	7.3128	0.2108	2919.2	3214.3	7.2376	0.18390	2916.5	3210.8	7.1719	380
400	0.3843	2959.7	3267.1	7.5716	0.3066	2957.3	3263.9	7.4651	0.2548	2954.9	3260.7	7.3774	0.2178	2952.5	3257.5	7.3026	0.19005	2950.1	3254.2	7.2374	400
420	0.3961	2992.6	3309.5	7.6336	0.3162	2990.4	3306.5	7.5275	0.2628	2988.2	3303.6	7.4401	0.2247	2986.0	3300.6	7.3658	0.19616	2983.8	3297.6	7.3009	420
440	0.4080	3025.6	3352.0	7.6941	0.3257	3023.6	3349.3	7.5883	0.2708	3021.6	3346.6	7.5012	0.2316	3019.5	3343.8	7.4272	0.20224	3017.5	3341.1	7.3627	440
460	0.4198	3058.9	3394.7	7.7531	0.3352	3057.0	3392.2	7.6476	0.2788	3055.1	3389.7	7.5609	0.2385	3053.2	3387.1	7.4871	0.20828	3051.4	3384.6	7.4229	460
480	0.4316	3092.3	3437.6	7.8108	0.3447	3090.6	3435.2	7.7055	0.2867	3088.8	3432.9	7.6190	0.2453	3087.1	3430.6	7.5456	0.21430	3085.3	3428.2	7.4816	480
500	0.4433	3126.0	3480.6	7.8673	0.3541	3124.4	3478.5	7.7622	0.2946	3122.8	3476.3	7.6759	0.2521	3121.1	3474.1	7.6027	0.2203	3119.5	3472.0	7.5390	500
540	0.4668	3194.0	3567.4	7.9767	0.3729	3192.6	3565.6	7.8720	0.3104	3191.2	3563.7	7.7861	0.2657	3189.8	3561.8	7.7132	0.2322	3188.4	3559.9	7.6499	540
580	0.4902	3263.0	3655.2	8.0820	0.3917	3261.8	3653.5	7.9776	0.3261	3260.6	3651.9	7.8920	0.2792	3259.3	3650.3	7.8194	0.2441	3258.1	3648.6	7.7563	580
620	0.5135	3333.0	3743.8	8.1836	0.4105	3332.0	3742.4	8.0794	0.3417	3330.9	3741.0	7.9940	0.2927	3329.8	3739.5	7.9217	0.2559	3328.7	3738.1	7.8588	620
660	0.5368	3404.1	3833.5	8.2818	0.4291	3403.1	3832.2	8.1778	0.3574	3402.1	3831.0	8.0926	0.3061	3401.2	3829.7	8.0204	0.2676	3400.2	3828.4	7.9578	660
700	0.5601	3476.2	3924.2	8.3770	0.4478	3475.3	3923.1	8.2731	0.3729	3474.4	3922.0	8.1881	0.3195	3473.6	3920.8	8.1160	0.2794	3472.7	3919.7	8.0535	700
750	0.5891	3567.8	4039.1	8.4921	0.4711	3567.0	4038.1	8.3883	0.3924	3566.2	4037.1	8.3034	0.3362	3565.5	4036.1	8.2315	0.2940	3564.7	4035.1	8.1692	750
800	0.6181	3661.1	4155.6	8.6033	0.4943	3660.4	4154.7	8.4996	0.4118	3659.7	4153.8	8.4148	0.3528	3659.0	4153.0	8.3431	0.3086	3658.3	4152.1	8.2808	800
850	0.6471	3756.1	4273.8	8.7109	0.5175	3755.5	4273.0	8.6074	0.4311	3754.9	4272.2	8.5227	0.3694	3754.2	4271.5	8.4510	0.3232	3753.6	4270.7	8.3888	850
900	0.6761	3852.8	4393.7	8.8153	0.5407	3852.2	4392.9	8.7118	0.4505	3851.6	4392.2	8.6272	0.3861	3851.1	4391.5	8.5556	0.3377	3850.5	4390.8	8.4935	900
950	0.7051	3951.1	4515.1	8.9167	0.5639	3950.5	4514.5	8.8133	0.4699	3950.0	4513.8	8.7287	0.4027	3949.5	4513.2	8.6571	0.3523	3948.9	4512.5	8.5951	950
1000	0.7340	4051.0	4638.2	9.0153	0.5871	4050.5	4637.6	8.9119	0.4892	4050.0	4637.0	8.8274	0.4192	4049.5	4636.4	8.7559	0.3668	4049.0	4635.8	8.6938	1000
1100	0.7919	4255.6	4889.1	9.2050	0.6335	4255.1	4888.6	9.1017	0.5278	4254.6	4888.0	9.0172	0.4524	4254.1	4887.5	8.9457	0.3958	4253.7	4887.0	8.8837	1100
1200	0.8497	4466.1	5145.9	9.3855	0.6798	4465.6	5145.4	9.2822	0.5665	4465.1	5144.9	9.1977	0.4855	4464.7	5144.4	9.1262	0.4248	4464.2	5143.9	9.0643	1200
1300	0.9076	4681.8	5407.9	9.5575	0.7261	4681.3	5407.4	9.4543	0.6051	4680.9	5407.0	9.3698	0.5186	4680.4	5406.5	9.2984	0.4538	4679.9	5406.0	9.2364	1300

Properties of H_2O

$P(t_{sat})$ Temp C	2×10^6 N/m^2 (212.42 C)				3×10^6 N/m^2 (233.90 C)				4×10^6 N/m^2 (250.40 C)				5×10^6 N/m^2 (263.99 C)				6×10^6 N/m^2 (275.64 C)				Temp C
	v m^3/kg	u kJ/kg	h kJ/kg	s kJ/kgK	v m^3/kg	u kJ/kg	h kJ/kg	s kJ/kgK	v m^3/kg	u kJ/kg	h kJ/kg	s kJ/kgK	v m^3/kg	u kJ/kg	h kJ/kg	s kJ/kgK	v m^3/kg	u kJ/kg	h kJ/kg	s kJ/kgK	
Sat liq	1.1767-3	906.44	908.79	2.4474	1.2165-3	1004.78	1008.42	2.6457	1.2522-3	1082.31	1087.31	2.7964	1.2859-3	1147.81	1154.23	2.9202	1.3187-3	1205.44	1213.35	3.0267	Sat liq
Evap	0.09845	1693.9	1890.7	3.8935	0.06546	1599.3	1795.8	3.5412	0.04853	1520.0	1714.1	3.2737	0.03815	1449.3	1640.1	3.0532	0.03112	1384.3	1571.0	2.8625	Evap
Sat vap	0.09963	2600.3	2799.5	6.3409	0.06668	2604.1	2804.2	6.1869	0.04978	2602.3	2801.4	6.0701	0.03944	2597.1	2794.3	5.9734	0.03244	2589.7	2784.3	5.8892	Sat vap
0	0.9992-3	-0.01	1.99	-0.0000	0.9987-3	0.01	3.01	-0.0000	0.9982-3	0.02	4.02	0.0001	0.9977-3	0.04	5.04	0.0001	0.9972-3	0.05	6.04	0.0001	0
40	1.0069-3	167.31	169.33	0.5717	1.0065-3	167.19	170.21	0.5713	1.0061-3	167.07	171.09	0.5709	1.0056-3	166.95	171.97	0.5705	0.9991-3	166.83	172.85	0.5701	40
80	1.0283-3	334.41	336.46	1.0740	1.0278-3	334.18	337.26	1.0734	1.0273-3	333.95	338.06	1.0727	1.0268-3	333.72	338.85	1.0720	1.0264-3	333.49	339.65	1.0714	80
120	1.0593-3	502.86	504.98	1.5260	1.0583-3	502.51	505.68	1.5251	1.0582-3	502.15	506.39	1.5242	1.0576-3	501.80	507.09	1.5233	1.0570-3	501.46	507.80	1.5224	120
160	1.1010-3	674.16	676.36	1.9410	1.0997-3	673.64	676.94	1.9398	1.0995-3	673.13	677.53	1.9387	1.0988-3	672.62	678.12	1.9375	1.0981-3	672.12	678.71	1.9358	160
200	1.1560-3	850.3	852.6	2.3302	1.1550-3	849.5	853.0	2.3286	1.1540-3	848.8	853.5	2.3271	1.1530-3	848.1	853.9	2.3255	1.1520-3	847.4	854.3	2.3239	200
210	1.1726-3	895.4	897.8	2.4247	1.1714-3	894.6	897.8	2.4229	1.1703-3	893.8	898.5	2.4212	1.1691-3	893.0	898.8	2.4195	1.1680-3	892.2	899.2	2.4179	210
220	0.10215	2617.4	2821.7	6.3861	1.1892-3	940.2	943.8	2.5185	1.1779-3	939.3	944.1	2.5146	1.1866-3	938.4	944.4	2.5128	1.1854-3	937.5	944.7	2.5110	220
230	0.10536	2638.9	2849.6	6.4423	1.2085-3	986.5	990.2	2.6095	1.2070-3	985.5	990.4	2.6075	1.2056-3	984.5	990.6	2.6055	1.2008-3	983.5	990.8	2.6035	230
240	0.10845	2659.6	2876.5	6.4952	0.06820	2619.7	2824.3	6.2265	1.2280-3	1032.5	1037.4	2.7001	1.2264-3	1031.4	1037.5	2.6979	1.2248-3	1030.3	1037.6	2.6957	240
250	0.11144	2679.6	2902.5	6.5453	0.07058	2644.0	2855.8	6.2872	1.2512-3	1080.4	1085.3	2.7927	1.2493-3	1079.1	1085.3	2.7902	1.2475-3	1077.9	1085.3	2.7878	250
260	0.11436	2699.0	2927.7	6.5931	0.07285	2667.0	2885.5	6.3435	0.05174	2629.4	2836.3	6.1362	1.2749-3	1127.9	1134.3	2.8830	1.2727-3	1126.5	1134.2	2.8803	260
270	0.11721	2717.9	2952.3	6.6388	0.07502	2688.9	2913.9	6.3963	0.05365	2655.5	2870.1	6.1990	0.04054	2616.2	2818.9	6.0189	1.3013-3	1176.6	1184.4	2.9736	270
280	0.12001	2736.4	2976.4	6.6828	0.07712	2709.9	2941.3	6.4462	0.05546	2680.0	2901.8	6.2568	0.04224	2645.6	2856.8	6.0880	0.03317	2605.2	2804.2	5.9252	280
290	0.12276	2754.6	3000.2	6.7253	0.07916	2730.3	2967.7	6.4936	0.05718	2703.2	2931.9	6.3108	0.04382	2672.7	2891.8	6.1507	0.03473	2637.7	2846.1	6.0003	290
300	0.12547	2772.6	3023.5	6.7664	0.08114	2750.1	2993.5	6.5390	0.05884	2725.3	2960.7	6.3615	0.04532	2698.0	2924.5	6.2084	0.03616	2667.2	2884.2	6.0674	300
320	0.13079	2807.9	3069.5	6.8452	0.08499	2788.4	3043.4	6.6245	0.06199	2767.4	3015.4	6.4553	0.04810	2744.7	2985.3	6.3125	0.03876	2720.0	2952.6	6.1846	320
340	0.13600	2842.6	3114.6	6.9201	0.08871	2825.5	3091.6	6.7045	0.06499	2807.3	3067.3	6.5413	0.05070	2788.0	3041.5	6.4058	0.04111	2767.4	3014.0	6.2866	340
360	0.14113	2877.0	3159.3	6.9917	0.09233	2861.7	3138.7	6.7801	0.06788	2845.7	3117.2	6.6215	0.05316	2828.9	3094.7	6.4912	0.04331	2811.2	3071.1	6.3782	360
380	0.14619	2911.2	3203.6	7.0606	0.09587	2897.4	3185.1	6.8522	0.07068	2883.2	3165.9	6.6971	0.05552	2868.3	3145.9	6.5708	0.04539	2852.8	3125.2	6.4623	380
400	0.15120	2945.2	3247.6	7.1271	0.09936	2932.8	3230.9	6.9212	0.07341	2919.9	3213.6	6.7690	0.05781	2906.6	3195.7	6.6459	0.04739	2892.9	3177.2	6.5408	400
420	0.15616	2979.3	3291.6	7.1915	0.10279	2967.9	3276.3	6.9878	0.07609	2956.2	3260.6	6.8379	0.06004	2944.2	3244.4	6.7172	0.04933	2931.8	3227.8	6.6148	420
440	0.16108	3013.4	3335.5	7.2540	0.10619	3002.9	3321.5	7.0520	0.07872	2992.2	3307.1	6.9041	0.06223	2981.3	3292.4	6.7855	0.05122	2970.0	3277.3	6.6853	440
460	0.16597	3047.6	3379.5	7.3147	0.10955	3037.9	3366.6	7.1144	0.08132	3028.1	3353.4	6.9680	0.06437	3018.0	3339.9	6.8511	0.05306	3007.7	3326.1	6.7527	460
480	0.17084	3081.8	3423.5	7.3740	0.11288	3072.9	3411.5	7.1749	0.08389	3063.8	3399.4	7.0299	0.06648	3054.6	3387.0	6.9145	0.05487	3045.1	3374.3	6.8176	480
500	0.17568	3116.2	3467.6	7.4317	0.11619	3108.0	3456.5	7.2338	0.08643	3099.5	3445.3	7.0901	0.06857	3091.0	3433.8	6.9759	0.05665	3082.2	3422.2	6.8803	500
520	0.18049	3150.8	3511.8	7.4882	0.11947	3143.1	3501.5	7.2913	0.08895	3135.3	3491.1	7.1486	0.07063	3127.3	3480.5	7.0355	0.05841	3119.2	3469.7	6.9410	520
540	0.18529	3185.6	3556.1	7.5434	0.12273	3178.4	3546.6	7.3474	0.09145	3171.1	3536.9	7.2056	0.07267	3163.7	3527.0	7.0934	0.06015	3156.1	3517.0	6.9999	540
560	0.19008	3220.5	3600.6	7.5974	0.12598	3213.8	3591.7	7.4022	0.09393	3206.9	3582.7	7.2612	0.07469	3200.0	3573.5	7.1499	0.06186	3193.0	3564.2	7.0573	560
580	0.19485	3255.6	3645.3	7.6504	0.12921	3249.3	3637.0	7.4559	0.09639	3242.9	3628.5	7.3156	0.07670	3236.5	3620.0	7.2050	0.06357	3229.9	3611.3	7.1132	580
600	0.19960	3290.9	3690.1	7.7024	0.13243	3285.0	3682.3	7.5085	0.09885	3279.1	3674.4	7.3688	0.07869	3273.0	3666.5	7.2589	0.06525	3266.9	3658.4	7.1677	600
640	0.20908	3362.2	3780.4	7.8035	0.13884	3357.0	3773.5	7.6106	0.10372	3351.8	3766.6	7.4720	0.08264	3346.4	3759.6	7.3632	0.06859	3341.0	3752.6	7.2731	640
680	0.21852	3434.5	3871.5	7.9011	0.14520	3429.8	3865.4	7.7091	0.10855	3425.1	3859.3	7.5714	0.08655	3420.4	3853.1	7.4634	0.07189	3415.6	3846.9	7.3743	680
720	0.2279	3507.6	3963.5	7.9956	0.15154	3503.5	3958.1	7.8043	0.11334	3499.3	3952.6	7.6673	0.09043	3495.0	3947.2	7.5601	0.07515	3490.7	3941.6	7.4716	720
760	0.2373	3581.8	4056.4	8.0874	0.15785	3578.0	4051.6	7.8966	0.11812	3574.2	4046.7	7.7601	0.09428	3570.4	4041.8	7.6535	0.07839	3566.5	4036.9	7.5656	760
800	0.2467	3657.0	4150.3	8.1765	0.16414	3653.5	4145.9	7.9862	0.12287	3650.0	4141.5	7.8502	0.09811	3646.6	4137.1	7.7440	0.08160	3643.1	4132.7	7.6566	800
850	0.2584	3752.4	4269.1	8.2847	0.17198	3749.3	4265.2	8.0948	0.12879	3746.1	4261.3	7.9593	0.10287	3743.0	4257.4	7.8535	0.08560	3739.8	4253.4	7.7666	850
900	0.2700	3849.3	4389.4	8.3895	0.17980	3846.5	4385.9	8.1999	0.13469	3843.6	4382.3	8.0647	0.10762	3840.7	4378.8	7.9593	0.08958	3837.8	4375.3	7.8727	900
950	0.2817	3947.9	4511.2	8.4912	0.18761	3945.2	4508.0	8.3019	0.14057	3942.5	4504.8	8.1669	0.11235	3939.8	4501.6	8.0618	0.09354	3937.2	4498.4	7.9755	950
1000	0.2933	4048.0	4634.6	8.5901	0.19541	4045.4	4631.6	8.4009	0.14645	4042.9	4628.7	8.2662	0.11707	4040.4	4625.7	8.1612	0.09749	4037.8	4622.7	8.0751	1000
1100	0.3166	4252.7	4885.9	8.7800	0.21098	4250.3	4883.3	8.5912	0.15817	4248.0	4880.6	8.4567	0.12648	4245.6	4878.0	8.3520	0.10536	4243.3	4875.4	8.2661	1100
1200	0.3398	4463.3	5142.9	8.9607	0.22652	4460.9	5140.5	8.7720	0.16987	4458.6	5138.1	8.6376	0.13587	4456.3	5135.7	8.5331	0.11321	4454.0	5133.3	8.4474	1200
1300	0.3631	4679.0	5405.1	9.1329	0.24206	4676.6	5402.8	8.9442	0.18156	4674.3	5400.5	8.8100	0.14526	4672.0	5398.2	8.7055	0.12106	4669.6	5396.0	8.6199	1300

Properties of H_2O

P(t_{sat})	7×10^6 N/m^2 (285.88 C)				8×10^6 N/m^2 (295.06 C)				9×10^6 N/m^2 (303.40 C)				10×10^6 N/m^2 (311.06 C)				12×10^6 N/m^2 (324.75 C)				
Temp	v	u	h	s	v	u	h	s	v	u	h	s	v	u	h	s	v	u	h	s	Temp
C	m^3/kg	kJ/kg	kJ/kg	kJ/kgK	m^3/kg	kJ/kg	kJ/kg	kJ/kgK	m^3/kg	kJ/kg	kJ/kg	kJ/kgK	m^3/kg	kJ/kg	kJ/kg	kJ/kgK	m^3/kg	kJ/kg	kJ/kg	kJ/kgK	C
Sat liq	1.3513-3	1257.55	1267.00	3.1211	1.3842-3	1305.57	1316.64	3.2068	1.4178-3	1350.51	1363.26	3.2858	1.4524-3	1393.04	1407.56	3.3596	1.5267-3	1473.0	1491.3	3.4962	Sat liq
Evap	0.02602	1323.0	1505.1	2.6922	0.02214	1264.2	1441.4	2.5364	0.01906	1207.3	1378.8	2.3914	16.574-3	1151.4	1317.1	2.2545	12.736-3	1040.7	1193.6	1.9962	Evap
Sat vap	0.02737	2580.5	2772.1	5.8133	0.02352	2569.8	2758.0	5.7432	0.02048	2557.8	2742.1	5.6772	18.026-3	2544.4	2724.7	5.6141	14.263-3	2513.7	2684.9	5.4924	Sat vap
0	0.9967-3	0.06	7.04	0.0002	0.9963-3	0.07	8.04	0.0002	0.9957-3	0.08	9.04	0.0002	0.9952-3	0.09	10.04	0.0002	0.9942-3	0.12	12.06	0.0003	0
40	1.0047-3	166.70	173.74	0.5698	1.0043-3	166.58	174.62	0.5694	1.0038-3	166.47	175.50	0.5690	1.0034-3	166.35	176.38	0.5686	1.0026-3	166.11	178.14	0.5678	40
80	1.0258-3	333.26	340.44	1.0707	1.0254-3	333.04	341.24	1.0701	1.0249-3	332.81	342.03	1.0694	1.0245-3	332.59	342.83	1.0688	1.0235-3	332.14	344.42	1.0675	80
120	1.0565-3	501.11	508.51	1.5215	1.0559-3	500.77	509.22	1.5207	1.0554-3	500.42	509.93	1.5198	1.0549-3	500.08	510.64	1.5189	1.0538-3	499.41	512.06	1.5171	120
160	1.0974-3	671.62	679.30	1.9352	1.0967-3	671.12	679.89	1.9340	1.0960-3	670.63	680.48	1.9329	1.0953-3	670.13	681.08	1.9317	1.0939-3	669.15	682.28	1.9294	160
200	1.1510-3	846.7	854.7	2.3224	1.1500-3	845.9	855.1	2.3208	1.1490-3	845.2	855.6	2.3193	1.1480-3	844.5	856.0	2.3178	1.1461-3	843.1	856.9	2.3148	200
240	1.2233-3	1029.2	1037.7	2.6936	1.2217-3	1028.1	1037.9	2.6914	1.2202-3	1027.0	1038.0	2.6893	1.2187-3	1026.0	1038.1	2.6872	1.2157-3	1023.9	1038.4	2.6831	240
280	1.3303-3	1226.4	1235.7	3.0648	1.3274-3	1224.5	1235.1	3.0614	1.3245-3	1222.7	1234.6	3.0581	1.3216-3	1220.9	1234.1	3.0548	1.3162-3	1217.5	1233.3	3.0485	280
290	0.02802	2596.6	2792.8	5.8501	1.3635-3	1277.7	1288.6	3.1572	1.3600-3	1275.6	1287.8	3.1533	1.3564-3	1273.4	1287.0	3.1495	1.3498-3	1269.4	1285.6	3.1422	290
300	0.02947	2632.2	2838.4	5.9305	0.02426	2590.9	2785.0	5.7906	1.4018-3	1331.0	1343.5	3.2515	1.3972-3	1328.4	1342.3	3.2469	1.3888-3	1323.5	1340.1	3.2363	300
310	0.03078	2663.9	2879.3	6.0013	0.02561	2629.0	2833.9	5.8752	0.021428	2588.3	2781.2	5.7446	1.4465-3	1386.6	1401.1	3.3485	1.4354-3	1382.6	1397.8	3.3379	310
320	0.03199	2692.9	2916.8	6.0650	0.02682	2662.7	2877.2	5.9489	0.022690	2628.5	2832.7	5.8322	0.019253	2588.8	2781.3	5.7103	1.4935-3	1441.9	1459.8	3.4433	320
330	0.03313	2719.8	2951.7	6.1234	0.02793	2693.1	2916.6	6.0147	0.02381	2663.6	2877.9	5.9078	0.020429	2630.4	2834.7	5.7996	0.015007	2546.8	2726.9	5.5624	330
340	0.03421	2745.2	2984.7	6.1776	0.02897	2721.3	2953.1	6.0747	0.02484	2695.3	2918.8	5.9751	0.021472	2666.6	2881.4	5.8763	0.016198	2598.4	2792.8	5.6708	340
360	0.03623	2792.5	3046.2	6.2763	0.03089	2772.7	3019.8	6.1819	0.02670	2751.6	2991.9	6.0924	0.02331	2729.1	2962.1	6.0060	0.018106	2678.4	2895.7	5.8361	360
380	0.03813	2836.6	3103.5	6.3655	0.03266	2819.7	3081.0	6.2769	0.02837	2801.9	3057.3	6.1941	0.02493	2783.2	3032.5	6.1154	0.019685	2742.6	2978.8	5.9654	380
400	0.03993	2878.6	3158.1	6.4478	0.03432	2863.8	3138.3	6.3634	0.02993	2848.4	3117.8	6.2854	0.02641	2832.4	3096.5	6.2120	0.02108	2798.3	3051.3	6.0747	400
420	0.04166	2919.1	3210.7	6.5248	0.03590	2905.9	3193.1	6.4437	0.03140	2892.4	3175.0	6.3691	0.02780	2878.4	3156.3	6.2995	0.02235	2848.9	3117.1	6.1711	420
440	0.04334	2958.5	3261.9	6.5976	0.03742	2946.7	3246.1	6.5190	0.03281	2934.6	3229.9	6.4471	0.02911	2922.1	3213.2	6.3805	0.02355	2896.1	3178.7	6.2586	440
460	0.04497	2997.2	3312.1	6.6670	0.03890	2986.5	3297.7	6.5904	0.03417	2975.5	3283.1	6.5207	0.03038	2964.3	3268.1	6.4564	0.02468	2941.1	3237.2	6.3396	460
480	0.04657	3035.5	3361.5	6.7335	0.04034	3025.7	3348.4	6.6586	0.03549	3015.7	3335.0	6.5906	0.03160	3005.4	3321.4	6.5282	0.02576	2984.4	3293.5	6.4154	480
500	0.04814	3073.4	3410.3	6.7975	0.04175	3064.3	3398.3	6.7240	0.03677	3055.2	3386.1	6.6576	0.03279	3045.8	3373.7	6.5966	0.02680	3026.6	3348.2	6.4871	500
520	0.04968	3111.0	3458.8	6.8594	0.04313	3102.7	3447.7	6.7871	0.03803	3094.2	3436.5	6.7219	0.03394	3085.6	3425.1	6.6622	0.02781	3068.0	3401.8	6.5555	520
540	0.05120	3148.5	3506.9	6.9193	0.04448	3140.8	3496.7	6.8481	0.03926	3132.9	3486.3	6.7839	0.03508	3125.0	3475.8	6.7254	0.02880	3108.7	3454.4	6.6209	540
560	0.05270	3185.9	3554.8	6.9775	0.04582	3178.7	3545.3	6.9072	0.04047	3171.5	3535.7	6.8440	0.03619	3164.1	3526.0	6.7864	0.02977	3149.0	3506.2	6.6840	560
580	0.05418	3223.3	3602.6	7.0342	0.04714	3216.6	3593.8	6.9646	0.04167	3209.8	3584.8	6.9022	0.03729	3203.0	3575.8	6.8455	0.03071	3189.0	3557.5	6.7448	580
600	0.05565	3260.7	3650.3	7.0894	0.04845	3254.4	3642.0	7.0206	0.04285	3248.1	3633.7	6.9589	0.03837	3241.7	3625.3	6.9029	0.03164	3228.7	3608.3	6.8037	600
620	0.05711	3298.1	3697.9	7.1433	0.04974	3292.2	3690.2	7.0751	0.04401	3286.3	3682.4	7.0141	0.03943	3280.3	3674.6	6.9587	0.03255	3268.2	3658.8	6.8608	620
640	0.05855	3335.6	3745.4	7.1960	0.05102	3330.1	3738.3	7.1283	0.04517	3324.5	3731.0	7.0679	0.04048	3318.9	3723.7	7.0131	0.03345	3307.5	3709.0	6.9164	640
660	0.05999	3373.1	3793.0	7.2476	0.05229	3368.0	3786.3	7.1804	0.04631	3362.7	3779.5	7.1204	0.04152	3357.5	3772.7	7.0661	0.03434	3346.8	3759.0	6.9705	660
680	0.06141	3410.8	3840.7	7.2981	0.05356	3405.9	3834.3	7.2313	0.04744	3401.0	3828.0	7.1718	0.04256	3396.1	3821.6	7.1180	0.03522	3386.0	3808.7	7.0233	680
700	0.06283	3448.5	3888.3	7.3476	0.05481	3443.9	3882.4	7.2812	0.04857	3439.3	3876.5	7.2221	0.04358	3434.7	3870.5	7.1687	0.03610	3425.2	3858.4	7.0749	700
720	0.06424	3486.4	3936.1	7.3961	0.05605	3482.1	3930.5	7.3301	0.04969	3477.7	3924.9	7.2714	0.04460	3473.3	3919.3	7.2183	0.03696	3464.5	3907.9	7.1253	720
740	0.06564	3524.5	3983.9	7.4438	0.05729	3520.4	3978.7	7.3782	0.05080	3516.2	3973.4	7.3197	0.04560	3512.1	3968.1	7.2670	0.03781	3503.7	3957.4	7.1746	740
760	0.06704	3562.7	4031.9	7.4907	0.05852	3558.8	4026.9	7.4253	0.05190	3554.8	4022.0	7.3672	0.04661	3550.9	4017.0	7.3148	0.03866	3543.0	4006.9	7.2230	760
780	0.06843	3601.0	4080.0	7.5368	0.05975	3597.3	4075.3	7.4717	0.05300	3593.6	4070.6	7.4138	0.04760	3589.8	4065.9	7.3617	0.03951	3582.3	4056.4	7.2704	780
800	0.06981	3639.5	4128.2	7.5822	0.06097	3636.0	4123.8	7.5173	0.05409	3632.5	4119.3	7.4596	0.04859	3628.9	4114.8	7.4077	0.04034	3621.8	4105.9	7.3170	800
850	0.07326	3736.7	4249.5	7.6926	0.06401	3733.5	4245.6	7.6282	0.05681	3730.3	4241.6	7.5710	0.05105	3727.2	4237.7	7.5196	0.04242	3720.8	4229.8	7.4299	850
900	0.07669	3835.0	4371.8	7.7991	0.06702	3832.1	4368.3	7.7351	0.05950	3829.2	4364.8	7.6783	0.05349	3826.3	4361.2	7.6272	0.04447	3820.6	4354.2	7.5382	900
950	0.08010	3934.5	4495.2	7.9021	0.07002	3931.8	4492.0	7.8384	0.06218	3929.2	4488.8	7.7818	0.05591	3926.5	4485.6	7.7311	0.04651	3921.2	4479.3	7.6426	950
1000	0.08350	4035.3	4619.8	8.0020	0.07301	4032.8	4616.9	7.9384	0.06485	4030.3	4614.0	7.8821	0.05832	4027.8	4611.0	7.8315	0.04853	4022.8	4605.3	7.7435	1000
1100	0.09027	4240.9	4872.8	8.1933	0.07896	4238.6	4870.3	8.1300	0.07016	4236.3	4867.7	8.0740	0.06312	4234.0	4865.1	8.0237	0.05256	4229.4	4860.1	7.9362	1100
1200	0.09703	4451.7	5130.9	8.3747	0.08489	4449.5	5128.5	8.3115	0.07544	4447.2	5126.2	8.2556	0.06789	4444.9	5123.8	8.2055	0.05656	4440.5	5119.2	8.1183	1200
1300	0.10377	4667.3	5393.7	8.5473	0.09080	4665.0	5391.5	8.4842	0.08072	4662.7	5389.2	8.4284	0.07265	4660.5	5387.0	8.3783	0.06055	4655.9	5382.6	8.2913	1300

Properties of H_2O

$P(t_{sat})$	14 x 10^6 N/m^2 (336.75 C)				16 x 10^6 N/m^2 (347.44 C)				18 x 10^6 N/m^2 (357.06 C)				20 x 10^6 N/m^2 (365.81 C)				22 x 10^6 N/m^2 (373.80 C)				
Temp C	v m^3/kg	u kJ/kg	h kJ/kg	s kJ/kgK	v m^3/kg	u kJ/kg	h kJ/kg	s kJ/kgK	v m^3/kg	u kJ/kg	h kJ/kg	s kJ/kgK	v m^3/kg	u kJ/kg	h kJ/kg	s kJ/kgK	v m^3/kg	u kJ/kg	h kJ/kg	s kJ/kgK	Temp C
Sat liq	1.6107-3	1548.6	1571.1	3.6232	1.7107-3	1622.7	1650.1	3.7461	1.8397-3	1698.9	1732.0	3.8715	2.036-3	1785.6	1826.3	4.0139	2.742-3	1961.9	2022.2	4.3110	Sat liq
Evap	9.874-3	928.2	1066.5	1.7485	7.595-3	809.0	930.5	1.4994	5.649-3	675.4	777.1	1.2329	3.798-3	507.4	583.4	0.9130	0.826-3	125.2	143.4	0.2217	Evap
Sat vap	11.485-3	2476.8	2637.6	5.3717	9.306-3	2431.7	2580.6	5.2455	7.489-3	2374.3	2509.1	5.1044	5.834-3	2293.0	2409.7	4.9269	3.568-3	2087.1	2165.6	4.5327	Sat vap
0	0.9933-3	0.14	14.05	0.0004	0.9923-3	0.16	16.04	0.0004	0.9914-3	0.17	18.03	0.0004	0.9904-3	0.19	20.01	0.0004	0.9894-3	0.22	22.01	0.0004	0
40	1.0017-3	165.88	179.88	0.5670	1.0009-3	165.64	181.65	0.5662	1.0000-3	165.40	100.75	0.2926	0.9992-3	165.17	185.16	0.5646	0.9983-3	164.94	186.90	0.5638	40
80	1.0226-3	331.70	346.01	1.0662	1.0218-3	331.26	347.61	1.0650	1.0209-3	330.83	349.20	1.0637	1.0199-3	330.40	350.80	1.0624	1.0190-3	329.97	452.39	1.0612	80
120	1.0527-3	498.74	513.48	1.5154	1.0517-3	498.07	514.90	1.5136	1.0506-3	497.42	515.61	1.5119	1.0496-3	496.76	517.76	1.5102	1.0485-3	496.13	519.19	1.5085	120
160	1.0925-3	668.19	683.49	1.9271	1.0911-3	667.23	684.69	1.9249	1.0898-3	666.29	685.90	1.9226	1.0885-3	665.35	687.12	1.9204	1.0872-3	664.43	688.35	1.9182	160
200	1.1442-3	841.7	857.8	2.3119	1.1424-3	840.4	858.6	2.3189	1.1406-3	839.1	859.5	2.3060	1.1388-3	837.7	860.5	2.3031	1.1370-3	836.4	861.5	2.3003	200
240	1.2128-3	1021.8	1038.8	2.6791	1.2100-3	1019.8	1039.2	2.6751	1.2073-3	1017.9	1039.6	2.6712	1.2046-3	1016.0	1040.0	2.6674	1.2020-3	1014.1	1040.5	2.6636	240
280	1.3110-3	1214.1	1232.5	3.0423	1.3060-3	1210.9	1231.8	3.0363	1.3011-3	1207.7	1231-2	3.0305	1.2965-3	1204.7	1230.6	3.0248	1.2920-3	1201.7	1230.2	3.0193	280
320	1.4792-3	1434.6	1455.3	3.4308	1.4664-3	1427.9	1451.4	3.4191	1.4546-3	1421.6	1447.8	3.4089	1.4437-3	1415.7	1444.6	3.3979	1.4337-3	1412.3	1441.7	3.3882	320
330	1.5491-3	1499.8	1521.5	3.5414	1.5258-3	1490.8	1514.9	3.5260	1.5135-3	1482.6	1509.9	3.5121	1.4987-3	1475.2	1505.2	3.4992	1.4856-3	1468.4	1501.1	3.4874	330
340	11.983-3	2503.3	2671.1	5.4266	1.6171-3	1561.4	1587.3	3.6444	1.5906-3	1549.8	1478.4	3.6247	1.5684-3	1539.7	1571.0	3.6075	1.5496-3	1530.8	1564.9	3.5924	340
350	13.221-3	2567.5	2752.6	5.5585	9.749-3	2459.7	2615.7	5.3020	1.7039-3	1628.1	1658.8	3.7547	1.6640-3	1612.3	1645.6	3.7281	1.6339-3	1599.7	1635.6	3.7067	350
360	14.218-3	2617.4	2816.5	5.6602	11.053-3	2539.0	2715.8	5.4614	8.090-3	2418.9	2564.5	5.1922	1.8226-3	1702.8	1739.3	3.8772	1.7589-3	1680.4	1719.1	3.8396	360
370	15.078-3	2659.5	2870.6	5.7450	12.038-3	2596.3	2788.9	5.5760	9.447-3	2514.3	2684.4	5.3801	6.898-3	2386.5	2524.4	5.1059	2.0188-3	1795.4	1839.9	4.0287	370
380	15.849-3	2696.5	2918.4	5.8188	12.867-3	2643.0	2848.9	5.6685	10.413-3	2578.5	2766.0	5.5060	8.255-3	2495.2	2660.3	5.3158	6.101-3	2368.3	2502.6	5.0522	380
400	17.216-3	2760.9	3001.9	5.9448	14.262-3	2719.4	2947.6	5.8175	11.901-3	2672.8	2887.0	5.6887	9.942-3	2619.3	2818.1	5.5540	8.253-3	2556.0	2737.6	5.4074	400
420	18.430-3	2817.2	3075.3	6.0521	15.452-3	2783.1	3030.3	5.9386	13.100-3	2745.9	2981.7	5.8274	11.183-3	2705.1	2928.8	5.7162	9.578-3	2660.0	2870.8	5.6025	420
440	19.543-3	2868.6	3142.2	6.1474	16.518-3	2839.4	3103.7	6.0429	14.143-3	2808.2	3062.8	5.9428	12.223-3	2774.9	3019.4	5.8450	10.632-3	2739.1	2973.0	5.7481	440
460	20.58-3	2916.7	3204.9	6.2340	17.500-3	2891.1	3171.1	6.1362	15.087-3	2864.1	3135.7	6.0436	13.144-3	2835.8	3098.7	5.9547	11.542-3	2805.9	3059.8	5.8681	460
480	21.57-3	2962.5	3264.5	6.3143	18.421-3	2939.7	3234.4	6.2215	15.961-3	2915.9	3203.2	6.1345	13.985-3	2891.2	3170.8	6.0518	12.360-3	2865.3	3137.2	5.9723	480
500	22.52-3	3006.8	3322.0	6.3897	19.296-3	2986.2	3294.9	6.3007	16.784-3	2964.9	3267.0	6.2181	14.768-3	2942.9	3238.2	6.1401	13.115-3	2920.0	3208.6	6.0658	500
520	23.43-3	3049.8	3377.8	6.4610	20.133-3	3031.1	3353.3	6.3752	17.566-3	3011.8	3328.0	6.2960	15.508-3	2992.0	3302.2	6.2218	13.821-3	2971.6	3275.6	6.1515	520
540	24.31-3	3092.0	3432.4	6.5289	20.941-3	3074.9	3410.0	6.4459	18.316-3	3057.3	3387.0	6.3695	16.213-3	3039.3	3363.5	6.2982	14.491-3	3020.8	3339.6	6.2311	540
560	25.17-3	3133.6	3486.0	6.5941	21.72-3	3117.8	3465.4	6.5132	19.039-3	3101.7	3444.4	6.4392	16.890-3	3085.2	3423.0	6.3705	15.131-3	3068.3	3401.2	6.3059	560
580	26.01-3	3174.7	3538.8	6.6567	22.48-3	3160.1	3519.8	6.5778	19.741-3	3145.2	3500.5	6.5058	17.544-3	3130.0	3480.9	6.4391	15.747-3	3114.5	3461.0	6.3768	580
600	26.83-3	3215.4	3591.1	6.7172	23.23-3	3201.8	3573.5	6.6399	20.42-3	3188.0	3555.6	6.5696	18.178-3	3174.0	3537.6	6.5048	16.341-3	3159.7	3519.2	6.4444	600
620	27.64-3	3255.8	3642.8	6.7758	23.95-3	3243.2	3626.5	6.6999	21.09-3	3230.4	3610.0	6.6311	18.795-3	3217.4	3593.3	6.5679	16.918-3	3204.2	3576.4	6.5090	620
640	28.43-3	3296.0	3694.1	6.8326	24.67-3	3284.2	3678.9	6.7580	21.74-3	3272.3	3663.6	6.6905	19.397-3	3260.2	3648.1	6.6286	17.480-3	3247.9	3632.5	6.5712	640
660	29.22-3	3336.0	3745.0	6.8878	25.37-3	3325.0	3730.9	6.8143	22.38-3	3313.9	3716.7	6.7480	19.986-3	3302.6	3702.3	6.6873	18.028-3	3291.2	3687.8	6.6311	660
680	29.99-3	3375.9	3795.7	6.9415	26.06-3	3365.6	3782.6	6.8691	23.01-3	3355.2	3769.3	6.8038	20.563-3	3344.6	3755.9	6.7442	18.565-3	3334.0	3742.4	6.6890	680
700	30.75-3	3415.7	3846.2	6.9939	26.74-3	3406.0	3833.9	6.9224	23.62-3	3396.3	3821.5	6.8580	21.13-3	3386.4	3809.0	6.7993	19.092-3	3376.4	3796.5	6.7451	700
720	31.50-3	3455.5	3896.5	7.0451	27.41-3	3446.4	3885.0	6.9744	24.23-3	3437.2	3873.4	6.9108	21.69-3	3428.0	3861.7	6.8529	19.609-3	3418.6	3850.0	6.7996	720
740	32.25-3	3495.2	3946.7	7.0952	28.08-3	3486.7	3935.9	7.0251	24.83-3	3478.0	3925.0	6.9623	22.24-3	3469.3	3914.1	6.9052	20.119-3	3460.5	3903.2	6.8526	740
760	32.99-3	3535.0	3996.8	7.1441	28.74-3	3526.9	3986.7	7.0747	25.43-3	3518.7	3976.5	7.0126	22.78-3	3510.5	3966.2	6.9561	20.621-3	3502.3	3955.9	6.9042	760
780	33.72-3	3574.7	4046.9	7.1921	29.39-3	3567.1	4037.3	7.1233	26.02-3	3559.4	4027.7	7.0617	23.32-3	3551.6	4018.1	7.0058	21.117-3	3543.8	4008.4	6.9545	780
800	34.45-3	3614.5	4096.9	7.2392	30.03-3	3607.3	4087.8	7.1708	26.60-3	3600.0	4078.8	7.1098	23.85-3	3592.7	4069.7	7.0544	21.61-3	3585.3	4060.6	7.0036	800
850	36.25-3	3714.4	4222.0	7.3531	31.63-3	3708.0	4214.1	7.2858	28.04-3	3701.5	4206.2	7.2258	25.16-3	3695.1	4198.3	7.1715	22.81-3	3688.6	4190.4	7.1218	850
900	38.03-3	3814.8	4347.2	7.4622	33.20-3	3809.0	4340.3	7.3957	29.45-3	3803.2	4333.3	7.3365	26.45-3	3797.5	4326.4	7.2830	23.99-3	3791.7	4319.5	7.2342	900
950	39.79-3	3915.9	4473.0	7.5672	34.76-3	3910.6	4466.8	7.5013	30.84-3	3905.3	4460.5	7.4427	27.71-3	3900.1	4454.3	7.3898	25.15-3	3894.8	4448.1	7.3416	950
1000	41.54-3	4017.9	4599.5	7.6685	36.30-3	4013.0	4593.8	7.6031	32.23-3	4008.0	4588.1	7.5449	28.97-3	4003.1	4582.5	7.4925	26.30-3	3998.2	4576.8	7.4448	1000
1100	45.02-3	4224.8	4855.1	7.8618	39.36-3	4220.3	4850.1	7.7969	34.96-3	4215.8	4845.1	7.7393	31.45-3	4211.3	4840.2	7.6874	28.57-3	4206.8	4835.4	7.6402	1100
1200	48.47-3	4436.0	5114.6	8.0442	42.40-3	4431.6	5110.0	7.9796	37.68-3	4427.2	5105.5	7.9222	33.91-3	4422.8	5101.0	7.8707	30.82-3	4418.5	5096.5	7.8238	1200
1300	51.91-3	4651.4	5378.1	8.2173	45.43-3	4646.9	5373.8	8.1528	40.39-3	4642.4	5369.4	8.0956	36.36-3	4638.0	5365.1	8.0442	33.06-3	4633.5	5360.8	7.9974	1300

Properties of H_2O

P	25 x 10^6 N/m^2				30 x 10^6 N/m^2				50 x 10^6 N/m^2				70 x 10^6 N/m^2				100 x 10^6 N/m^2				
Temp	v	u	h	s	v	u	h	s	v	u	h	s	v	u	h	s	v	u	h	s	Temp
C	m^3/kg	kJ/kg	kJ/kg	kJ/kgK	m^3/kg	kJ/kg	kJ/kg	kJ/kgK	m^3/kg	kJ/kg	kJ/kg	kJ/kgK	m^3/kg	kJ/kg	kJ/kg	kJ/kgK	m^3/kg	kJ/kg	kJ/kg	kJ/kgK	C
0	0.9880-3	0.24	24.95	0.0003	0.9856-3	0.25	29.82	0.0001	0.9766-3	0.20	49.03	-0.0014	0.9682-3	-0.06	67.71	-0.0042	0.9566-3	-0.69	94.97	-0.0101	0
40	0.9971-3	164.60	189.52	0.5626	0.9951-3	164.04	193.89	0.5607	0.9872-3	161.86	211.21	0.5527	0.9796-3	159.80	228.38	0.5447	0.9688-3	156.94	253.82	0.5326	40
80	1.0178-3	329.34	354.78	1.0592	1.0156-3	328.30	358.77	1.0561	1.0073-3	324.34	374.70	1.0440	0.9994-3	320.66	390.61	1.0322	0.9882-3	315.59	414.41	1.0152	80
120	1.0470-3	495.16	521.33	1.5060	1.0445-3	493.59	524.93	1.5108	1.0348-3	487.65	539.39	1.4857	1.0259-3	482.17	553.98	1.4704	1.0134-3	474.66	576.00	1.4486	120
160	1.0853-3	663.06	690.19	1.9149	1.0821-3	660.82	693.28	1.9096	1.0703-3	652.41	705.92	1.8891	1.0595-3	644.72	718.89	1.8698	1.0448-3	634.30	738.78	1.8429	160
200	1.1344-3	834.5	862.8	2.2961	1.1302-3	831.4	865.3	2.2893	1.1146-3	819.7	875.5	2.2634	1.0009-3	809.3	886.4	2.2397	1.0827-3	795.4	903.7	2.2071	200
240	1.1982-3	1011.3	1041.3	2.6580	1.1920-3	1006.9	1042.6	2.6490	1.1702-3	990.7	1049.2	2.6158	1.1515-3	976.7	1057.3	2.5863	1.1279-3	958.5	1071.3	2.5470	240
280	1.2856-3	1197.5	1229.6	3.0113	1.2755-3	1190.7	1229.0	2.9986	1.2415-3	1167.2	1229.3	2.9537	1.2144-3	1147.8	1232.8	2.9157	1.1818-3	1123.8	1242.0	2.8672	280
320	1.4200-3	1402.4	1437.9	3.3746	1.3997-3	1390.7	1432.7	3.3539	1.3388-3	1353.3	1420.2	3.2868	1.2956-3	1325.0	1415.7	3.2348	1.2478-3	1292.3	1417.1	3.1728	320
360	1.6955-3	1655.9	1698.3	3.7986	1.6265-3	1626.6	1675.4	3.7494	1.4838-3	1556.0	1630.2	3.6291	1.4064-3	1512.1	1610.5	3.5525	1.3312-3	1465.8	1598.9	3.4694	360
370	1.8470-3	1742.8	1789.0	3.9407	1.7252-3	1698.1	1749.9	3.8661	1.5324-3	1610.6	1687.2	3.7184	1.4408-3	1560.9	1661.7	3.6327	1.3556-3	1510.2	1645.7	3.5428	370
380	2.2135-3	1879.1	1934.5	4.1649	1.8691-3	1781.4	1837.5	4.0012	1.5884-3	1667.2	1746.6	3.8101	1.4786-3	1610.6	1714.1	3.7135	1.3818-3	1555.0	1693.2	3.6160	380
390	4.6126-3	2275.7	2391.0	4.8585	2.1270-3	1890.2	1954.0	4.1782	1.6536-3	1726.2	1808.9	3.9047	1.5203-3	1661.4	1767.8	3.7951	1.4098-3	1600.3	1741.3	3.6891	390
400	6.004-3	2430.1	2580.2	5.1418	2.790-3	2067.4	2151.1	4.4728	1.7309-3	1788.1	1874.6	4.0031	1.5664-3	1713.2	1822.8	3.8775	1.4399-3	1646.0	1790.0	3.7620	400
410	6.885-3	2517.2	2689.3	5.3029	3.972-3	2275.6	2394.8	4.8324	1.8243-3	1853.5	1944.7	4.1064	1.6178-3	1766.1	1879.4	3.9609	1.4722-3	1692.2	1839.4	3.8349	410
420	7.577-3	2581.7	2771.1	5.4218	4.924-3	2407.4	2555.1	5.0655	1.9393-3	1923.1	2020.0	4.2159	1.6751-3	1820.3	1937.5	4.0454	1.5068-3	1738.9	1889.5	3.9077	420
430	8.166-3	2634.5	2838.7	5.5186	5.641-3	2495.8	2665.0	5.2229	2.083-3	1997.5	2101.7	4.3329	1.7394-3	1875.6	1997.4	4.1311	1.5440-3	1785.9	1940.3	3.9840	430
440	8.687-3	2680.1	2897.3	5.6013	6.228-3	2563.4	2750.3	5.3434	2.264-3	2076.9	2190.1	4.4577	1.8116-3	1932.3	2059.1	4.2182	1.5838-3	1833.4	1991.7	4.0531	440
450	9.162-3	2720.7	2949.7	5.6744	6.735-3	2619.3	2821.4	5.4424	2.486-3	2159.6	2284.0	4.5884	1.8931-3	1990.2	2122.7	4.3068	1.6266-3	1881.1	2043.8	4.1256	450
460	9.600-3	2757.7	2997.7	5.7403	7.187-3	2667.7	2883.3	5.5274	2.745-3	2242.8	2380.0	4.7203	1.9849-3	2049.2	2188.2	4.3986	1.6724-3	1929.3	2096.5	4.1979	460
470	10.011-3	2792.1	3042.4	5.8008	7.600-3	2710.8	2938.8	5.6026	3.028-3	2322.7	2474.1	4.8478	2.0881-3	2109.3	2255.5	4.4879	1.7215-3	1977.6	2149.8	4.2701	470
480	10.400-3	2824.4	3084.4	5.8570	7.982-3	2750.1	2989.6	5.6705	3.321-3	2396.8	2562.9	4.9665	2.2030-3	2169.9	2324.1	4.5797	1.7741-3	2026.2	2203.6	4.3421	480
490	10.769-3	2855.0	3124.3	5.9096	8.340-3	2786.5	3036.7	5.7327	3.611-3	2464.3	2644.9	5.0747	2.3293-3	2230.5	2393.6	4.6713	1.8303-3	2075.0	2258.0	4.4138	490
500	11.123-3	2884.3	3162.4	5.9592	8.678-3	2820.7	3081.1	5.7905	3.892-3	2525.5	2720.1	5.1726	2.466-3	2290.6	2463.2	4.7619	1.8903-3	2123.8	2312.8	4.4852	500
520	11.793-3	2939.8	3234.7	6.0515	9.309-3	2884.0	3163.3	5.8955	4.418-3	2631.8	2852.7	5.3420	2.761-3	2406.3	2599.6	4.9361	2.0218-3	2221.4	2423.5	4.6266	520
540	12.422-3	2992.2	3302.8	6.1363	9.891-3	2942.5	3239.2	5.9900	4.895-3	2722.5	2967.3	5.4847	3.071-3	2513.7	2728.7	5.0969	2.1681-3	2318.0	2534.8	4.7651	540
560	13.019-3	3042.4	3367.8	6.2154	10.437-3	2997.5	3310.6	6.0768	5.332-3	2802.5	3069.1	5.6084	3.382-3	2611.7	2848.4	5.2424	2.327-3	2412.7	2645.4	4.8995	560
580	13.589-3	3090.8	3430.5	6.2897	10.953-3	3050.0	3378.6	6.1574	5.735-3	2875.0	3161.7	5.7183	3.684-3	2701.2	2959.1	5.3737	2.496-3	2504.5	2754.1	5.0284	580
600	14.137-3	3137.9	3491.4	6.3602	11.446-3	3100.5	3443.9	6.2331	6.112-3	2942.0	3247.6	5.8178	3.976-3	2783.4	3061.7	5.4925	2.671-3	2592.7	2859.8	5.1509	600
620	14.667-3	3184.0	3550.7	6.4274	11.918-3	3149.6	3507.1	6.3047	6.467-3	3004.9	3328.2	5.9091	4.255-3	2859.7	3157.5	5.6010	2.849-3	2677.0	2961.9	5.2665	620
640	15.181-3	3229.3	3608.8	6.4917	12.374-3	3197.5	3568.7	6.3729	6.803-3	3064.6	3404.8	5.9939	4.521-3	2931.1	3247.6	5.7008	3.027-3	2757.4	3060.1	5.3752	640
660	15.681-3	3273.8	3665.8	6.5536	12.815-3	3244.3	3628.8	6.4380	7.124-3	3121.8	3478.1	6.0732	4.776-3	2998.6	3332.9	5.7933	3.203-3	2834.1	3154.4	5.4774	660
680	16.169-3	3317.8	3722.0	6.6131	13.243-3	3290.4	3687.7	6.5004	7.432-3	3177.0	3548.6	6.1480	5.021-3	3062.9	3414.4	5.8796	3.376-3	2907.5	3245.1	5.5736	680
700	16.646-3	3361.3	3777.5	6.6707	13.661-3	3335.8	3745.6	6.5606	7.727-3	3230.5	3616.8	6.2189	5.256-3	3124.5	3492.4	5.9606	3.546-3	2977.7	3332.3	5.6642	700
720	17.114-3	3404.5	3832.3	6.7265	14.069-3	3380.6	3802.6	6.6186	8.013-3	3282.5	3683.1	6.2863	5.482-3	3183.7	3567.5	6.0370	3.712-3	3045.2	3416.4	5.7497	720
740	17.574-3	3447.3	3886.6	6.7806	14.468-3	3424.9	3858.9	6.6747	8.290-3	3333.3	3747.8	6.3508	5.701-3	3241.0	3640.1	6.1094	3.875-3	3110.1	3497.6	5.8306	740
760	18.027-3	3489.8	3940.5	6.8332	14.859-3	3468.8	3914.6	6.7291	8.558-3	3383.0	3810.9	6.4125	5.913-3	3296.6	3710.5	6.1782	4.033-3	3172.8	3576.1	5.9074	760
780	18.472-3	3532.1	3993.9	6.8845	15.244-3	3512.3	3969.6	6.7819	8.820-3	3431.8	3872.8	6.4718	6.118-3	3350.6	3778.9	6.2438	4.187-3	3233.5	3652.2	5.9803	780
800	18.912-3	3574.3	4047.1	6.9345	15.623-3	3555.5	4024.2	6.8332	9.076-3	3479.8	3933.6	6.5290	6.318-3	3403.4	3845.7	6.3066	4.338-3	3292.3	3726.1	6.0499	800
850	19.991-3	3678.9	4178.7	7.0543	16.548-3	3662.6	4159.0	6.9560	9.692-3	3596.7	4081.3	6.6636	6.797-3	3530.3	4006.1	6.4528	4.700-3	3432.5	3902.5	6.2105	850
900	21.045-3	3783.0	4309.1	7.1680	17.448-3	3768.5	4291.9	7.0718	10.283-3	3710.3	4224.4	6.7882	7.252-3	3651.6	4159.2	6.5862	5.044-3	3564.3	4068.6	6.3553	900
950	22.080-3	3886.9	4438.9	7.2763	18.329-3	3873.8	4423.6	7.1817	10.854-3	3821.3	4364.0	6.9048	7.688-3	3768.5	4306.7	6.7093	5.373-3	3689.4	4226.7	6.4872	950
1000	23.10-3	3990.9	4568.5	7.3802	19.196-3	3978.8	4554.7	7.2867	11.411-3	3930.5	4501.1	7.0146	8.110-3	3882.1	4449.8	6.8240	5.690-3	3809.2	4378.2	6.6087	1000
1100	25.12-3	4200.2	4828.2	7.5765	20.903-3	4189.2	4816.3	7.4845	12.496-3	4145.7	4770.5	7.2184	8.927-3	4102.4	4727.3	7.0338	6.299-3	4037.1	4667.1	6.8272	1100
1200	27.11-3	4412.0	5089.9	7.7605	22.589-3	4401.3	5079.0	7.6692	13.561-3	4359.1	5037.2	7.4058	9.722-3	4317.3	4997.8	7.2240	6.889-3	4254.7	4943.6	7.0216	1200
1300	29.10-3	4626.9	5354.4	7.9342	24.266-3	4616.0	5344.0	7.8432	14.616-3	4572.8	5303.6	7.5808	10.509-3	4530.0	5265.6	7.3999	7.469-3	4466.6	5213.6	7.1989	1300

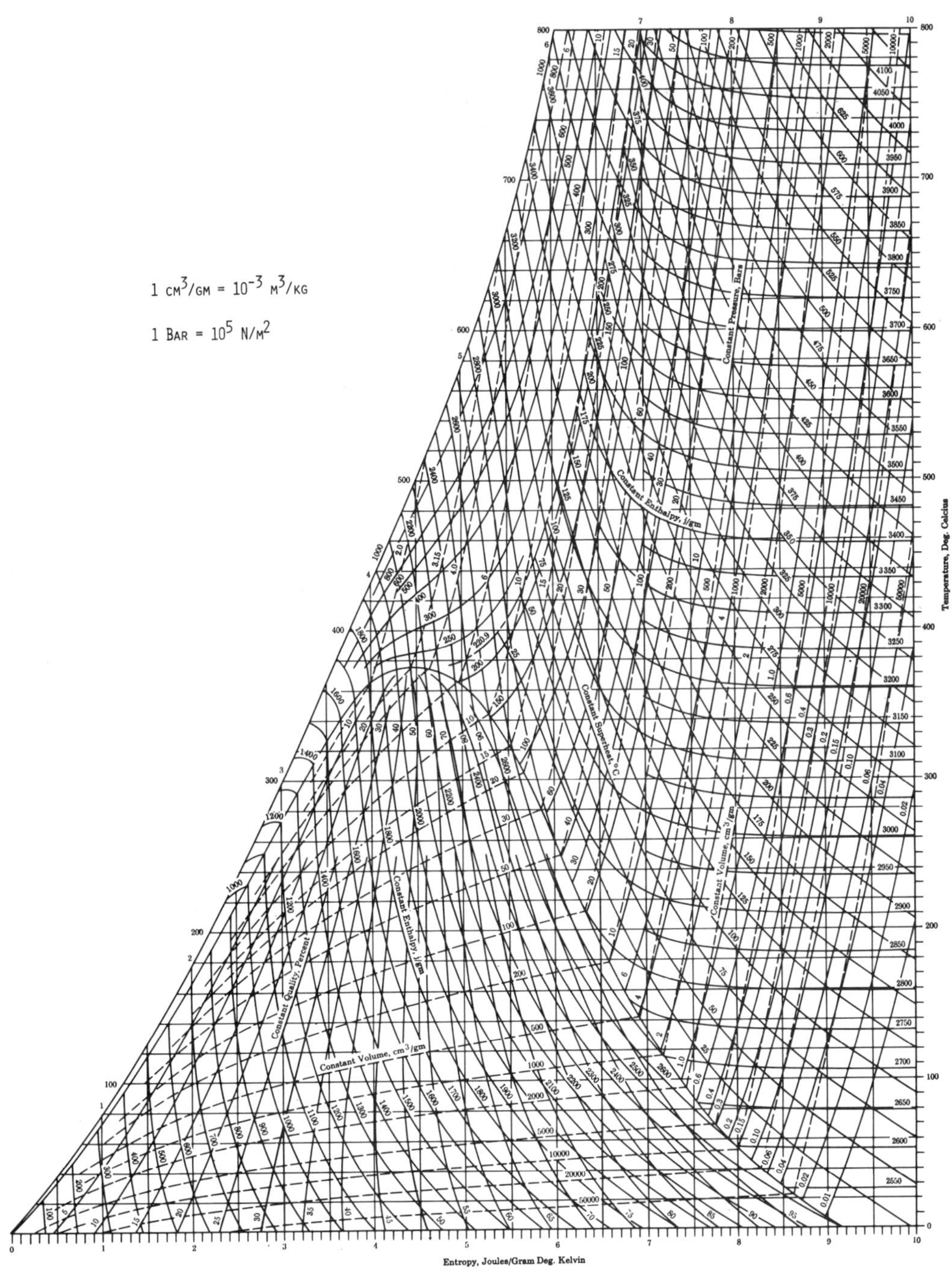

Temperature-Entropy Diagram for H_2O

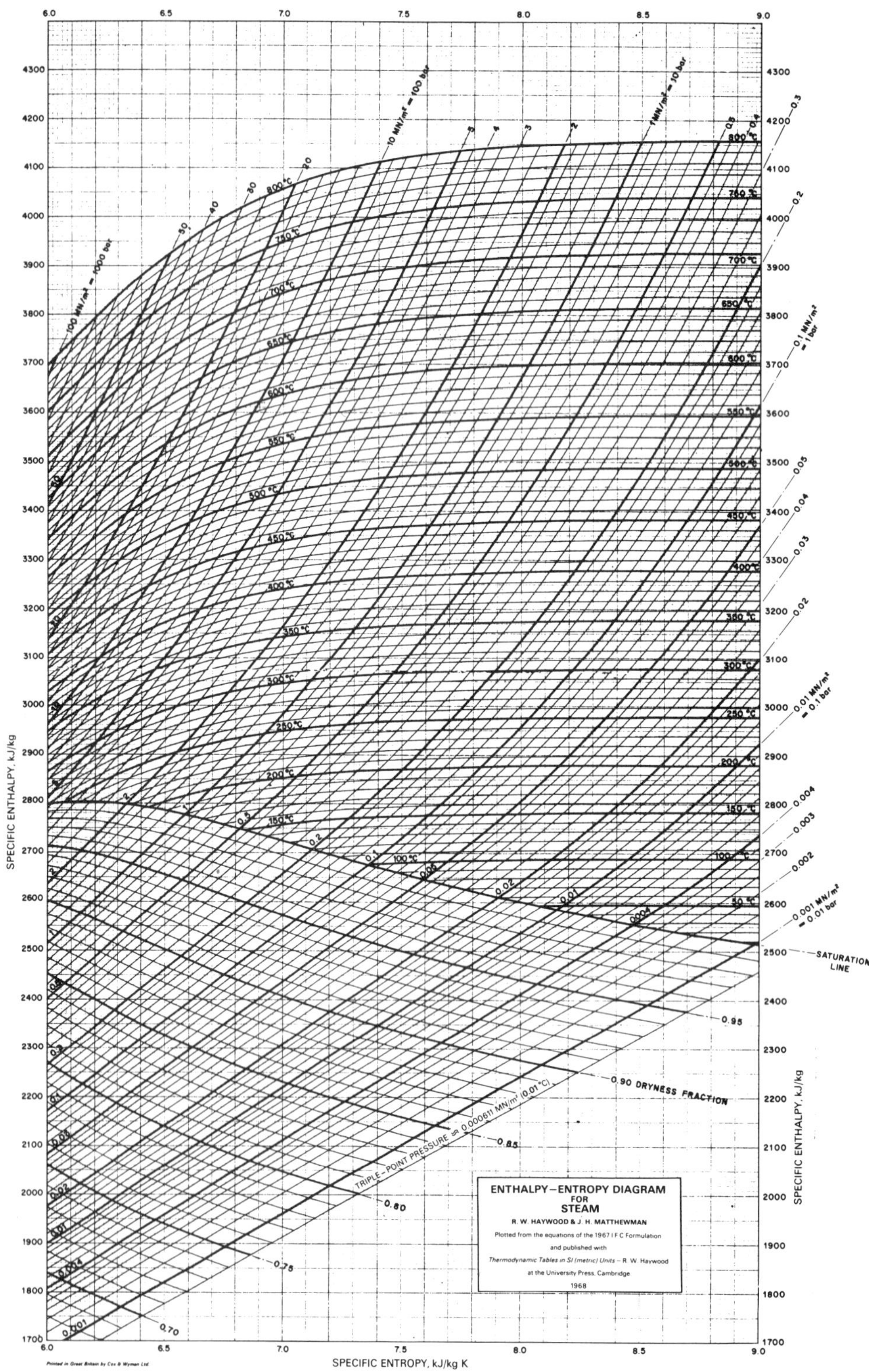

ENTHALPY–ENTROPY DIAGRAM
FOR
STEAM
R. W. HAYWOOD & J. H. MATTHEWMAN
Plotted from the equations of the 1967 I F C Formulation
and published with
Thermodynamic Tables in SI (metric) Units – R. W. Haywood
at the University Press, Cambridge
1968
SPECIFIC ENTROPY, kJ/kg K
SPECIFIC ENTHALPY, kJ/kg
SATURATION LINE
0.90 DRYNESS FRACTION
TRIPLE-POINT PRESSURE ≈ 0.000611 MN/m² (0.01 °C)
100 MN/m² = 1000 bar
10 MN/m² = 100 bar
1 MN/m² = 10 bar
0.1 MN/m² = 1 bar
0.01 MN/m² = 0.1 bar
0.001 MN/m² = 0.01 bar

Properties of Saturated N_2

T	P	v_f	v_{fg}	u_g	u_f	u_{fg}	u_g	h_f	h_{fg}	h_g	s_f	s_{fg}	s_g
K	N/m^2	m^3/kg	m^3/kg	m^3/kg	kJ/kg	kJ/kg	kJ/kg	kJ/kg	kJ/kg	kJ/kgK	kJ/kgK	kJ/kgK	kJ/kgK
63.17	12.53+3	1.153-3	1.480	1.481	-150.4	196.7	46.31	-150.4	215.3	64.90	2.429	3.412	5.841
64.	14.55+3	1.157-3	1.290	1.291	-148.7	195.6	46.88	-148.7	214.3	65.66	2.454	3.353	5.807
65.	17.37+3	1.162-3	1.096	1.098	-146.7	194.3	47.56	-146.7	213.3	66.60	2.485	3.286	5.770
66.	20.59+3	1.167-3	937.1-3	938.3-3	-144.7	192.9	48.24	-144.7	212.2	67.54	2.515	3.219	5.735
67.	24.28+3	1.172-3	805.3-3	806.5-3	-142.7	191.6	48.91	-142.7	211.1	68.46	2.546	3.155	5.700
68.	28.47+3	1.178-3	695.6-3	696.7-3	-140.7	190.2	49.57	-140.6	210.0	69.37	2.575	3.092	5.667
69.	33.20+3	1.184-3	603.7-3	604.8-3	-138.6	188.9	50.23	-138.6	208.9	70.27	2.605	3.031	5.636
70.	38.54+3	1.189-3	526.3-3	527.5-3	-136.6	187.5	50.88	-136.6	207.7	71.16	2.634	2.971	5.605
71.	44.52+3	1.195-3	460.8-3	462.0-3	-134.6	186.1	51.52	-134.5	206.5	72.03	2.663	2.912	5.576
72.	51.21+3	1.201-3	405.1-3	406.3-3	-132.5	184.7	52.15	-132.5	205.4	72.89	2.692	2.855	5.547
73.	58.64+3	1.208-3	357.5-3	358.7-3	-130.5	183.2	52.78	-130.4	204.1	73.74	2.720	2.800	5.520
74.	66.89+3	1.214-3	316.6-3	317.8-3	-128.4	181.8	53.39	-128.3	202.9	74.57	2.748	2.745	5.493
76.	86.05+3	1.227-3	250.9-3	252.1-3	-124.3	178.9	54.59	-124.2	200.4	76.18	2.803	2.640	5.443
78.	109.1+3	1.241-3	201.2-3	202.5-3	-120.2	176.0	55.75	-120.1	197.8	77.71	2.856	2.539	5.395
80.	136.7+3	1.256-3	163.2-3	164.5-3	-116.1	173.0	56.86	-115.9	195.1	79.18	2.908	2.441	5.350
82.	169.2+3	1.271-3	133.7-3	135.0-3	-112.0	169.9	57.93	-111.7	192.3	80.55	2.959	2.348	5.307
84.	207.2+3	1.287-3	110.5-3	111.8-3	-107.8	166.8	58.95	-107.6	189.4	81.84	3.009	2.257	5.267
86.	251.2+3	1.304-3	92.03-3	93.34-3	-103.7	163.6	59.91	-103.3	186.4	83.04	3.058	2.170	5.228
88.	301.9+3	1.322-3	77.20-3	78.53-3	-99.50	160.3	60.82	-99.10	183.2	84.13	3.106	2.084	5.191
90.	359.7+3	1.341-3	65.17-3	66.51-3	-95.30	157.0	61.66	-94.81	179.9	85.11	3.153	2.001	5.155
92.	425.4+3	1.360-3	55.32-3	56.68-3	-91.06	153.5	62.44	-90.48	176.5	85.97	3.200	1.920	5.120
94.	499.4+3	1.381-3	47.18-3	48.56-3	-86.79	149.9	63.15	-86.10	172.8	86.71	3.246	1.840	5.086
96.	582.3+3	1.403-3	40.41-3	41.81-3	-82.48	146.3	63.77	-81.66	169.0	87.30	3.292	1.761	5.053
98.	674.9+3	1.427-3	34.73-3	36.15-3	-78.12	142.4	64.31	-77.15	164.9	87.75	3.337	1.684	5.021
100.	777.7+3	1.452-3	29.93-3	31.38-3	-73.70	138.5	64.76	-72.57	160.6	88.03	3.382	1.607	4.988
102.	891.3+3	1.479-3	25.84-3	27.32-3	-69.21	134.3	65.11	-67.89	156.0	88.14	3.426	1.530	4.957
104.	1.016+6	1.508-3	22.34-3	23.85-3	-64.65	130.0	65.33	-63.12	151.2	88.04	3.471	1.454	4.925
106.	1.154+6	1.540-3	19.32-3	20.86-3	-60.00	125.4	65.43	-58.23	146.0	87.72	3.515	1.377	4.892
108.	1.304+6	1.575-3	16.70-3	18.27-3	-55.26	120.6	65.38	-53.20	140.4	87.16	3.560	1.300	4.860
110.	1.467+6	1.613-3	14.41-3	16.02-3	-50.39	115.6	65.16	-48.02	134.3	86.30	3.605	1.221	4.826
112.	1.645+6	1.655-3	12.39-3	14.04-3	-45.38	110.1	64.73	-42.66	127.8	85.11	3.651	1.141	4.792
114.	1.838+6	1.703-3	10.59-3	12.29-3	-40.19	104.2	64.06	-37.06	120.6	83.52	3.698	1.057	4.755
116.	2.046+6	1.758-3	8.973-3	10.73-3	-34.76	97.83	63.07	-31.17	112.6	81.43	3.746	0.970	4.716
118.	2.271+6	1.824-3	7.501-3	9.324-3	-29.02	90.68	61.66	-24.88	103.6	78.70	3.796	0.877	4.673
120.	2.513+6	1.904-3	6.130-3	8.035-3	-22.80	82.48	59.68	-18.02	93.10	75.09	3.850	0.776	4.626
122.	2.774+6	2.012-3	4.809-3	6.821-3	-15.79	72.60	56.80	-10.21	80.36	70.14	3.910	0.659	4.569
124.	3.055+6	2.180-3	3.431-3	5.611-3	-7.101	59.36	52.26	-.4422	63.18	62.74	3.985	0.510	4.494
126.	3.358+6	4.167-3	7.451-6	4.174-3	+34.05	14.08	48.13	+48.04	116	48.15	4.365	0.001	4.366
126.22	3.400+6	3.087-3	0	3.087-3	18.19	0	18.19	28.71	0	28.71	4.212	0	4.212

Data from computer program supplied by R. D. McCarty, Cryogenics Division, National Bureau of Standards, Boulder, Colorado..

Properties of N_2

$P(T_{sat})$	10.0×10^3 N/m^2 (61.988 K)				100×10^3 N/m^2 (77.251 K)				200×10^3 N/m^2 (83.644 K)				500×10^3 N/m^2 (94.016 K)				1.00×10^6 N/m^2 (103.748 K)			
Temp K	v m^3/kg	u kJ/kg	h kJ/kg	s kJ/kgK	v m^3/kg	u kJ/kg	h kJ/kg	s kJ/kgK	v m^3/kg	u kJ/kg	h kJ/kg	s kJ/kgK	v m^3/kg	u kJ/kg	h kJ/kg	s kJ/kgK	v m^3/kg	u kJ/kg	h kJ/kg	s kJ/kgK
SAT LIQ	1.147-3	-152.6	-152.6	2.392	1.236-3	-121.7	-121.6	2.836	1.284-3	-108.6	-108.3	3.000	1.381-3	-86.76	-86.07	3.246	1.505-3	-65.23	-63.72	3.465
EVAP	1.824	198.1	216.3	3.495	0.2183	177.1	198.8	2.576	0.1142	167.3	189.9	2.273	0.04712	149.9	172.8	1.839	0.02275	130.5	151.8	1.464
SAT VAP	1.825	45.49	63.73	5.887	0.2195	55.32	77.15	5.412	0.1155	58.77	81.62	5.274	0.04850	63.15	86.71	5.086	0.02426	65.31	88.07	4.929
64.	1.886	46.99	65.85	5.921	1.156-3	-148.7	-148.6	2.454	1.156-3	-148.7	-148.5	2.453	1.156-3	-148.8	-148.3	2.452	1.155-3	-149.0	-147.8	2.449
65.	1.916	47.74	66.90	5.937	1.162-3	-146.7	-146.6	2.484	1.161-3	-146.8	-146.5	2.484	1.161-3	-146.9	-146.3	2.482	1.160-3	-147.0	-145.9	2.480
66.	1.946	48.49	67.95	5.953	1.167-3	-144.7	-144.6	2.515	1.167-3	-144.8	-144.5	2.514	1.166-3	-144.9	-144.3	2.513	1.165-3	-145.0	-143.9	2.510
67.	1.976	49.24	69.00	5.969	1.172-3	-142.7	-142.6	2.545	1.172-3	-142.8	-142.5	2.545	1.171-3	-142.9	-142.3	2.543	1.170-3	-143.0	-141.9	2.540
68.	2.006	49.99	70.05	5.985	1.178-3	-140.7	-140.6	2.575	1.178-3	-140.7	-140.5	2.575	1.177-3	-140.8	-140.3	2.573	1.176-3	-141.0	-139.9	2.570
69.	2.036	50.74	71.10	6.000	1.183-3	-138.7	-138.5	2.605	1.183-3	-138.7	-138.5	2.604	1.183-3	-138.8	-138.2	2.602	1.181-3	-139.0	-137.8	2.600
70.	2.066	51.49	72.15	6.015	1.189-3	-136.6	-136.5	2.634	1.189-3	-136.7	-136.4	2.633	1.188-3	-136.8	-136.2	2.632	1.187-3	-137.0	-135.8	2.629
71.	2.096	52.24	73.20	6.030	1.195-3	-134.6	-134.5	2.663	1.195-3	-134.6	-134.4	2.662	1.194-3	-134.8	-134.2	2.661	1.193-3	-135.0	-133.8	2.658
72.	2.126	52.99	74.25	6.045	1.201-3	-132.5	-132.4	2.692	1.201-3	-132.6	-132.3	2.691	1.200-3	-132.7	-132.1	2.689	1.199-3	-132.9	-131.7	2.686
73.	2.156	53.74	75.30	6.059	1.208-3	-130.5	-130.4	2.720	1.207-3	-130.5	-130.3	2.719	1.206-3	-130.7	-130.1	2.717	1.205-3	-130.9	-129.7	2.714
74.	2.186	54.49	76.35	6.073	1.214-3	-128.4	-128.3	2.748	1.214-3	-128.5	-128.2	2.747	1.213-3	-128.6	-128.0	2.745	1.211-3	-128.9	-127.7	2.742
75.	2.216	55.23	77.39	6.087	1.221-3	-126.4	-126.3	2.775	1.220-3	-126.4	-126.2	2.775	1.219-3	-126.6	-126.0	2.773	1.218-3	-126.8	-125.6	2.770
80.	2.366	58.97	82.62	6.155	0.2283	57.38	80.22	5.451	1.256-3	-116.1	-115.9	2.908	1.254-3	-116.3	-115.7	2.906	1.252-3	-116.6	-115.3	2.902
85.	2.515	62.70	87.85	6.218	0.2442	61.31	85.73	5.518	0.1178	59.64	83.21	5.292	1.294-3	-105.9	-105.3	3.032	1.291-3	-106.3	-105.0	3.027
90.	2.664	66.42	93.06	6.278	0.2600	65.19	91.19	5.581	0.1262	63.74	88.98	5.358	1.339-3	-95.42	-94.75	3.152	1.336-3	-95.85	-94.51	3.147
95.	2.813	70.15	98.28	6.334	0.2755	69.04	96.60	5.639	0.1344	67.76	94.64	5.420	0.04927	63.37	88.01	5.099	1.388-3	-85.13	-83.75	3.264
100.	2.962	73.87	103.5	6.388	0.2910	72.87	102.0	5.694	0.1425	71.71	100.2	5.477	0.05305	67.88	94.41	5.165	1.449-3	-74.00	-72.55	3.379
105.	3.111	77.59	108.7	6.438	0.3063	76.68	107.3	5.746	0.1505	75.63	105.7	5.531	0.05669	72.23	100.6	5.225	0.02486	65.19	90.05	4.948
110.	3.260	81.30	113.9	6.487	0.3216	80.47	112.6	5.796	0.1584	79.51	111.2	5.581	0.06021	76.45	106.6	5.281	0.02711	70.41	97.51	5.017
115.	3.409	85.02	119.1	6.533	0.3369	84.25	117.9	5.843	0.1662	83.37	116.6	5.630	0.06364	80.59	112.4	5.333	0.02919	75.27	104.5	5.079
120.	3.557	88.73	124.3	6.577	0.3521	88.02	123.2	5.888	0.1740	87.21	122.0	5.676	0.06702	84.65	118.2	5.382	0.03117	79.90	111.1	5.135
125.	3.706	92.45	129.5	6.620	0.3672	91.78	128.5	5.931	0.1817	91.03	127.4	5.719	0.07034	88.67	123.8	5.428	0.03307	84.36	117.4	5.187
130.	3.855	96.16	134.7	6.661	0.3824	95.54	133.8	5.973	0.1894	94.83	132.7	5.761	0.07362	92.64	129.5	5.472	0.03491	88.70	123.6	5.236
135.	4.003	99.88	139.9	6.700	0.3975	99.29	139.0	6.012	0.1971	98.63	138.0	5.802	0.07686	96.58	135.0	5.514	0.03670	92.94	129.6	5.281
140.	4.152	103.6	145.1	6.738	0.4125	103.0	144.3	6.050	0.2048	102.4	143.4	5.840	0.08008	100.5	140.5	5.555	0.03846	97.12	135.6	5.324
145.	4.301	107.3	150.3	6.774	0.4276	106.8	149.5	6.087	0.2124	106.2	148.7	5.877	0.08327	104.4	146.0	5.593	0.04018	101.2	141.4	5.365
150.	4.449	111.0	155.5	6.810	0.4426	110.5	154.8	6.123	0.2200	110.0	154.0	5.913	0.08644	108.3	151.5	5.630	0.04188	105.3	147.2	5.404
155.	4.598	114.7	160.7	6.844	0.4576	114.3	160.0	6.157	0.2276	113.7	159.2	5.948	0.08959	112.1	156.9	5.666	0.04356	109.3	152.9	5.442
160.	4.746	118.4	165.9	6.877	0.4726	118.0	165.2	6.190	0.2352	117.5	164.5	5.982	0.09273	115.9	162.3	5.700	0.04523	113.3	158.5	5.478
165.	4.895	122.1	171.1	6.909	0.4876	121.7	170.5	6.223	0.2428	121.2	169.8	6.014	0.09585	119.8	167.7	5.733	0.04687	117.3	164.1	5.512
170.	5.044	125.9	176.3	6.940	0.5026	125.4	175.7	6.254	0.2503	125.0	175.1	6.045	0.09896	123.6	173.1	5.765	0.04850	121.2	169.7	5.545
175.	5.192	129.6	181.5	6.970	0.5176	129.2	180.9	6.284	0.2579	128.7	180.3	6.076	0.1021	127.4	178.4	5.796	0.05012	125.1	175.3	5.578
180.	5.341	133.3	186.7	6.999	0.5325	132.9	186.2	6.314	0.2654	132.5	185.6	6.105	0.1052	131.2	183.8	5.826	0.05174	129.0	180.8	5.609
185.	5.489	137.0	191.9	7.028	0.5475	136.6	191.4	6.342	0.2730	136.2	190.8	6.134	0.1082	135.0	189.1	5.856	0.05334	132.9	186.3	5.639
190.	5.638	140.7	197.1	7.055	0.5624	140.4	196.6	6.370	0.2805	140.0	196.1	6.162	0.1113	138.8	194.4	5.884	0.05493	136.8	191.7	5.668
195.	5.786	144.4	202.3	7.082	0.5774	144.1	201.8	6.397	0.2880	143.7	201.3	6.189	0.1144	142.6	199.8	5.912	0.05652	140.7	197.2	5.696
200.	5.935	148.1	207.5	7.109	0.5923	147.8	207.0	6.423	0.2955	147.4	206.5	6.216	0.1174	146.3	205.1	5.939	0.05810	144.5	202.6	5.724
210.	6.231	155.5	217.9	7.159	0.6222	155.2	217.5	6.474	0.3105	154.9	217.0	6.267	0.1236	153.9	215.7	5.990	0.06124	152.2	213.4	5.776
220.	6.528	163.0	228.2	7.208	0.6520	162.7	227.9	6.523	0.3255	162.4	227.5	6.316	0.1296	161.4	226.2	6.039	0.06436	159.8	224.2	5.826
230.	6.825	170.4	238.6	7.254	0.6818	170.1	238.3	6.569	0.3405	169.8	237.9	6.362	0.1357	168.9	236.8	6.086	0.06746	167.4	234.9	5.874
240.	7.122	177.8	249.0	7.298	0.7116	177.5	248.7	6.613	0.3555	177.3	248.4	6.407	0.1418	176.4	247.3	6.131	0.07055	175.0	245.6	5.920
250.	7.419	185.2	259.4	7.340	0.7414	185.0	259.1	6.656	0.3704	184.7	258.8	6.449	0.1478	183.9	257.8	6.174	0.07363	182.6	256.2	5.963
260.	7.716	192.6	269.8	7.381	0.7712	192.4	269.5	6.697	0.3853	192.2	269.2	6.490	0.1539	191.4	268.3	6.215	0.07670	190.1	266.8	6.005
270.	8.013	200.1	280.2	7.420	0.8009	199.8	279.9	6.736	0.4003	199.6	279.7	6.529	0.1599	198.9	278.8	6.255	0.07976	197.7	277.5	6.045
280.	8.310	207.5	290.6	7.458	0.8307	207.3	290.3	6.774	0.4152	207.0	290.1	6.567	0.1659	206.4	289.3	6.293	0.08281	205.2	288.0	6.083
290.	8.607	214.9	301.0	7.495	0.8605	214.7	300.8	6.810	0.4301	214.5	300.5	6.604	0.1719	213.8	299.8	6.330	0.08586	212.8	298.6	6.120
300.	8.904	222.3	311.4	7.530	0.8902	222.1	311.2	6.846	0.4450	221.9	310.9	6.639	0.1779	221.3	310.3	6.365	0.08890	220.3	309.2	6.156

Data from computer program supplied by R. D. McCarty, Cryogenics Division, National Bureau of Standards, Boulder, Colorado.

Properties of N_2

$P(T_{sat})$	2.00×10^6 N/m^2 (115.571 K)				3.00×10^6 N/m^2 (123.620 K)								10.0×10^6 N/m^2				20.0×10^6 N/m^2			
Temp K	v m^3/kg	u kJ/kg	h kJ/kg	s kJ/kgl	v m^3/kg	u kJ/kg	h kJ/kg	s kJ/kgK	v m^3/kg	u kJ/kg	h kJ/kg	s kJ/kgK	v m^3/kg	u kJ/kg	h kJ/kg	s kJ/kgK	v m^3/kg	u kJ/kg	h kJ/kg	s kJ/kgK
SAT LIQ	1.746-3	-35.95	-32.46	3.735	2.139-3	-8.981	-2.563	3.969												
EVAP	9.306-3	99.26	114.4	0.989	3.707-3	62.30	67.01	0.542												
SAT VAP	0.01105	63.31	81.92	4.725	5.846-3	53.32	64.44	4.511												
64.	1.153-3	-149.3	-147.0	2.444	1.152-3	-149.6	-146.2	2.439	1.149-3	-150.2	-144.4	2.430	1.142-3	-151.5	-140.1	2.408	1.129-3	-153.7	-131.2	2.371
65.	1.158-3	-147.3	-145.0	2.475	1.157-3	-147.7	-144.2	2.470	1.153-3	-148.3	-142.5	2.460	1.146-3	-149.7	-138.2	2.438	1.132-3	-151.9	-129.3	2.400
66.	1.163-3	-145.4	-143.1	2.505	1.162-3	-145.7	-142.2	2.500	1.158-3	-146.3	-140.5	2.490	1.150-3	-147.8	-136.3	2.467	1.136-3	-150.2	-127.4	2.428
67.	1.169-3	-143.4	-141.1	2.535	1.167-3	-143.7	-140.2	2.530	1.163-3	-144.4	-138.6	2.520	1.154-3	-145.9	-134.3	2.497	1.139-3	-148.3	-125.6	2.456
68.	1.174-3	-141.4	-139.1	2.565	1.172-3	-141.8	-138.2	2.559	1.168-3	-142.4	-136.6	2.549	1.159-3	-144.0	-132.4	2.525	1.143-3	-146.5	-123.7	2.484
69.	1.179-3	-139.4	-137.0	2.594	1.177-3	-139.8	-136.2	2.589	1.173-3	-140.5	-134.6	2.578	1.163-3	-142.1	-130.5	2.554	1.147-3	-144.7	-121.8	2.512
70.	1.185-3	-137.4	-135.0	2.623	1.183-3	-137.8	-134.2	2.618	1.178-3	-138.5	-132.6	2.607	1.168-3	-140.2	-128.5	2.582	1.150-3	-142.9	-119.9	2.539
71.	1.191-3	-135.4	-133.0	2.652	1.188-3	-135.8	-132.2	2.646	1.184-3	-136.5	-130.6	2.635	1.173-3	-138.3	-126.5	2.610	1.154-3	-141.1	-118.0	2.566
72.	1.196-3	-133.4	-131.0	2.680	1.194-3	-133.8	-130.2	2.674	1.189-3	-134.6	-128.6	2.663	1.178-3	-136.4	-124.6	2.637	1.158-3	-139.3	-116.1	2.592
73.	1.202-3	-131.3	-128.9	2.708	1.200-3	-131.8	-128.2	2.702	1.195-3	-132.6	-126.6	2.691	1.183-3	-134.5	-122.6	2.664	1.163-3	-137.5	-114.2	2.618
74.	1.208-3	-129.3	-126.9	2.736	1.206-3	-129.8	-126.2	2.730	1.200-3	-130.6	-124.6	2.718	1.188-3	-132.5	-120.7	2.691	1.167-3	-135.7	-112.4	2.644
75.	1.215-3	-127.3	-124.9	2.763	1.212-3	-127.8	-124.1	2.757	1.206-3	-128.6	-122.6	2.745	1.193-3	-130.6	-118.7	2.717	1.171-3	-133.9	-110.5	2.669
80.	1.248-3	-117.2	-114.7	2.895	1.245-3	-117.7	-114.0	2.888	1.237-3	-118.8	-112.6	2.874	1.221-3	-121.1	-108.9	2.843	1.194-3	-124.9	-101.0	2.791
85.	1.286-3	-107.0	-104.4	3.019	1.281-3	-107.6	-103.8	3.011	1.272-3	-108.9	-102.5	2.996	1.251-3	-111.6	-99.13	2.962	1.218-3	-116.0	-91.66	2.905
90.	1.329-3	-96.68	-94.02	3.138	1.323-3	-97.47	-93.50	3.129	1.311-3	-98.95	-92.40	3.112	1.285-3	-102.2	-89.32	3.074	1.245-3	-107.2	-82.28	3.012
95.	1.378-3	-86.15	-83.40	3.253	1.370-3	-87.12	-83.01	3.242	1.354-3	-88.89	-82.12	3.223	1.322-3	-92.66	-79.44	3.181	1.274-3	-98.39	-72.90	3.113
100.	1.436-3	-75.29	-72.41	3.365	1.424-3	-76.48	-72.21	3.353	1.404-3	-78.64	-71.62	3.331	1.362-3	-83.09	-69.46	3.283	1.305-3	-89.61	-63.51	3.210
105.	1.506-3	-63.89	-60.88	3.478	1.490-3	-65.42	-60.95	3.463	1.461-3	-68.11	-60.81	3.436	1.408-3	-73.41	-59.33	3.382	1.338-3	-80.84	-54.07	3.302
110.	1.597-3	-51.64	-48.44	3.594	1.571-3	-53.72	-49.01	3.574	1.529-3	-57.19	-49.54	3.541	1.459-3	-63.59	-49.01	3.478	1.374-3	-72.06	-44.59	3.390
115.	1.726-3	-37.73	-34.28	3.720	1.678-3	-40.94	-35.90	3.691	1.613-3	-45.69	-37.63	3.647	1.516-3	-53.59	-38.43	3.572	1.412-3	-63.28	-35.04	3.475
120.	0.01260	66.93	92.14	4.811	1.847-3	-25.79	-20.25	3.824	1.722-3	-33.31	-24.70	3.757	1.582-3	-43.38	-27.56	3.665	1.454-3	-54.50	-25.43	3.557
125.	0.01403	73.46	101.5	4.888	6.644-3	53.28	73.21	4.581	1.880-3	-19.28	-9.883	3.878	1.659-3	-32.90	-16.31	3.757	1.498-3	-45.71	-15.74	3.636
130.	0.01529	79.20	109.8	4.953	8.280-3	65.33	90.18	4.715	2.176-3	-1.339	9.542	4.030	1.751-3	-22.10	-4.592	3.848	1.546-3	-36.91	-5.991	3.712
135.	0.01644	84.48	117.4	5.010	9.423-3	73.38	101.6	4.801	3.018-3	26.11	41.20	4.269	1.862-3	-10.90	7.720	3.941	1.598-3	-28.13	3.830	3.786
140.	0.01752	89.46	124.5	5.062	0.01038	80.04	111.2	4.871	4.237-3	49.25	70.43	4.482	2.001-3	0.727	20.74	4.036	1.654-3	-19.37	13.71	3.858
145.	0.01855	94.22	131.3	5.110	0.01124	85.96	119.7	4.930	5.178-3	62.85	88.74	4.610	2.175-3	12.73	34.47	4.132	1.715-3	-10.65	23.65	3.928
150.	0.01955	98.83	137.9	5.155	0.01204	91.43	127.5	4.984	5.923-3	72.50	102.1	4.701	2.388-3	24.81	48.69	4.229	1.781-3	-1.974	33.64	3.996
155.	0.02051	103.3	144.3	5.197	0.01279	96.59	134.9	5.032	6.559-3	80.34	113.1	4.773	2.636-3	36.52	62.88	4.322	1.852-3	6.626	43.66	4.061
160.	0.02145	107.7	150.6	5.236	0.01350	101.5	142.0	5.077	7.128-3	87.17	122.8	4.835	2.907-3	47.49	76.56	4.409	1.928-3	15.13	53.68	4.125
165.	0.02237	112.0	156.7	5.274	0.01419	106.3	148.8	5.119	7.652-3	93.36	131.6	4.889	3.189-3	57.54	89.43	4.488	2.009-3	23.50	63.67	4.187
170.	0.02327	116.2	162.8	5.310	0.01485	110.9	155.5	5.159	8.144-3	99.11	139.8	4.938	3.472-3	66.68	101.4	4.559	2.095-3	31.71	73.61	4.246
175.	0.02415	120.4	168.7	5.345	0.01550	115.4	161.9	5.196	8.611-3	104.5	147.6	4.983	3.750-3	74.99	112.5	4.624	2.186-3	39.74	83.45	4.303
180.	0.02503	124.5	174.6	5.378	0.01614	119.8	168.2	5.232	9.058-3	109.7	155.0	5.025	4.020-3	82.61	122.8	4.682	2.280-3	47.57	93.17	4.358
185.	0.02589	128.6	180.4	5.410	0.01676	124.2	174.4	5.266	9.489-3	114.7	162.2	5.064	4.282-3	89.67	132.5	4.735	2.378-3	55.17	102.7	4.410
190.	0.02675	132.7	186.2	5.441	0.01737	128.5	180.6	5.298	9.907-3	119.6	169.1	5.101	4.535-3	96.27	141.6	4.784	2.479-3	62.54	112.1	4.460
195.	0.02760	136.7	191.9	5.470	0.01797	132.7	186.6	5.330	0.01031	124.3	175.9	5.136	4.779-3	102.5	150.3	4.829	2.581-3	69.68	121.3	4.508
200.	0.02844	140.7	197.6	5.499	0.01857	136.9	192.6	5.360	0.01071	128.9	182.5	5.170	5.016-3	108.4	158.6	4.871	2.685-3	76.59	130.3	4.553
210.	0.03010	148.7	208.9	5.554	0.01973	145.2	204.4	5.417	0.01149	137.9	195.3	5.232	5.470-3	119.6	174.3	4.947	2.895-3	89.78	147.7	4.638
220.	0.03173	156.6	220.0	5.606	0.02088	153.3	215.9	5.471	0.01223	146.6	207.8	5.291	5.903-3	130.0	189.0	5.016	3.106-3	102.2	164.3	4.716
230.	0.03336	164.4	231.1	5.655	0.02201	161.3	227.4	5.522	0.01296	155.2	220.0	5.345	6.319-3	139.9	203.1	5.079	3.316-3	114.0	180.3	4.787
240.	0.03496	172.2	242.1	5.702	0.02312	169.3	238.7	5.570	0.01368	163.6	232.0	5.396	6.722-3	149.5	216.7	5.136	3.524-3	125.2	195.6	4.852
250.	0.03656	179.9	253.0	5.747	0.02422	177.2	249.9	5.616	0.01438	171.9	243.8	5.444	7.112-3	158.7	229.9	5.190	3.728-3	135.9	210.5	4.913
260.	0.03814	187.6	263.9	5.789	0.02531	185.1	261.0	5.660	0.01507	180.1	255.4	5.489	7.494-3	167.8	242.7	5.241	3.929-3	146.2	224.8	4.969
270.	0.03972	195.3	274.7	5.830	0.02638	192.9	272.1	5.701	0.01575	188.2	266.9	5.533	7.867-3	176.6	255.3	5.288	4.127-3	156.3	238.8	5.022
280.	0.04128	203.0	285.5	5.870	0.02746	200.7	283.3	5.741	0.01642	196.2	278.3	5.574	8.234-3	185.3	267.6	5.333	4.322-3	166.0	252.5	5.071
290.	0.04284	210.6	296.3	5.907	0.02852	208.4	294.0	5.780	0.01709	204.2	289.6	5.614	8.594-3	193.9	279.8	5.376	4.514-3	175.5	265.8	5.118
300.	0.04440	218.2	307.0	5.944	0.02958	216.2	304.9	5.817	0.01775	212.1	300.8	5.652	8.950-3	202.3	291.8	5.416	4.704-3	184.9	278.9	5.163

Properties of Saturated Freon-12 (Refrigerent 12)

t	T	P	v_{liq}	v_{vap}	u_{liq}	u_{vap}	h_{liq}	h_{evap}	h_{vap}	s_{liq}	s_{vap}
C	K	N/m^2	m^3/kg	m^3/kg	kJ/kg	kJ/kg	kJ/kg	kJ/kg	kJ/kg	kJ/kgK	kJ/kgK
-70	203.15	12.34+3	0.0006234	1.1259	359.38	525.70	359.39	179.99	539.59	3.9377	4.8240
-65	208.15	16.88+3	0.0006289	0.8413	363.19	527.61	363.20	178.60	541.81	3.9565	4.8149
-60	213.15	22.70+3	0.0006349	0.6394	367.08	530.06	367.09	177.18	544.28	3.9752	4.8067
-55	218.15	30.06+3	0.0006406	0.4930	371.05	531.93	371.07	175.67	546.75	3.9937	4.7992
-50	223.15	39.22+3	0.0006468	0.3854	375.07	534.11	375.09	174.12	549.22	4.0120	4.7925
-45	228.15	50.50+3	0.0006527	0.3050	379.12	536.29	379.15	172.53	551.69	4.0300	4.7865
-40	233.15	64.24+3	0.0006592	0.2441	383.26	538.48	383.30	170.86	554.16	4.0480	4.7810
-35	238.15	80.79+3	0.0006658	0.1973	387.43	540.69	387.48	169.14	556.63	4.0658	4.7762
-30	243.15	100.5+3	0.0006725	0.1613	391.69	542.89	391.75	167.34	559.10	4.0835	4.7719
-25	248.15	123.7+3	0.0006793	0.1331	396.02	545.10	396.11	165.46	561.57	4.1010	4.7679
-20	253.15	151.0+3	0.0006868	0.1107	400.36	547.28	400.46	163.53	564.00	4.1183	4.7645
-15	258.15	182.6+3	0.0006940	0.09268	404.82	549.50	404.94	161.48	566.43	4.1356	4.7614
-10	263.15	219.1+3	0.0007018	0.07813	409.31	551.74	409.46	159.39	568.86	4.1528	4.7586
- 5	268.15	260.9+3	0.0007092	0.06635	413.84	553.89	414.03	157.17	571.20	4.1698	4.7561
0	273.15	308.6+3	0.0007173	0.05667	418.45	556.06	418.68	154.87	573.55	4.1868	4.7539
5	278.15	362.4+3	0.0007257	0.04863	423.10	558.22	423.36	152.48	575.85	4.2036	4.7519
10	283.15	423.0+3	0.0007342	0.04204	427.83	560.33	428.14	149.97	578.11	4.2204	4.7501
15	288.15	491.1+3	0.0007435	0.03648	432.63	562.41	432.99	147.33	580.33	4.2371	4.7484
20	293.15	566.7+3	0.0007524	0.03175	437.47	564.47	437.89	144.57	582.46	4.2537	4.7469
25	298.15	650.8+3	0.0007628	0.02773	442.34	566.47	442.83	141.68	584.51	4.2702	4.7455
30	303.15	743.4+3	0.0007734	0.02433	447.28	568.39	447.86	138.62	586.48	4.2867	4.7441
35	308.15	846.0+3	0.0007849	0.02136	452.26	570.21	452.92	135.35	588.28	4.3031	4.7425
40	313.15	958.2+3	0.0007968	0.01882	457.31	572.05	458.07	132.01	590.08	4.3194	4.7410
45	318.15	1.081+6	0.0008104	0.01656	462.43	573.81	463.31	128.40	591.72	4.3357	4.7393
50	323.15	1.215+6	0.0008244	0.01459	467.66	575.67	468.67	124.72	593.39	4.3520	4.7379
55	328.15	1.360+6	0.0008410	0.01316	473.01	577.17	474.15	120.91	595.07	4.3681	4.7369
60	333.15	1.518+6	0.0008568	0.01167	478.38	578.86	479.68	116.89	596.57	4.3844	4.7357
65	338.15	1.688+6	0.0008741	0.01036	483.85	580.46	485.33	112.62	597.95	4.4007	4.7340
70	343.15	1.873+6	0.0008936	0.00919	489.39	581.87	491.07	108.01	599.08	4.4171	4.7319
75	348.15	2.072+6	0.0009149	0.00814	495.03	583.14	496.93	103.07	600.01	4.4338	4.7298
80	353.15	2.284+6	0.0009398	0.00723	500.81	584.12	502.96	97.67	600.63	4.4501	4.7269
85	358.15	2.512+6	0.0009680	0.00639	506.72	584.79	509.15	91.69	600.84	4.4669	4.7231
90	363.15	2.756+6	0.0010009	0.00564	512.72	584.88	515.47	84.95	600.42	4.4799	4.7181
95	368.15	3.018+6	0.0010416	0.00497	518.90	584.17	522.05	77.12	599.17	4.4857	4.7110
100	373.15	3.296+6	0.0010952	0.00437	525.43	582.25	529.04	67.61	596.66	4.5192	4.7005
105	378.15	3.595+6	0.0011736	0.00359	532.23	579.57	536.45	56.01	592.47	4.5389	4.6871
110	383.15	3.910+6	0.0013513	0.00266	545.02	571.10	550.31	31.19	581.50	4.5707	4.6520
115.5 (crit.)	388.65	4.009+6	0.0017934	0.00179	556.98	556.98	564.17	0	564.17	4.6122	4.6122

Properties of Freon-12 (Refrigerent 12)

P(t_{sat}) Temp C	9.807 x 10^3 N/m^2 (-73.3 C) v m^3/kg	u kJ/kg	h kJ/kg	s kJ/kgK
Sat liq	0.6194-3	356.901	356.907	3.9255
Evap	1.3411	168.569	181.721	0.9048
Sat vap	1.3417	525.470	538.628	4.8303
-70	1.419	525.51	539.43	4.8412
-65	1.454	527.72	541.98	4.8538
-60	1.489	529.98	544.58	4.8659
-55	1.524	532.22	547.17	4.8780
-50	1.559	534.52	549.81	4.8898
-45	1.594	536.86	552.49	4.9015
-40	1.629	539.23	555.21	4.9132
-35	1.664	541.61	557.93	4.9245
-30	1.699	544.04	560.70	4.9358
-25	1.734	546.50	563.50	4.9471
-20	1.769	548.96	566.31	4.9584
-15	1.804	551.46	569.15	4.9693
-10	1.839	553.97	572.00	4.9802
- 5	1.875	556.46	574.85	4.9911
0	1.911	559.00	577.74	5.0020
5	1.947	561.58	580.67	5.0124
10	1.983	564.15	583.60	5.0229
15	2.019	566.77	586.57	5.0334
20	2.054	569.40	589.54	5.0438
25	2.089	572.07	592.56	5.0543
30	2.124	574.78	595.61	5.0648
35	2.159	574.50	598.67	5.0748
40	2.194	580.25	601.77	5.0849
45	2.229	583.01	604.87	5.0949
50	2.264	585.81	608.01	5.1045
55	2.299	588.64	611.19	5.1142
60	2.333	591.53	614.41	5.1238
65	2.367	594.47	617.68	5.1330
70	2.401	597.44	620.99	5.1422
75	2.434	600.42	624.29	5.1514
80	2.467	603.41	627.60	5.1606
85				
90				
95				
100				
105				
110				
115				
120				
125				

Temp C	29.42 x 10^3 N/m^2 (-55.4 C) v m^3/kg	u kJ/kg	h kJ/kg	s kJ/kgK
Sat liq	0.6401-3	370.739	370.758	3.9922
Evap	0.5027	161.024	175.812	0.8076
Sat vap	0.5033	531.763	546.570	4.7998
-55	0.5077	531.73	546.67	4.8018
-50	0.5190	534.12	549.39	4.8136
-45	0.5304	536.51	552.11	4.8249
-40	0.5420	539.41	554.83	4.8362
-35	0.5536	541.31	557.60	4.8475
-30	0.5654	543.77	560.40	4.8588
-25	0.5772	546.23	563.21	4.8701
-20	0.5891	548.68	566.01	4.8814
-15	0.6010	551.14	568.82	4.8927
-10	0.6129	553.59	571.62	4.9036
- 5	0.6248	556.09	574.47	4.9145
0	0.6367	558.63	577.36	4.9254
5	0.6486	561.21	580.29	4.9362
10	0.6605	563.83	583.26	4.9471
15	0.6724	566.46	586.24	4.9576
20	0.6843	569.12	589.25	4.9681
25	0.6962	571.78	592.26	4.9785
30	0.7081	574.49	595.32	4.9886
35	0.7200	577.20	598.38	4.9986
40	0.7318	579.95	601.48	5.0087
45	0.7436	582.69	604.57	5.0187
50	0.7554	585.49	607.71	5.0283
55	0.7672	588.33	610.90	5.0380
60	0.7790	591.20	614.12	5.0476
65	0.7908	594.12	617.39	5.0572
70	0.8026	597.08	620.69	5.0669
75	0.8145	600.04	624.00	5.0765
80	0.8264	603.00	627.31	5.0857
85	0.8383	605.96	630.62	5.0949
90	0.8502	608.91	633.92	5.1041
95	0.8621	611.87	637.23	5.1133
100	0.8740	614.87	640.58	5.1225
105	0.8859	617.87	643.93	5.1313
110	0.8978	620.87	647.28	5.1401
115	0.9097	623.87	650.63	5.1489

Temp C	58.84 x 10^3 N/m^2 (-41.8 C) v m^3/kg	u kJ/kg	h kJ/kg	s kJ/kgK
Sat liq	0.6569-3	381.755	381.794	4.0416
Evap	0.2635	155.993	171.500	0.7415
Sat vap	0.2642	537.748	553.294	4.7831
-40	0.2705	538.33	554.25	4.7863
-35	0.2762	540.80	557.05	4.7981
-30	0.2820	543.27	559.86	4.8102
-25	0.2878	545.78	562.71	4.8215
-20	0.2936	548.27	565.55	4.8328
-15	0.2994	550.78	568.40	4.8441
-10	0.3053	553.29	571.25	4.8550
- 5	0.3112	555.83	574.14	4.8659
0	0.3171	558.36	577.02	4.8768
5	0.3231	560.95	579.96	4.8877
10	0.3291	563.57	582.93	4.8981
15	0.3351	566.18	585.90	4.9086
20	0.3411	568.85	588.92	4.9187
25	0.3471	571.55	591.97	4.9287
30	0.3531	574.25	595.03	4.9392
35	0.3591	577.00	598.13	4.9488
40	0.3651	579.74	601.22	4.9593
45	0.3711	582.48	604.32	4.9689
50	0.3771	585.27	607.46	4.9787
55	0.3831	588.10	610.64	4.9886
60	0.3891	590.97	613.87	4.9990
65	0.3951	593.88	617.13	5.0083
70	0.4011	596.84	620.44	5.0179
75	0.4071	599.80	623.75	5.0271
80	0.4131	602.75	627.06	5.0371
85	0.4191	605.70	630.36	5.0464
90	0.4251	608.66	633.67	5.0556
95	0.4311	611.65	637.02	5.0644
100	0.4371	614.65	640.37	5.0740
105	0.4431	617.65	643.72	5.0837
110	0.4491	620.64	647.07	5.0916
115	0.4551	623.64	650.42	5.1008

Temp C	98.07 x 10^3 N/m^2 (-30.6 C) v m^3/kg	u kJ/kg	h kJ/kg	s kJ/kgK
Sat liq	0.6717-3	391.191	391.257	4.0814
Evap	0.1645	151.437	167.572	0.6913
Sat vap	0.1652	542.628	558.829	4.7724
-30	0.1650	542.84	559.02	4.7734
-25	0.1684	545.39	561.91	4.7847
-20	0.1717	547.96	564.80	4.7960
-15	0.1750	550.53	567.69	4.8069
-10	0.1783	553.13	570.62	4.8178
- 5	0.1816	555.74	573.55	4.8291
0	0.1849	558.35	576.48	4.8395
5	0.1882	560.99	579.45	4.8504
10	0.1915	563.65	582.43	4.8617
15	0.1948	566.30	585.40	4.8713
20	0.1981	568.98	588.41	4.8818
25	0.2014	571.72	591.47	4.8919
30	0.2047	574.49	594.57	4.9023
35	0.2080	577.27	597.67	4.9120
40	0.2112	580.05	600.76	4.9220
45	0.2144	582.83	603.86	4.9321
50	0.2176	585.70	607.04	4.9425
55	0.2208	588.62	610.27	4.9521
60	0.2240	591.56	613.53	4.9626
65	0.2272	594.52	616.80	4.9722
70	0.2304	597.47	620.07	4.9819
75	0.2335	600.43	623.33	4.9915
80	0.2366	603.48	626.68	5.0007
85	0.2397	606.52	630.03	5.0099
90	0.2428	609.57	633.38	5.0196
95	0.2459	612.62	636.73	5.0288
100	0.2490	615.66	640.08	5.0376
105	0.2522	618.70	643.43	5.0468
110	0.2554	621.73	646.78	5.0560
115	0.2586	624.77	650.13	5.0648

Temp C	196.1 x 10^3 N/m^2 (-13.1 C) v m^3/kg	u kJ.kg	h kJ/kg	s kJ/kgK
Sat liq	0.6791-3	406.519	406.656	4.1422
Evap	0.08610	143.802	160.689	0.6181
Sat vap	0.08680	550.321	567.345	4.7603
-10	0.0880	551.94	569.20	4.7667
- 5	0.0897	554.58	572.17	4.7776
0	0.0915	557.19	575.14	4.7884
5	0.0934	559.79	578.11	4.7993
10	0.0954	562.42	581.13	4.8102
15	0.0974	565.04	584.14	4.8207
20	0.0994	567.70	587.20	4.8311
25	0.1014	570.41	590.30	4.8416
30	0.1034	573.12	593.40	4.8521
35	0.1054	575.87	596.54	4.8621
40	0.1073	578.63	599.68	4.8722
45	0.1092	581.40	602.82	4.8822
50	0.1111	584.21	606.00	4.8923
55	0.1130	587.10	609.26	4.9023
60	0.1149	589.99	612.53	4.9124
65	0.1168	592.88	615.79	4.9220
70	0.1187	595.82	619.10	4.9321
75	0.1206	598.76	622.41	4.9413
80	0.1225	601.69	625.72	4.9509
85	0.1244	604.67	629.07	4.9605
90	0.1262	607.67	632.42	4.9702
95	0.1280	610.66	635.77	4.9798
100	0.1298	613.70	639.16	4.9890
105	0.1316	616.78	642.59	4.9982
110	0.1334	619.85	646.02	5.0074
115	0.1352	622.94	649.46	5.0166
120	0.1370	626.10	652.97	5.0254
125	0.1388	629.59	656.81	5.0342
130	0.1406	632.55	660.13	5.0430

Properties of Freon-12 (Refrigerent 12)

P(t_{sat}) Temp C	294.2 x 10^3 N/m^2 (-1.5 C) v m^3/kg	u kJ/kg	h kJ/kg	s kJ/kgK	490.3 x 10^3 N/m^2 (15 C) v m^3/kg	u kJ/kg	h kJ/kg	s kJ/kgK	686.5 x 10^3 N/m^2 (27 C) v m^3/kg	u kJ/kg	h kJ/kg	s kJ/kgK	980.7 x 10^3 N/m^2 (41 C) v m^3/kg	u kJ/kg	h kJ/kg	s kJ/kgK	1.373 x 10^6 N/m^2 (55.6 C) v m^3/kg	u kJ/kg	h kJ/kg	s kJ/kgK	Temp C
Sat liq	0.7151-3	417.088	417.298	4.1818	0.7434-3	432.581	432.946	4.2369	0.7669-3	444.310	444.837	4.2769	0.7993-3	458.280	459.064	4.3225	0.8423-3	473.450	474.607	4.3694	Sat liq
Evap	0.05864	138.307	155.561	0.5727	0.03579	129.813	147.360	0.5115	0.02553	123.004	140.531	0.4680	0.01758	114.099	131.340	0.4182	0.01220	103.780	120.586	0.3674	Evap
Sat vap	0.05936	555.395	572.859	4.7545	0.03653	562.394	580.306	4.7484	0.02630	567.314	585.368	4.7449	0.01838	572.379	590.404	4.7407	0.01304	577.230	595.133	4.7368	Sat vap
0	0.0596	556.23	573.76	4.7575																	0
5	0.0610	558.82	576.77	4.7683																	5
10	0.0624	561.47	579.83	4.7788																	10
15	0.0638	564.12	582.89	4.7893																	15
20	0.0651	566.83	585.98	4.8002	0.0377	564.98	583.47	4.7591													20
25	0.0664	569.56	589.08	4.8106	0.0385	567.77	586.65	4.7696													25
30	0.0677	572.30	592.22	4.8211	0.0393	570.57	589.84	4.7801	0.0268	568.84	587.24	4.7512									30
35	0.0690	575.10	595.40	4.8316	0.0401	573.36	593.02	4.7905	0.0274	571.74	590.55	4.7621									35
40	0.0703	577.90	598.59	4.8420	0.0409	576.22	596.28	4.8006	0.0280	574.64	593.86	4.7721									40
45	0.0716	580.70	601.77	4.8521	0.0417	579.10	599.55	4.8111	0.0286	577.53	597.16	4.7830	0.01877	574.90	593.31	4.7495					45
50	0.0729	583.50	604.95	4.8621	0.0425	581.98	602.82	4.8215	0.0292	580.42	600.47	4.7935	0.01935	577.72	596.70	4.7604					50
55	0.0742	586.39	608.22	4.8722	0.0433	584.93	606.16	4.8316	0.0297	583.47	603.86	4.8039	0.01989	580.67	600.18	4.7709					55
60	0.0755	589.30	611.52	4.8822	0.0441	587.87	609.51	4.8420	0.0303	586.45	607.25	4.8144	0.02039	583.65	603.65	4.7813	0.01315	580.28	598.34	4.7466	60
65	0.0768	592.23	614.83	4.8923	0.0449	590.84	612.86	4.8255	0.0309	589.43	610.64	4.8244	0.02087	586.70	607.17	4.7918	0.01358	583.29	601.94	4.7579	65
70	0.0780	595.19	618.14	4.9019	0.0457	593.80	616.21	4.8625	0.0315	592.42	614.04	4.8345	0.02135	589.75	610.69	4.8023	0.01400	586.36	605.58	4.7688	70
75	0.0793	598.12	621.45	4.9115	0.0465	596.76	619.56	4.8722	0.0321	595.43	617.47	4.8445	0.02182	592.80	614.20	4.8127	0.01441	589.48	609.26	4.7792	75
80	0.0806	601.04	624.75	4.9212	0.0473	599.72	622.91	4.8818	0.0327	598.45	620.90	4.8546	0.02228	595.87	617.72	4.8232	0.01482	592.68	613.03	4.7897	80
85	0.0819	604.05	628.15	4.9308	0.0481	602.76	626.35	4.8919	0.0333	601.52	624.38	4.8646	0.02274	598.98	621.28	4.8328	0.01523	595.89	616.80	4.8002	85
90	0.0831	607.09	631.54	4.9404	0.0489	605.80	629.78	4.9011	0.0339	604.58	627.85	4.8743	0.02320	602.09	624.84	4.8429	0.01563	599.11	620.57	4.8102	90
95	0.0843	610.13	634.93	4.9496	0.0497	608.84	633.21	4.9107	0.0345	607.69	631.37	4.8839	0.02365	605.25	628.44	4.8546	0.01601	602.36	624.34	4.8207	95
100	0.0855	613.21	638.36	4.9588	0.0505	611.88	636.64	4.9199	0.0352	610.73	634.89	4.8935	0.02409	608.41	632.04	4.8625	0.01637	605.62	628.10	4.8303	100
105	0.0868	616.25	641.79	4.9681	0.0511	615.02	640.08	4.9295	0.0358	613.82	638.40	4.9027	0.02453	611.62	635.68	4.8718	0.01673	608.90	631.87	4.8408	105
110	0.0881	619.31	645.23	4.9777	0.0520	618.09	643.59	4.9387	0.0364	616.97	641.96	4.9120	0.02497	614.88	639.37	4.8818	0.01708	612.23	635.68	4.8505	110
115	0.0894	622.40	648.70	4.9869	0.0528	621.26	647.15	4.9480	0.0372	6]0.06	645.60	4.9216	0.02541	618.13	643.05	4.8910	0.01743	615.56	639.49	4.8605	115
120	0.0907	625.58	652.26	4.9957	0.0536	624.52	650.80	4.9572	0.0377	623.41	649.29	4.9304	0.02585	621.38	646.73	4.9006	0.01776	618.92	643.30	4.8701	120
125	0.0920	628.79	655.86	5.0045	0.0543	620.53	654.44	4.9660	0.0383	626.68	652.97	4.9396	0.02629	624.72	650.50	4.9099	0.01810	622.26	647.11	4.8797	125
130	0.0932	632.04	659.46	5.0133	0.0551	631.06	658.08	4.9743	0.0389	629.96	656.66	4.9484	0.02673	628.06	654.27	4.9195	0.01844	625.60	650.92	4.8889	130
135	0.0944	635.29	663.06	5.0221	0.0559	634.31	661.72	4.9835	0.0395	633.22	660.34	4.9572	0.02717	631.39	658.04	4.9287	0.01877	628.96	654.73	4.8986	135
140	0.0956	638.53	666.66	5.0304	0.0567	637.57	665.37	4.9923	0.0401	636.54	664.07	4.9660	0.02761	634.73	661.81	4.9379	0.01911	632.30	658.54	4.9082	140
145	0.0967	641.86	670.31	5.0392	0.0575	640.85	669.05	5.0011	0.0407	639.85	667.79	4.9748	0.02805	638.07	665.58	4.9467	0.01945	635.74	662.44	4.9174	145
150					0.0583	644.15	672.74	5.0099	0.0413	643.17	671.52	4.9840	0.02849	641.40	669.34	4.9555	0.01979	639.20	666.37	4.9266	150
155					0.0591	647.44	676.42	5.0183	0.0418	646.55	675.25	4.9928	0.02893	644.78	673.15	4.9647	0.02013	642.67	670.31	4.9358	155
160					0.0600	650.68	680.10	5.0271	0.0424	649.91	679.02	5.0020	0.02936	648.17	676.96	4.9739	0.02046	646.15	674.24	4.9450	160
165					0.0608	653.98	683.79	5.0355	0.0430	653.26	682.78	5.0099	0.02979	651.60	680.82	4.9823	0.02078	649.65	678.18	4.9542	165
170													0.03022	655.07	684.71	4.9911	0.02110	653.19	682.16	4.9635	170
175													0.03065	658.54	688.60	4.9995	0.02142	656.72	686.13	4.9722	175
180													0.03107	662.07	692.54	5.0087	0.02174	660.30	690.15	4.9815	180
185													0.03149	665.59	696.47	5.0175	0.02206	663.88	694.17	4.9902	185
190																	0.02238	667.46	698.19	4.9990	190
195																	0.02270	671.12	702.29	5.0083	195
200																	0.02301	674.89	706.48	5.0175	200

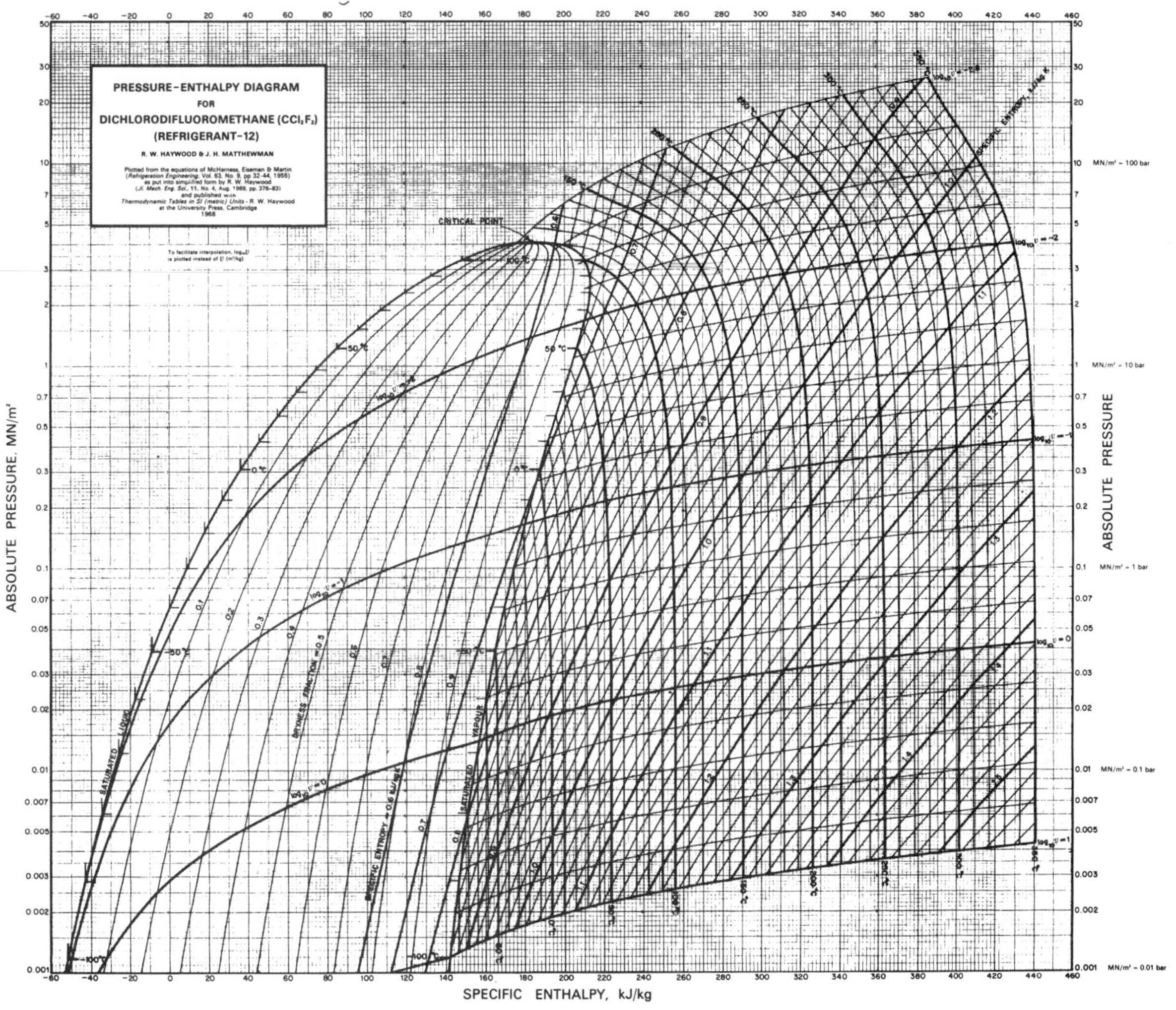

For this chart

$h = 0$ for saturated liquid at -40 C

$s = 0$ for saturated liquid at -40 C

To convert to same zero state as tables 5 and 6

$h_{table} = h_{chart} + 383.30$ kJ/kg

$s_{table} = s_{chart} + 4.0480$ kJ/kgK

Properties of Saturated CO_2

t	T	P	v_{liq}	v_{vap}	u_{liq}	u_{vap}	h_{liq}	h_{evap}	h_{vap}	s_{liq}	s_{vap}	t
C	K	N/m^2	m^3/kg	m^3/kg	kJ/kg	kJ/kg	kJ/kg	kJ/kg	kJ/kg	kJ/kgK	kJ/kgK	C
-56.6	216.55	517.8+3	0.000849	0.0722	301.01	611.98	301.45	347.92	649.37	3.7200	5.3273	-56.6
-55	218.15	555.1+3	0.000853	0.0676	303.90	612.26	304.38	345.41	649.79	3.7334	5.3172	-55
-50	223.15	683.5+3	0.000867	0.055407	313.45	613.46	314.05	337.28	651.34	3.7765	5.2883	-80
-47.5	225.65	752.2+3	0.000873	0.050250	318.29	614.21	318.95	333.06	652.01	3.7974	5.2745	-47.5
-45	228.15	832.6+3	0.000881	0.045809	322.90	614.54	323.64	329.04	652.68	3.8184	5.2607	-45
-42.5	230.65	915.0+3	0.000889	0.041780	327.51	615,03	328.32	324.93	653.26	3.8393	5.2477	-42.5
-40	233.15	1.005+6	0.000897	0.038164	332.32	615.49	333.22	320.62	653.85	3.8594	5.2348	-40
-37.5	235.65	1.098+6	0.000905	0.034900	336.96	615.98	337.95	316.55	654.31	3.8795	5.2222	-37.5
-35	238.15	1.202+6	0.000913	0.032008	341.38	617.29	342.48	311.95	655.77	3.8996	5.2096	-35
-32.5	240.65	1.309+6	0.000922	0.029480	346.33	616.55	347.54	307.60	655.15	3.9197	5.1975	-32.5
-30	243.15	1.427+6	0.000931	0.027001	351.15	616.95	352.48	302.99	655.48	3.9389	5.1854	-30
-27.5	245.65	1.546+6	0.000940	0.024850	355.89	617.33	357.34	298.39	655.73	3.9607	5.1728	-27.5
-25	248.15	1.681+6	0.000950	0.022885	360.68	617.48	362.28	293.66	655.94	3.9779	5.1615	-25
-22.5	250.65	1.832+6	0.000960	0.021070	365.54	617.47	367.30	288.76	656.07	3.9984	5.1489	-22.5
-20	253.15	1.967+6	0.000971	0.019466	370.41	618.11	372.32	283.82	656.40	4.0168	5.1380	-20
-17.5	255.15	2.129+6	0.000982	0.017950	375.47	617.93	377.56	278.59	656.15	4.0377	5.1259	-17.5
-15	258.15	2.289+6	0.000994	0.016609	380.56	618.05	382.84	273.23	656.07	4.0570	5.1154	-15
-12.5	260.65	2.461+6	0.001006	0.015320	385.85	618.15	388.32	267.53	655.86	4.0779	5.1029	-12.5
-10	263.15	2.647+6	0.001019	0.014194	391.23	618.08	393.93	261.71	655.65	4.0976	5.0924	-10
-7.5	265.65	2.844+6	0.001033	0.013120	396.81	617.96	399.75	255.52	655.27	4.1177	5.0807	-7.5
-5	268.15	3.045+6	0.001048	0.012141	402.55	617.88	405.74	249.11	654.85	4.1407	5.0698	-5
-2.5	270.65	3.257+6	0.001063	0.011230	408.43	617.69	411.89	242.37	654.27	4.1625	5.0585	-2.5
0	273.15	3.485+6	0.001081	0.010383	414.91	617.49	418.68	235.00	653.68	4.1868	5.0472	0
2.5	275.65	3.722+6	0.001100	0.009584	422.29	616.71	426.38	226.00	652.38	4.2077	5.0334	2.5
5	278.15	3.972+6	0.001120	0.008850	427.21	615.68	431.65	219.17	650.83	4.2299	5.0179	5
7.5	280.65	4.236+6	0.001142	0.008175	433.85	614.65	438.69	210.59	649.28	4.2517	5.0041	7.5
10	283.15	4.506+6	0.001166	0.007519	440.64	613.35	445.89	201.34	647.23	4.2781	4.9894	10
12.5	285.65	4.789+6	0.001193	0.006910	447.29	611.46	453.01	191.54	644.55	4.3015	4.9718	12.5
15	288.15	5.093+6	0.001223	0.006323	454.73	609.09	460.96	180.32	641.29	4.3292	4.9551	15
17.5	290.65	5.403+6	0.001253	0.005774	461.73	606.32	468.50	169.02	637.52	4.3543	4.9362	17.5
20	293.15	5.733+6	0.001297	0.005269	469.85	602.41	477.29	155.33	632.62	4.3827	4.9128	20
22.5	295.65	6.065+6	0.001346	0.004753	478.34	597.09	486.50	139.42	625.92	4.4141	4.8843	22.5
25	298.15	6.432+6	0.001409	0.004232	488.32	589.62	497.39	119.44	616.84	4.4497	4.8504	25
27.5	300.65	6.801+6	0.001501	0.003679	500.58	580.18	510.79	94.41	605.20	4.4924	4.8094	27.5
30	303.15	7.192+6	0.001680	0.002979	515.03	568.70	527.11	63.01	590.12	4.5444	4.7524	30
31 (crit.)	304.15	7.351+6	0.002156	0.002156	543.08	543.08	558.93	0	558.93	4.6465	4.6465	31

Index